INDUSTRIAL ENGINEERING AND MANAGEMENT

A NEW PERSPECTIVE

McGraw-Hill Series in Industrial Engineering and Management Science

CONSULTING EDITORS

Kenneth E. Case, *Department of Industrial Engineering and Management, Oklahoma State University*
Philip M. Wolfe, *Department of Industrial and Management Systems Engineering, Arizona State University*

Barnes: Statistical Analysis for Engineers and Scientists: A Computer-Based Approach
Bedworth, Henderson, and Wolfe: Computer-Integrated Design and Manufacturing
Black: The Design of the Factory with a Future
Blank: Statistical Procedures for Engineering, Management, and Science
Denton: Safety Management: Improving Performance
Dervitsiotis: Operations Management
Hicks: Industrial Engineering and Management: A New Perspective
Huchingson: New Horizons for Human Factors in Design
Juran and Gryna: Quality Planning and Analysis: From Product Development through Use
Khoshnevis: Discrete Systems Simulation
Law and Kelton: Simulation Modeling and Analysis
Lehrer: White-Collar Productivity
Moen, Nolan, and Provost: Improving Quality through Planned Experimentation
Niebel, Draper, and Wysk: Modern Manufacturing Process Engineering
Polk: Methods Analysis and Work Measurement
Riggs and West: Engineering Economics
Riggs and West: Essentials of Engineering Economics
Taguchi, Elsayed, and Hsiang: Quality Engineering in Production Systems
Wu and Coppins: Linear Programming and Extensions

INDUSTRIAL ENGINEERING AND MANAGEMENT

A NEW PERSPECTIVE

SECOND EDITION

Philip E. Hicks

Hicks & Associates, Inc.

McGRAW-HILL, INC.

New York St. Louis San Francisco Auckland Bogotá
Caracas Lisbon London Madrid Mexico City
Milan Montreal New Delhi San Juan Singapore Sydney Tokyo Toronto

This book was set in Times Roman by Publication Services.
The editors were Eric M. Munson and John M. Morriss;
The production supervisor was Leroy A. Young.
The cover was designed by Rafael Hernandez.
R. R. Donnelley & Sons Company was printer and binder.

INDUSTRIAL ENGINEERING AND MANAGEMENT

A New Perspective

This book is printed on recycled, acid-free paper containing a minimum of 50% recycled de-inked fiber.

1 2 3 4 5 6 7 8 9 0 **DOH DOH** 9 0 9 8 7 6 5 4 3

ISBN 0-07-028807-0

Library of Congress Cataloging-in-Publication Data

Hicks, Philip E.
Industrial engineering and management: a new perspective / Philip E. Hicks. — 2nd ed.
p. cm. — (McGraw-Hill series in industrial engineering and management science)
Rev. ed. of: Introduction to industrial engineering and management science. c1977
Includes bibliographic references and index.
ISBN 0-07-028807-0
1. Industrial engineering. I. Hicks, Philip E. Introduction to industrial engineering and management science. II. Title. III. Series.
T56.H47 1994
658.5—dc20 93-38101

ABOUT THE AUTHOR

Philip E. Hicks is an internationally recognized leader in the fields of industrial engineering and management. His career has progressed from practicing manufacturing engineering and industrial engineering in industry, to industrial manager, to professor and department head, to full-time consultant.

He received his Ph.D. in Industrial Engineering, with a minor in Management, from the Georgia Institute of Technology. Over the last 15 years of consulting he has continuously developed and taught the Institute of Industrial Engineers one-week seminar *Introduction to Industrial Engineering* throughout the United States.

Dr. Hicks has consulted for a broad range of organizations, including the IBM World Trade Corporation in Europe, Tropicana Products, United States Naval Shipyards, Digital Equipment Corporation, and YHY Foods Limited in Hong Kong. The many clients he has served from various industries have provided him with an opportunity to see, feel, and practice industrial engineering and management in the "trenches," where all battles are ultimately won or lost.

To My Dear Wife and Best Friend, Vermont

CONTENTS

PREFACE

Any student considering a professional discipline should take an introductory course that surveys that discipline early in his or her academic program. This book is intended to provide such a student with sufficient insight into the nature of industrial engineering so that he or she can make a commitment to pursue the field as a life's work. With that commitment made, he or she can get on with the serious business of preparing for a successful professional career to follow. For those students that like the challenge and opportunity that an engineering program offers, and who want day-to-day association with technical staff professionals, line and support management, and numerous employees in an organization generally, an industrial engineering degree provides that opportunity.

Undergraduate industrial engineering curricula often include a course, typically taught by the Department Chairperson, entitled Senior Seminar, or its equivalent. It offers a last chance to discuss topics that may have "fallen through the cracks" in a curriculum, before granting students a diploma and "turning them loose" on the public. It is a last opportunity to prepare them for professional practice. This text can help serve that need as well.

Business programs, both undergraduate and graduate, may also use this text to introduce their students to production system design, control, and management. In an Operations Management course, for example, it is desirable not only to provide students with an introductory understanding of the relationships and quantitative tools of production systems design, control, and management, but also to provide them with a sense of the industrial work environment in which they will practice. This text attempts to "shed some light" on that environment.

Response to the first edition suggests that in a lifetime learning environment, this text has also served other lifelong "students" as well (e.g., engineers, technicians, first line foremen, supervisors, managers, upper management, entrepreneurs, union leaders, etc.).

Prior to the original edition of this text, more than a decade ago, a major shift in the field of industrial engineering had just occurred from traditional industrial engineering (i.e., the stopwatch era) to Operations Research and Systems Engineering. Today the field of industrial engineering is undergoing another major shift. The shift today is toward participative management and continuous improvement. Industrial engineers are ideally suited to make major contributions in implementing and improving this "kinder and gentler" and highly effective management approach.

Until very recently, industrial engineering had been practiced almost exclusively in the "expert" mode, whereas today industrial engineers are learning to form, guide, and participate in improvement teams at all levels in organizations. So much change has occurred, especially in management philosophy and approach, that nearly half of the present content of this text is new in this edition. Industrial engineers have gone from solely being "technical experts" to being coaches, advisers, counselors, team builders, a supporting resource, and, when needed, experts. In many organizations today, industrial engineers participate in a dual role of "production system expert" concerning the design and control of a productive system, while also teaching, encouraging, and participating in top management visioning, and as team building and development facilitators at all levels in an organization. Participative management is rapidly becoming the industrial management reality because it works; and it will continue to be the reality of the future. Industrial engineers are intimately involved in shaping and perfecting this improved management culture shift.

Practicing industrial engineers today are on the forefront of recognizing the need for "operational behavior modification," especially for managers, as a necessary element of long-term successful development and implementation of production systems and improvement processes that will work and endure in the future. William T. Morris's 1979 text *Implementation Strategies for Industrial Engineers,* small in size but large in contribution, laid a foundation for this revolution. There is increased recognition today of the need for an improved *psychology of implementation* in industrial engineering—a very promising industrial engineering practice and research area in years to come.

It is the opinion of the author that students should not only be introduced to their discipline early in their studies, but that they should be provided with a historical perspective as well. Chapter 1, therefore, offers a brief, and, it is hoped, relevant history of industrial engineering.

Chapter 2, the largest chapter, surveys various elements of the design of productive systems. The word "productive" throughout this text will be used to connote both "production" (e.g., manufacturing) and "services" (e.g., hospitals, insurance operations, banking, railroading). Ergonomics has been given greater attention compared to the earlier edition, because of the inherent importance of ergonomics in the development of any truly effective human-machine system.

It should be noted that whereas Chapter 2 is devoted solely to the design of productive systems, most of this text is concerned with the design of continuous improvement processes for such systems. What remains are control and management issues and techniques.

Chapter 3 surveys production system control approaches. Production and inventory control have gone through a minirevolution of their own during the last decade as a result of the introduction of productivity and quality improvement techniques and JIT (Just-in-Time). Originally developed in Japan, both are being aggressively implemented, typically very successfully, in the United States today. Those that understand JIT recognize that productivity and quality improvement, and JIT, are overlapping and inseparable.

Chapter 4 surveys forty years of management theory and practice in the United States. The chapter stresses the need for improvement in this critical, related discipline to industrial engineering. The future utilization and successful implementation of industrial engineering is dependent on motivated and highly competent management focused on continuous productivity and quality improvement and responsiveness. Industrial engineers will play an important role in assisting management in the development of effective management approaches that will be right for a twenty-first century workforce, just ahead.

Chapter 5 reviews the development of Total Quality Control, a significant and important redirection of quality control and management efforts in organizations. It offers considerable opportunity for the effective development of productive systems and those who must manage them.

Chapter 6 reviews the general topic of productivity. Industrial engineering started with a focus on direct labor. While direct labor may only represent 5 to 10 percent of the manufactured cost in many operations today, it is still very important to assure that those few remaining direct labor people are truly effective, because they essentially determine what gets shipped today and what does not. Those that do not represent direct labor—indirect and overhead labor—typically representing as much as half of manufactured cost, obviously must be effective as well. This chapter suggests that significant efforts which encompass a vision which leads to continuous productivity and quality improvement must be effectively planned if the opportunities they represent are to be realized.

Chapter 7 offers a brief introduction to Operations Research. A more thorough introduction is constrained by the limited mathematical development of freshman and sophomore students and their lack of familiarity with production systems design and control at that point in their degree programs. Three software sources are referenced in this chapter—LINDO, Quant Systems, and STORM—as supplementary resources to this text (see Chapter 7 references for more details). Each provides quantitative software capability, and end-of-chapter problems, related to topics reviewed in this text. Their parallel use is highly encouraged, particularly with respect to topics reviewed in Chapter 7.

Chapter 8 reviews various aspects of decision sciences; specifically engineering economy, decision theory, and cost engineering. Engineering economy in particular is a subdiscipline of industrial engineering with which all industrial engineers, and engineers in general, must become familiar in preparing for professional practice in their respective fields.

Chapter 9 surveys aspects of systems engineering. It is important that students develop a systems perspective of the field of industrial engineering; otherwise they might accumulate all the subdiscipline pieces in their education, but fail to understand the importance, need, and techniques for system synthesis and integration.

Chapter 10 looks at where industrial engineering is today and some aspects of practice that need to be improved to enhance the "batting average" effectiveness of future industrial engineers. Industrial engineering faculty will need to upgrade continuously their teaching skills, focus, and materials to address the special needs of relating to the changing role of the industrial engineer. Industrial engineers of the future will need to acquire an increasing level of "bedside manner" never before assumed as a requirement in their educational development.

Laboratory examples are provided in addition to problems at the end of each chapter. The laboratory examples are offered to provide students with team or group laboratory exercises relating to the material reviewed in this text. They will give students an opportunity to experiment with some of the participative team-based management approaches discussed in this book and industrial engineering techniques as well. It is assumed that some laboratory examples might be assigned from the supplementary texts referenced in Chapter 7, and that they be utilized as well relative to Chapters 2 and 3.

At various places throughout the text, attention is often directed to productivity generally; after all, this is what industrial engineering is all about. Until recently, Americans enjoyed the highest standard of living and sense of national security in the world. We still enjoy a high sense of national security, but our standard of living has dropped from first place in recent years. The present and future war or competition is and will likely continue to be economic, not military; therefore, we must improve the way we do business if we hope to compete successfully internationally to regain our previous first-place position.

Thanks are due to Ginny Miller, Olivia Garcia, and Mary Ann Westbrook, all of New Mexico State University, and to Gwendolyn Fuller, at North Carolina A & T State University, who helped me in the original edition of this text. Special thanks is given to Karen Perry, my exceptionally capable executive secretary at Hicks & Associates, for her consistent and faithful dedication to this project.

McGraw-Hill and I would also like to thank the following reviewers for their many helpful comments and suggestions: Suraj M. Alexander, University of Louisville; William F. Girouard, Cal Poly, Pamona; Ali A. Housmand, University of Cincinnati; Gary P. Maul, The Ohio State University; Roderick J. Reasor, Virginia Tech; Clarence L. Smith, North Carolina State University; and L. Jackson Turvaville, Tennessee Technological University.

Philip E. Hicks

INDUSTRIAL ENGINEERING AND MANAGEMENT

A NEW PERSPECTIVE

1

A BRIEF HISTORY OF INDUSTRIAL ENGINEERING

Human history becomes more and more a race between education and catastrophe.

H. G. Wells

One of the first engineers in this world may well have been a fellow by the name of Joe Ogg. He is the lead character in a cartoon film [1] produced for the Institute of Industrial Engineers depicting primitive attempts at industrial organization. The film introduces concepts such as specialization of labor, methods study, material handling, and quality control with respect to the production of arrows and hides. Ogg's son, Junior, emerges as a hero because his radical concepts are readily accepted by his peers, and the fruits of their efforts immediately validate his approach. This could only happen in the movies.

In contrast, Sprague de Camp states [27, p. 13], "The story of civilization is, in a sense, the story of engineering—that long and arduous struggle to make the forces of nature work for man's good." In this sense, it is obvious that engineering is as old as civilization itself. What may not be readily apparent is that although industrial engineering is the newest of the major fields of engineering (i.e., civil, mechanical, electrical, chemical, and industrial engineering), fundamental principles of industrial engineering were employed in the age of Ogg.

Throughout history, engineering works have often been taken for granted. In 1514 Pope Paul III was faced with the problem of replacing the architect Bramante after his death during the rebuilding of St. Peter's Cathedral. An artist and engineer by the name of Michelangelo Buonarroti, known to us today as Michelangelo, was selected to see the project through to conclusion. How many of us today think of Michelangelo as an engineer? His work in completing St. Peter's Cathedral is well known. It is less known, however, that in Florence, and again in Rome, he was called on to design fortifications for the city. After building them, since he was convinced the fortifications would not hold because of the incompetence of the defenders, he slipped through the lines of the attacking enemy. This shows that in addition to being one of our heroes he was very human. Michelangelo was a stubborn individualist, and his face was disfigured by a broken nose he received in a fight with a fellow sculptor.

Another of Michelangelo's many enemies was Leonardo da Vinci. Like Michelangelo, da Vinci is best known for his artistic endeavors; however, he was an active, almost continuously absorbed scholar. He tried to master astronomy, anatomy, aeronautics, botany, geology, geography, genetics, and physics. His studies of physics represented a broad coverage of what was known at that time. He had a scientific curiosity that got him into trouble on occasion. He was dismissed by Pope Leo X when the Pope was informed that da Vinci was learning human anatomy by dissecting the real thing. From a purely scientific point of view, what better way is there to learn human anatomy?

In 1483 da Vinci moved to Milan, and he submitted the following résumé to Duke Lodovico Sforza in the hope of gaining employment [27, pp. 363–364]:[1]

> Having, My Most Illustrious Lord, seen and now sufficiently considered the proofs of those who consider themselves masters and designers of instruments of war and that the design and operation of said instruments is not different from those in common use, I will endeavor without injury to anyone to make myself understood by your Excellency, making known my own secrets and offering thereafter at your pleasure, and at the proper time, to put into effect all those things which for brevity are in part noted below—and many more, according to the exigencies of the different cases.
>
> I can construct bridges very light and strong, and capable of easy transportation, and with them pursue or on occasion flee from the enemy, and still others safe and capable of resisting fire and attack, and easy and convenient to place and remove; and have methods of burning and destroying those of the enemy.
>
> I know how, in a place under siege, to remove the water from the moats and make infinite bridges, trellis work, ladders, and other instruments suitable to the said purposes.
>
> Also, if on account of the height of the ditches, or of the strength of the position and the situation, it is impossible in the siege to make use of bombardment, I have means of destroying every fortress or other fortification if it be not built of stone.
>
> I have also means of making cannon easy and convenient to carry, and with them throw out stones similar to a tempest; and with the smoke from them cause great fear to the enemy, to his grave damage and confusion.
>
> And if it should happen at sea, I have the means of constructing many instruments capable of offense and defense and vessels which will offer resistance to the attack of the largest cannon, powder, and fumes.
>
> Also, I have means by tunnels and secret and tortuous passages, made without any noise, to reach a certain and designated point, even if it be necessary to pass under ditches or some river.
>
> Also, I will make covered wagons, secure and indestructible, which entering with their artillery among the enemy, will break up the largest body of armed men. And behind these can follow infantry unharmed and without any opposition. Also, if the necessity occurs, I will make cannon, mortars, and field pieces of beautiful and useful shapes, different from those in common use.
>
> Where cannon cannot be used, I will contrive mangonels, dart throwers, and machines for throwing fire, and other instruments of admirable efficiency not in com-

[1] Excerpt from *The Ancient Engineers* by L. Sprague de Camp. Reprinted by permission of Doubleday & Company, Inc., 1963, and Barthold Fles, Literary Agent.

mon use; and in short, according as the case may be, I will contrive various and infinite apparatus for offense and defense.

In times of peace I believe that I can give satisfaction equal to any other in architecture, in designing public and private edifices, and in conducting water from one place to another.

Also, I can undertake sculpture in marble, in bronze, or in terra cotta; similarly in painting, that which it is possible to do I can do as well as any other, whoever he may be.

Furthermore, it will be possible to start work on the bronze horse, which will be to the immortal glory and eternal honor of the happy memory of your father, My Lord, and of the illustrious House of Sforza.

And if to anyone the above-mentioned things seem impossible or impracticable, I offer myself in readiness to make a trial of them in your park or in such a place as may please your Excellency; to whom as humbly as I possibly can, I commend myself.

Duke Lodovico Sforza evidently was not impressed and did not hire da Vinci after reading his résumé. Da Vinci was later commissioned by the duke as the result of an association da Vinci had with another artist. However, the duke had a habit of paying late, if at all, which resulted in da Vinci's quitting once, but he reconsidered later.

Leonardo da Vinci was one of the great geniuses of all time. He anticipated many engineering developments that were to follow, such as the steam engine, machine gun, camera, submarine, and helicopter. However, he probably had little influence on engineering thought at the time. His research was an unpublished mishmash of thoughts and sketches. He was an impulsive researcher and never summarized his research for the benefit of others through publication. His research was recorded from right to left in his notebooks, possibly for ease of writing because he was left-handed.

Many years later in 1795, Napoleon authorized the establishment of the École Polytechnique in Paris, which became the first engineering school. Rensselaer Polytechnic Institute, founded in 1824, was the first engineering school in the United States.

Until 1880 engineering was either civil or military and for all but the last 100 years was both. In 1880 the American Society of Mechanical Engineers was founded, followed by the American Society of Electrical Engineers in 1884 and the American Institute of Chemical Engineers in 1908. The American Institute of Industrial Engineers, representing the last major field of engineering to become organized, was incorporated in 1948.

It is difficult to say when industrial engineering began. Certainly in the age of Ogg there were production problems associated with making arrows that have their parallel today. If the individual in a toy factory today most concerned with how to make arrows is an industrial engineer, does that mean that when Ogg was deciding how to make arrows he was doing industrial engineering? The basic *what, how, where*, and *when* questions of production analysis have characterized this approach for centuries.

Adam Smith's *Wealth of Nations* [26] published in 1776 was one of the first works promoting "specialization of labor" to improve productivity. He observed in pin making that the division of the task into four separate operations increased output by a factor of almost five. Whereas one worker performing all the operations produced 1000 pins per day, ten workers employed on four more specialized tasks could produce 48,000 pins per day. The concept of designing a process to use the work force efficiently had arrived.

It should be noted, however, that what worked for one process (e.g., pin manufacture) in 1776 may not work well for a similar process today. Manufacturing cells, for example, in common use today represent a reversal of this same concept (i.e., de-departmentalization of processes), whereby a manufacturing cell permits the integrated processing of all materials for a product within a single area of the plant. Such de-departmentalization greatly reduces the material handling cost of the product during its manufacture, inventory costs, and throughput time, and creates a sense of ownership among those producing the product. Determining the best balance of costs in the manufacture of a product is what industrial engineering is all about.

Around 1800, Matthew Boulton and James Watt, Jr., sons of prominent steam engine developers in England, attempted organizational improvements in their Soho foundry that were well ahead of their time. Their efforts were pioneering prototypes for industrial engineering techniques to follow. At about this time, an increasing number of mechanical improvements, such as Arkwright's spinning jenny, were making a considerable impact on productivity. The industrial revolution of this period was freeing humans and beasts from being sources of power in industry. The development of water and steam power and other mechanical devices is the usual primary connotation given to the term *industrial revolution*.

In 1832 Charles Babbage, a self-made mathematician, again suggested division of labor for improved productivity in his book *On the Economy of Machinery and Manufacturers*. In fact, his *difference engine*, the prototype of the modern mechanical calculator, was conceived after he heard about French attempts to produce handbook tables by dividing the calculation task into small steps requiring simple operations. Later, he created his *analytical engine*, which was a mechanical prototype of our modern computers. His difference engine was never completed in his lifetime; the British government abandoned the project after he had spent £17,000 in development. Babbage, like Leonardo da Vinci, was a tireless researcher who had little patience in completing what he had already conceived. Babbage was also aware of the need for improved organization in industry; he toured a number of plants in England and the Continent in the hope of improving his knowledge of the "mechanical art."

After the American Revolution, there was a considerable demand in the United States for muskets, and independence made it possible to produce manufactured goods. Eli Whitney found backers to support the concept of manufacturing interchangeable parts in producing muskets. However, his backers became quite impatient when, after a considerable time had elapsed and much money had been spent, they learned he was still making tools to make parts. Eventually, however, his efforts did produce cheap, interchangeable parts in large quantities. The concept, which is readily accepted today, of producing a set of dies to produce a million parts cheaply was simply not understood at the time. Whitney's invention of the cotton gin typifies many highly significant mechanical improvements of the day, but there is little question that his concept of tooling up for interchangeable parts was the major innovation of that period.

Around the turn of this century, Henry Ford, on observing carcasses on a moving slat conveyor in a slaughterhouse, got the idea for progressive assembly of automobiles by use of conveyors. Conveyors are so much a part of our industrial heritage that it becomes necessary in an industrial engineering course dealing with materials handling to offer a case study for which the use of conveyors is a poor choice of approach. This shock seems necessary to convince students that conveyors help most of the time but not all of the time; in fact, in many Just-in-Time (JIT) installations today, conveyor removal becomes part of the plan. There is little question that the mass production of Ford automobiles gave considerable impetus to the mass production concept in the United States.

FREDERICK W. TAYLOR

In 1886, Henry Towne of the Yale and Towne Company published a paper in the *Transactions of the American Society of Mechanical Engineers* called "The Engineer as Economist." In it, Towne stressed the need for engineers to be concerned with the economic effects of their decisions. Until that time engineers were primarily battling the elements, and costs were assumed to be a relatively uncontrollable necessity for winning the battle against nature. Another member of the American Society of Mechanical Engineers (ASME), much impressed by the concepts offered by Towne, was Frederick W. Taylor. Taylor has often been referred to as the "father of industrial engineering." Based on the accomplishments he made, in light of the times in which they were made, the title seems most appropriate. Whereas the industrial revolution brought new sources of power that made widespread industrialization possible, Taylor offered the concept that it was an engineering responsibility to design, measure, plan, and schedule work.

Frederick Winslow Taylor was from a well-to-do Philadelphia family. Following attendance in French and German schools, he entered Phillips Exeter Academy to prepare for entrance into Harvard Law School. Two things happened at Phillips Exeter Academy that affected his life and subsequently the development of industrial engineering. First, he was influenced by George Wentworth, a professor of mathematics at the academy, who determined the time to solve math problems for homework by timing students solving problems in class. And second, his eyesight became impaired; although Taylor passed his Harvard entrance exams, instead of entering law school he became a machinist's apprentice in the Enterprise Hydraulic Works in 1874. Nine years later he was married, had received a mechanical engineering degree from Stevens Institute a year earlier, and had just been promoted to chief engineer at the Midvale Steel Company. His efforts at Midvale Steel led the way to what came to be known as "scientific management."

To appreciate Taylor's accomplishments, it is necessary to understand the prevailing work environment of the times. The owner-manager, along with the sales and office staff, typically had little direct contact with the production activity. In most cases, a superintendent was given full responsibility for producing the products demanded by the sales staff. All planning and staff functions were informally performed by the superintendent, who had to deal with journeyman mechanics in attempting to get work done. There were no recognized staff functions, and work methods were determined by the individual mechanics on the basis of personal experience, preference, and what tools were readily available. Taylor, influenced by both Towne and Wentworth, developed the concept that work design, work measurement, production scheduling, and other staff functions were engineering responsibilities. His attempts to implement his concepts revolutionized industrial productivity.

In 1881 Taylor began a study of metal cutting that went on for 25 years, ending with publication in 1907 of the longest paper (more than 200 pages) ever published in the *Transactions of the American Society of Mechanical Engineers* [28]. Before this study the geometry of metal-cutting tools and speeds and feeds for metal cutting were determined by experience or rules of thumb. Taylor, along with other experimenters who assisted him, turned metal cutting into a science. Phenomenal increases in the rate of metal cutting have resulted from these initial experiments.

Later, at Bethlehem Steel, Taylor made an analysis of shoveling tasks, the most prevalent task performed in a steel mill in those days. He was not the first to study shoveling, however; Leonardo da Vinci had studied the art of shoveling and calculated the rate at which dirt could be removed as part of one of his engineering projects [9, p. 3]. Taylor noted that although there was a considerable variety of shoveling tasks performed at the mill, the same type of shovel was used for all tasks. A shovelful of rice coal weighed only $3\frac{1}{2}$ pounds, whereas a shovelful of iron ore weighed 38 pounds. Taylor reasoned that with this degree of variability in shovel-load weight, the same type of shovel could not be ideal for all tasks. After some experimentation he found that $21\frac{1}{2}$ pounds seemed to represent an ideal weight for a shovel load and then designed shovels of different sizes for different tasks such that in all cases a shovel load would weigh approximately the ideal weight. Shoveling productivity in the mill improved dramatically. In a period of $3\frac{1}{2}$ years the number of workers performing shoveling tasks was reduced from 500 to 140.

Even more important than the improvement in shoveling productivity was the concept of applying engineering analysis to human work situations. Taylor initiated the practice of performing an engineering analysis of work requirements specifying the exact method, tools, and equipment to be employed, and then training the worker to perform the operations as specified. It is ironic that inattention to this practice generally today represents a major source of industrial inefficiency. In fact, an engineer describing a most efficient method for performing a task has become an unacceptable practice in many plants today; the worker is supposed to discover the best method. It is difficult to explain why such a concept that worked so well for so long is effectively ignored if not discouraged by many industrial managers today. Inattention to the details of workplace design is commonplace in much of American industry and represents an inexcusable drain on our international competitiveness.

Another classic work situation in which Taylor performed a pioneering study was the handling of pig iron. He eventually determined an optimum method of handling, optimum pace, and optimum work and rest periods. He then carefully selected workers to perform this task and carefully trained them to perform the task exactly as he specified. As a result of his analysis, dramatic changes in pig-iron handling productivity resulted.

Analysis of the work requirements and specifications for a method to perform an operation is now called *work design* or *methods study.* The shoveling and pig-iron studies were primarily concerned with the design of work. However, Taylor also pioneered the activity we now generally call *work measurement.* This activity is concerned with determining the amount of time an operator should be allowed for performing an operation. The reciprocal of such a time value permits determination of the amount of production that is expected of an employee in a given time period. Taylor very carefully timed employees performing a task exactly as he specified, which established a basis for determining the amount they should produce in a day. Taylor invented stopwatch timestudy, which is still used extensively to determine the time to perform an operation, commonly referred to today as a time standard.

Before Taylor, labor control was attempted through direct supervision. His development of timestudy led to time standards, which are the underlying basis for control of labor costs and are a necessary input to scheduling and pricing activities in industry.

In June 1895, Taylor presented his first significant paper, entitled "A Piece Rate System," at a meeting of the ASME. His paper was not well received, primarily because it was assumed to be another attempt at devising yet another wage payment plan, as the title suggests. A number of wage payment plans had been attempted, often embodying unsound and unethical rate-cutting practices. Rate cutting is the practice of reducing a time standard once it has been reached in the hope of increasing output at the same pay level, when such a reduction in time is not justified. Apparently, the general distaste for wage payment plans in the engineering community drew attention away from the significant breakthroughs in management thought embodied, but for the most part missed, within the paper. Taylor was disappointed at the reception his paper received, but he concluded that he had not properly presented his findings and thoughts and decided to present an improved paper at a later date.

In June 1903, at the Saratoga, New York, meeting of the ASME, he presented a second paper, entitled "Shop Management" [29]. Again, the reaction was less than enthusiastic; in fact, the mechanical engineering academic community generally regarded it as an insignificant work. The paper was later well received, however, by many plant management personnel, who in everyday manufacturing practice work closer to the situations that would be most affected by Taylor's innovations. Henry Towne, a past president of the ASME, drew attention to what he considered to be interesting concepts in the paper, and a short time thereafter it was the center of controversy of management thought. When one considers the content of the educational experience of the mechanical engineer during this period, it is easier to understand why plant personnel perceived the value of the innovations suggested, whereas the mechanical engineering academics of the day seemed to miss their significance. Mechanical engineers then typically received little education in production management. In fact, the need for engineers trained in production management was the basic justification for initiating the first options in industrial engineering within mechanical engineering departments at universities, and later separate departments of industrial engineering. When one reviews the "Shop Management" paper today, it is interesting to note the number and diversity of concepts embodied within this one paper. Here are some examples:

1 Methods study
2 Time study
3 Standardization of tools
4 A planning department
5 The exception principle of management
6 Instruction cards for workers
7 Slide rules for metal cutting
8 Mnemonic classification systems for parts and products
9 A routing system
10 Costing methods
11 Employee selection in relation to the job
12 Task idea permitting a bonus if the job is completed in the specified time

To understand the significance of the above concepts it is essential that one consider the environment of scientific thought at the time. In the latter half of the nineteenth century, mechanical engineering had become established in Europe. In the United States, however, during this same period Copley [6] states, "The very idea that there could be a true science of mechanical devices continued to be generally scorned."

Industrial engineering, which represents the science of operations, has not been embraced by many responsible for operations today; they choose to deny its existence a century after its initiation. They choose to believe that operations simply require common sense. The history of engineering is replete with such attitudes. There are firms just today discovering that industrial engineering effectively deals with problems they have been unable to solve effectively throughout their past. In a sense, even today, in some firms, doing industrial engineering work is a bit like doing missionary work.

Copley offers the following quotation from a statement by Taylor which typifies the state of the mechanical art in the United States during this period:

> I can remember distinctly the time when an educated scientific engineer was looked upon with profound suspicion by practically the whole manufacturing community.
>
> The successful engineers of my boyhood were mostly men who were endowed with a fine sense of proportion—men who had the faculty of carrying in their minds the size and general shape of parts of machinery, for instance, which had proved themselves suc-

> cessful, and who through their intuitive judgment were able to make a shrewd guess at the proper size and strength of the parts required for a new machine.
>
> It was my pleasure and honor to know intimately one of the greatest and one of the last of this school of empirical engineers—Mr. John Fritz—who had such an important part in the development of the Bessemer process, as well as almost all of the early elements of the steel industry of this country.
>
> When I was a boy and first saw Mr. Fritz, most of the drawings which he made for his new machinery were done with a stick on the floor of the blacksmith shop, and in many cases the verbal description of the parts of the machines which he wished to have made were more important than his drawings. Time and again he himself did not know just what he wanted until after the pattern or model was made and he had an opportunity of seeing the shape of the piece which he was designing. One of his favorite sayings whenever a new machine was finished was, "Now, boys, we have got her done, let's start her up and see why she doesn't work."
>
> The engineer of his day confidently expected that the first machine produced would fail to work, but that by studying its defects he might be able to make a success of the second machine.

In today's computer-aided design environment it may be incomprehensible for students today to imagine what my first job was as an industrial engineer, "co-oping" at Northeastern University in Boston in 1951. One of my tasks was making drawings for parts used in the manufacture of automotive switches for which there were no drawings. I deciphered handwritten notes with part numbers scribbled on them; *reverse engineering* was the order of the day. Reverse engineering refers to making part drawings by measuring parts to produce the drawing. In fact, we made products for which there were no drawings or bills of material; a production manager on the production floor knew which parts to put together to produce a particular customer order. Only he and the customer knew what constituted the product. He liked it that way, too; it provided him some measure of job security.

SCIENTIFIC MANAGEMENT

In 1909, Taylor attempted again to shed light on his concepts by offering another paper to the ASME, entitled "Principles of Scientific Management" [29]. By this time his approach had received a mixture of fame and notoriety, particularly as the result of railroad rate case hearings before the Interstate Commerce Commission in which his concepts represented the center of controversy. Louis Brandeis, representing eastern industrialists, contended that the rate increase was unjustified because the railroads had failed to take advantage of the Taylor system. Because of the controversy *Scientific Management* had created, 18 months passed without action on his paper submitted to ASME. Taylor, feeling obliged to clarify much of the confusion surrounding his techniques, withdrew the paper from ASME, published it privately, and sent it to all ASME members at his own expense. "Principles of Scientific Management," in contrast to his earlier papers, was philosophical in nature. For the most part, it attempted to describe and justify his approach.

In 1911 and 1912 Taylor was questioned at length by a special committee of the U.S. House of Representatives concerned about the Taylor system. In looking back over Taylor's efforts, it is not difficult to understand why his approach was met with considerable criticism at the time; his approach was both a departure from traditional practice and successful.

As will be discussed later in this text, one burden that industrial engineers must bear in doing their work is that if they are successful in identifying potential productivity improvement concerning present operations, someone responsible for present operations will often try to defend present operations. Dealing with wounded managerial egos is a fact of life for practicing industrial engineers; it may well be the single largest barrier to the implementation of productivity improvements in industry today.

Taylor increased output concurrently with a reduction in overall labor cost, while paying higher wages. He taught workers how to work and then expected them to work almost to capacity for higher wages. In some applications, Taylor produced a fourfold increase in production. He had no shortage of workers wanting to receive the increased wages and was never "struck" by labor. It is doubtful that a union today, however, would permit the above distribution of the value of increased productivity.

Another source of criticism against Taylor was his frankness with respect to human behavior, as indicated in the following quotation from *Scientific Management* [29, p. 59]:

> Now one of the very first requirements for a man who is fit to handle pig iron as a regular occupation is that he shall be so stupid and so phlegmatic that he more nearly resembles in his mental make-up the ox than any other type. The man who is mentally alert and intelligent is for this very reason entirely unsuited to what would, for him, be the grinding monotony of work of this character. Therefore the workman who is best suited to handling pig iron is unable to understand the real science of doing this class of work. He is so stupid that the word "percentage" has no meaning to him, and he must consequently be trained by a man more intelligent than himself into the habit of working in accordance with thc laws of this science before he can be successful.

In 1913, concerned about the effects of the Taylor system, Congress added an amendment to the government appropriation bill stipulating that no part of the appropriation should be made available for the pay of any person engaged in timestudy work. Later, a law was passed making it illegal to use a stopwatch in a post office; it contained both a possible fine and imprisonment for violation. As late as 1947, the Military Establishment Appropriation Act and the Navy Department Appropriation Act specified that wages for performing a stopwatch timestudy could not be paid, nor incentive wages for employees, from these funds. In July 1947 a bill was passed in the House of Representatives allowing the War Department to use timestudy; and in 1949 all federal restrictions against the use of stopwatches were discontinued.

Taylor joined Bethlehem Steel in 1898 and created considerable controversy among managers in the twelve years that followed; in 1910, Robert P. Lindeman, then president of Bethlehem Steel, summarily dismissed him. Not until around 1910 had he received any preponderant acceptance of his concepts, in the midst of considerable heated controversy. Following his dismissal from Bethlehem Steel he divided his time between consulting and lecturing in the hope of explaining his concepts. He died of pneumonia five years later.

FRANK B. GILBRETH

Another industrial engineering pioneer—Frank B. Gilbreth—was born on July 7, 1868, in Fairfield, Maine. At college age, after moving to Boston with his family, he wanted to enter the Massachusetts Institute of Technology; however, the family budget was already strained by the education of an older sister, so he took a job as a bricklayer's helper instead.

He had a questioning attitude about his work, but it would seem that he received too few satisfying answers. Bricklaying was an evolved art, and under detailed analysis there was considerable room for improvement. To make a long story short, before he was 30 years of age he was the owner of a profitable construction firm with offices throughout the world.

In the past, 120 bricks laid per worker per hour was normal; Gilbreth's innovations resulted in an average production rate of 350 bricks per worker per hour. These rate increases were not gained by making bricklayers work faster, but by a more effective method. This is what Drucker [7, p. 271] later called "working smarter, not harder." In analyzing the *standard method,* a concept that Gilbreth had initiated, for laying exterior brickwork, Gilbreth reduced the number of motions from 18 to 5. Traditionally, a bricklayer would bend over and pick up a brick from a pile of bricks on a relatively unadjustable scaffold, rotate the brick to find the best side, and then lay the brick by tapping with mortar of often poor consistency. Gilbreth proposed otherwise. Consider that bricks brought to a site are densely arranged, all touching one another, on a pallet. Also consider that to pick up one brick pressing up against another brick, one of the two bricks has to be pushed away from the other brick before you can get your hand around one of the bricks to pick it up. Gilbreth wanted master bricklayers to be able to pick up a brick most efficiently; therefore, he had minimum-cost labor people arranging the bricks on a pallet for ease of pickup by the master bricklayer. He then provided adjustable scaffolds, the proper location of bricks and mortar, and mortar of proper consistency. The result was a vast improvement in productivity with less fatigue. Gilbreth was always in search of the one best way.

Frank Gilbreth married Lillian Moller Gilbreth, a Phi Beta Kappa psychology graduate of the University of California, who later received a Ph.D. from Brown University. The Gilbreths worked very closely together; the combination of engineer and psychologist made significant inroads into the analysis of human work behavior.

Of particular interest to Frank Gilbreth was the analysis of fundamental motions of human activity. He classified the basic motions into what he called *therbligs* (which is almost Gilbreth spelled backwards), such as search, find, transport empty, preposition, grasp, and so forth. In an attempt to analyze motions in more detail he employed industrial motion picture cameras in a technique he labeled *micromotion study* at an ASME meeting in 1912. Because cameras were generally hand-cranked and not of constant speed, he included a clock called a microchronometer, graduated to 1/2000 of a minute, to provide a time dimension to filmed activity. Today, micromotion filming is done at 1000 frames per minute, which automatically spaces successive frames of activity 0.001 minutes apart. Time is determined by counting frames, and a counter is built into the projector.

Gilbreth also studied the motions of parts of the human body, typically the hands. Using an open lens, he filmed a light attached to a point on the body in a darkened room. The photo produced by this technique is called a cyclegraph. By adding "blips" in the light path at fixed time intervals he was able to add a time dimension to the motion path photograph.The photograph thus produced was called a chronocyclegraph. These techniques and others were used to study fundamental motions in human activity to determine average times for such motions under various conditions.

Gilbreth was influenced greatly by Taylor, but whereas Taylor applied his methods almost exclusively to the industrial shop, Gilbreth brought out the generality of these techniques by applying them to fields such as construction, canal building, education, medicine, and the military. His wife added the "human factor" dimension to their work, which led eventually to organization theory and analysis of management practice.

By 1924 Frank Gilbreth had become internationally famous for his contributions. He died three days before he was to leave to give invited papers at the World Power Conference in England and the International Management Conference in Czechoslovakia. Lillian Gilbreth continued the work after his death and became the most distinguished woman engineer in the United States to date. In her later years, she possessed an enthusiasm and charm that served as an inspiration to many. I heard her speak while I was a graduate student at Georgia Tech; she was the most inspiring speaker I have ever heard then or since. She had just returned from Australia and was challenging everyone in the room to go there to do good works. Lillian Gilbreth has become known as both the "first lady of engineering" and "the first ambassador of management" [9, p. 25]. The most prestigious award bestowed by the Institute of Industrial Engineers is the Frank and Lillian Gilbreth Industrial Engineering Award. Two of the Gilbreths' children later described their family in the book *Cheaper by the Dozen,* which was made into a movie of the same name.

One of Taylor's associates at Midvale Steel was Carl Barth, a mathematician, who had started his career in the drafting room at Midvale Steel. Barth became deeply involved in the metal-cutting experiments of Taylor and developed slide rules, such as the one illustrated in Fig. 1-1, to be used by workers to quickly calculate feed and speed parameters for a particular operation.

Barth also performed some early fatigue studies in an effort to establish appropriate fatigue allowances in timestudy. Taylor had more than a dozen associates helping him with different aspects of the study. Proponents of operations research today suggest that one unique feature of operations research as compared to earlier approaches is the use of the team approach. This differs little in concept from the

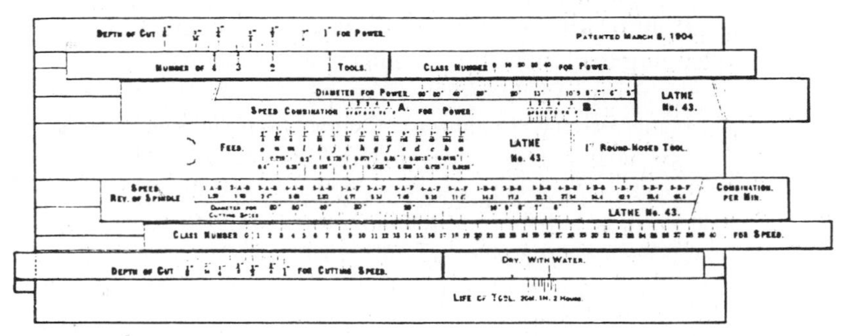

FIGURE 1-1 A Barth slide rule. (From Charles D. Flagle, W. H. Huggins, and R. H. Roy, Operations Research and Systems Engineering, Johns Hopkins Press, Baltimore, 1960, p. 18.)

approach in Taylor's metal-cutting study, as indicated in the following quotation [28, p. 35]:

> Mr. White [Mansel White] is undoubtedly a much more accomplished metallurgist than any of the rest of us; Mr. Gantt [H. L. Gantt] is a better all around manager, and the writer of this paper has perhaps the faculty of holding on tighter with his teeth . . . . Mr. Barth [Carl G. Barth], who is a very much better mathematician than any of the rest of us, has devoted a large part of his time . . . to carrying on the mathematical work.

Another associate of Taylor was Henry Laurence Gantt, who is best known today for the type of chart, which is used for scheduling production equipment, that bears his name. The Gantt chart, illustrated in Fig. 1-2, is used to show graphically the work that has been scheduled ahead for each machine and the progress of those jobs to date.

FIGURE 1-2 A Gantt chart.

Lathes	5 M	6 T	7 W	8 TH	9 F	12 M	13 T	14 W	15 TH
#261	#106	#108		#211		#212			
#263		#316			#227		#87		
#268			#251				#26	#301	
#273	#247				#248				
#281		#11						#304	
#286	#26			#28		#108			
#294									
#303									

Gantt charts are, therefore, a means of planning production and observing and planning the utilization of equipment. Variations of the Gantt chart are in common use in many of the smaller job shop production plants today.

Gantt also developed a wage incentive plan in 1901, which paid workers a bonus if they worked above the standard rate of activity. The plan encouraged cooperation between management and labor and softened the tougher approach used by Taylor. Taylor did not believe in retaining a person who could not work above the standard rate; he preferred keeping only a first-class person and paying a bonus for work above standard.

INDUSTRIAL ENGINEERING ACADEMICS

The first industrial engineering course taught in the United States was taught by Professor Hugo Diemer at the University of Kansas in 1902. The course description read as follows: "*Factory Economics*—Factory design, equipment, and organization. Selection and arrangement of machinery for given conditions. Shop plans. Study of organization and management of industrial establishments. Senior, 2nd term. Daily, at 9, Professor Diemer" [9, p. 43]. In 1904 Professor Dexter Kimball offered an elective industrial engineering course for mechanical engineering seniors at Cornell University. In 1908, Professor William Kent initiated an industrial engineering curriculum at Syracuse University. At this same time the Harvard School of Business Administration incorporated the Taylor system into its shop management courses.

At the recommendation of Frederick W. Taylor, Professor Diemer was hired by what is now Pennsylvania State University to teach an industrial engineering approach to mechanical engineering, which in 1908 produced the first continuously taught curriculum in industrial engineering, offered as an option in mechanical engineering. Because of the interest immediately shown in the course of study, the next year it became a separate four-year curriculum, with Professor Diemer heading the program. Diemer then went on to write the first industrial engineering text in 1910 entitled *Factory Organization and Administration.* Three years later Professor Kimball at Cornell wrote the text *Principles of Industrial Organization,* which was well received and became a standard text in the emerging field for many years.

A major subcommittee presentation entitled "The Present State of the Art of Industrial Management" was given at the annual ASME meeting, on December 6, 1912, which may have been the single greatest assemblage of industrial engineering pioneers in industrial engineering history. In attendance were Taylor, Gilbreth, Gantt, Towne, Diemer, Kent, Kimball, Thompson, and Barth. One notable pioneer was not in attendance that day—Lillian Gilbreth—because women were not allowed at ASME meetings. How outrageous historical facts can be in retrospect—but we are making progress today.

INDUSTRIAL ENGINEERING PRACTICE

From 1912 to 1913 a number of leading United States industrial firms also initiated industrial engineering programs in their plants: Armstrong Cork (now Armstrong Industries), Dow Chemical, Eastman Kodak, and Eli Lilly, to name a few.

Whereas most attention in the early stages of the development of industrial engineering in the United States was directed at the production floor, Henri Fayol in France was concerned with the application of the principles of management throughout an organization.

About the same time as Fayol, Taylor was developing the concept of *functional foremanship* in which the labor force was divided into operators and planners; he was in effect inventing staff in manufacturing organizations. Before this development staff did not exist. The typical plant superintendent had no designated staff for planning his operations; he did it when he could find the time.

Harrington Emerson attempted to use Taylor's approach and some ideas of his own in analyzing labor efforts in the Santa Fe Railroad system. Emerson reorganized the management of the company and employed improved shop practices, standard costing, and tabulating machines for accounting purposes. His improvements resulted in reported annual savings well in excess of $1 million per year for the line. He later wrote a book entitled *Twelve Principles of Efficiency* [8], in which he attempted to elucidate his approach. Emerson's success precipitated modernization efforts in a wide range of industrial firms because his approach had applicability across both commercial and industrial fields.

Morris L. Cooke attempted to employ scientific management in city governments. Later, he and Philip Murray, president of the Congress of Industrial Organizations, published "Organized Labor and Production," a pamphlet that brought out the desirability of a common goal of optimum productivity for both labor and management.

Dwight V. Merrick, following the timestudy work of Taylor, performed a study of elemental times, which was published in the journal *American Machinist*. Merrick, like Gantt, also developed a wage incentive plan as a hoped-for improvement over those offered by Taylor and Gantt.

Times of war have always provided a stimulus to improve technology. Franklin D. Roosevelt, through the Department of Labor, recommended the use of time standards during World War II, and a significant improvement resulted from their use. In fact, Regional War Labor Board III in the middle east coast area encouraged wage incentives and issued guidelines for their use.

An interesting study of human performance was started in 1927 at the Hawthorne Works of the Western Electric Company (now AT&T Technologies), the manufacturing subsidiary of the Bell Telephone System. One part of the study was concerned with the effect of illumination on productivity. One area in the plant was provided with increased illumination, and an increase in productivity was noted. However, it was later realized that the increased production was related to management interest in the study (i.e., frequent visits to the study area by top-level management) and, to a much lesser degree, to the increased illumination. It became a classic example of the necessity to include a control group in a study of this type. The control group should have experienced all effects, except the increased illumination, experienced by the other group (i.e., including management attention). The productivity of the control group would have increased also. The erroneous assumption of the causal relationship of an effect when a control group has not been employed has acquired the name "the Hawthorne effect."

F. W. Harris was one of the earliest to convert a graphical description of the simplest of inventory models, the Economic Order Quantity (EOQ) model, into mathematical terms. Unfortunately for Harris, another gentleman made extensive application of the Harris formula, and it later became generally known as the Wilson formula. There may be a message here: If you have a good idea, let the system know (e.g., write an article) or someone else just might get the credit for it.

CLASSIC INDUSTRIAL ENGINEERING TEXTS

In 1931 F. E. Raymond wrote the first book on inventory control, in which he endeavored to detail the utility of inventory control in manufacturing.

In 1924 W. A. Shewhart of the Bell Telephone Laboratories offered the first description of a control chart and in 1931 published the first text on statistical quality control [24]. Professor Eugene Grant of Stanford University in 1948 published the text *Statistical Quality Control* [11], which has been a mainstay in the field; the Grant and Leavenworth text is now in its sixth edition. Earlier, in 1930, Grant [10] published the text *Principles of Engineering Economy,* which was a pioneering effort in this subdiscipline; this Grant and Leavenworth text is now in its eighth edition and is a mainstay in this subdiscipline as well.

Texts by Barnes [3] and later Niebel [19] and Mundel [17] extended the methods and timestudy efforts of Gilbreth and Taylor. Charting methods are still effective for performing production analysis and often defy useful mathematical description. Such a technique as work sampling, however, is applied statistical sampling and depends on mathematical description for its solution.

The objective rating method of Mundel, in which the inherent difficulty of a method is a factor to be considered, suggested the complexity of this area more clearly than earlier performance rating methods. Harold Smalley, who had been a student of Mundel, in a lengthy article [25] on work measurement in the *Journal of Industrial Engineering,* highlighted the controversial nature of many of the underlying concepts in this field. What has always seemed ironic to this author is the lack of attention being given to this subdiscipline in light of its importance with respect to productivity nationally. There are generally recognized inherent inconsistencies between competing approaches in this field, which suggests that research is needed, and yet research support has been minimal over the years. Krick's text [12] identifies many of these theoretical inconsistencies, as uncomfortable as they may seem to the typical sophomore industrial engineering student, and draws attention to the need for continued research in this area.

A number of consultants were offering advances in the state of the art in the mid-1900s. Charles Bedaux, in wage incentives; Phil Carroll, in time and motion study and cost analysis; H. B. Maynard, in predetermined time systems; and Allen Mogensen, in work simplification. The self-directed team approaches of today suggest that we have come full circle in the sense that the work simplification approaches taught in Mogensen's summer seminars at the Lake Placid Club in Lake Placid, New York, were in many ways similar to the latest thing being espoused today.

As mentioned earlier, one of the earliest texts in the field of capital investment analysis was authored by Eugene Grant in 1930 entitled *Principles of Engineering Economy*. Another pioneer in the field was H. G. Thuesen, who formed the industrial engineering department at Oklahoma State University in 1925 and authored *Engineering Economy* in 1950.

Mallick and Gaudreau [13], Muther [18], and Apple [2] offered early texts in the area of plant layout. Apple's text was a joy to use because it was obvious that it was written by a man who had successfully practiced his profession. It provided a very logical presentation of the material and a product example that ran throughout his text. Plant layout has been, and for the most part in application still is, dictated by graphical and conceptual techniques. CAD/CAE/CAM (i.e., computer-aided design, computer-aided engineering, and computer-aided manufacturing) techniques deal very effectively with the graphical aspects of the field today. Other texts, such as those by Reed [21] and Moore [16], were early attempts to include a more mathematical point of view. The most comprehensive text in the field today is Tompkins and White [30]. However, in light of the number of variables to be considered, it is likely that much of the basic workload of plant design will continue to be done by the traditional techniques developed to date. That is not to say that modern techniques will not be employed. Digital computer simulation and queuing theory have been both successfully and extensively employed in this field and will continue to be employed. Other quantitative techniques are making numerous inroads into the field as well.

INDUSTRIAL ENGINEERING ORGANIZATIONS

There is little question that industrial engineering developed as an offshoot of mechanical engineering, and the American Society of Mechanical Engineers was the first major technical society to represent industrial engineering interests. Symbols for flow process charting commonly employed in production analysis are still defined by ASME standards. Also, a few industrial engineering programs are still options within mechanical engineering departments at universities in the United States.

In 1911, when the controversy over scientific management was at its height, the Taylor system was being discussed in the railroad rate case before the Interstate Commerce Commission, and Taylor himself was giving testimony before a committee of the U.S. House of Representatives, Morris Cooke and Harlow Persons organized a conference on scientific management at the Amos Tuck School of Dartmouth College.

A year later, the Efficiency Society was formed in New York City, and the Society to Promote the Science of Management was initiated, which in 1915 became the Taylor Society. In 1917 the Society of Industrial Engineers (SIE) was formed to specifically represent the interests of production specialists and managers, as compared to the focus on general management philosophy that had developed within the Taylor Society.

A number of individuals who wished to develop corporate training programs for managerial staff formed the American Management Association (AMA) in

1922. This is still the major organization in the United States representing the art and science of management.

In 1932, the Society of Manufacturing Engineers (SME) was formed in Detroit. SME is "dedicated to the advancement of scientific knowledge in the field of manufacturing engineering and to applying its resources to research, writing, publishing, and disseminating information." SME has 80,000 members in 70 countries and sponsors over 300 senior chapters and 200 student chapters and units.

With Frank Gilbreth, Morris Cooke, and Robert Kent in attendance, the first society representing industrial engineering, the Taylor Society, came into existence on November 11, 1910, at the Athletic Club in New York City. Monthly meetings followed at Keene's Chop House. In 1917 the Western Efficiency Society held a meeting in Chicago in which the Society of Industrial Engineers, a San Francisco Bay organization, was formed. In 1936 the Taylor Society and the Society of Industrial Engineers merged to form the Society for Advancement of Management (SAM). This society combined the interests of production specialists, production managers, and those interested in general management philosophy. SAM was later absorbed by AIIE, which will be discussed later, and was combined with its Management Division.

Two strong institutions representing the interests of manufacturers came into being during this period: the American Management Association (AMA), founded in 1922 as the Association of Corporation Schools; and the American Manufacturers Association, founded in 1929.

INSTITUTE OF INDUSTRIAL ENGINEERS

Wyllys G. Stanton, a professor of industrial engineering at Ohio State University, was a man of vision. In 1948 he called a meeting of a number of his associates to consider the formation of a new professional organization to represent industrial engineers. In August 1948, 12 members formed the Columbus Chapter of the American Institute of Industrial Engineers (AIIE), with Eldon Raney as their president. A month later, AIIE was incorporated under the laws of the state of Ohio. One year later there were student chapters at the University of Alabama, Columbia University, Georgia Tech, Northeastern University, Ohio State University, Oklahoma State University, the University of Pittsburgh, Syracuse University, Texas Tech, and Washington University. AIIE's first headquarters was Stanton's office at Ohio State University, and shortly thereafter it was moved to Columbus. In 1960 the office was moved to the United Engineering Center in New York City and later was moved once again to a suburb of Atlanta.

The first official journal of AIIE was the *Journal of Industrial Engineering*. Colonel Frank Groseclose, now deceased, formerly professor emeritus and former director of the School of Industrial Engineering at Georgia Tech, was its first editor. He produced the first issue of the journal in June 1949. By 1969 a dichotomy of interests had developed within the AIIE, which led in that year to the offering of two journals—*Industrial Engineering* for the practitioners and *AIIE Transactions* for the academicians.

In recent years, AIIE has fostered the development of divisions within the institute. Typical of these is the Facilities Planning and Design Division. It is interesting to note that the development of the division concept within the ASME (e.g., the Management Division) may have been one of the factors that led to the formation of industrial engineering as a separate engineering discipline.

In 1981, with chapters in over 70 countries throughout the world, the American Institute of Industrial Engineers (AIIE) elected to change its name to the Institute of Industrial Engineers (IIE) to better represent the worldwide industrial engineering community.

The Institute of Industrial Engineers, along with other engineering disciplines, is a member of the National Society of Professional Engineers (NSPE). The practice of all engineering fields is regulated by state law. "Registered professional engineer" is a designation reserved for those graduate engineers who pass both an Engineer-In-Training (EIT) examination, typically taken in their senior year in college, and then the professional engineering examination in their respective fields after they have acquired and documented a minimum of five years of professional engineering practice. Such designation is typically considered prerequisite for serving as an expert witness in legal proceedings (e.g., suits, rate cases, NLRB [National Labor and Relations Board] hearings) in which expert testimony is considered necessary by one of the litigants. Documents such as drawings involving safety to the public typically require "sealing" by a professional engineer, as a check of their adequacy before construction. The professional engineer seals a drawing or calculation by affixing his seal to the drawing using a hand-held machine similar to a notary public stamp, and thus accepts professional responsibility for the design.

INTERNATIONAL INDUSTRIAL ENGINEERING

It is doubtful that anyone has contributed more to the industrial engineering profession than Frederick W. Taylor. How tragic it seems that a man who contributed so much should have been arguing for acceptance of his ideas three years before his death. His efforts, ultimately, did not go unnoticed. In 1918 Georges Clemenceau of the French Ministry of War referred to Taylor's methods as [4, p. 14] "the employment in every kind of work of the minimum of labor through scientific research into the most advantageous methods of procedure in each particular case," and encouraged their use. Lenin wrote the following in *Pravda* [4, pp. 14–15]:

> We should try out every scientific and progressive suggestion of the Taylor System. . . . To learn how to work—this problem the Soviet authority should present to the people in all its comprehensiveness. The last word of capitalism in this respect, the Taylor System, as well as all progressive measures of capitalism, combined the refined cruelty of bourgeois exploitation and a number of most valuable scientific attainments in the analysis of mechanical motions during work, in dismissing superfluous and useless motions, in determining the most correct methods of work, the best systems of accounting and control, etc. The Soviet Republic must adopt valuable and scientific technical advance in this field. The possibility of socialism will be determined by our success in

> combining the Soviet rule and the Soviet organization of management with the latest progressive measures of capitalism.

It is interesting to note that the Soviet people in the early 1990s have decided to move to a market economy and more western industrial practices as a means to meeting their needs. In many ways they are relearning "how to work."

Another example of the never-ending international development of industrial engineering is the following recent summary of the development of industrial engineering in the People's Republic of China, provided by Zhang Shuwu, Senior Engineer, China International Engineering Consulting Corporation, Beijing [32]:

> Industrial engineering is not as yet an independent engineering discipline in China; however, the application of IE methods can be traced back to the early 1950s. The evolution of IE in China to date is as follows:
>
> Scientific management was introduced during the 1950s and 60s. In the 1950s, new China implemented an overall rehabilitation of both its national economy and industrial construction. The newly established industrial and economic system utilized the Soviet system as a model. All enterprises adopted Soviet enterprise management methods, including scientific management, thereby indirectly introducing the Taylor system into China. Time study was widely used by enterprises under the name "labor quota management." During the period of the Cultural Revolution (1966–1976), further development was delayed because time study and methods study were viewed as "capitalist tricks."
>
> In the period from 1978 to the mid-1980s, modern management techniques were applied which, in fact, belong to IE. After the Cultural Revolution, the State initiated reform policies and "opened up" to the outside world. A new upsurge of science and technology development and industrial construction followed. Emphasis was placed on enterprise management, and encouragement was offered for learning advanced technology and management skills from developed countries. Most enterprises resumed the quota management system, and numerous modern management techniques were introduced in production and operations management. Examples were: Total Quality Control (TQC), Value Engineering (VE), Systems Engineering (SE), Operations Research (OR), Management by Objectives (MBO), Group Technology (GT), Material Requirements Planning (MRP), Management Information Systems (MIS), Just-in-Time (JIT)—Kanban management, and Computer Aided Process Planning (CAPP). A "management fever" developed in attempting to meet the need for management professionals. Around 1980, most of the engineering colleges, in succession, established Industrial Management Engineering (IME) Departments which offered both business management and engineering courses.
>
> From the mid-1980s to the present, a new stage of IE development has emerged, encouraged by the following major events:
>
> **1** IIE's official definition of industrial engineering has been adopted in China, and IE knowledge has been spread widely throughout China.
>
> **2** The application of IE has been extended, is comprehensive and is becoming systematized. Some manufacturing plants employ work study (i.e., methods engineering and work measurement) and the Japanese IE method, "the 5 S's," to improve productivity. Adoption of such practices has lead to cost reductions, and productivity and quality improvements.
>
> **3** An IE forum was held by the Chinese Mechanical Engineering Society (CMES) in the summer of 1989, leading to the first Chinese IE organization—the Chinese Institute of Indus-

trial Engineering—founded in June 1990, which marks the emergence of the IE profession in China.

4 Because of the urgent need for IE personnel, the College of Continuing Engineering Education for Mechanical Engineers has prepared an IE professional training program for practicing engineers, and some universities are planning to provide formal IE education to students.

Much work has already been done on various aspects of operation and management skills, such as production and inventory control, factory layout, and material handling, etc. Today, more and more people are interested in learning and applying IE; the trend is a popularization of IE knowledge. Efforts are underway now to better understand the nature and full potential of IE, and to determine how IE can best serve the unique needs of China. In this respect, some major organizations are taking steps to stimulate enterprise technological improvements aimed at overall optimization of integrated systems. The expectation is that there will be both considerable and rapid future development of IE in China.

Industrial-engineering-type programs have been in existence in China for over 40 years under the title of Management Engineering (e.g., at Northeast University of Technology, Shenyang, Liaoning Province, People's Republic of China). It may be of interest to note that Management Engineering is a title that would probably better serve industrial engineers in the United States today, as compared to "industrial engineering," because of the broadening of the field beyond the industrial sector. More than half of industrial engineering graduates today enter nonmanufacturing work environments with a field name that erroneously connotes an industrial association. *Management engineer* is not an unfamiliar title in the United States; practicing industrial engineers in the healthcare industry commonly employ that title. All industrial engineers and the public would probably be better served if the title *management engineer* were substituted for *industrial engineer* in the future.

It is interesting to note that throughout the western world and China, all are adopting industrial engineering as a means to quality and productivity improvements in their respective national economies. In the last 15 years, for example, Mexico has been actively developing industrial engineering as an engineering field to enhance and support its economic development. The single largest of 10,000+ industrial parks in the world is in Juarez, Mexico. The Parque Industrial Antonio J. Bermudez, Ciudad Juarez, with 48 *maquiladora* plants employs 35,000 workers [5]. Maquiladora plants permit materials to be brought into Mexican plants from the United States so that Mexican labor can be added to the material, with international agreement that all products produced must be shipped back to the United States; this prevents the products produced from competing with Mexican products produced in Mexican-owned plants. Present trends suggest that the former Soviet Union, and other eastern European countries, will increasingly embrace industrial engineering as a means to quality and productivity improvement in their respective emerging free enterprise economies.

In the first half of this century industrial engineers were concerned with the design of manufacturing plants and controls for operating them. In the latter half of the century the emergence of operations research, management science, systems engineering, and computer science has greatly broadened the scope of industrial engineering. It is increasingly more difficult to describe the domain of industrial engineering in simple terms. It is hoped that the following chapters will offer insight into the kinds of work industrial engineers perform. Industrial engineers receive a unique education today, by comparison not only with other engineers but with students in other colleges as well; it prepares them to analyze a broad spectrum of productive activity. The key to understanding the breadth of this activity is the economist's definition of the word *production.* In economics, production refers to the creation of either a product or service. The present-day technology of industrial engineering is sufficiently universal with respect to application to analyze production in such diverse areas as manufacturing, banking, hospitals, defense systems, distribution, retailing, shipbuilding, construction, the chemical industry, insurance, Goodwill Industries, dental offices, and mail-order houses. The work of an industrial engineer today can be so diversified that the following definition, adopted by IIE, is about as specific as is possible:

> Industrial engineering is concerned with the design, improvement, and installation of integrated systems of people, materials, and equipment; drawing upon specialized knowledge and skill in the mathematical, physical, and social sciences together with the principles and methods of engineering analysis and design, to specify, predict, and evaluate the results to be obtained from such systems.

This text endeavors not to define industrial engineering but to describe the activities of industrial engineers, so that the reader can acquire an understanding of the capability and, consequently, the role of industrial engineers.

RELATED DISCIPLINES

Today, more than ever, industrial engineering means different things to different people. In fact, one of the ways to develop an understanding of modern industrial engineering is by gaining an understanding of both its subdisciplines and how it relates to other fields. It would be convenient, for purposes of explanation, if there were clearly defined boundaries between subdisciplines of and related fields to industrial engineering; unfortunately, this is not the case. The fields most commonly referred to today as subdisciplines of or related to industrial engineering are management, statistics, operations research, management science, ergonomics, manufacturing engineering, and systems engineering. There are those in each of these disciplines who believe their field is separate and distinct from industrial engineering.

The education of the modern industrial engineer involves some combination of content from all the disciplines just mentioned. In any particular instance, the combination depends on the industrial engineering academic department and the company in which individuals gain work experience. What may or may not be apparent at this point is the diversity of course offerings in industrial engineering. Whereas depth in a single discipline is the primary strength of an electrical, mechanical, or civil engineering degree program, breadth of understanding across a broad range of related topic areas, both within and outside the college of engineering, as well as depth in industrial engineering subjects, is the primary strength of an industrial engineering degree program.

The following introduction to each of these sub- and related disciplines is intended to offer both relevant history and a limited comparative understanding of the present nature of each discipline.

Management

Of all the disciplines mentioned above, management was one of the earliest to emerge in human history. If management is the art and science of directing human effort, then it must have begun when one person attempted to get another person to work. There is considerably less than unanimity of opinion today as to how best to do that.

Recognition of the need for planning, organizing, and controlling human effort can be traced back at least as far as early Egyptian times. The execution of these functions is essential if, for example, one is to build a pyramid in a reasonable amount of time.

With the possible exception of an introductory statement or paragraph about pre-twentieth-century management thought, most modern texts in management begin their development with a discussion of the scientific concepts of Frederick W. Taylor. Many authors refer to Taylor as the "father of scientific management," whereas others call him the "father of industrial engineering."

There is little question that the subdivision of management commonly referred to as production management has a great deal in common with industrial engineering. In most business colleges, production management is a sequence of one to two courses at the undergraduate level that attempt to familiarize management students with concepts and techniques specific to the analysis and management of production activity. Industrial engineering, on the other hand, is an engineering degree curriculum concerned with the analysis, design, and control of *productive* systems. A productive system is any system that produces either a product or a service. Production management courses are often primarily concerned with teaching management students how to manage (i.e., direct human efforts) in a production environment, with less attention paid to the analysis and design of productive systems.

Industrial engineering students, on the other hand, are taught primarily how to analyze and design productive systems and the control procedures for efficiently operating such systems. Except for a possible course or two concerned with fundamental understanding of management concepts for directing the human effort associated with such systems, it is generally assumed that industrial engineers will not operate the systems they design. The training of a race car driver is analogous to management education; the designing of the car is analogous to industrial engineering education. The race car driver wants to know first and foremost how to drive the car and is less concerned with a detailed understanding of how it works. The industrial engineer designs the car with a driver in mind but with no intention of getting behind the wheel on the day of the race. The engineer does intend to be there, however, to observe the performance of the car and assist with appropriate adjustments. The engineer's concern after the initial design is with design improvements or the continued development of procedures that result in optimum performance.

Chapter 4 provides a chronology of recent historical developments in the field of management.

Operations Research

When World War II began, a small group of military researchers, headed by A. P. Rowe, was interested in the military use of a technique called *radiolocation,* which was developed by civilian scientists. Some historians consider this research to be the most readily identifiable starting point for operations research. Others believe that studies with the characteristics of operations research work can be traced further back in time. Archimedes' analysis and solution of the naval blockade of Syracuse for the king of Syracuse in the third century is considered by some to be the beginning. Just before World War I, F. W. Lanchester in England developed mathematical relationships (i.e., differential equations) representing the firepower of opposing forces, which when solved with respect to time, could determine the final outcome of a military engagement. Confederate General Nathan Bedford Forrest was once asked how to win a war. Using applied Lanchester equation reasoning, his answer was essentially "Get there fustest with the mostest." In retrospect, 1991's Operation Desert Storm seems to owe something to this line of reasoning. Thomas Edison made studies of antisubmarine warfare. Neither the studies of Lanchester nor those of Edison had any immediate impact; along with Archimedes, they were early examples of the employment of scientists for determining the optimum conduct of war.

Not long after the outbreak of World War II, the Bawdsey Research Station, under Rowe, became involved in devising optimum-use policies for a new early-warning detection system called radar. Shortly thereafter, this effort developed into an analysis of all phases of night operations, and the study became a model for operational research studies to follow.

In August 1940, a research group was organized under the direction of P. M. S. Blackett of the University of Manchester to study the use of a new radar-controlled antiaircraft system. The research group came to be known as "Blackett's circus." The name does not seem unreasonable in light of their diverse backgrounds. The group was composed of three physiologists, two mathematical physicists, one astrophysicist, one Army officer, one surveyor, one general physicist, and two mathematicians. The formation of this group seems to be commonly accepted as the beginning of operations research.

In 1941 Blackett and part of his group became involved in the problems associated with the detection of ships and submarines by airborne radar. This study led to Blackett's becoming director of Naval Operational Research of the British Admiralty. The remaining part of his group later became the Operational Research Group of the Air Defense Research and Development Establishment, and then splintered again, forming the Army Operational Research Group. Within two years after the beginning of the war, all three major services had operational research groups.

Here is an example of these earlier studies. The Coastal Command was having difficulty sinking enemy submarines with a newly developed antisubmarine bomb. The bombs were triggered to explode at depths of not less than 100 feet. After detailed study, a Professor Williams concluded that maximum kill likelihood would occur at a depth setting of 20 to 25 feet. The bombs were then set for the minimum possible depth setting of 35 feet, and kill rate increases from different estimates ranged from 400 to 700 percent. Research was immediately undertaken to develop a firing mechanism that could be set to the optimum depth of 20 to 25 feet.

Another problem considered by the Admiralty was the relative merits of large and small convoys. The results indicated that large convoys fared much better.

Within a few months after the United States entered the war, operations research activities were initiated in the Army Air Force and the Navy. By D-day (the allied invasion of France) 26 operations research groups, with approximately 10 scientists per group, had developed in the Air Force. A similar development had also occurred in the Navy. Philip M. Morse of the Massachusetts Institute of Technology headed a group in 1942 to analyze the sea and air attack data against German U-boats. Another study was undertaken later to determine the best maneuvering policy for ships in convoy attempting to evade enemy planes, including the effects of antiaircraft accuracy. The results of the study showed that large ships should attempt severe changes in direction, whereas small ships should change direction gradually.

One final example of early military operations research was a study headed by Ellis A. Johnson. The effort involved computer-simulation war gaming to determine optimum policies for the deployment of sea mines. The efforts of this group ultimately culminated in Operation Starvation. The study showed that B-29s could be extremely effective in mining if employed at night at an altitude of 5000 feet. This approach to mining reduced loss rates to one-tenth of the previous rate. Operation Starvation cost the Japanese war effort 1,200,000 tons of shipping with B-29 loss rates of less than 1 percent.

What is surely obvious by now is that operations research developed to meet the need for analysis of military operational systems during World War II. The first chapter of the text by McCloskey and Trefethen [14] is an excellent account of this early development of operations research.

Operational research in Britain and *operations research* in the United States in post-World War II days had different characteristics and rates of acceptance in the industrial sector of the respective economies. In the United States, private consulting and industrial engineering were familiar improvement activities before the war. Industry experimented with operations research as a possible successor to consulting and industrial engineering. Although U.S. firms had become flexible in experimenting with new approaches, they often jealously guarded study results from their competitors, limiting dissemination of the experience gained.

Prewar British industry, by comparison, had a more traditional trade orientation and had not integrated improvement activities into the industrial structure to the extent that U.S. industry had. Consequently, operational research, particularly in light of its wartime successes and the backlog of opportunities for improvement in British industry, was welcomed as a needed, if not a somewhat all-encompassing, improvement activity. Much of what had been done in the United States by industrial engineers and consultants was being sold as operations research in Britain, with the one exception possibly being *work study.* Work study is the equivalent terminology in Britain for what is now called *methods engineering* and was previously called *motion and time study* in the United States.

One of the first courses in operations research was offered in 1948 at the Massachusetts Institute of Technology. The following year a lecture series was presented at University College, London. Case Western Reserve became the first university to offer a degree program, and shortly thereafter universities such as Johns Hopkins and Northwestern were staffing departments of Operations Research.

The group of scientists who had initiated the operations research lecture series at University College had formed the Operational Research Club a year earlier. The club later initiated the first periodical in operations research, entitled *The Operational Research Quarterly.* The first operations research publication in the United States was *Operations Research with Special Reference to Non-Military Applications,* published by the Committee on Operations Research, which was formed in 1949 by the National Research Council. A year later the Operations Research Society of America was chartered, with Philip M. Morse as its first president. The quarterly *Journal of the Operations Research Society of America* was first published in November 1952.

Systems Engineering

Two highly significant works were published in 1948; one was Norbert Wiener's *Cybernetics, or Control and Communication in the Animal and the Machine* [31], and the other was Claude Shannon's *The Mathematical Theory of Communication* [23]. Wiener derived the word *cybernetics* from a Greek word meaning *steersman,* and his subject was the generality of negative feedback in systems spanning the biological and physical world.

The most commonly used example of negative feedback is the thermostat. When the temperature drops sufficiently below some desired value, the thermostat initiates the heating portion of the cycle, and heat is added until a temperature is reached that is greater than the desired temperature. Heating is then stopped to permit cooling to negate the overheating. Negative feedback means that some action is taken to oppose or negate an unacceptable difference.

Figure 1-3 is a conceptual model of negative feedback in management systems. An apparent condition is compared with a goal, and if a sufficient difference (i.e., error) exists, management action is taken to reduce the difference. The action should result in a change in the apparent condition so that later comparison of the apparent condition with the goal will cause the controlling action to cease. Assume that a manufacturer wishes to have 100 units of inventory on hand. After reviewing his inventory status he notes that he has only 80. If 20 is a sufficient difference, he would probably perform the management action of ordering more

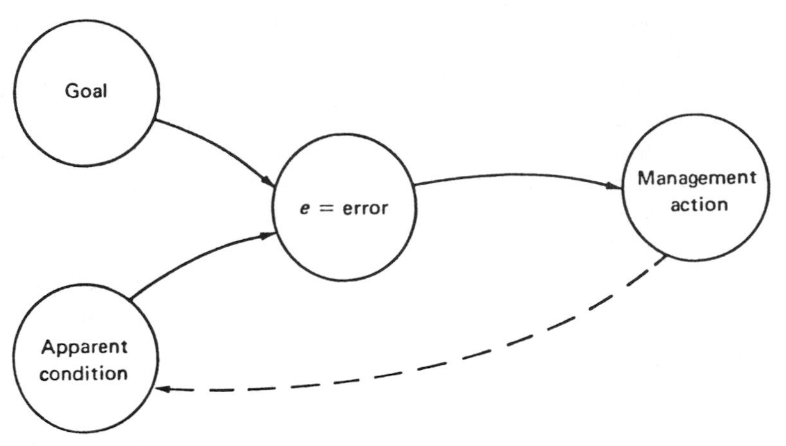

FIGURE 1-3 Negative feedback in management systems.

material in an attempt to raise his inventory level closer to the desired level. The order, after a purchase delay, should bring the apparent condition closer to the goal, removing the need for additional management action. This concept is consistent with the exception principle of managerial control, which says that management attention should be directed to situations in which abnormal values are known to exist. It is management's job to cause an undesirable value to return to a normal or steady-state level.

It is the generality of this concept and other characteristics of systems that makes Wiener's text significant. *Homeostasis* is a word commonly employed in the biological sciences in connection with the regulatory processes in living organisms. Analogous regulation can be identified in such diverse systems as water flow in irrigation and current flow in electrical networks.

Wiener's work is generally considered to be the starting point of what is now commonly referred to as general systems theory. Schlager [22] reported in 1956, on the basis of a review of a nationwide survey, that the first known use of the term *systems engineering* was in the Bell Telephone Laboratories in the early 1940s. Considering the problems the Bell System faced at that time in expanding its system, it is understandable that the term might well have been conceived there. The RCA Corporation, in the years just preceding this, had recognized the need for a systems engineering point of view in the development of a television broadcasting system.

In 1946 the newly created RAND Corporation developed a methodology that they labeled *systems analysis*. Quade and Boucher, in *Systems Analysis and Policy Planning*, defined systems analysis as "a systematic approach to helping a decisionmaker choose a course of action by investigating his full problem, searching out objectives and alternatives, and comparing them in the light of their consequences, using an appropriate framework—in so far as possible analytic—to bring expert judgment and intuition to bear on the problem" [20, p. 2].

Some fairly clear differences have emerged over the years between operations research and systems engineering. Although the early philosophers of operations research believed it to be the beginning of an analytical attack, via mathematics, on large-scale problems, a review of the operations research literature shows that for most problems, the number and complexity of representations must be limited if analytically sound solutions are to be reached. Some operations research problems involve a large number of equations—some linear programming solutions, for example—but the complexities of representation in any one of the many equations may, and often do, make the entire set of equations unsolvable. For many problems today the techniques of operations research offer solutions that were unavailable in the recent past.

Systems engineering seems to have developed with less dependence on "hard" mathematical representation of all aspects of a system. Digital simulation is a much more frequently employed technique in systems engineering, particularly if the system cannot be tightly represented and solved analytically because there is no appropriate analytical technique or the data are not in the form required for a specific operations research technique.

As will become clear in Chapter 9, *systems* demands that a macro perspective be attained in effectively dealing with any significant problem. There is a considerable danger to attempting to solve a problem without first getting the big picture of the total system in which the problem is embedded. You may mess up the system in the process of fixing the problem—it is commonly called "winning the battle, but losing the war."

Statistics

Statistics has been and will continue to be distinct from industrial engineering. However, the approach of industrial engineering has changed significantly; the world around us is viewed as probabilistic in nature rather than deterministic. By deterministic it is meant that all actions under consideration in a particular study situation are assumed to be certain. Probabilistic implies that at least one aspect of the study situation has a probability of occurrence associated with it that must be considered.

In a deterministic problem you may assume, for example, that the cost of a used car is $2000. All calculations concerning buying the car would assume the fixed $2000 cost. In a similar probabilistic problem, you may assume that there is an 80 percent chance that the car can be purchased for $2000, and a 20 percent chance that it can be bought for $1500.

The probabilistic view of the world has so pervaded industrial engineering practice and education that a beginning course in probability and statistics has now become the most important prerequisite in a typical industrial engineering degree program. Industrial engineering has been leading the way for other engineering disciplines in this development, and it seems likely that the improved insight it offers to problems will ultimately result in all disciplines shifting toward a more probabilistic view of the world.

Management Science

Management science is a field that developed closely allied to operations research in the 1960s. The underlying techniques were typically the same as those employed in operations research. What was different was the background of the management scientist and the area of application of the discipline. Management science, for the most part, was an outgrowth of the desire in many business administration and industrial management programs to offer degree options of a quantitative nature employing the discovered techniques of operations research. Because these programs developed in business colleges, the area of application was most often the one with which they were most familiar—management. Whereas an operations research study might offer a linear programming solution to a problem involving refueling submarines, a management science study might well involve a linear programming solution to a portfolio selection problem.

In many instances the leaders in the operations research movement were also leaders in the management science movement. In the affluent 1960s, two closely allied societies with independent publications coexisted with little difficulty. Since then, the merger of Management Science with the Operations Research Society resulted in the initiation of the joint publication *Interfaces*. In even more recent years the two societies have held joint conferences and have a joint business office, which suggests that a merger into one society at some future time is likely.

Ergonomics

Ergonomics, previously called *human factors*, is a subdiscipline of industrial engineering, closely associated with both industrial and experimental psychology. The field of psychology has produced a wealth of information and theory about the human body and mind that is readily available to human factors engineers. Industrial engineering systems by nature are often human-machine systems, in contrast to hardware systems in electrical engineering, for example. The design of human-machine systems involves determining the best combination of human and machine elements. A typical course summarizes the considerable research that has been performed to date in national ergonomics laboratories (e.g., Wright Patterson Air Force Base) and in universities (e.g., Texas Technological University and Virginia Polytechnic Institute). These ergonomics accomplishments assist in familiarizing the industrial engineering student with human-machine systems design. The text *Human Factors Engineering* by McCormick [15] has been used extensively for this purpose. A significant amount of ergonomics research is now being performed in industrial engineering departments, complementing the continuing research that has been underway for many years in industrial psychology.

The author recalls a research effort for which he was responsible at White Sands Missile Range. One of the key issues became, when is a forward observer first capable of recognizing that the black "glob" coming at him in the distance is a Soviet tank? Data collected during World War II, often referred to as the Blackwell Experiments, were concerned with visibility—what factors determine when you are first able to recognize something? Contrast ratio, for example, is a major factor. These old data, collected aboard ship in many instances years ago, still work, and are available if you know where to find them.

Manufacturing Engineering

As mentioned earlier, the Society of Manufacturing Engineers has represented manufacturing engineers in the United States since 1932. Manufacturing engineering is a familiar industrial function name in manufacturing organizations but has never been as well established as an academic degree program in U.S. universities as industrial engineering, for example. There are more than 70 industrial engineering and 10 manufacturing engineering academic programs in the United States.

Industrial engineering and manufacturing engineering are distinct and typically complementary functions in a manufacturing organization. Most firms need both functions represented in their organizations to be truly effective. If one tries to substitute one function for the other, the function omitted typically represents a weakness in that manufacturing organization that will likely limit the overall capability of the technical effort in that organization.

A typical manufacturing engineering department is composed of numerous technical professionals (mechanical engineers, electrical engineers, chemical engineers, structural engineers, mathematicians, thermodynamicists, materials engineers, computer scientists, etc.). Each professional represents some part of the technical processes in use at that manufacturing plant. For example, the thermodynamicist may concern herself with product fin design for dissipating heat, an electrical engineer may worry about test sets and related procedures, and the chemical engineer may concern himself with solution concentrations and related specifications for plating processes. The processes function as they were intended to function because the manufacturing engineering department represents the assemblage of technical expertise necessary to keep all the manufacturing processes under control.

If that is what the manufacturing engineering department does, why do we need an industrial engineering department as well? The core of a typical industrial engineering department is a more homogeneous collection of professionals, typically made up of industrial engineers, with and without degrees, and technicians/technologists. It may well include other specialists, however, with degrees or experience in psychology, management, computer science, and statistics, as well as other engineering disciplines. The smallest entity that an industrial engineer typically deals with is a machine. The machine to an industrial engineer is a black box that has a production rate, yield rate, required operator skills, process capabilities, and other production system attributes. The industrial engineer is concerned with developing a production system that produces the required quantity of products at an appropriate cost and quality. If a machine does not work properly, he may refer the problem to maintenance, and if they cannot fix it, he may refer it to the manufacturing engineering department. If they cannot fix it, they should design a machine that will work for the required step in the production system under design.

The industrial engineering responsibility involves the integration of workers, machines, materials, information, capital, and managerial know-how into a producing system that will produce the right product, at the right cost, at the right time. Manufacturing engineering technical talent is one of the underlying technical plant-supporting resources that guarantees success of that production system.

In summary, it is necessary to know the technical details of each of the processes (i.e., manufacturing engineering) and then to integrate all the elements of a producing system (workers, materials, equipment, information, etc.) so that a quality product is made at the right time and cost (i.e., industrial engineering).

What, then, are modern industrial engineers? First and foremost, they are engineers. They take the same science, mathematics, and engineering core courses as other engineering students. In addition, they take courses in their own discipline, have an interest in management, and acquire a capability in computer science, operations research, systems engineering, and ergonomics while developing a probabilistic view of the world. Industrial engineering students, more than any other engineering students, are in classes in other departments in other colleges in various remote corners of a campus. Industrial engineers have a far broader training than students in other engineering disciplines. That training is probably their greatest asset when it comes time to leave campus, as most students must sooner or later, and go to work.

SUMMARY

Joe Ogg was concerned about how to make arrows and hides; it's a jungle out there, and he may well have needed more arrows than the other guy if he hoped to live a long life. Basically, nothing has changed; we are still trying to figure out how to make widgets, if not arrows and hides, and there still is a jungle (the corporate world) out there. Everyday people like Michelangelo, Leonardo da Vinci, and you and I get our noses bent as Michelangelo did and put bread on the table by doing whatever it is we do, and occasionally join the enemy (a competing corporation) as Michelangelo did. Some also become famous, like Michelangelo and Leonardo da Vinci, not for their engineering, of course, but for their art; no one has ever gotten famous doing engineering. Industrial engineering is a search, in each instance, for the best way to do something.

Nothing stays the same, however. In Adam Smith's time, breaking the process of pin making down to four distinct steps (i.e., specialists or departments), called specialization of labor, resulted in labor efficiencies. Today, with a Just-in-Time production philosophy, to be discussed in a subsequent chapter, combining different operations required for a product into a manufacturing cell that eliminates sequential departmental manufacturing often creates considerable improvement in quality, cost, and delivery of the product.

About a century ago Taylor's papers "Shop Management" [29] and "Principles of Scientific Management" [29] initiated a management approach that was first called *scientific management* and later evolved into what we now call *industrial engineering*. Taylor in the United States and Henri Fayol in France both initiated aspects of what is now called *organizational development*, in trying to determine what organizational structure would best serve a particular organization. Management consulting firms have been performing this function ever since.

The first industrial engineering course taught in 1902 at the University of Kansas by Diemer, later courses at Cornell, and courses in shop management at the Harvard Business School and at Rensselaer initiated what is now more than 70 accredited degree programs in industrial engineering throughout the United States today.

Shortly following the introduction of these courses, progressive firms such as Armstrong Cork (now Armstrong Industries), Eastman Kodak, Dow Chemical, and Eli Lilly initiated the practice of industrial engineering in major U.S. corporations. Hundreds of thousands of industrial engineers now practice across a broad range of organizations throughout the United States.

An Ohio State University professor, Wyllys Stanton, had the audacity (and foresight) to initiate the first chapter of the American Institute of Industrial Engineers, in Columbus. Here is a suggestion: develop the ability to be selectively audacious, like Stanton, for "good purpose."

Industrial engineering, in a historical sense, is a new field. For that reason it is not uncommon for an industrial engineer to tour a plant that has not embraced industrial engineering and note that they are doing what can best be described as dumb things, simply out of ignorance, of course. In history, as in cultures, things change slowly. In a topic area called technological forecasting, for example, it was discovered that it took 25 years for adoption of the diesel locomotive to replace the steam locomotive, even though the diesel locomotive was vastly superior in performance.

Students often express the view that there is nothing left for them to change; everything has been done. Not true. Management has been chasing a behavioral management approach for the past 40 years, and more recently golden parachutes (their own personal and generous retirement programs). Behavioral management has real strengths but also some very definite weaknesses, to be discussed later. In the meantime, management has neglected the workplace. We know in the United States today that management must improve if it hopes to compete in the world economy of today and tomorrow.

As proof of the rate of change that is going on now, consider that the Chinese economy, representing one billion people—one-third of the world population—has until recently been following Soviet enterprise management methods, and the Soviet economy has collapsed. It was only in 1990 that the Chinese formally initiated industrial engineering with the creation of the Chinese Institute of Industrial Engineering. Students typically think of history as being about the past, which it is, but it is now as well (i.e., history in the making). At few times in history has there been more need for change in the economies of the world, and more likelihood that significant change will occur, with the recent collapse of outdated economic systems of the past. Students worldwide entering industrial engineering will have a limitless opportunity to make significant changes in the way all businesses around the world can and will do business in the future, and how people will live at work.

REFERENCES

1 "The Story of Joe Ogg," a 16-mm sound film produced at Kansas State University for the Institute of Industrial Engineers, Norcross, GA, 1969.
2 Apple, James M.: *Plant Layout and Materials Handling*, 2nd ed., The Ronald Press Company, New York, 1963.
3 Barnes, Ralph M.: *Motion and Time Study*, 6th ed., John Wiley & Sons, Inc., New York, 1968.
4 Blair, Raymond N., and C. Wilson Whitston: *Elements of Industrial Systems Engineering*, Prentice-Hall, Inc., Englewood Cliffs, NJ, 1971.
5 del Campo, Martha G., "Ciudad Juarez," *Twin Plant News*, Vol. 6, No. 6, 1991.
6 Copley, Frank Barkley: *Frederick W. Taylor*, Harper & Row Publishers, Inc., New York, 1923.
7 Drucker, Peter: *The Age of Discontinuity*, Harper & Row Publishers, Inc., New York, 1968.
8 Emerson, Harrington: *Twelve Principles of Efficiency,* The Engineering Magazine Co., New York, 1912.
9 Emerson, Howard P., and Douglas C. E. Naehring: *Origins of Industrial Engineering*, Industrial Engineering and Management Press, Institute of Industrial Engineers, Norcross, GA, 1988.
10 Grant, Eugene L.: *Principles of Engineering Economy*, Ronald Press, New York, 1930.
11 Grant, Eugene L.: *Statistical Quality Control*, McGraw-Hill Book Company, New York, 1948.
12 Krick, Edward V.: *Methods Engineering*, John Wiley & Sons, Inc., New York, 1962.
13 Mallick, Randolph W., and A. T. Gaudreau: *Plant Layout Planning and Practice*, John Wiley & Sons, Inc., New York, 1951.
14 McCloskey, Joseph F., and Florence N. Trefethen (eds.): *Operations Research for Management*, The Johns Hopkins Press, Baltimore, MD, 1954.
15 McCormick, Ernest J.: *Human Factors Engineering*, McGraw-Hill Book Company, New York, 1957.
16 Moore, James M.: *Plant Layout and Design*, The Macmillan Company, New York, 1962.
17 Mundel, Marvin E.: *Motion and Time Study*, 3rd ed., Prentice-Hall, Inc., Englewood Cliffs, NJ, 1960.
18 Muther, Richard: *Practical Plant Layout*, McGraw-Hill Book Company, New York, 1955.
19 Niebel, Benjamin W.: *Motion and Time Study*, 5th ed., Richard D. Irwin, Inc., Homewood, IL, 1972.
20 Quade, E. S., and W. S. Boucher (eds.): *Systems Analysis and Policy Planning*, American Elsevier Publishing Co., New York, 1968.
21 Reed, Ruddell, Jr.: *Plant Layout*, Richard D. Irwin, Inc., Homewood, IL, 1961.
22 Schlager, K. J.: "Systems Engineering—Key to Modern Development," *IRE Transactions on Engineering Management,* Vol. 3, no. 12, 1956, pp. 64–66.
23 Shannon, Claude: *The Mathematical Theory of Communication*, University of Illinois Press, Urbana, IL, 1948.
24 Shewhart, Walter A.: *Economic Control of Quality of Manufactured Product*, D. Van Nostrand Co., New York, 1931.
25 Smalley, Harold E.: "Another Look at Work Measurement," *Journal of Industrial Engineering*, Vol. 18, No. 3, 1967.
26 Smith, Adam: *An Inquiry into the Nature and Causes of the Wealth of Nations*, Edwin Cannan (ed.), Random House, Inc., New York, 1937 (originally published in 1776).

27 Sprague de Camp, L.: *The Ancient Engineers*, Doubleday & Company, Inc., Garden City, NY, 1963.
28 Taylor, Frederick W.: "On the Art of Cutting Metals," *Transactions of the ASME*, Vol. 28, paper 1119, pp. 31–350, 1907.
29 Taylor, Frederick W.: *Scientific Management*, Harper & Row Publishers, Inc., New York, 1947 (comprises "Shop Management," "The Principles of Scientific Management," and "Testimony Before the Special House Committee").
30 Tompkins, James A., and John A. White: *Facilities Planning*, John Wiley & Sons, Inc., New York, 1984.
31 Wiener, Norbert: *Cybernetics, or Control and Communication in the Animal and the Machine*, John Wiley & Sons, Inc., New York, 1948.
32 Zhang Shuwu: China International Engineering Consulting Corporation, Beijing.

REVIEW QUESTIONS AND PROBLEMS

1 What was Eli Whitney's main contribution to engineering development?
2 What new role did Towne believe an engineer should accept as part of his responsibility?
3 Who initiated an approach that came to be known as scientific management?
4 Differentiate between methods study and work measurement.
5 What is rate cutting?
6 What is reverse engineering?
7 Describe the changing attitude of government toward stopwatch timestudy over the first half of the twentieth century.
8 What are therbligs, and what relationship do they have to Gilbreth?
9 What is micromotion study?
10 Was Taylor a lone researcher like Leonardo da Vinci or was he a team player?
11 In what industry did Harrington Emerson make extensive use of the Taylor system?
12 What is meant by the Hawthorne effect?
13 Who initiated the field of quality control?
14 From what major engineering discipline did industrial engineering emerge?
15 Who was primarily responsible for initiating the American Institute of Industrial Engineers?
16 Has interest in the Taylor system been confined solely to the United States?
17 What is a *maquiladora* plant?
18 In what year was the first Chinese industrial engineering organization created?
19 In what way does the economist's definition of production suggest an expanded role for industrial engineering?
20 How does production management as an area of study differ from industrial engineering?
21 What is the difference in connotation between operational research and operations research?
22 What was the name of the first operations research periodical?
23 Give an example of negative feedback.
24 How does industrial engineering compare with other engineering disciplines in the extent to which a probabilistic view has replaced the more classical deterministic view of natural phenomena?
25 How does management science basically differ from operations research?
26 What characteristic of industrial engineering systems often makes ergonomics an unavoidable consideration in the design of such systems?

2

PRODUCTION SYSTEMS DESIGN

Any truly labor-saving device will win out. All that you have to do to find proof of this is to look at the history of the industrial world. And, gentlemen, scientific management is merely the equivalent of a labor-saving device.

F. W. Taylor

The design of a production system starts with the design of the product to be manufactured. Figure 2-1 describes a typical sequence of steps starting with a product design concept that culminates in a final product design for manufacture. Product engineers are those individuals in a manufacturing organization most familiar with the function of a product and the customers' changing needs relative to that product.

Arts-Way Manufacturing is a manufacturer of farm machinery in Armstrong, Iowa. As soon as the beet-harvesting season comes to an end at the end of summer, Arts-Way product engineering and marketing personnel evaluate their most recent harvester design successes and any unique conditions or problems that affected the performance of their equipment. There is nothing like a harvest to bring to light the strengths and weaknesses of a harvester design. As soon as the harvester performance information is collected and evaluated and customer and dealer inputs have been reviewed, it is likely that design improvements and related engineering tests will begin immediately for an improved harvester product to be available for next summer. The more mature is the design, the fewer design changes that will

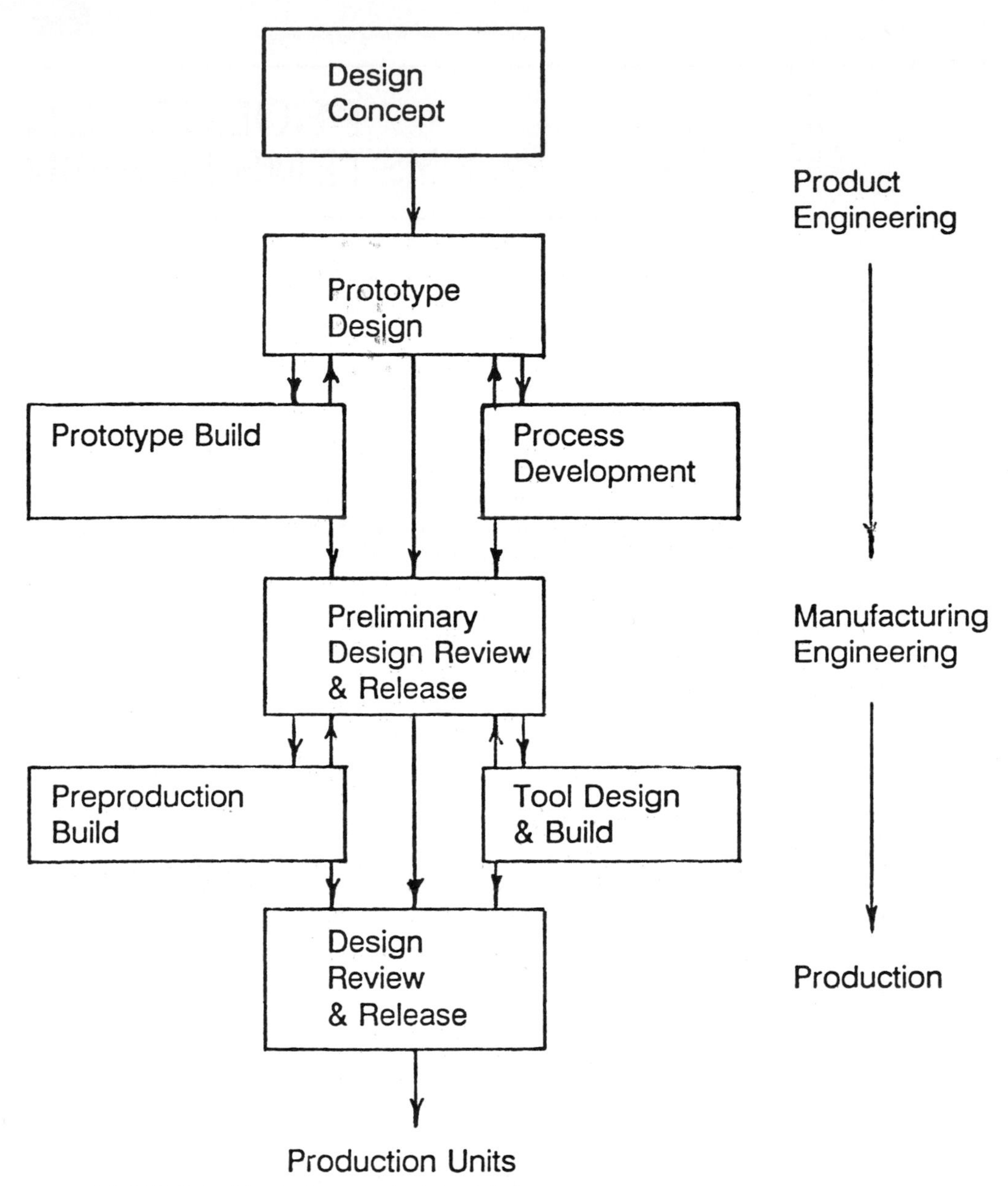

FIGURE 2-1 The design-manufacturing interface.

likely be developed during the next improvement cycle. This is the kind of interest a small manufacturer has in its products that no amount of central planning in the Soviet economy could have ever successfully duplicated. The Russians should talk to the people in Armstrong, Iowa, and watch them analyze, design, and make harvesters if they want to learn how to make farm machinery the American way.

If all goes well, the nine months between harvests will provide sufficient time to make the desired engineering tests, and to add design improvements needed before the release of material requisitions, so that next year's improved harvesters will be ready in time. It is not uncommon that, as the time approaches to make material releases for the improved designs, product engineering will beg for more time to run one more test. A saying in the Architectural and Engineering (A&E) business often applies in this situation: "Sooner or later you have to shoot the engineer and build the building." The best compromise often is for some product improvements to wait for next year's product design.

It will be noted in reviewing Fig. 2-1 that much of the above effort lies in interactions between product engineering, manufacturing engineering, and production. The manufacture of a new design is always a process of discovery. In Frederick W. Taylor's day, the new design for a machine element might be a line drawn on the molding shop floor, which the master mold maker would then use to produce a stronger machine element for the next test of the machine. When the stronger part was molded, it would be installed on the machine, the machine would run for some seconds, minutes, or hours, and would fail again, often in a different place, and the process would be repeated. Ultimately, a machine design would evolve that would produce a machine that would run all the time. These machines were sold to customers. With today's perfected engineering knowledge it is much more likely that the design will perform as designed, if not the first time, with far fewer prototype redesigns. Expert systems and Taguchi methods, both to be discussed later in this text, provide the means for doing a much better job today of optimizing both the product design in its underlying function and identifying the best means of manufacture for producing a high-quality, reliable, and cost-effective product.

COMPUTER-AIDED DRAFTING AND DESIGN (CADD)

Whereas design may have been accomplished with a stick on the molding shop floor in Taylor's time, CAD/CAE/CAM (computer-aided design, computer-aided engineering, computer-aided manufacturing) is becoming the preferred means today for producing designs. Most people think of CADD (computer-aided drafting and design) as simply electronic drafting, which greatly understates the computer revolution associated with the tasks implied in Fig. 2-1. The following excerpt from a paper by Floyd [11] concerning the use of CADD in the automotive industry provides some insight as to the overall comprehensiveness of the computer revolution in engineered product design today.

CAD/CAE/CAM FOR THE AUTOMOTIVE INDUSTRY

Bryan Floyd
Executive Manager,
Mechanical Design/Engineering/Manufacturing, Intergraph Corp.

Fully integrated design, engineering, and manufacturing. Automotive manufacturing is a complex business that integrates the efforts of many departments and disciplines. Tools that promote the integration of design, engineering, and manufacturing processes yield the greatest productivity benefits. Intergraph offers automakers the master model concept: a single intelligent product definition that drives all aspects of development, from concept through production. Intergraph's tightly integrated systems eliminate intermediate transfer or re-entry of data between design, analysis, and manufacturing phases. Additionally, all product development capabilities are simultaneously accessible through a single user interface, allowing engineers to combine functions, as needed, without changing environments.

Conceptual design and styling. Intergraph systems provide advanced tools for conceptual design and automotive styling, with high-performance graphics for concept visualization and communication. I/DESIGN, Intergraph's industrial design system, includes high-precision modeling and photo-realistic rendering capabilities that aid in developing a functional, ergonomic, and aesthetic design.

Precision geometric modeling. Automotive engineers require CAD/CAE/CAM modeling that can precisely describe complex surfaces and completely model intricate assemblies. Intergraph meets these demands with the Engineering Modeling System (I/EMS), which is based on highly accurate non-uniform rational B-spline (NURBS) mathematics. Intergraph is distinguished from other CAD/CAE/CAM vendors by offering advanced geometric modeling as a foundation for analysis and manufacturing.

Solid modeling. When designing an automobile, engineers must know critical geometric properties including masses and displacement volumes that are only available with solid modeling techniques. Property calculations, such as volume, cross sectional area, radius of gyration, moments of inertia, mass density, and others are included in I/EMS as standard functions supporting the software's solid modeling capability.

Assembly design and configuration management. Automotive development depends on a wealth of application data for thousands of components. Intergraph provides Product Data Manager (I/PDM) as a complete system for controlling and managing access to the product database. Without regard for physical storage locations, file names, or operating system platforms, engineers can locate and retrieve data from any location on a heterogeneous network.

Structural analysis. By simulating performance characteristics of designs before products are built, automakers complete designs in less time and reduce overdesign. Finite element analysis techniques help ensure compliance with performance standards and reduce the risk of failure in the field. These benefits are achieved with automatic and interactive meshing, h-adaptive refinement technology, an integrated solver, and full postprocessing functions, all available with Intergraph's Finite Element Modeling (I/FEM) system.

Plastics design and analysis. When integrated into the mechanical design process, plastics design and analysis functions can improve the quality of plastic components, increase yield, and reduce manufacturing cycle times. Plastics engineers can predict plastics behavior under molding conditions using the Injection Flow Analysis (I/FLOW) package. The I/FLOW model can then be used in conjunction with the Plastics Cooling Analysis (I/COOL) software to analyze heat transfer in cooling circuit layouts. By analyzing temperature distribution, engineers can reduce distortion and cooling times for injected plastic parts.

Mechanism and kinematic analysis. Engineers designing mechanical systems must determine how forces and motions vary over time to achieve performance goals and eliminate part-to-part interference. With Mechanical Systems Modeler (I/MSM), engineers analyze motions and part-to-part interactions and conduct kinematic and kinetostatic analyses with the built-in solution program. To conduct static equilibrium and dynamic analyses, engineers have access to I/MSM's modeling and post-processing functions and direct interface to third-party programs, including ADAMS and DADS.

Manufacturing capabilities. The diversity of processes required in automotive production demands a versatile set of manufacturing tools. Intergraph manufacturing solutions include the industry's broadest range of NC programming and fabrication tools. I/NC, Intergraph's off-line programming environment, supports machining capabilities for multiple-axis milling, thermal cutting, wire EDM, turret punching, and turning. Integrated fabrication software addresses flat pattern development and nesting processes.

Integrated design and manufacturing. To minimize production lead times, increase equipment and material yields, and reduce errors, Intergraph offers automakers complete CAD-to-CAM integration. Manufacturing processes are developed directly from the design model. An intelligent database structure automatically maintains the relationships between component geometries, toolpaths, machine and tool characteristics, and other variables to greatly reduce the input required to generate, maintain, and verify manufacturing data.

Electronic design and analysis. The increasing electronic content of automobiles demands coordination of electronic and mechanical design processes. To satisfy this demand, Intergraph provides a full suite of integrated electronics design applications. The Design Engineer series of products works in conjunction with mechanical design applications to promote a concurrent engineering environment.

Facilities management. To operate at peak efficiency, manufacturing facilities must optimize spatial and functional relationships. Designers can avoid trial-and-error space planning and factory layout with Project Planner, a software package that models facilities, simulates manufacturing scenarios, and determines an optimal "fit" within the building envelope.

Whereas Intergraph, Computervision, and AutoCad are often required CADD formats for the Department of Defense and other government entities, a myriad of other CADD software packages have been used by smaller users. Relatively inexpensive 486 operating system platforms, connected either to a plotter or by modem to a CADD service center, let the small business entrepreneur compete effectively in many application areas today. Some enhancements to AutoCad, such as FactoryCad [42], to be discussed later in this chapter, offer specialized CADD software capability to assist industrial facility planners and designers.

Fully integrating software systems into the design of both processes and plants is commonly referred to as computer-integrated manufacturing (CIM). All plants, every day of the week, are inching toward this ultimate goal, and all are progressing at different rates. Figure 2-2 provides a generic overview of the elements of a CIM technology approach. This figure might be compared with the two articles in Chapter 9, one authored by Parks and the other by Ellis-Brown, to note that we are quite rapidly moving toward that goal.

Since the days of Taylor, industrial engineers have been concerned with the design of manufacturing plants. At first, attention was centered on activity within an employee's workplace, and this type of analysis was first called "time and motion study" and later became known as "methods engineering." Later, attention was also given to methods of handling materials between workstations and the relative spatial arrangement of all entities within a plant. These two areas of anal-

FIGURE 2-2 Generic CIM architecture. (From Hodson [15, p. 7.147]).

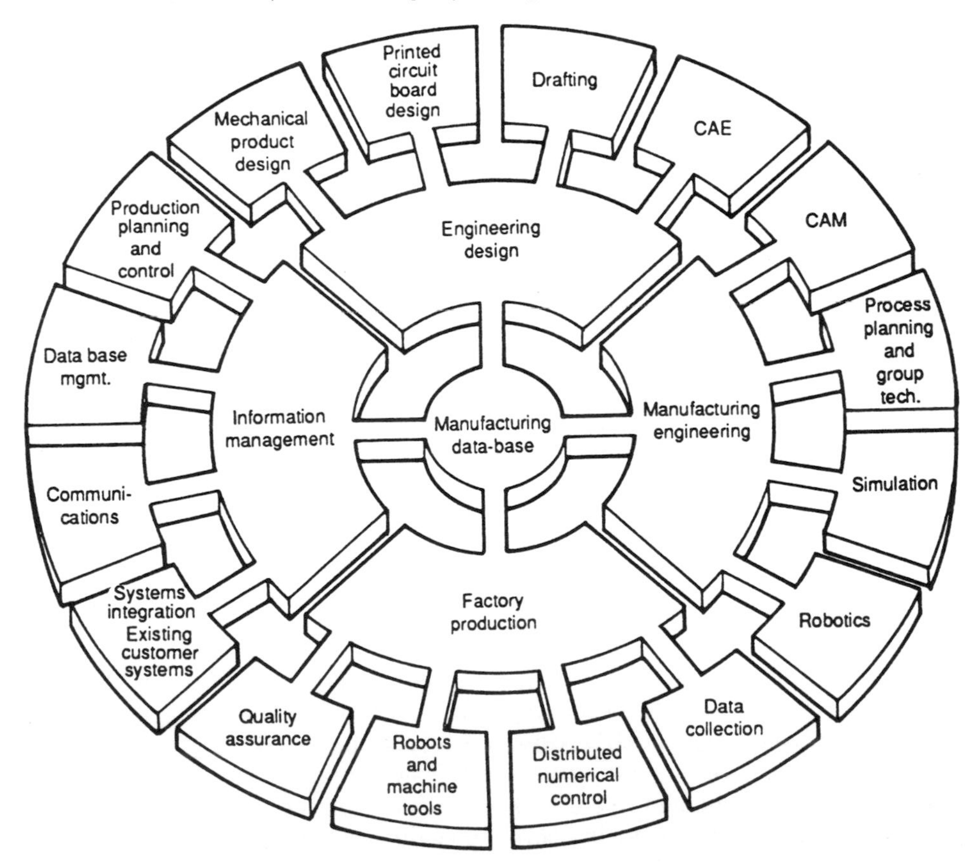

ysis are commonly referred to as materials handling and plant layout, respectively. All three activities—methods engineering, materials handling, and plant layout—and others are employed in what is referred to in industrial engineering circles as plant design. In Architectural and Engineering (A&E) circles, this preliminary activity would typically be included in Phase I of an A&E assignment under the heading Programming. This is not computer programming, but rather a first phase of a facility design concerned with determining the client's needs and wishes relative to a production system to be housed in a new, renovated, or expanded facility that will be designed by the architects and engineers. Industrial engineers first must design the production system to be housed in the facility. Obviously, a production system can be no better than its workstation designs because employees predominantly perform their work functions at workstations (that is where the action is).

METHODS ENGINEERING

A typical text on methods engineering covers both methods study and work measurement. Methods study is primarily concerned with design, and work measurement with control. The focus of this chapter is design; however because of the high degree of interdependence between methods study and work measurement, both are reviewed in this chapter.

Methods Study

The study of the detailed design of workstations, and to a lesser extent the relationships between workstations, is called *methods study.* In the planning stage, an estimate is made of the time it will take a typical employee to perform a given task at a workstation. Later, when the employee has learned the task and the conditions affecting the task have stabilized (e.g., tooling, material, method, and conditions are available and are consistently applied), management normally requires a detailed restudy of the job. Through observation and analysis, an industrial engineer or technician defines and documents the standard method and determines the time standard for performing the task, including nonproductive allowances. This time becomes management's official benchmark of how long it should take a typical trained employee at a normal rate of activity to perform the required operation, per unit of product, including prorated allowances. The sum of all operations for a product represents the expected direct labor time for a completed unit of product. This time, therefore, serves the purpose of providing management with a basis for determining employee performance by comparing the actual number of units produced by an employee, or employees, over a period of time and the number of units that the employee(s) would have produced based on standard time. The process of determining the standard time for an operation is called *work measurement*. The term *methods engineering* connotes both methods study and work measurement, which respectively attempt to answer the questions, How should a task be performed? and How much time should it take to perform the task, including allowances?

Charting Techniques Many of the traditional approaches in methods engineering can be classified as charting methods. These methods have changed very little in the last 50 years, but they still possess a general utility that has not been replaced by many of today's more sophisticated approaches. Here is a hint for students and managers alike: be concerned less with how sophisticated an approach is, and more concerned with what it produces. In general, charting methods offer a graphic dimension to a problem and also channel data collection with respect to it.

The operation process chart has been in use for many years to display in one figure the operations and inspections, and their sequence, for the manufacture of a complete product. Figure 2-3 is a typical operation process chart.

FIGURE 2-3 Typical operation process chart.

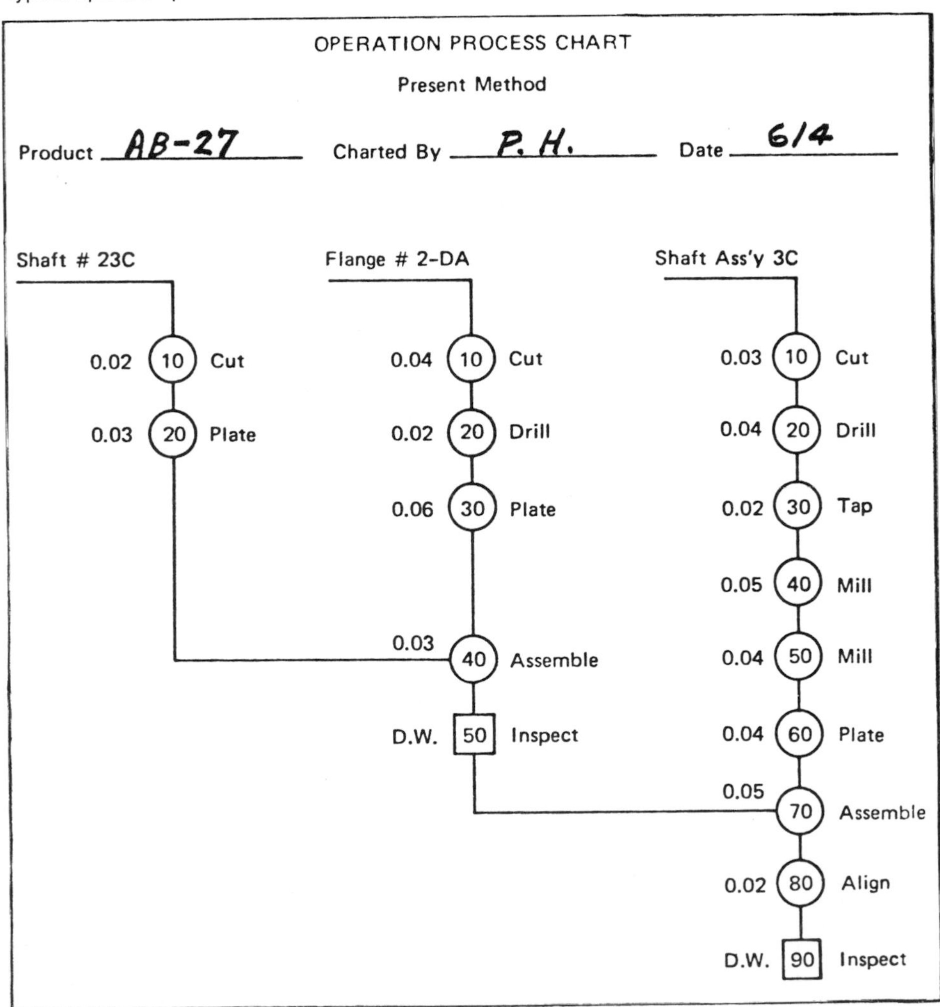

If the plant has a product volume requirement and other attributes that dictate a production line layout, the operation process chart offers a possible first hint as to a likely relative arrangement of processing operations. If you took the roof off a plant and looked at the operations from a helicopter, Fig 2-3 could represent the general product flows in the plant. In fact, we rarely have an opportunity to take the roof off a plant; however, plant personnel can readily view detailed equipment arrangements within limited areas with little difficulty. This is one of the reasons why areas in a plant are more likely to be poorly located with respect to one another, as compared to the equipment arrangements within areas. The traditional definition of an operation is that it includes all work performed by a worker or a crew at one location at one time.

After the creation of an operation process chart, the next step typically is to perform a more detailed analysis of each part or assembly of the total product. Whereas analysis was restricted only to operations and inspections in using the operation process chart, the flow process chart includes additional consideration of moves, delays, and storages. The boundary between delay and storage is the degree of control production has over the object. *Storage* connotes that the product has left production's control and is under the control of another entity (e.g., the stockroom or warehouse), and that retrieval of the product will require some form of authorization (i.e., paperwork). *Delay* implies that the product remains under production's direct control. Figure 2-4 is a typical flow process chart.

In effect, analysis has shifted from intraoperation handling to interoperation handling. Methods study is predominantly the study of transportation within a workstation, whereas the typical connotation of materials handling concerns the handling of materials between workstations.

Although the focus of most methods engineering work is individual workstations, the analysis extends to a series of workstations, as is obvious in the use of the flow process chart. Figure 2-5 is an example of a flow diagram that was used in conjunction with a flow process chart, to indicate on a layout the transportations involved in "comb inspection after sawing."

It may be instructive to reflect on the fact that a typical production process embodies only two main functions: transformations and moves. The transformations cause the product to change in nature and value as combined materials offer increased utility; for instance, an assembled bicycle is worth more than the sum of its parts. (If you do not believe this, ask the parents who have to assemble a tricycle on Christmas Eve if they would prefer to purchase it already assembled.) In a plant, material moves typically represent a necessary evil in terms of additional manufacturing cost, which therefore should be minimized. Movement of parts and assemblies within a plant does not add to a product's functional utility. For this reason, the single main criterion in plant layout is typically minimum material handling cost. The flow process chart is particularly effective in tracing the sometimes incredible distance a part travels in a plant, particularly if the plant layout has been allowed to evolve over time with little detailed analysis and improvement.

NO. 1
PAGE 1 OF 1

FLOW PROCESS CHART

JOB Receive air freight package and bring to outgoing freight area
☒ MAN OR ☐ MATERIAL Baggage handler
CHART BEGINS At receiving dock
CHART ENDS Outgoing freight area
CHARTED BY A.S. DATE 9/26/–

SUMMARY

	PRESENT		PROPOSED		DIFFERENCE	
	NO.	TIME	NO.	TIME	NO.	TIME
○ OPERATIONS	50	6.6				
⇨ TRANSPORTATIONS	43	21.3				
☐ INSPECTIONS	17	21.9				
D DELAYS	1	5.5				
▽ STORAGES	-	-				
DISTANCE TRAVELED	1471	FT.		FT.		FT.

	DETAILS OF (PRESENT) METHOD	DISTANCE IN FEET	QUANTITY	MIN. TIME	ANALYSIS: WHY? WHAT?	WHERE?	WHEN?	WHO?	HOW?	NOTES	ACTION: ELIMINATE	COMBINE	CHNGE: SEQUE.	PLACE	PERSON	IMPROVE
1	Other duties															
2	Goes to equipment area for hand truck	62		1.0	.	√				PLACE NEAR USE AREA				√		
3	Grasps hand truck and returns to receiving dock	62		1.0	√					"	√					
4	Loads packages on H.T.		4	.2	.	.	.	.	√	USE SEMI-LIVE SKID						√
5	Pushes H.T. to receiving dock scale	21		.5	.	.	.	.	√	"						√
6	Tips packages off H.T. onto scale		4	-	√					PAINT WEIGHT ON SKID	√					
7	Checks weight of each package		4	.8	√					"	√					
8	Checks packages for cond.		4	1.8	.	.	√			CHECK AS LOADED ON SKID			√			
	Loads packages on H.T.		4	.2	√						√					
10	Pushes to check-in area	32		.3	.	.	.		.							
11	Tips packages off H.T.		4	-	√					LEAVE ON SKID	√					
12	Returns with H.T. to receiving dock	26		.3	.	.	.	.	√							√
	Items 4-11 repeated 7 times Item 12 repeated 6 times															
75	Rec. air bill from truck driver and checks no. of pkgs. with A.B.		32	1.1	.	.	.	.	.							
76	Goes with driver to billing office	48		.65	√					USE WIRE BASKET ON OVERHEAD CABLE	√					
77	Waits while bill is processed and lot labels prepared			5.5	√						√					
78	Returns to pkgs. with processed copy of bill and lot labels	48		.65	√						√					
79	Pastes lot labels to each pkg. and air bill to one		32	1.8	.	.	.	.	√	USE STAMPING MACHINE						√
80	Loads packages on H.T.		4	.2	√						√					
81	Pushes H.T. to outgoing freight area	41		.6	.	.	.	.	√							√
82	Tips packages off H.T.		4	-	√						√					
83	Returns with H.T. to check in area	41		.6	.	.	.	.	√							√
	Items 80-82 repeated 7 times Item 83 repeated 6 times															
111	Returns H.T. to equip. area	30		.5	√						√					

FIGURE 2-4 Typical flow process chart. (From Hodson [15, p. 3.8].)

There is a class of charts commonly referred to as multiple activity charts. These charts can be effective in analyzing situations in which at least two resources are employed within an operation. The object typically is to group or arrange the sequence of events with respect to the resources employed so that a minimum unit production time is determined.

One such chart, commonly referred to as a man-machine chart, is illustrated in Fig. 2-6. This chart involves the analysis of four resources—an operator, a lead pot, a cooler, and a knockoff bar. The general approach in such an analysis is to attempt to make the best use of the key resource or resources. In this example the

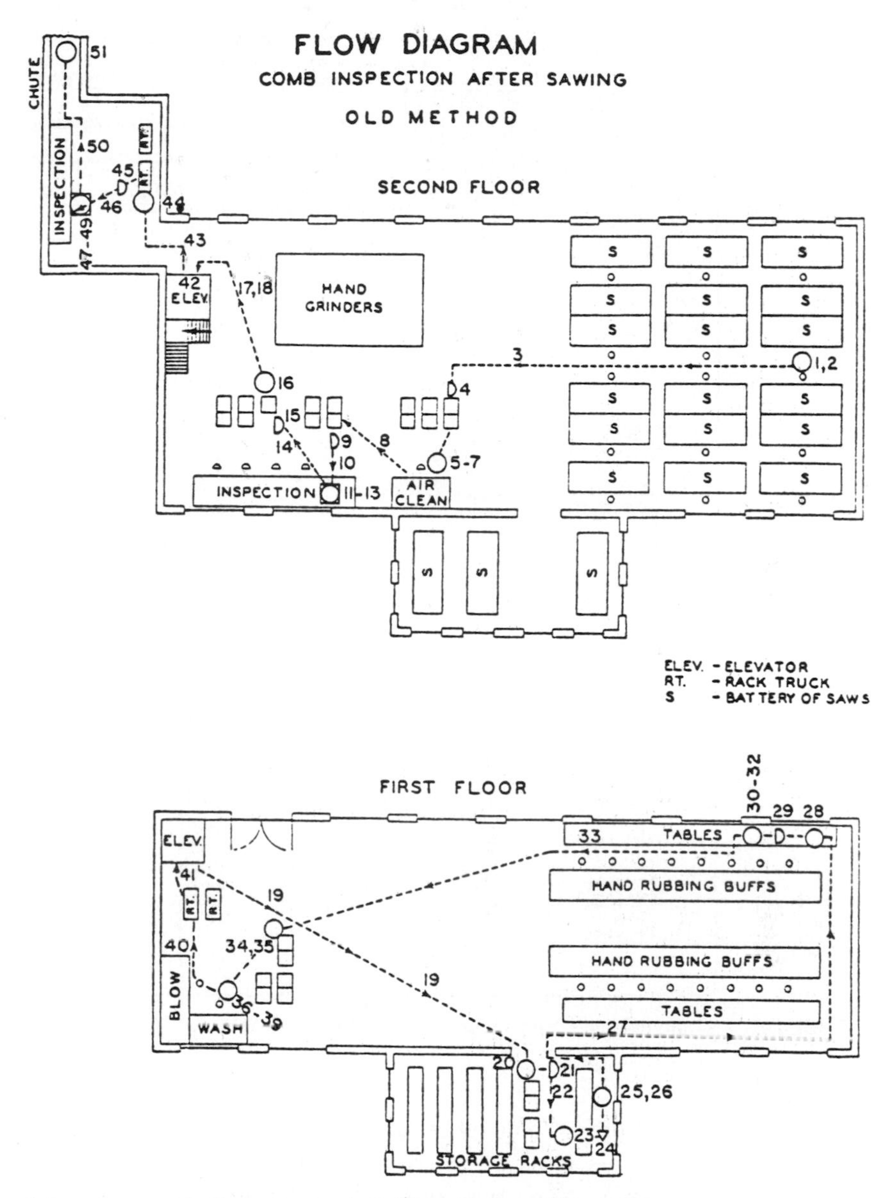

FIGURE 2-5 A flow diagram of a flow process chart. (From Hodson [15, p. 3.9].)

operator is the key resource and is busy continuously. The focus of the analysis is to minimize the cycle time of the operator. Because of extreme competitiveness in the garment industry, multiple activity analysis is common and accepted in that industry. There are test operations in other, less profit-margin-sensitive industries, for example, in which a test employee sits idle for long periods of time waiting for

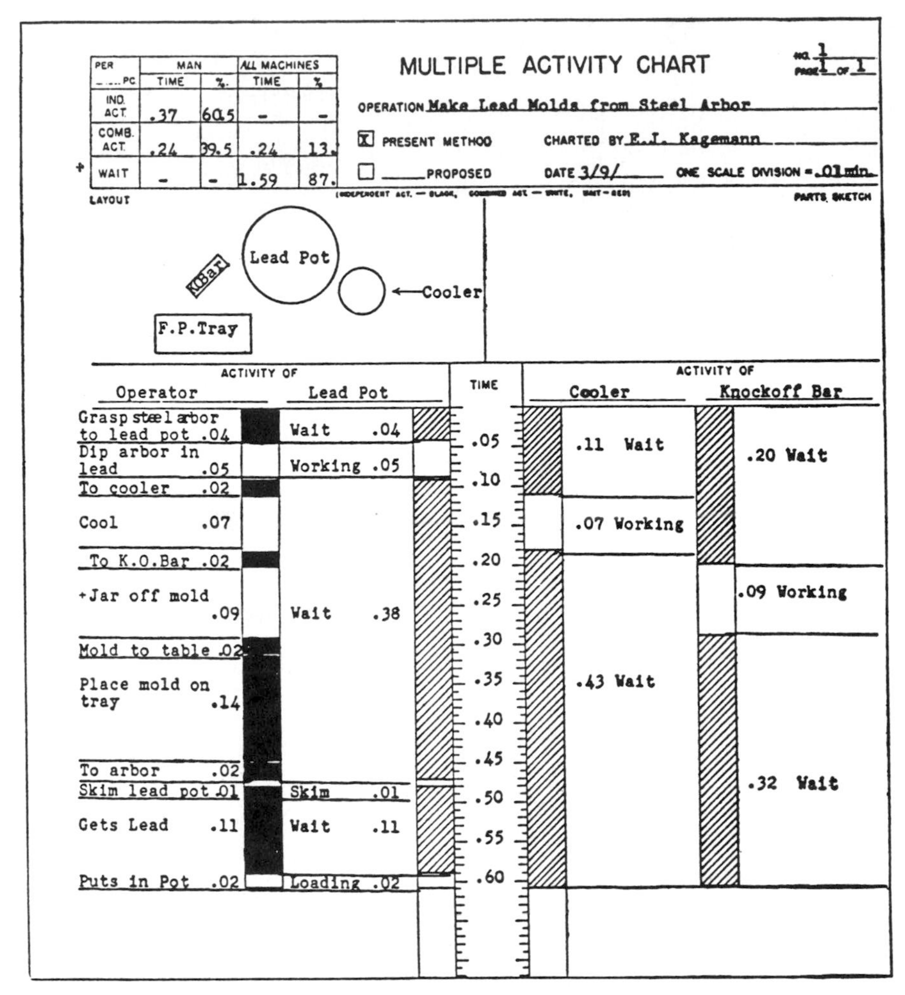

FIGURE 2-6 A man-machine chart. (From Hodson [15, p. 3.20].)

temperature or some other variable to stabilize. In such instances, multiple test stations are often not designed or collocated to utilize the multiple activity approach, forgoing the efficiencies provided by this approach.

The "before" and "after" gang charts in Fig. 2-7 and 2-8 illustrate how the elemental activities of workers acting as a gang or crew can be analyzed to determine a minimum cycle time for a unit of production. Note that both charts have a *working time* (a cycle time) in the lower left corner of each chart of 1 minute, and yet the before and after crews are composed of six and four crew members, respectively. Note in reviewing the before chart (Fig. 2-7), that there is considerable idle time for four of the six gang members. A significant part of the solution involves eliminating this idle time.

GANG PROCESS CHART OF PRESENT METHOD

HYDRAULIC EXTRUSION PRESS DEPT. 11 BELLEFONTE PA. PLANT

CHARTED BY B.W.N. 4-15- CHART NO. G-85

MACHINE

OPERATION	TIME
Elevate Billet	.07
Position Billet	.08
Position Dummy	.04
Build Pressure	.05
Extrude	.45
Unlock Die	.06
Loosen & Push Out Shell	.10
Withdraw Ram & Lock Die in Head	.15

PRESS OPERATOR

OPERATION	TIME
Elevate Billet	.07
Position Billet	.08
Position Dummy	.04
Build Pressure	.05
Extrude	.45
Unlock Die	.06
Loosen & Push Out Shell	.10
Withdraw Ram & Lock Die in Head	.15

ASSISTANT PRESS OPERATOR

OPERATION	TIME
Grease Die & Fosition Back in Die Head	.12
Idle Time	.68
Run Head & Shell Out	.11
Shear Rod from Shell	.04
Pull Die Off End of Rod	.05

FURNACE MAN

OPERATION	TIME
Rearrange Billets in Furnace	.20
Idle Time	.51
Open Furnace Door & Remove Billet	.19
Ram Billet from Furnace & Close Furnace Door	.10

DUMMY KNOCKER

OPERATION	TIME
Position Shell on Small Press	.10
Press Dummy Out of Shell	.12
Dispose of Shell	.18
Dispose of Dummy and Lay Aside Tongs	.12
Idle	.43
Grab Tongs & Move to Position	.05

ASSISTANT DUMMY KNOCKER

OPERATION	TIME
Move Away from Small Press and Lay Aside Tongs	.12
Idle Time	.68
Guide Shell from Shear to Small Press	.20

PULL-OUT MAN

OPERATION	TIME
Pull Rod toward Cooling Rack	.20
Walk Back toward Press	.15
Grab Rod with Tongs and Pull Out	.45
Straighten Rod End with Mallet	.11
Hold Rod while Die Removed at Press	.09

	MACHINE	PRESS OPERATOR	ASSISTANT PRESS OPERATOR	FURNACE MAN	DUMMY KNOCKER	ASSISTANT DUMMY KNOCKER	PULL-OUT MAN
WORKING TIME	1.00 MIN.	1.00 MIN.	.32 MIN.	.49 MIN.	.57 MIN.	.32 MIN.	1.00 MIN.
IDLE TIME	0 "	0 "	.68 "	.51 "	.43 "	.68 "	0 "

IDLE TIME = 2.30 MAN-MINUTES PER CYCLE = 18.4 MAN-HOURS PER EIGHT-HOUR DAY

FIGURE 2-7 A gang chart before analysis. (Reproduced with permission from Niebel [36, p. 156].)

GANG PROCESS CHART—PROPOSED METHOD

Hydraulic Extrusion Press Dept. 11 Bellefonte, Pa. Plant

Charted by B.W.N. 4-15 Chart G-85

MACHINE OPERATION	TIME	PRESS OPERATOR OPERATION	TIME	ASSISTANT PRESS OPERATOR OPERATION	TIME	DUMMY KNOCKER OPERATION	TIME	PULL-OUT MAN OPERATION	TIME
Elevate Billet	.07	Elevate Billet	.07	Grease Die & Position Back in in Die Head	.12	Position Shell on Small Press	.10	Pull Rod toward Cooling Rack	.20
Position Billet	.08	Position Billet	.08	Walk to Furnace	.05	Press Dummy Out of Shell	.12	Walk Back toward Press	.15
Position Dummy	.04	Position Dummy	.04	Rearrange Billets in Furnace	.20	Dispose of Shell	.18	Grab Rod with Tongs and Pull Out	.45
Build Pressure	.05	Build Pressure	.05	Return to Press	.05	Dispose of Dummy and Lay Aside Tongs	.12		.11
Extrude	.45	Extrude	.45	Idle Time	.09	Idle Time	.23		.09
Unlock Die	.06	Unlock Die	.06		.19	Grab Tongs & Move to Position	.05		
Loosen & Push Out Shell	.10	Loosen & Push Out Shell	.10		.10	Guide Shell from Shear to Small Press	.20		
Withdraw Ram & Lock Die in Head	.15	Withdraw Ram & Lock Die in Head	.15	Run Head & Shell Out	.11				
				Shear Rod from Shell	.04				
				Pull Die Off End of Rod	.05				
Working Time	1.00 Min.		1.00 Min.		.91 Min.		.77 Min.		1.00 Min.
Idle Time	0		0		.09 Min.		.23 Min.		0

FIGURE 2-8 A gang chart after analysis. (Reproduced with permission from Niebel [36, p. 157].)

A wide variety of charts have been and are still in general use for analyzing the sequence of events of parts and product movement both within and between workstations. Figure 2-9 is a right- and left-hand chart. Note that the left hand is holding the bolt throughout the assembly. By employing a holding fixture that orients and holds two bolts extending up, bolts can be picked up by both hands and placed into the fixture two at a time; and both hands can proceed to add the other parts, two at a time, almost doubling the output of the operation. Most operators employing such fixtures initially have difficulty assembling with their left hand until they acquire dexterity with that hand. Without the employment of such fixtures, assembly demands are not placed on the left hand, and therefore the required dexterity is not acquired.

Workplace Design In a real sense, a typical production employee works in a workstation, not a plant. Therefore, it is the sum effect of well-designed workstations that results in a productive plant. For this reason, most methods engineering is performed in relation to workstations. Figure 2-10 represents a typical work chair today, particularly for office workers, because most office employees sit while they work. Figure 2-11 offers basic biometric data for a "sit or stand" workbench that was popular in years past and less common today. Sit or stand

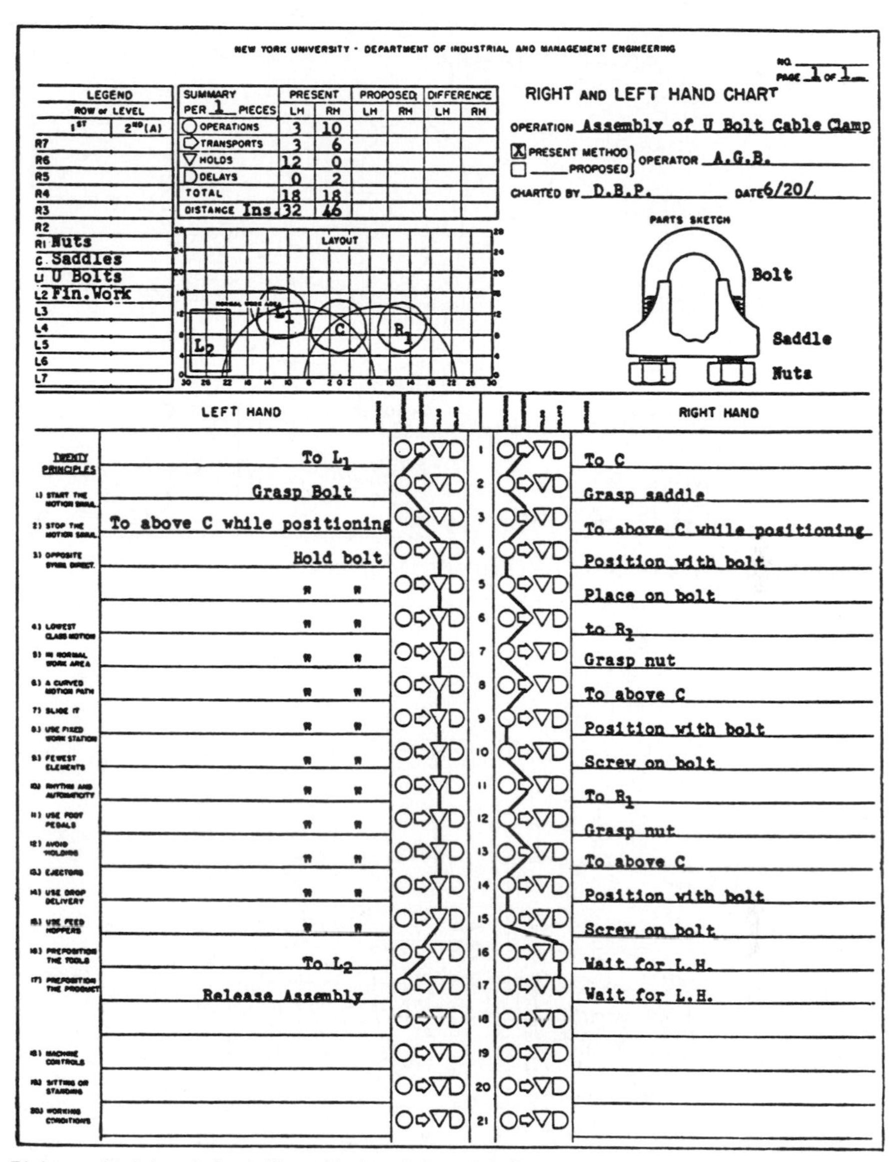

NEW YORK UNIVERSITY - DEPARTMENT OF INDUSTRIAL AND MANAGEMENT ENGINEERING

NO. ____ PAGE 1 OF 1

RIGHT AND LEFT HAND CHART

OPERATION Assembly of U Bolt Cable Clamp

[X] PRESENT METHOD / [ ] PROPOSED — OPERATOR A.G.B.

CHARTED BY D.B.P. DATE 6/20/

SUMMARY PER 1 PIECES	PRESENT LH	PRESENT RH	PROPOSED LH	PROPOSED RH	DIFFERENCE LH	DIFFERENCE RH
OPERATIONS	3	10				
TRANSPORTS	3	6				
HOLDS	12	0				
DELAYS	0	2				
TOTAL	18	18				
DISTANCE Ins.	32	46				

LEGEND — ROW or LEVEL: R7, R6, R5, R4, R3, R2, R1 Nuts, C Saddles, U U Bolts, L2 Fin. Work, L3, L4, L5, L6, L7

LAYOUT — L_1, C, R_1, L_2

PARTS SKETCH — Bolt, Saddle, Nuts

	LEFT HAND		RIGHT HAND
	To L_1	1	To C
	Grasp Bolt	2	Grasp saddle
	To above C while positioning	3	To above C while positioning
	Hold bolt	4	Position with bolt
	" "	5	Place on bolt
	" "	6	to R_1
	" "	7	Grasp nut
	" "	8	To above C
	" "	9	Position with bolt
	" "	10	Screw on bolt
	" "	11	To R_1
	" "	12	Grasp nut
	" "	13	To above C
	" "	14	Position with bolt
	" "	15	Screw on bolt
	To L_2	16	Wait for L.H.
	Release Assembly	17	Wait for L.H.
		18	
		19	
		20	
		21	

TWENTY PRINCIPLES: 1) Start the motion simul. 2) Stop the motion simul. 3) Opposite symm. direct. 4) Lowest class motion 5) In normal work area 6) A curved motion path 7) Slide it 8) Use fixed work station 9) Fewest elements 10) Rhythm and automaticity 11) Use foot pedals 12) Avoid holding 13) Ejectors 14) Use drop delivery 15) Use feed hoppers 16) Preposition the tools 17) Preposition the product 18) Machine controls 19) Sitting or standing 20) Working conditions

FIGURE 2-9 Right- and left-hand chart. (From Hodson [15, p. 3.21].)

workbenches are typically used in situations where employee trunk movement is expected in performing the work, such as in a packing station. Employees often prefer such an opportunity of alternate sitting and standing. The goal in methods study typically is to develop a best work design for a workstation such that a best sequence of movements of the hands of an employee result in a minimum operation time (i.e., the product transformation is accomplished at a minimum labor cost). Figure 2-12 is an operation sheet that specifies the workplace arrangement

FIGURE 2-10 Typical work chair. (From UAW [44, p. 28].)

FIGURE 2-11 Dimensions of normal and maximum work areas. (From Barnes [6, p. 260].)

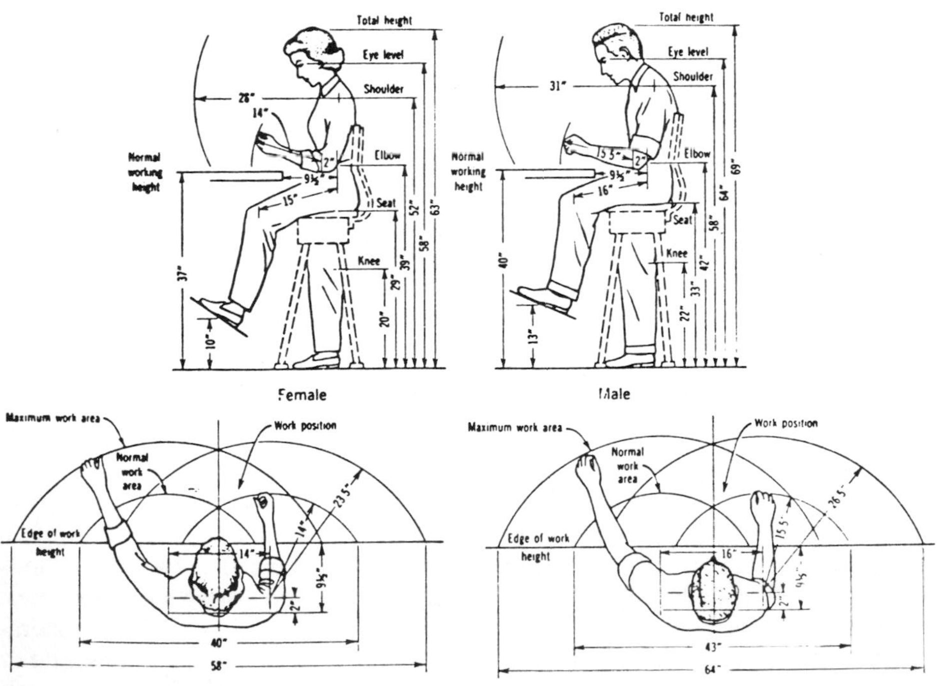

HIXY CLAMP COMPANY

Operation Sheet

Oper. Desc. BONDING OF HOLDER Oper. No. 125

Drawn By P. H. Dept. 4-12 Date 4/7/ Sheet 1 of 1

#2257
#2256
#1858
#S-23-C
#T-12B
#T-127

Parts	Supplies	Tools
#2256	#23-C EPOXY	#12B CLAMP
#2257		#127 SPREADER
#1858		

Operation

Step	Description
1	LOCATE #2257 IN CLAMP.
2	ALIGN #2256 WITH RESPECT TO #2257.
3	SPREAD EPOXY ON #2257 & #2256 PER SPEC. NO. 116-2.
4	DRY FOR 10 MINUTES.
5	SCREW #1858 TO 1/4" DEPTH.
6	RELEASE ASSEMBLY FROM CLAMP AND DISPOSE.

FIGURE 2-12 An operation sheet.

and sequence of motions to be employed. An ideal workstation minimizes intraoperation handling through detailed analysis and specification of the exact method to be employed.

Motion Economy Principles A number of common principles of good design have evolved over the years and are commonly referred to as "principles of motion economy." Twenty-two motion economy principles from Barnes' text, which have been in common use for a number of years, are as follows [7, pp. 174–236]:

1 The two hands should begin as well as complete their motions at the same time.

2 The two hands should not be idle at the same time except during rest periods.

3 Motions of the arms should be made in opposite and symmetrical directions, and should be made simultaneously.

4 Hand and body motions should be confined to the lowest classification with which it is possible to perform the work satisfactorily.

5 Momentum should be employed to assist the worker wherever possible, and it should be reduced to a minimum if it must be overcome by muscular effort.

6 Smooth continuous curved motions of the hands are preferable to straight-line motions involving sudden and sharp changes in direction.

7 Ballistic movements are faster, easier, and more accurate than restricted (fixation) or "controlled" movements.

8 Work should be arranged to permit easy and natural rhythm wherever possible.

9 Eye fixations should be as few and as close together as possible.

10 There should be a definite and fixed place for all tools and materials.

11 Tools, materials, and controls should be located close to the point of use.

12 Gravity feed bins and containers should be used to deliver the material close to the point of use.

13 Drop deliveries should be used wherever possible.

14 Materials and tools should be located to permit the best sequence of motions.

15 Provisions should be made for adequate conditions for seeing. Good illumination is the first requirement for satisfactory visual perception.

16 The height of the work place and the chair should preferably be arranged so that alternate sitting and standing at work are easily possible.

17 A chair of the type and height to permit good posture should be provided for every worker.

18 The hands should be relieved of all work that can be done more advantageously by a jig, a fixture, or a foot-operated device.

19 Two or more tools should be combined wherever possible.

20 Tools and materials should be pre-positioned wherever possible.

21 Where each finger performs some specific movement, such as in typewriting, the load should be distributed in accordance with the inherent capacities of the fingers.

22 Levers, hand wheels, and other controls should be located in such positions that the operator can manipulate them with the least change in body position and with the greatest speed and ease.

Principle 4 above, as an example, refers to the "lowest (motion) classification." The five motion classifications are (1) a finger movement, (2) a wrist movement, (3) movement from the elbow, (4) movement from the shoulder, and (5) a trunk movement. Note in Fig. 2-11 that the *normal work area* does not exceed the level

3 motion classification. The maximum work area does not exceed a level 4 motion. Much less time and energy are required for hand motions of level 3 or less, so good bench assembly workstation design attempts to limit motions to this reach envelope. It takes considerably more energy to reach from the shoulder (level 4), than from the elbow (level 3) with the upper arm acting only as a vertical link. Recalling your statistics, consider the difference in relative moment generated from the shoulder compared to the elbow, resulting from the difference in their respective moment lengths. Energy minimization (i.e., fatigue) is an important variable to consider in developing efficient bench assembly operations. Table 2-1 is a list of typical workstation design considerations.

The use of these principles in combination with a generally questioning attitude has traditionally been very productive in developing efficient workstation designs. A challenge if you are in a plant now (or will be tomorrow): randomly select six employees as a sample from the plant telephone directory, and review their workstations with them using the list above. The productivity potential is often enormous. There is still considerable need today to improve workstation designs so that workers can be more productive and take more pleasure and pride in their work.

CONCURRENT ENGINEERING

In recent years there has been a growing recognition of the need to change the traditional product design process. A new approach generally referred to as *concurrent engineering* or *simultaneous engineering* is improving the design process.

The traditional approach involved a sequence of steps, each performed in different departments, which culminated in a final product design. The departmentalization of the effort into relatively independent sequential steps produced rivalries between functions, limited communication, and an inability of the design to evolve with full simultancous participation of all design-related participants. Sequential consideration of the design essentially requires each player to provide his or her

TABLE 2-1 TYPICAL WORKSTATION DESIGN CONSIDERATIONS

Proper equipment
Proper tools
Proper hand or elbow height relative to tasks
Minimization of reach distances
Adequate task lighting
Proper seating (back/foot rest, seat angle or "sit-stand" if appropriate for task, etc.)
Proper seat height for focal length relative to task
Adequate space for material and tools
Location and positioning of tools (e.g., overhead tension reels)
Proper distance and orientation of monitoring equipment
Proper distance and orientation of input data
Efficient means for identifying place on input data
Proper placement of incoming materials to eliminate or minimize bending or walking to materials
Work surface position and orientation relative to task

best professional input based on whatever combined knowledge is available concerning the project up until that time. Once a technical decision is made by a function in an organization (e.g., marketing), the function is less inclined at a later date to revise its earlier recommendations when new and better information becomes available, because it would require starting the whole process over again. Human nature is such that professionals dislike having to admit that their earlier opinions were incorrect; therefore, they have the bad habit of defending past positions, even in the face of new information.

Concurrent engineering utilizes the concept of a team approach to product design. All professionals assigned to the design team jointly and simultaneously initiate the design; thus, if a design concept offered by one member of the team will potentially cause problems in another discipline, that design concept can be dismissed as unacceptable when it is in its formative stage, without excessive bruising of the other professional's ego. Professional egos are one of the greatest barriers to progress in many projects involving interaction between numerous professional players in a major design effort.

There is general recognition today that the preponderance of quality problems in manufacturing and in product use have as their origin the product design; in some cases the quality problem arose during general conceptual development (e.g., in not adequately understanding underlying customer needs) and in others it occurred in the product engineering design that followed conceptual development. There is so much accumulated process knowledge available today that it would be unreasonable to assume that a product engineer, familiar with the product and its operating environment, would also be fully knowledgeable about all the processing choices and variables involved. Three texts that provide a representative storehouse of manufacturing engineering knowledge concerning processes are Koenig [19], Ludema [23], and Tanner [45]. These texts demonstrate the need for teamwork between product engineering and manufacturing engineering in what ideally should be the simultaneous development and design of both the product and its associated manufacturing process. A sequential design process involving first design of the product followed by design of the process is a flawed, if not doomed, development process in these times. Product engineering and manufacturing are only two of typically many members of a product/process development process that would work best today. The team representation should be primarily a function of the products and processes involved.

Another development in recent years is the concept of "producibility" or "manufacturability."

Producibility

The following lists highlight the limitations of methods engineering, the traditional approach, as compared to a producibility approach:

The Methods Engineering Approach

- Product design is given (i.e., relatively frozen).
- Manufacturing and methods engineering follow the product design.

- Equipment and tools are selected and arranged to best accommodate the manufacturing process for the given product design.
- Equipment is typically selected from what is presently available.
- Tools are purchased or designed to accommodate the equipment selected.

Focus: To create the best workstation design for a given product design and available equipment and tools.

The Producibility Engineering Approach

- The product design can be changed to accommodate final intended product function at lowest cost.
- The producibility engineer is a member of the product design team, advising the product designer of the manufacturing alternatives available and their relative effects and costs.
- Product design and process design are simultaneous and dependent activities.
- The product may be modified to accommodate tooling requirements to reduce cost of manufacture.
- The producibility engineer optimizes at lowest cost the product design in relation to intended function, concerning the determination of design features affecting equipment capabilities/limitations, tolerances, material selection, and process controls.
- The team may specify modifications to existing or purchased equipment or tools, or provide design criteria for the purchase or development of new equipment or tools.

Focus: To produce the most manufacturable product design; employ existing or purchased equipment and tools or build new ones as required, and then provide the best workstation design.

With the earlier traditional approach of methods engineering, one means to changing the product design was through the use of "value engineering." This approach typically was a team approach to reevaluating a product design after its completion. Unfortunately, since the design was at least initially complete, those who designed it were placed in the position of having to defend their original designs. A value engineering effort can be very effective in uncovering design weaknesses before those shortcomings show up in the field, but the process can be disheartening for those who are forced to defend their design decisions after the fact. A process, such as producibility engineering, that disposes of flawed design concepts during team deliberations, before they become part of a hardened design, simply saves everyone involved the necessity of extricating the design weaknesses after they have become a formal part of the design.

Manufacturability Tanner [45] authored a five-part series in *Automation* starting in May 1989 concerning manufacturability. Figure 2-13 offers one example of the numerous manufacturability concepts provided in this series.

One of the recognized problems in the traditional product design process, particularly in the United States, has been the practice of allowing recent graduates of

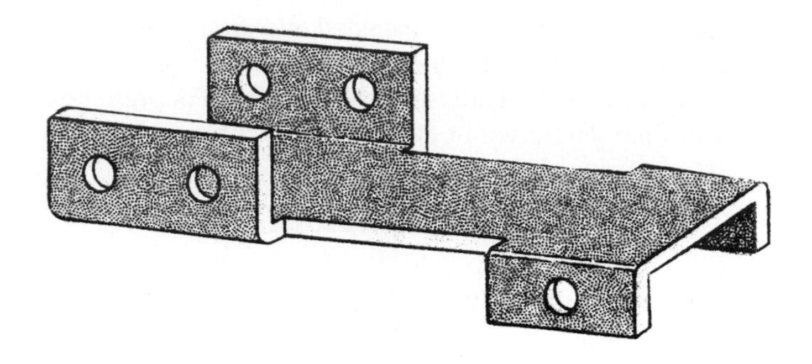

Not Recommended - Two Forming Operations

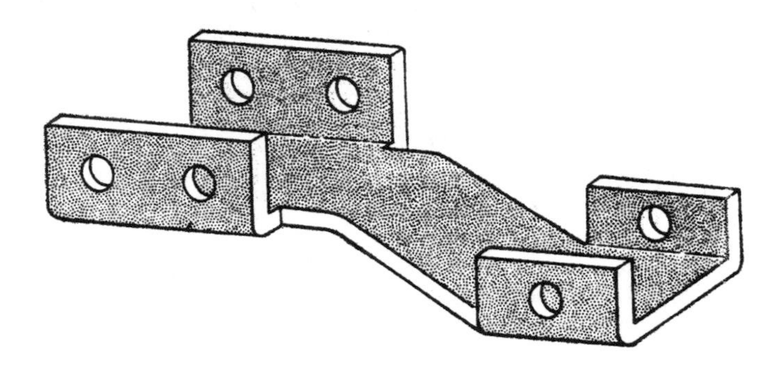

Recommended - One Forming Operation

FIGURE 2-13 A manufacturability example. (From Tanner [45].)

engineering programs (e.g., a mechanical engineer) to perform product design with insufficient acquired experience(or none at all) in the manufacturing processes required for making the product. It is a well-known fact in industry that numerous product designs are issued to production that can be made only with great difficulty, if at all, because product designers often do not know the capabilities and limitations of the processing equipment. This condition has actually worsened in recent years as engineers receive an increasingly theoretical engineering education, which supplants previous, less theoretical courses that attempted to provide some understanding of machining and other processing knowledge. The

Germans and Japanese have done a better job of requiring that product designers acquire shop-floor experience before they are allowed to produce product designs. As will be discussed later in this text, most product quality problems have as their source a design element that made their existence possible. The preponderance of quality problems can best be eliminated within the design process.

Design for Assembly (DFA) is a product design concept that concentrates on utilizing a thorough understanding of assembly operations in the design of a product. A video, produced by a major United States manufacturer for in-house use and viewed by the author, effectively describes and demonstrates the assembly of a computer printer developed by a cross-functional design team (product design, quality control, manufacturing, etc.). The final assembly of the printer is performed in 3.5 minutes with a labor-cost-effective, high-quality, and reliable design that does not utilize any traditional fasteners (bolts, nuts, screws, etc.) or springs. Instead, parts are designed to snap together or be assembled in such a way that they maintain their required relative positions. The video offers a dramatic example of how design for manufacturing can significantly simplify the assembly process.

ROBOTICS

A development that reached its peak in the 1980s was the introduction of the robot in industry. Four basic robot configurations are shown in Fig. 2-14. Figure 2-15 indicates degrees of freedom in a robot wrist.

Robots work well for tasks in which the same product is always available in the same place and in the same orientation. Industry abounds with excellent examples of such tasks: welding parts together, punching holes in parts, forging parts, painting parts and assemblies, and so on. Note that welding, punching, forging, and painting all involve relatively hostile environments. Robots do not perceive, because they are unable to perceive, hostile environments as being hostile. They go on welding or forging or painting unaffected (at least psychologically) by the hostile nature of their surroundings as if they were on a Sunday afternoon picnic. On rare occasions a robot may reach into a press, which no longer requires the expensive and limiting safety devices for humans, and because of some inappropriate timing get a finger chopped off. All that is required is to go to the finger drawer, get a new finger, and screw it on and the robot is good as new. Industry abounds with applications yet to be analyzed in which the robot is a far superior choice to a worker for both a safer and a more cost-effective means for performing a task. Figure 2-16 is a conceptual design for a robot palletizing four different products being delivered for palletization.

What if the products received in the same general location differ (have different plan-view outlines), and their orientation (their specific 3-dimension rotational position) is variable? On the leading edges of robot technology today are robots that can see (have vision systems). The robot that Dr. Kok Meng Lee of the Georgia Institute of Technology is adjusting in Fig. 2-17 has vision capabilities. The different products are placed in general locations, separated by carton

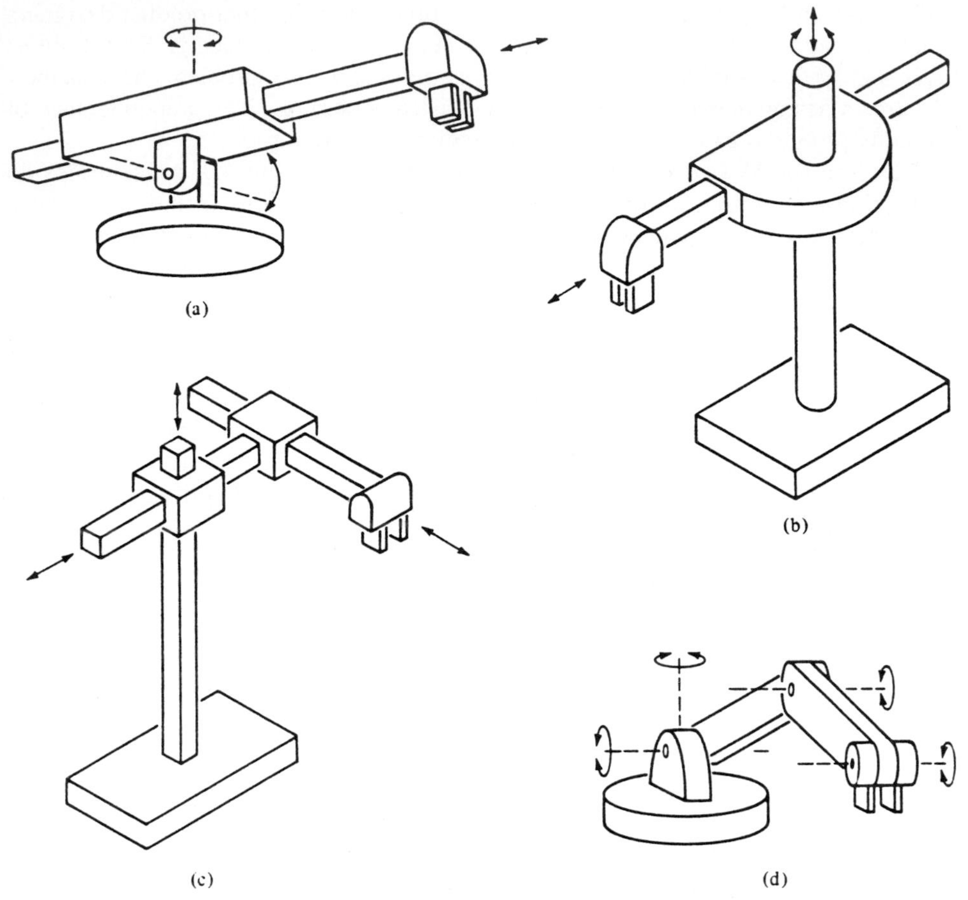

FIGURE 2-14 Four basic robot configurations: (a) polar, (b) cylindrical, (c) cartesian, and (d) jointed arm. (From Hodson [15, p. 7.173].)

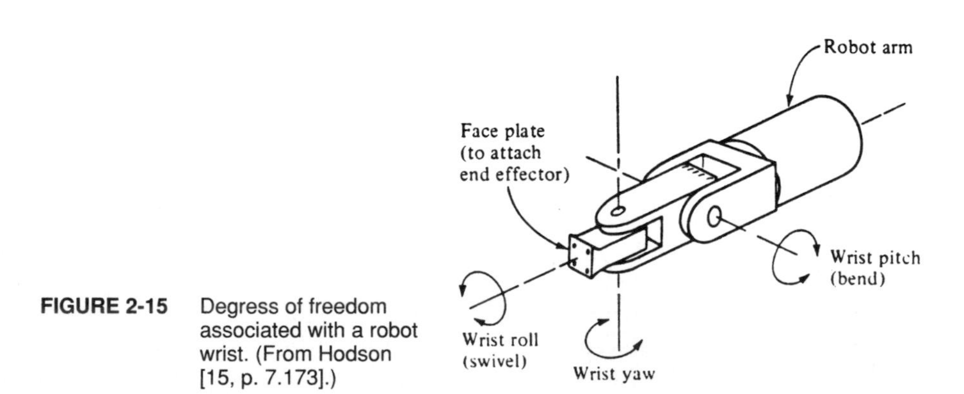

FIGURE 2-15 Degress of freedom associated with a robot wrist. (From Hodson [15, p. 7.173].)

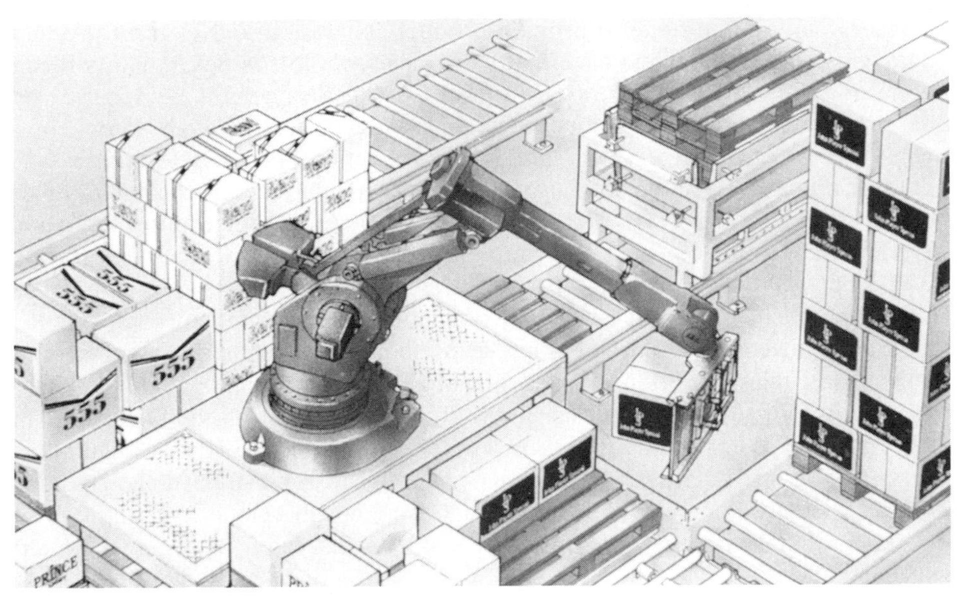

FIGURE 2-16 A robot concept for multiproduct palletization. (Source: ABB Robotics Inc., New Berlin, Wisconsin.)

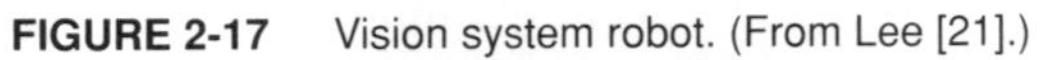
FIGURE 2-17 Vision system robot. (From Lee [21].)

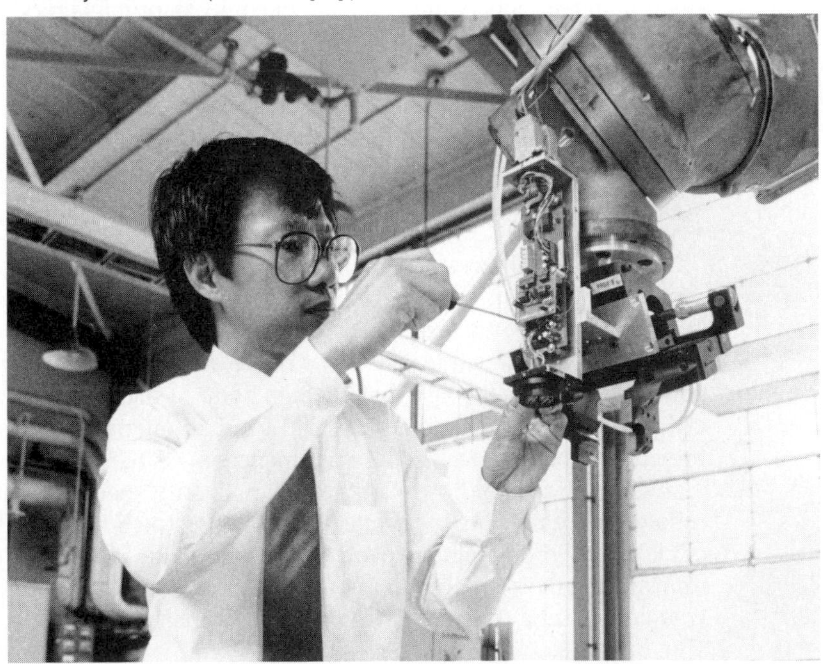

dividers, as is often done to provide product separation and to limit product damage. Here, the dividers have a reflective surface; this is not typically being done now but could be. The robot first moves to the general *x-y* location of the product tray, scans the product at the location with its reflective vision, compares the product with its stored library of product outlines to determine which product it is looking at, determines its rotational orientation, and then determines how best to orient its arm to pick it up. Note that it is performing the same eye/hand coordination functions you would perform if you were attempting to pick it up.

One problem with robots is that they often mimic human motions (i.e., complex rotational movements). The author has seen robots employed in practice for functions that could have been performed much more efficiently, and with less maintenance cost with in-line transfer machines. Robots have their place in industry, but there are also places in industry where a robot is a bad choice. That is what industrial engineering is all about—distinguishing between good and bad choices. GM some years ago got onto the robot bandwagon and bought a lot of robots, many of which have since been replaced. It was a classic example of "silver bullet" management at work.

Robots became such a fad in the 1980s that every plant had to have one. There are a lot of plants in the United States, as elsewhere, that make needed products with operations that have little if anything particularly interesting about them; they use the same equipment everyone else uses, and touring these plants can be a relatively dull event; hence the need for at least one robot to serve as something interesting for visitors to look at. In most plants, what the plant manager wants the plant manager gets, which has taught some plant managers to be cautious about expressing wants.

A story of admittedly questionable validity was circulated about a plant manager who insisted that the manufacturing engineering department in his plant determine the best application in the plant for a robot, and that they procure one without delay. As the story goes, manufacturing engineering found Harold lifting product off of a monorail conveyor one unit at a time and placing it on a belt conveyor.

A robot was procured without delay, and the robot replaced Harold lifting the product off hangers on the monorail conveyor and placing it on the belt conveyor. When visitors toured the plant with the plant manager, he typically brought them by the robot to show them how the robot transferred product from the monorail conveyor to the belt conveyor (the high point of the tour) without any expenditure of human labor; the manual task had been automated. Everyone involved in the project was happy because the plant manager was happy.

As time went on, however, the cheap, hastily specified, inherently unreliable robot began damaging the product when it attempted to grasp the product and then lift it off the monorail conveyor. Occasionally, the robot would grasp the product improperly and drop the product on the floor on the way to the belt conveyor. It was determined that if the product was properly located for the robot where it formed its grasp, the robot could successfully perform the task. The decision was made to have Harold position the product for the robot to grasp it. As a result, Harold positioned the product for the robot, the robot carried and placed the product onto the belt conveyor, and the robot maintenance technician spent a significant amount of his time attempting to prevent further deterioration of the robot. No one ever mentioned the robot to the plant manager after that, and no additional robots were purchased until he retired shortly thereafter as plant manager. In the year following his retirement as plant manager, Harold had his old job back lifting product off the monorail conveyor and placing it on the belt conveyor. This story is not intended as an indictment of robots; it is an indictment of doing engineering in haste and with possibly limited justification.

The lesson of this story is that a robot, or any machine for that matter, added to a process can be a plus or a minus depending on the application attributes in relation to the functionality, availability, selection, cost, and other attributes of the equipment. Industry abounds with opportunities for rationally implementing robotics, and industrial engineers are playing an important role in identifying these opportunities and bringing them to the attention of management and overseeing their implementation.

WORK MEASUREMENT

Methods study and work measurement are subsets of methods engineering. Because of the central role work measurement has played in the field of industrial engineering concerning the concept of labor control, it constitutes a considerable theoretical and practical technique development area over the years. It represents a considerable body of developed technology in the field of industrial engineering, and it has been and still is important as an element in attaining the overall primary objective of industrial engineering—productivity. It is unfortunate that many practicing industrial engineers, who have been over influenced by academicians during their educational development choose to treat it as old (which it is) and irrelevant (which it is not). The "math types," in particular, thought that by mathematization it would be possible somehow to avoid the messy necessity of dealing directly with workers to determine how long it takes workers to perform their tasks. Mathematization will never eliminate the need to determine operation times, and without operation times, production systems development will not go anywhere.

Time has always been one of the most important variables in engineering and science, including manufacturing. Galileo's experiments with falling bodies, for example, depended greatly on measurements of distance and time. Although time has been an important variable throughout history, it was Taylor who offered the concept of measuring the time of human activity as a means of monitoring labor performance in industry. A watch is a device that, by means of gears powered by a spring through an escapement mechanism, rotates hands that record and display elapsed time. Because a watch measures time and accomplishes nothing else, it is understandable that stopwatch time-study was the first work measurement technique developed.

Stopwatch Time-study

In stopwatch time-study, the analyst breaks down an operation into elements, as suggested by Taylor. Then an operator performs the operation a number of times while the analyst observes elapsed time at the end of each element for a number of cycles of the study. The analyst also observes the rate of activity of the operator and records a "performance rating factor," which is the observed pace of the operator compared to the analyst's concept of normal pace for the operation under study, considering applicable allowances for the operation. Figure 2-18 is a representative time-study sheet on which the observed times and rating factor or factors have been recorded. Additional descriptive information about the operation is added either at the workplace or elsewhere. The time-study analyst then calculates the standard time for the operation. The calculation, with reference to Fig. 2-18, is as follows:

$$ST = NT + A$$

$$NT = \sum_{i=1}^{n}(RF_i \times \overline{T}_i) \qquad i = 1,2,3,4$$

where ST = standard time
NT = normal time
A = allowances (14.3% of NT in this example)
RF_i = rating factor for the ith element
$\overline{T}_i$ = average observed time for the ith element

Therefore,

$$\begin{aligned} NT &= (0.90 \times 0.07) + (1.05 \times 0.15) + (1.00 \times 0.24) + (0.90 \times 0.10) \\ &= 0.06 + 0.16 + 0.24 + 0.09 \\ &= 0.55 \text{ min.} \end{aligned}$$

$$\begin{aligned} ST &= 0.55 + (0.143 \times 0.55) \\ &= 0.63 \text{ min.} \end{aligned}$$

The above calculation is typical of the way in which time-studies are performed today.

Between 1910 and 1940, there was considerable misuse of time-study by individuals neither trained in its use nor ethically concerned with the employee's best interests; consultants are hired by management, not the employees. Far too many managements allowed and even encouraged misapplication, which resulted in a practice known as "rate cutting." For example, if an employee assigned to the task in Fig. 2-18 readily performed at a rate consistently equal to or above the standard

Time Study Observation Sheet

Identification of operation	Assemble 24" x 36" chart blanks	Date 10/9

Began timing: 9:26 Ended timing: 9:32	Operator 109	Approval PML	Observer MWS

	Element Description and Breakpoint		Cycles										Summary			
			1 (0.00)	2	3	4	5	6	7	8	9	10	ΣT	$\overline{T}$	RF	NT
1	Fold over end (grasp stapler)	T	.07	.07	.05	.07	.09	.06	.05	.08	.08	.06	.68	.07	.90	.06
		R	.07	.61	.14	.67	.24	.78	.33	.88	.47	.09				
2	Staple five times (drop stapler)	T	.16	.14	.14	.15	.16	.16	.14	.17	.14	.15	1.51	.15	1.05	.16
		R	.23	.75	.28	.82	40	.94	.47	4.05	.61	.24				
3	Bend and insert wire (drop pliers)	T	.22	.25	22	.25	.23	.23	.21	.26	.25	.24	2.36	.24	1.00	.24
		R	.45	1.00	.50	2.07	.63	3.17	.68	.31	.86	.48				
4	Dispose of finished chart (touch next sheet)	T	.09	.09	.10	.08	.09	.11	.12	.08	.17	.08	1.01	.10	.90	.09
		R	.54	.09	.60	.15	.72	.28	.80	.39	1.03	5.56				
5		T														0.55
		R														normal minute
6		T														for cycle
		R														
7		T														
		R														
8		T														
		R														
9		T														
		R														
10		T														
		R														

Normal cycle time 0.55 + Allowance (0.55 x 0.143) or 0.08 = Std. time 0.63 min./pc.

FIGURE 2-18 A representative time-study sheet. (From Krick [20, p. 246].)

in the chart, management might have assumed that it was "too loose" (i.e., provided too much time), in which case they would have the operation restudied, which would have typically resulted in the standard being lowered (e.g., 0.58 min.).

Additionally complicating the issue was the desire of the operator, typically on incentive earnings (daily earnings based on output), to perform improvements in the method (e.g., moving material closer to the machine, or increasing the revolutions per minute of the rotating head of the machine) to permit higher earnings employing the existing standard. Improvements initiated by the employee often resulted in operational improvement, thereby increasing take-home pay, which then often lead to restudy of the standard to reflect the improved method. The restudy often had the effect of reducing the previous time standard, thereby effectively reducing take-home pay at the previous rate of output. Each reduction in the standard required the employee to work faster or employ an increasingly more effective method for the same take-home pay; the company was benefiting from the improvements but the employee was not. Rate cutting understandably led to much labor and union dissatisfaction with time-study.

Today, a permanent standard is normally guaranteed by management, which means that it cannot be changed as long as the job and method remain in use. The specifics of such arrangements are typically spelled out in the union contract. The author recalls a union contract that stated that a significant number of changes had to occur before a time standard could be restudied. "Significant" was later judged to mean "at least two." Therefore, if a new machine was purchased that reduced the cycle time by half, the employee still had to be paid at the old rate because a significant number of changes had not occurred.

Previous judgments such as this and numerous others—holdovers from many years of negotiations between management and the unions—have embedded numerous inefficiencies and disincentives into present American industrial practice. That is one advantage that a Japanese firm, for example, has in setting up shop in the United States; they do not have to hold to the history of inefficient work rules that hampers traditional American firms. American firms do have a legitimate gripe on this issue, but their management can change the rules. Moreover, it is their duty to change those rules that need to be changed and thus earn their pay.

Too many times in the past, managements have not bitten the bullet and have acquiesced to unions, and the legacy is work rules, and wages, in traditional American plants that make it more difficult, if not impossible, to compete in a global economy. An average annual salary of an auto worker (a good example of a low-skilled, highly paid worker) of $55,000 per year including fringes in an Arlington, Texas, automobile assembly plant, recently reported in the press, is indefensible in light of the relative educational requirements and responsibilities for performing the work involved, which is fundamentally putting nut A on bolt B. Teachers, with a much higher educational requirement and the responsibility for molding the minds of our children and grandchildren, typically make a lower salary. This is typical of the inequities that need to be corrected in our own national self-interest.

It became apparent in the practice of methods engineering that methods improvements should be implemented first, and when the best method becomes known, learned, and stabilized, a standard should then be established as a basis for labor performance measurement. Firms have also learned to install "temporary standards" in the initial stage of developing new operations, to test out equipment and methods without having to guarantee the temporary standard. Unions typically limit the period of time a temporary standard can be kept in effect, however, for obvious reasons.

For time-studies involving a large number of possible elements, a number of hand-held data collection devices, and often supporting software, are from time to time advertised in *Industrial Engineering* [18] for collecting and summarizing the data. This technique permits capturing the time-study data within the device during the study followed by downloading of the data to a personal computer for later processing of the data.

Micromotion Study

In seeking to identify basic motions employed in human work, Gilbreth invented the technique known as micromotion study. This method originally utilized a constant speed (1000 frames per minute) 16-millimeter industrial camera; therefore, the time between successive frames was 0.001 minutes. By examining and counting successive frames, usually with the aid of a frame counter attached to the projector, one could break down detailed human activity, such as high-volume bench work, into detailed timed work elements. Figure 2-19 is a "simo chart," which is used for displaying a micromotion study analysis. By filming various alternative ways of performing an operation and analyzing their respective charts, it is possible to devise combinations of motion sequences that minimize unit cycle times. The employee is then trained to perform the operation using the superior method.

Micromotion study can involve a considerable amount of film; therefore, a technique known as memomotion study (similar to time-lapse photography) was developed, which employs a longer time interval between successive frames. It is often more feasible than micromotion study for medium- and long-cycle operations.

With the introduction of the video camera in recent years, it has become common practice to record worker performances on videotape as part of the time-study process. Subsequent performance rating for establishing a standard for that operation is thereby documented on the videotape. Any future question of the efficacy of the prior performance rating process for establishing the time standard can be reevaluated by reviewing the videotaped worker performance. Before videotaping worker performances, time study performance rating was done on the spot by the time-study person. Since this rating performance could not be restudied at a later date, it was often a source of contention in questioning the standard produced.

Predetermined Time Systems (PTS)

Another type of work measurement technique that involves a level of detail similar to that of micromotion study is commonly known as "predetermined time systems." Nine of these systems are listed in Fig. 2-20. A more complete list of predetermined time systems is provided in Fig. 2-21.

By far, the predetermined time system in most common use to date has been MTM (Methods Time Measurement). Figure 2-22, which is a typical table in the MTM system, gives times in TMUs (Time Measurement Units), for example, reaching defined distances under various conditions. A TMU is equal to 0.0006 minutes or 0.00001 hours. A predetermined time system, therefore, is a classification system that provides times for detailed human motions utilized in the performance of work under various conditions. An analyst trained in MTM breaks down an operation into a sequence of elemental activities as classified by the MTM system. The sum of the times for these detailed activities becomes the basis for

FIGURE 2-19 A simo chart. (From Krick [20, p. 100].)

Simo chart

Operation:	Assemble Tab Shaft		
Part:	NA 37124	Operator:	R. Rees
Department:	Assembly	Date:	7/10
Analysis:	Smalley	Film No.:	16 - 48

Left-Hand Description	Time in Minutes	Symbol	Time Scale	Symbol	Time in Minutes	Right-Hand Description
To shaft	0.007	R		D	0.007	
Shaft	0.016	G	0.010 0.020	R	0.016	To key
	0.006	D	0.030	G	0.006	Key
To assembly point	0.014	M	0.040	M	0.014	To shaft
Support assembly		H	0.050	P	0.009	To shaft
				RL	0.002	
			0.060	R	0.009	To collar
			0.070	G	0.007	Collar
			0.080	M	0.010	To assembly
			0.090	P	0.009	To assembly
				RL	0.003	
			0.100	R	0.008	To screwdriver

Publication Containing Information about System	How Data Were Originally Obtained	System Developed by
"Motion-Time-Analysis" by A. B. Segur, in *Industrial Engineering Handbook*, H. B. Maynard, editor, McGraw-Hill Book Co., New York, pp. 4-101 to 4-118, 1956	Motion pictures, micromotion analysis, kymograph	A. B. Segur
Applied Time and Motion Study by W. G. Holmes, Ronald Press Co., New York, 1938	Not known	W. G. Holmes
Motion and Time Study: Design and Measurement of Work, 6th ed., by Ralph M. Barnes, John Wiley & Sons, New York, 1968, Ch. 30	Time study, motion pictures of factory operations, laboratory studies	Harold Engstrom and H. C. Geppinger of Bridgeport Plant of General Electric Co.
Work-Factor Time Standards, by Joseph H. Quick, James H. Duncan, and James A. Malcolm, Jr., McGraw-Hill Book Co., New York, 1962 *Ready Work-Factor Time Standards*, by J. A. Malcolm, Jr. et al., Haddonfield, N. J., 1966.	Time study, motion pictures of factory operations, study of motions with stroboscopic light unit	J. H. Quick W. J. Shea R. E. Koehler
"Establishing Time Values by Elementary Motions" by M. G. Schaefer, *Proceedings Tenth Time and Motion Study Clinic*, IMS, Chicago, November, 1946. Also "Development and Use of Time Values for Elemental Motions" by M. G. Schaefer, *Proceedings Second Time Study and Methods Conference*, SAM-ASME, New York, April, 1947	Kymograph studies, motion pictures of industrial operations, and electric time-recorder studies (time measured to 0.0001 minute)	Western Electric Co.
Methods-Time Measurement by H. B. Maynard, G. J. Stegemerten, and J. L. Schwab, McGraw-Hill Book Co., New York, 1948	Time study, motion pictures of factory operations	H. B. Maynard G. J. Stegemerten J. L. Schwab
Basic Motion Timestudy by G. B. Bailey and Ralph Presgrave, McGraw-Hill Book Co., New York, 1958	Laboratory studies	Ralph Presgrave G. B. Bailey J. A. Lowden
Dimensional Motion Times by H. C. Geppinger, John Wiley & Sons, New York, 1955	Time study, motion pictures, laboratory studies	H. C. Geppinger
"Synthesized Standards from Basic Motion Times," *Handbook of Industrial Engineering and Management*, W. G. Ireson and E. L. Grant, editors, Prentice-Hall, Englewood Cliffs, N. J., pp. 373–378, 1955	Motion pictures of factory operations	Irwin P. Lazarus

FIGURE 2-20 Review of some predetermined time systems. (From Barnes [pp. 364–365].)

Name of System	First Applied Date	First Publication Describing System
Motion-Time Analysis (MTA)	1924	Data not published, but information concerning MTA published in *Motion-Time Analysis Bulletin*, a publication of A. B. Segur & Co.
Body Member Movements	1938	*Applied Time and Motion Study* by W. G. Holmes, Ronald Press Co., New York, 1938
Motion-Time Data for Assembly Work (Get and Place)	1938	*Motion and Time Study*, 2nd ed., by Ralph M. Barnes, John Wiley & Sons, New York, 1940, Chs. 22 and 23
The Work-Factor System	1938	"Motion-Time Standards" by J. H. Quick, W. J. Shea, and R. E. Koehler, *Factory Management and Maintenance*, Vol. 103, No. 5, pp. 97–108, May, 1945
Elemental Time Standard for Basic Manual Work	1942	"Establishing Time Values by Elementary Motion Analysis" by M. G. Schaefer, *Proceedings Tenth Time and Motion Study Clinic*, IMS, Chicago, pp. 21–27, November, 1946
Methods-Time Measurement (MTM)	1948	*Methods-Time Measurement* by H. B. Maynard, G. J. Stegemerten, and J. L. Schwab, McGraw-Hill Book Co., New York, 1948
Basic Motion Timestudy (BMT)	1950	Manuals by J. D. Woods & Gordon, Ltd., Toronto, Canada, 1950
Dimensional Motion Times (DMT)	1952	"New Motion Time Method Defined" by H. C. Geppinger, *Iron Age*, Vol. 171, No. 2, pp. 106–108, January 8, 1953
Predetermined Human Work Times	1952	"A System of Predetermined Human Work Times" by Irwin P. Lazarus, Ph.D. thesis, Purdue University, 1952

FIGURE 2-20 *Continued*

MTS	General Electric Company
MOST	H. B. Maynard & Co. Ltd.
MODAPTS	Heyde Dynamics Pty. Ltd.
MANPRO	Methods Management
CUE	General Analysis Inc.
MICRO	Standards International, Inc.
MSD	Sirge A. Burn Company
WORK FACTOR	Science Management Corporation
UNIVEL	Management Science, Inc.
AM COST ESTIMATOR	Costcom, Inc.
ADAM	MTM Association
FAST	Royal J. Dossett Corporation
4M	MTM Association
Taskmaster	Artifacts Software
Labour Standard Builder	Applied Computer Services, Inc.
EASE	Ease Inc.
MODCAD	R. Wygant and R. Dawood
MTM	generally public domain or MTM association
MACRO	Standards International, Inc.

FIGURE 2-21 A list of predetermined time systems. (From Hodson [15, p. 4.91].)

establishing the standard time for the operation. Figure 2-23 is an example of MTM analysis for a simple operation.

In recent years another predetermined time system, developed by the H. B. Maynard organization in Sweden and labeled MOST (Maynard Operational Sequence Technique), is replacing MTM in many application environments. This technique is especially popular in work environments involving medium- to long-cycle operations, particularly where standards are utilized to measure overall departmental accomplishment. Whereas the MTM system involves 2 weeks of training before application, training in MOST involves only 3 days of initial training. Figure 2-24 illustrates a MOST analysis for an electronic assembly.

Methods Time Measurement and other predetermined time systems have their greatest utility in repetitive short-cycle operations. Yet a larger proportion of activity today has shifted from short-cycle repetitive work, such as an assembly operation on an electronics assembly line, to less repetitive medium- and long-cycle activity. A much greater proportion of workers today perform servicing or monitoring functions as equipment designs reach higher levels of automation. One technique that has been developed for analyzing nonrepetitive work is called "work sampling."

Work Sampling

Work sampling is nothing more than statistical sampling theory applied to industrial situations. In a basic course in probability theory it is often convenient to

TABLE 1 REACH-R

Distance moved, in.	Time, TMU				Hand in motion	
	A	B	C or D	E	A	B
3/4 or less	2.0	2.0	2.0	2.0	1.6	1.6
1	2.5	2.5	3.6	2.4	2.3	2.3
2	4.0	4.0	6.9	3.8	3.5	2.7
3	5.3	5.3	7.3	5.3	4.5	3.6
4	6.1	6.4	8.4	6.8	4.9	4.3
5	6.5	7.8	9.4	7.4	5.3	5.0
6	7.0	8.6	10.1	8.0	5.7	5.7
7	7.4	9.3	10.8	8.7	6.1	6.5
8	7.9	10.1	11.5	9.3	6.5	7.2
9	8.3	10.8	12.2	9.9	6.9	7.9
10	8.7	11.5	12.9	10.5	7.3	8.6
12	9.6	12.9	14.2	11.8	8.1	10.1
14	10.5	14.4	15.6	13.0	8.9	11.5
16	11.4	15.8	17.0	14.2	9.7	12.9
18	12.3	17.2	18.4	15.5	10.5	14.4
20	13.1	18.6	19.8	16.7	11.3	15.8
22	14.0	20.1	21.2	18.0	12.1	17.3
24	14.9	21.5	22.5	19.2	12.9	18.8
26	15.8	22.9	23.9	20.4	13.7	20.2
28	16.7	24.4	25.3	21.7	14.5	21.7
30	17.5	25.8	26.7	22.9	15.3	23.2

Case	Case and description
A	Reach to object in fixed location, or to object in other hand or on which other hand rests
B	Reach to single object in location, which may vary slightly from cycle to cycle
C	Reach to object jumbled with other objects in a group so that search and select occur
D	Reach to a very small object or where accurate grasp is required
E	Reach to indefinite location to get hand in position for body balance, next motion, or out of way

FIGURE 2-22 An MTM reach table. (From Maynard [26, pp. 5–18].)

consider drawing a sample of colored balls from an urn, called a Shewhart bowl, containing balls of known characteristics (e.g., 400 white and 100 black balls). In work sampling, instead of drawing balls from an urn, random minutes are drawn from a population of possible minutes in a day. Based on the composition of activity of the sampled minutes, it is possible to make statistical inferences about the composition of activity for that day. Instead of black and white balls in a Shewhart bowl, activity may have been classified, for example, as either working or idle.

The proportion of observed minutes of a particular class of activity to the total sample drawn is a best estimate of the percentage occurrence of that type of activity over the period of the study. If an employee is observed for 100 randomly selected minutes and for 90 of those observations is busy, our best estimate of the percentage of total time the employee is busy is 90/100 = 90 percent.

If the observer wanted to be more sure that the percent calculated was an accurate reflection of the true underlying percent occurrence, he or she could increase the sample size. The following equation is commonly employed to relate the level

MTM ELEMENT ANALYSIS

PART 1/2" Pipe Union Knurled Ends Hex Nut PART NO 652 DEPT 20
OPERATION Assemble DATE 12/1

DESCRIPTION — LEFT HAND	•	MOTION	TMU	MOTION	•	DESCRIPTION — RIGHT HAND
C Reach Forward Swivel		R12C	14.2	R12C	✓	Reach to Swivel
Grasp			7.3	G4A	✓	Grasp Swivel
Grasp Swivel	✓	G4A	7.3			
Swivel to Nut		M12C	15.2	M12C	✓	Swivel to Nut
Preposition Swivel		G2	--	G2		Preposition Swivel
		P1SE	5.6	P1SE	✓	Engage Swivel in Nut
		RL1	2.0	RL1	✓	Release Swivel
			51.6			

• — LIMITING MOTION SHEET ___ OF ___ SHEETS

FIGURE 2-23 An MTM analysis. (From Maynard [26, pp. 5–33].)

of confidence with how well a sample represents the underlying percent of occurrence:

$$Sp = K_{\alpha}\sqrt{\frac{p(1-p)}{n}} \tag{2-1}$$

where S = relative accuracy
p = percent occurrence
K_{α} = number of standard deviations from the mean for a given confidence level assuming a normal distribution
n = sample size

Part of the justification for Eq. (2-1) is derived from what is called the "central limit theorem" in statistics. The theorem basically says that a distribution of means of samples, irrespective of the distribution of the population from which they are drawn, will approach a normal distribution as a limit as sample size increases. A distribution of proportions from samples behaves in a similar fashion.

MOST-calculation

ELECTRONIC ASSEMBLY

Code	
Date	7/29/87
Sign.	A.A.
Page	1/1

Activity: INSTALL CONNECTOR ON PC-BOARD

Conditions: EDGE CONNECTORS ONLY

No.	Method
1	POSITION EDGE CONNECTOR TO BOARD
2	ALIGN CONNECTOR TO ACCURATE LOCATION
3	PLACE SCREW TO HOLE IN CONNECTOR F2
4	MOVE WASHER TO SCREW ON BOARD F4
5	FASTEN NUTS 2 SPINS USING FINGERS F2
6	FASTEN 2 SCREWS 5 SPINS USING SCREWDRIVER
7	PLACE BOARD TO RACK

No.	Sequence Model	Fr	TMU
1	$A_1\ B_0\ G_1\ A_1\ B_0\ P_6\ A_0$		90
3	$A_1\ B_0\ G_1\ A_1\ B_0\ P_3\ A_0$	2	120
4	$A_1\ B_0\ G_1\ A_1\ B_0\ P_1\ A_0$	4	160
7	$A_1\ B_0\ G_1\ A_1\ B_0\ P_3\ A_0$		60
	A B G A B P A		
	A B G A B P A		
	A B G A B P A		
	A B G A B P A		
	A B G A B P A		
	A B G A B P A		
	A B G A B P A		
	A B G A B P A		
	A B G A B P A		
	A B G A B P A		
	A B G A B P A		
	A B G A B P A		
	A B G A B P A		
	A B G A B P A		
2	$A_0\ B_0\ G_0\ M_3\ X_0\ I_{16}\ A_0$		190
	A B G M X I A		
	A B G M X I A		
	A B G M X I A		
	A B G M X I A		
	A B G M X I A		
	A B G M X I A		
	A B G M X I A		
5	$A_1\ B_0\ G_1\ A_1\ B_0\ P_1\ F_3\ A_0\ B_0\ P_0\ A_0$	2	140
6	$A_1\ B_0\ G_1\ A_1\ B_0\ P_1\ (L_3\ A_1\ F_{10})\ A_1\ B_0\ P_1\ A_0$	(2)	330
	A B G A B P A B P A		
	A B G A B P A B P A		
	A B G A B P A B P A		
	A B G A B P A B P A		
	A B G A B P A B P A		
	A B G A B P A B P A		
	A B G A B P A B P A		
	A B G A B P A B P A		
	A B G A B P A B P A		
	A B G A B P A B P A		
	A B G A B P A B P A		
	A B G A B P A B P A		

TIME = .65 ~~millihours (mh)~~/minutes (min.) 1090

FIGURE 2-24 A MOST analysis. (From Hodson [15, p. 4.98].)

Let us assume that a particular activity does involve 90 percent busy, 10 percent idle activity. The central limit theorem says then that if a number of samples were taken from that activity and "percent busy" was estimated from each, the estimates of the proportion would be distributed as shown in Fig. 2-25.

In statistics it can be shown that

$$\sigma_p = \sqrt{\frac{p(1-p)}{n}}$$

Note in the right half of Fig. 2-25 that the distance $K_\alpha \sigma_p$ equals the distance Sp. This constitutes the derivation of Eq. (2-1). An example will illustrate its use, but first Eq. (2-1) must be solved for n:

$$n = \frac{K_\alpha^2(1-p)}{S^2 p} \tag{2-2}$$

Assume that a sample of size 50 was taken, and for 40 of the minutes the employee was observed to be working; how large should the sample size be to provide a relative accuracy of ±10 percent at a confidence level of approximately 95 percent? A K_α of ±2 contains 95.4 percent of the area under a normal

FIGURE 2-25 A distribution of percent occurrences.

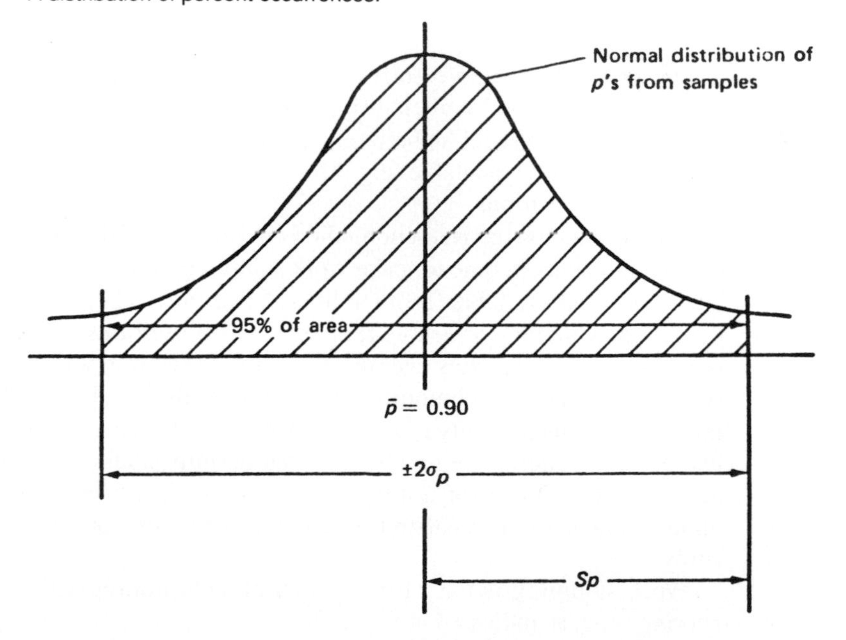

distribution between these limits and is close to the required 95 percent; therefore, it is often assumed that $K_\alpha = 2$. Our estimate of n, therefore, with $p = 40/50 = 0.80$, and $S = 0.10$, is:

$$n = \frac{2^2(1 - 0.80)}{0.10^2(0.80)} = 100$$

This calculation indicates that the sample of 50 was not large enough for the desired level of accuracy and confidence level; therefore, sampling should continue. When 100 samples have been taken, the calculation should be repeated until the estimated size of n required as calculated is less than the total size of the accumulated sample taken thus far.

A number of means have been developed to assist the work sampling analyst in performing work sampling calculations. Table 2-2 is a table used to simplify the determination of sample size. Figure 2-26 is a nomograph for determining sample size as well. In utilizing these two sources, however, note that the "absolute error" in Table 2-2 and the "precision interval" in Fig. 2-26 are identical in meaning, simply involving different terminology used by different practitioners, therefore:

$$A = Sp$$

where A = absolute error (or precision interval) = absolute accuracy
S = relative error = relative accuracy
p = percent occurrence

In the previous example following Eq. (2-2) the absolute error was equal to 0.10(0.8) = 0.08, or 8 percent (10 percent of 80 percent = 8 percent, so p, 95 percent of the time, should lie between 72 and 88 percent).

Software is also available (e.g., from IIE) for generating random times for work sampling observations.

One of the most convenient methods in practice of discretely signaling the successive moments in time to make work sampling observations is by use of a device developed and available from Divilbiss Electronics [10], shown in Fig. 2-27. The device is available in various ranges of sampling rates. Each device provides a number of sampling rates over the range for which it is set, employing DIP switch settings. Inexpensively modified custom models are also available for sampling rates outside the typically available ranges. The device permits a vibrator signaling choice, as an alternative to a beeper, that permits silent signaling to the analyst—an excellent feature for not disturbing employees being reviewed. The device either clips on (to a waistband) or can be carried in a pocket or handbag during a study.

Work sampling is used for analysis of such nonrepetitive activities as engineering time, as indicated in Fig. 2-28.

TABLE 2-2

NUMBER OF OBSERVATIONS REQUIRED FOR A GIVEN ABSOLUTE ERROR FOR VALUE OF p AT A 95 PERCENT CONFIDENCE LEVEL

Percent occurrence	±1.0%	±1.5%	±2.0%	±2.5%	±3.0%	±3.5%
1	396	176	99	63	44	32
2	784	348	196	125	87	64
3	1164	517	291	186	129	95
4	1536	683	394	246	171	125
5	1900	844	475	304	211	155
6	2256	1103	564	361	851	184
7	2604	1157	651	417	239	213
8	2944	1308	736	471	327	240
9	3276	1456	819	524	364	267
10	3600	1690	900	576	400	294
11	3916	1740	979	627	435	320
12	4224	1577	1056	656	469	344
13	4524	2011	1131	724	503	369
14	4816	2140	1204	771	535	393
15	5100	2267	1275	816	567	416
16	5376	2389	1344	860	597	439
17	5644	2508	1411	903	627	461
18	5904	2624	1476	945	656	482
19	6156	2736	1539	985	684	502
20	6400	2844	1000	1024	711	522
21	6636	2949	1059	1062	737	542
22	6864	3050	1716	1098	763	560
23	7084	3148	1771	1133	797	578
24	7296	3243	1824	1167	811	596
25	7500	3333	1875	1200	833	612
26	7696	3420	1924	1231	855	628
27	7834	3504	1971	1261	876	644
28	8064	3584	2016	1290	896	658
29	8236	3660	2059	1318	915	672
30	8400	3733	2100	1344	933	686
31	8556	3803	2139	1369	951	698
32	8704	3868	2176	1393	967	710
33	8844	3931	2211	1415	983	722
34	8976	3989	2244	1436	997	733
35	9100	4044	2275	1456	1011	743
36	9216	4096	2304	1475	1024	753
37	9324	4144	2331	1492	1036	761
38	9424	4188	2356	1508	1047	769
39	9516	4229	2379	1523	1057	777
40	9600	4266	2400	1536	1067	784
41	9676	4300	2419	1548	1075	790
42	9744	4330	2436	1559	1083	795
43	9804	4357	2451	1569	1089	800
44	9856	4380	2464	1577	1095	804
45	9900	4400	2475	1584	1099	808
46	9936	4416	2484	1590	1104	811
47	9964	4428	2491	1594	1107	813
48	9984	4437	2496	1597	1109	815
49	9996	4442	2499	1599	1110	816
50	10000	4444	2500	1600	1111	816

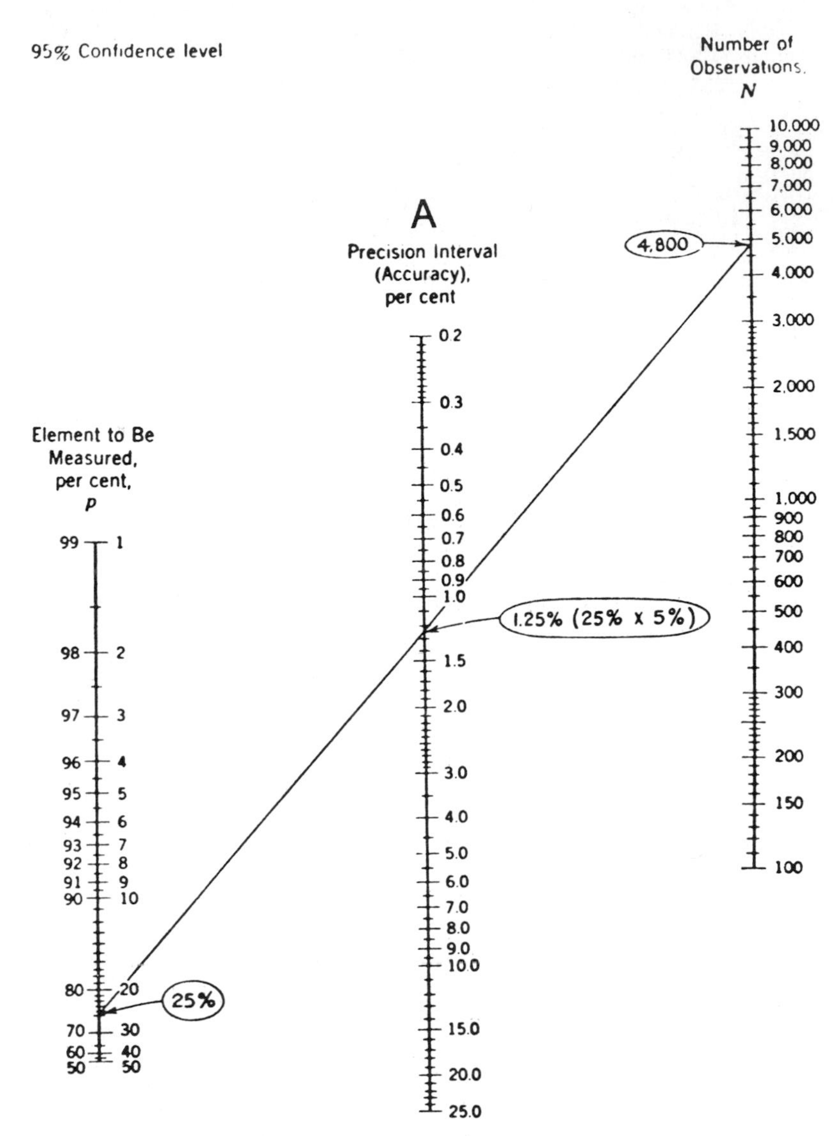

FIGURE 2-26 A work sampling nomograph. (Courtesy Manville Corporation.)

Service functions such as maintenance work, or long-cycle tasks such as manual washing of an airplane, can sometimes be analyzed best by employing work sampling. For work sampling to qualify as a work measurement technique, it is necessary not only to classify the nature of the activity during observations, but also to rate the pace of activity observed. The rated observations, along with a production count for a defined period of time, offer a means of determining a standard time for a unit of production. Such techniques have been successfully

FIGURE 2-27 The JD-7 Random Reminder from Divilbiss. (From Divilbiss Electronics [10].)

employed, for example, in determining how long it should take to replace a broken pane of glass, or replace a valve in a factory. Because of the increasing proportion of indirect labor activities such as maintenance, as compared to direct labor activities, the ability to measure work in these indirect areas is becoming increasingly more important, and in many companies will be the key to effective labor cost monitoring and reporting in the future.

Here is an example of how work sampling will be increasingly employed in indirect labor areas in the future. Assume that a maintenance worker has been assigned as one of her key tasks the repair of operating conveyors in a manufacturing plant. As soon as a conveyor goes down, her beeper is beeped, and she attends to the conveyor repair as her highest priority task. When she completes the

Work Sampling Observation Sheet

Study of: *Utilization of Engineering Time*

Schedule of Observations. Observe at:	Engr. to be observed	Idle	Unaccounted for	Nondelegable											Delegable								
				Drawing	Computation	Measurement	Construction	Paperwork	Acquiring info.	Thinking	Reading	Consultation	Supervision	Misc.	Drawing	Computation	Measurement	Construction	Paperwork	Acquiring info.	Acquiring mater.	Misc.	
Mon. 8:02	C			✓																			
8:17	M										✓												
8:18	B	✓																					
8:31	F															✓							
8:48	H	✓																					
9:34	C																					✓	
9:36	G												✓										
9:57	D							✓															
10:07	M															✓							
10:41	A					✓																	
11:12	C					✓																	
11:25	F															✓							
1:09	E		✓																				
1:28	B											✓											
1:33	H			✓																			
2:28	K	✓																					
2:43	G		✓																				
2:57	B									✓													
3:10	M									✓													
3:18	F																	✓					
3:52	D																			✓			
3:55	A				✓																		
4:00	F							✓															
4:21	E												✓										
4:48	B			✓																			
Daily total		3	2	3	1	2	0	2	0	2	1	1	2	0	0	3	0	1	0	1	0	1	

FIGURE 2-28 A work sampling observation sheet. (From Krick [20, p. 290].)

repair and the production process is working again, she then returns to completing any interrupted lower priority task. During a previous typical week, she had been one of the subjects of an ongoing work sampling project, and 200 random observations had been made of her activities during the 40 hours that she worked that week. Of the 200 observations taken of her during the week, 20 of the observations involved her performing some aspect of conveyor repair (working on a conveyor, going to the tool room to get a tool to fix a conveyor, walking to a conveyor needing repair, etc.). The following represent the rated observations (which not only classify her activity by work sampling category, but also are estimates of her performance pace at the time of observation) taken of her performing the conveyor repairs:

80	80	95	100
80	85	110	105
100	110	105	80
85	95	100	100
105	110	85	80

The average of the above 20 ratings is 94.5 percent. Because 20 of the 200 total observations involved conveyor repair, the best estimate of the fraction of time she performed conveyor repairs is 20/200 = 0.10, or 10 percent of her time. She worked 40 hours last week; therefore, a best estimate of the time she spent performing conveyor repairs is 10 percent of 40 hours, or 4 hours. If she performed 10 conveyor repairs last week, the average time spent per conveyor repair is estimated to be 4 hours/10 = 0.4 hours, or 24 minutes. More important than how long she spent to complete a conveyor repair on average is the question of how long should it have taken her to complete a conveyor repair. *Normal time = observed time × performance rating factor,* so the time it should have taken her to complete an average conveyor repair is 0.4 hours × 0.945 = 0.378 hours/#. It is assumed in this example that she is employing proper methods when performing conveyor repairs.

Next week when tasks are being assigned to this maintenance worker, if 10 conveyor repairs are expected to occur, 10 × 0.378 hours/# can be employed to estimate that she will need 3.78 hours of time to complete conveyor repairs. It is assumed that the result of a sufficient number of samples of conveyor repair provides the estimate of 0.378 hours/#. Other tasks will be included in her duties to utilize her total expected available work time. Over time, as more task time estimates are generated and combined, improved estimating times will be generated for all repeated tasks in the maintenance activity, and maintenance labor performance can be evaluated over time in reference to these standards.

Whereas unmeasured maintenance activities in a typical plant may produce labor performance of 50 percent, maintenance performance with standards developed in a work measurement process, as stated in the above example, may result in maintenance labor performance in the 90 percent to 100 percent range. The effect on maintenance labor productivity can be dramatic. At a time when indirect labor cost is becoming a larger segment of the overall cost of operations, and in fact the largest in some operations, such techniques offer considerable promise in making future operations more cost-effective.

It should be noted that the cost of direct or indirect labor is often not the primary management concern. One misconception in industry is that if direct labor is only 7 percent of manufactured cost, it is not important. It should be noted that those individuals whose efforts represent 7 percent of the manufactured cost of the product determine to a considerable extent what is produced in their production facility. If they work more efficiently, more product flows through the plant. It is not inconceivable that a motivated and efficient direct labor force can produce twice as much product as an unmotivated and inefficient direct labor force. In such an instance, twice as much product is produced and 93 percent of the remaining costs of the operation are spread over twice as much product (i.e., unit overhead cost drops dramatically, almost in half). Therefore, in many instances it is not the cost of direct labor as much as the effect of direct labor that is the primary concern. Maximizing throughput is the primary focus, and motivated and efficient direct labor workers are therefore critical to a cost-effective production operation.

Indirect labor personnel (e.g., maintenance or material-handling workers) often greatly impact throughput of production operations. The focus, therefore, is the effect they have on throughput, not necessarily their cost.

Standard Data

Over the years, a work measurement technique known as *standard data* has evolved. Standard data employs "lowest common denominator" work elements from previous time-studies, or other work measurement sources, thereby making it unnecessary to restudy work elements that have been timed sufficiently in the past. For example, the use of a 100-watt soldering iron to solder a number 1 lug with standard solder might have been a commonly occurring element in previous operations. If a new operation includes soldering a number 1 lug employing the same solder, it is not necessary to time-study that element again. Note that standard data employs common work elements, whereas predetermined time systems employ common human motions. Each system accumulates different element types to represent the overall operation. Figure 2-29 offers an example of standard data used for molding operations.

Generally, time-studies should be taken in such a way that the elemental data they generate will be useful in meeting future standard data needs. For that reason, it is typically critical that element endpoints be uniformly defined for future standard data collection purposes. It is not uncommon in a time-study function to have one person control endpoint definitions to ensure that in the future, when summarization of like elements occurs, that standard definitions have been employed. Otherwise, the data cannot be combined in a single element classification because the elements have different endpoints.

Standard data is a very efficient method for setting time standards. The work measurement analyst simply reviews a drawing of the part to be made, visualizes the machining elements and their sequence, and simply extracts the appropriate elemental times from the data available, thereby producing a time standard for the operation.

It needs to be stated that not everyone today supports work measurement. There has been a decline in support of work measurement, in particular from management circles in the United States. One of the most vocal in opposition to time standards is W. Edwards Deming [8, 9], an author to be reviewed in more detail in following chapters of this book. In a chapter entitled "What Top Management Must Do to Improve Productivity" [8], Deming listed 14 points that he wished to stress. Point 11a states, "Eliminate work standards (quotas) on the factory floor. Substitute leadership." With rare exceptions, the author is generally in agreement with, and supports, Deming's efforts to improve productivity in the United States, as will be clear in subsequent discussions in this text. His point on eliminating work measurement,

SQUEEZER MOLDING SPECIFICATION

DATE *May 13 -* BY *P Gr* PART NO *47943*

PATTERN *Metal* SIZE *12 × 14*

LOOSE ___ MOUNTED ✓ PART NAME *Gear*

SPLIT ___ FOLLOW BOARD ___. WEIGHT *4½* CASTING PER MOLD *1*

METAL	FLASK SIZE	COPE DRAG	DEPTH OF DRAW IN INCHES								RAISE DEPTH	STAND.
			0	1/2	1	1 1/2	2	2 1/2	3	3 1/2	r	
IRON	12" X 12"	COPE	30	.50	.65	.80	(.90)	1.00	1.10	1.20	.30	.90
	12" X 14"	DRAG	—	4.06	4.18	(4.29)	4.40	4.46	4.57	4.68		4.29
BRASS	12" X 12"	COPE	.35	.60	.70	.80	.90	1 00	1.10	1.20	.30	
	12" X 14"	DRAG	—	3.05	3.15	3.25	3.35	3.40	3.50	3.60		
	10" X 19"	COPE	.45	.70	.80	.90	1.00	1.10	1.20	1.30	.40	
		DRAG	—	4.05	4.15	4.25	4.35	4.40	4.50	4.60		

MISCELLANEOUS DATA	PER	1	2	3	4	5	6	STAND.
SET SOLDIERS	SOLDIER	.15	.25	.40	.50	.60	.68	
SET JOB NAILS	NAIL	.07	.14	.20	(.25)	.30	.35	.25
VENT (COPE, DRAG, OR SIDE)	VENT	.08	(.15)	.22	.30	.35	.40	15
BLACKEN MOLD	100 SQ IN	.06	.10	.15	.20	.23	.26	
WATER MOLD AFTER DRAW	10 LIN. IN.	.05	.12	.16	.20	.25	.29	
CUT GATE	GATE	.90	1.50	2.00	2.50	2.90	3.25	
SET, CUT AND REAM RISERS	RISER	.45	.90	1.30	1.70	2.00	2.30	
SPECIAL RAM (DRAG OR COPE)	100 CU. IN.	.30	.55	(.80)	1.00	1.15	1.30	.80
PUT PASTE ON CORES	SQ. INCH.	.05	.10	.15	.18	.20	.22	

CORE SETTING CLASSIFICATION	SKETCH OF TYPE	VOL RANGE CU IN.	CORES SET PER MOLD								STAND.
			1	2	3	4	5	6	7	6	
STOCK CORES		0-10	.10	.20	.25	.30	.35	.40	.45	.50	
		11-20	.15	.25	.30	.35	.40	.45	.50	.55	
		21-50	(.20)	.30	.35	.40	.45	.50	.55	60	.20
REGULAR BLOCK OR CYLINDER CORES		0-50	.20	.30	.40	.45	.50	.55	.60	65	
		51-100	.25	.40	.45	.50	.55	.60	65	.70	
		101-200	.35	.45	.50	.55	.60	.65	.70	.75	
IRREGULAR BLOCK OR CYLINDER CORES		0-50	.30	.35	.45	.50	.55	60	.65	.70	
		51-100	.40	.45	.50	.55	60	.65	.70	.75	
		101-200	.50	.55	.60	.65	.70	.75	.80	.85	

TOTAL STANDARD PER MOLD *6.59*

DIVIDED BY *1* CASTINGS = STANDARD PER CASTING *6.6*

FIGURE 2-29 Standard data for squeezer molding. (From Maynard [26, pp. 3–157].)

however, is one of those exceptions. Mundel, who also disagrees with Deming on this point, offers an example in an *Industrial Engineering* article entitled "Now Is the Time to Speak Out in Defense of Time Standards" [32, p. 51]:

> One more case. Deming suggests that records be kept, following a statistical procedure, for a year, and these data be used in lieu of the usual time standards. This was exactly the procedure used for years by the Social Security Administration (SSA) of the United States in setting "standards" for its 70 odd thousand employees. During the period 1978 to 1980 a massive work measurement [work sampling] project was undertaken covering the entire field staff in the 1570 offices of the SSA. The study required the efforts of roughly 50 SSA employees and two consultants. The data obtained indicated that the SSA had about 23,000 excess employees; about 32 percent too many. The Government Accounting Office and the Office of Management and Budget gave SSA six years to allow attrition to reduce their workforce by 23,000. At the end of the six years they whacked them another 3000. Total reduction was 35 percent; productivity rose 54 percent.
>
> This discussion could keep on going for many pages but it would save time to summarize. It is common experience, that when workers work without time standards of performance, they complete about 65 percent of the 130 they could do without excessive effort, or 65 of the 100 when no incentive system is employed. Is this the sort of data that would support the exhortation to abandon standards?

The two consultants mentioned in the above passage were Mundel and the late Robert Newsom, both past presidents of the Institute of Industrial Engineers. The passage does not mention that data for the study were collected not by the consultants, but by SSA employees, following the work sampling training they received from the consultants. It should also be noted that a work sampling study, by itself, does not improve the methods being used by those performing work; it simply determines the percentage of their time devoted to performing specified categories of activity. A methods study of the work being performed would probably have provided another significant productivity improvement.

Why not consider a third study category as well? What do government employees do, and what are the costs and benefits of what they do? Excerpts from a *USA Today* article [14] announcing an upcoming "massive audit to identify waste and inefficiency throughout the government" offered some interesting information, as follows:

> The all-out audit would be like a Texas audit in 1991 that found $4.2 billion savings in a $30 billion budget....
>
> Bob Davis, a Library of Congress manager, says if the government was run efficiently "you could eliminate 40% of the people we have."

This is not a criticism of government employees per se. It is a criticism of government systems, and their managements' ineffectiveness and relative inactivity and lack of skill in determining how few (instead of how many) employees it needs to provide necessary government services.

There are too many government operations (city, county, state, and federal) in which all the employees would be happier if some fraction of them did not work there. Those extra employees could be contributing to the commercial, potentially exportable, gross domestic product and reducing both the trade deficit and national debt of this nation, while probably feeling better about themselves. It is demanding, frustrating, tiring, and demeaning to work in a place that has too many employees and too little work. It obviously has a negative effect on one's self-image as well. Industrial engineers and system managers have a responsibility to ensure that employees are not required to function in such self-image-degrading and unmotivating environments. At the same time, there are also government employees who have inordinate case loads that make it essentially impossible for them to perform a credible job. Such inequities in assignment of government personnel need to be improved if the nation is to function effectively.

The author acknowledges that there are government employees, managers, and industrial engineers today performing valuable services. The point is that the industrial engineers in government today probably represent one-tenth to one-hundredth of the number of industrial engineers that should be there today, "leaning down" the operation while creating the necessary services. One can only wonder how much other opportunity for improvement lies in the huge federal, state, county, and city government employment base that exists today, which typically has yet to have experienced serious across-the-board methods study and work measurement (and other) review. One might be tempted to ask, "Should we unemploy 23,000 SSA workers?" Consider that if it weren't for the study performed by Mundel and Newsom, those 23,000 government employees would still be there today. Is it not better for them, and for us, that those who would have replaced the ones who left through attrition are now gainfully employed somewhere else, and we as a society are surely more internationally competitive because of it? Can we as a nation afford to have 23,000 employees employed in the SSA that are not needed? What does it do for an employee's sense of accomplishment and morale when there are 70,000 employees in SSA and only enough work for 47,000 employees? Obviously, the opportunity for this kind of improvement is not restricted in any way to government operations.

The author is not aware of any credible hard evidence to date that indicates that methods engineering and work measurement, properly performed, do not produce results.

FACILITIES PLANNING AND DESIGN

For most industrial engineers, methods engineering connotes workstation design and the work measurement associated with such designs. Facilities planning and design, however, extends the analysis to include the design of the entire productive system. Plant layout, a familiar term in industrial engineering, connotes design of a plant or other productive facility. However, corporate industrial engineering responsibilities today often extend this analysis to the determination of what facilities are needed, where, and in what sizes, to meet corporate objectives. This broader responsibility is called facilities planning and design, and it obviously includes plant sizing and location. Plant layout, a subset of facilities planning and design, is concerned with determining the best arrangement of the appropriate number of

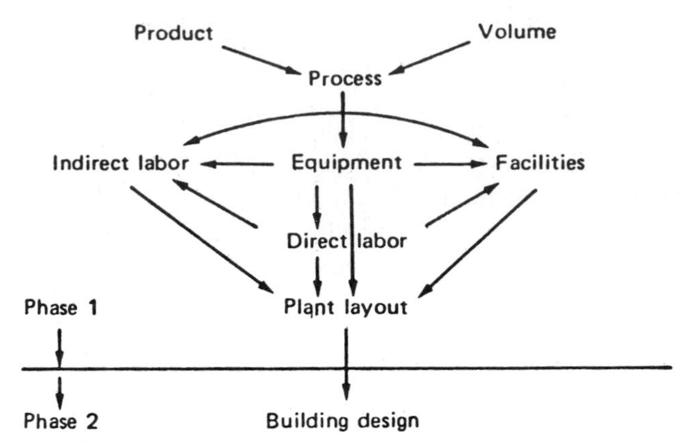

FIGURE 2-30 Major steps in the design of a plant.

various entities needed in the design of a productive facility, and it is closely associated with material handling and storage. It must, because material handling cost is typically the key criterion in evaluating the success of a plant layout design.

A typical sequence of steps in the analysis for the design of a plant is indicated in Fig. 2-30. The starting point for an industrial engineer assigned to such a study is typically a given product design at a stated rate of production (e.g., monthly volumes) and product production life. The design of the product is normally the responsibility of a product engineer, and the rate of production is normally determined at a level of management that bridges both the production and marketing functions of the company. The responsibility of the industrial engineer, then, is to design a production facility that will produce the specified product at the stated production rate at a minimum cost.

Occasionally, it is discovered that a change could be made in the material or design of the product that would improve the product while reducing its cost, or at least not increasing its cost intolerably. That is, product designs change somewhat as a plant design is developed, but for the most part, the overall goal of the industrial engineer is the design of a process to produce a specified product at a given rate of production at a minimum cost.

For a given product at a stated rate of production, it is normally assumed that there is a most appropriate process for producing it. As product production life and annual volumes increase, higher levels of mechanization become feasible, whereas at low levels of production, general-purpose equipment is often employed. With higher levels of mechanization and consequent specialization, process methods change and materials change to accommodate those processes. For example, molded plastic parts might be justified if sufficient volume is demanded, but if not, machined metal parts may be more appropriate.

The process then establishes the requirement for equipment. The number of pieces of equipment of a particular type required in a specific area of the plant is a function of scrap rates, operation times, equipment utilization, machine operator performance, number and length of shifts, and other related factors, such as the type of plant layout.

The sum of operation times for equipment, along with an appropriate allowance for expected unproductive time, establishes a concurrent requirement for direct labor. Operation time refers to the time an employee takes to perform an operation, whereas unproductive time refers to all the remaining time the employee expends (e.g., discussing the World Series game played the day before).

At this point, product and volume have led to a process, which dictates equipment, which in turn requires operators. What remains are the people necessary to support these activities. Equipment and machine operators need maintenance support, janitors, material handlers, and many other indirect employees. Consider also that all the equipment and laborers thus far mentioned establish a need for staff personnel to plan and manage the total facility; this includes accountants, quality control engineers, and production supervisors. And finally, equipment, personnel, and material need to be covered, or at least controlled. The determination of the correct number and arrangement of all these entities, indicated as phase 1 in Fig. 2-30, is called plant layout, and is normally performed by industrial engineers. At this point, the industrial engineer often engages the services of an architect, or the construction division of the company, to design the "cake cover" (i.e., the building) to house the designed production system.

The previous discussion might suggest that plant design is a set of steps, for example, 26 steps, such that having completed all the steps to step 13, the job is half done. Unfortunately, this is not the case. Although the sequence of events described above is typical, it is at step 13, for example, that one gains the insight and perspective to improve steps 2, 5, and 7, and so one goes back to do the process again. In general, plant design is first macro, then micro, and ultimately macro. That is, the first concern is the relative spatial sizing and arrangement of major plant areas, including aisle planning. Then detailed layout of entities within an area is undertaken, and then ultimately area layouts are combined and adjusted to arrive at a total layout contained within a predefined overall outer shape.

Layout Types

One of the first decisions in designing a plant layout is what plant layout type to employ. There are three main plant layout types: (1) process, (2) product, and (3) fixed location, as depicted in Fig. 2-31.

The process layout is typically employed where a large number of products are being processed in generic processes, particularly if the volume of each is relatively low. A production facility for books, for example, is typical of this job-shop type of layout. By grouping similar types of equipment together, it is possible to attain equipment type utilizations that are relatively high (e.g., if 18.6 cutters are needed, 19 cutters will result in little overall cutter nonutilization).

The product layout, however, offers the advantage that the types of equipment a product needs are arranged in sequence to suit the product. This results in much

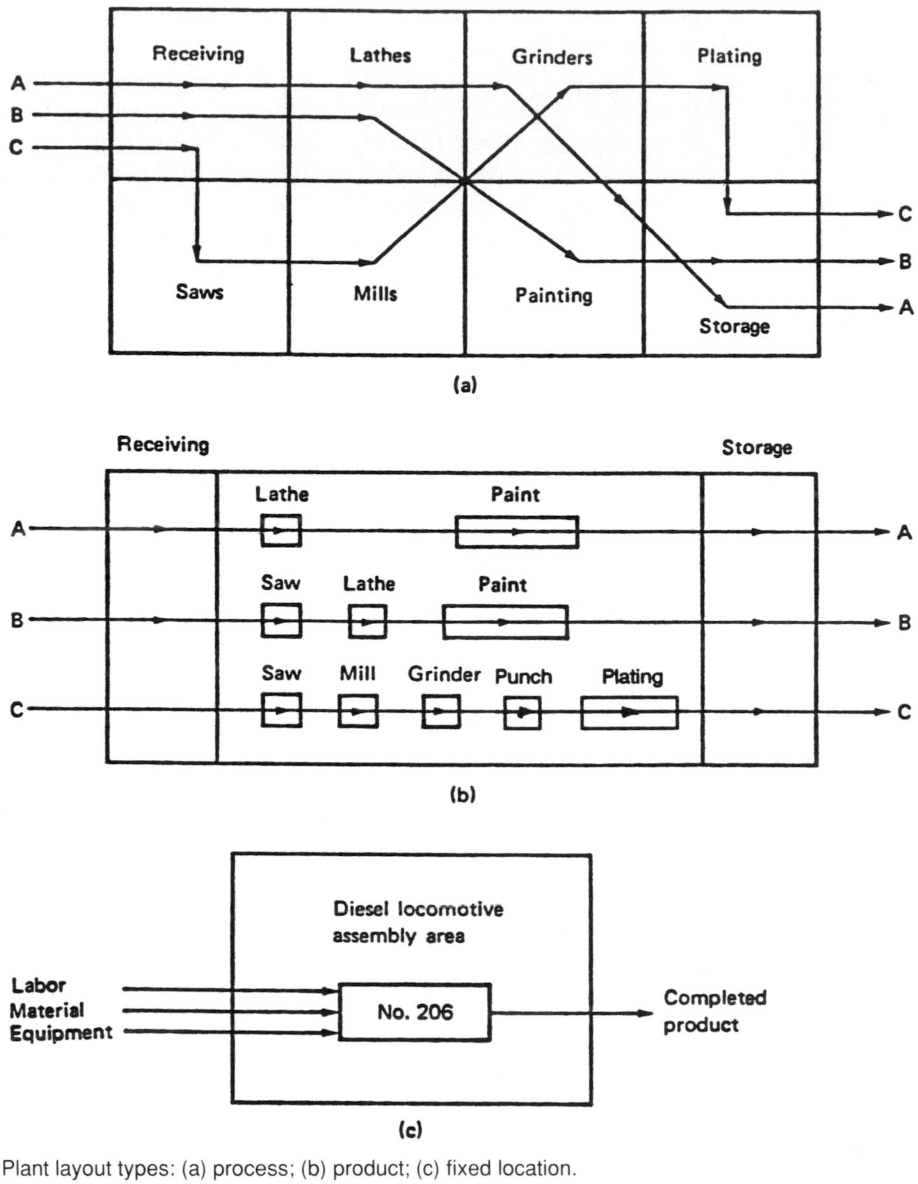

FIGURE 2-31 Plant layout types: (a) process; (b) product; (c) fixed location.

shorter material handling distances traveled, because the equipment is arranged in the desired processing sequence. Also, in the product layout the material stays in one area, which eliminates the need to expedite it from one area to another. Because such production lines are paced at some designed rate, control of production is limited to seeing that the expected amount of production is coming off the end of the line. If it is, the entire production line can be assumed to be producing as it should. Of course, such inflexibility offers the danger that if any part of the line has problems, the entire production line may be affected.

It would seem, therefore, that the main cost trade-off in choosing between the process and product types of layout involves comparing the cost effects of the utilization of equipment and the cost of material handling, also including consideration of control effects. Parts in a process layout that utilizes general area material handling equipment tend to get lost, and an activity (i.e., expediting) is typically necessary to continuously find such parts; in a product layout, material follows along fixed path devices (e.g., conveyors) and therefore tends not to get lost. If the volume is sufficient to permit acceptable utilization of equipment, the product layout becomes superior to the process layout. In most plants there is a combination of both types. In plants involving parts fabrication and assembly, fabrication tends to employ the process layout, whereas assembly areas often employ the product type of layout. In the final analysis, the combination that produces the desired volume of product at least total cost is preferred. Marketing is concerned with maximizing income, industrial engineering is concerned with maximizing cost, and management is attempting to ensure that there is a net positive difference (i.e., profit) to its credit.

Many plants are hybrids of the process, product, and fixed location layouts. An aircraft factory is a hybrid between product and fixed location layout. An airplane stays in a bay for some time while its fuselage is constructed on a fixed location basis, but at some time it is moved one bay position forward to have its wings attached. After moving five or six times, it eventually goes out the end of the building as a completed airplane. It often will then immediately enter a modification line to incorporate recent engineering changes that could not be readily incorporated in the basic processing line.

In shipbuilding, the hull of the ship will typically be built in a dry dock, but as soon as the hull is complete, it will often be moved out of the dry dock to be "finished out" at a pier, utilizing cranes to add remaining assemblies to complete the ship. Pier space is much less expensive than dry dock space; therefore, ship production is often a sequential process containing at least two fixed position stations.

As a historical note, Dry Dock No. 1 at the Norfolk Naval Shipyard actually precedes the formation of the United States. Passageways under the dry dock show how old it really is; rings in passageway walls under the dry dock that remain today were used in the early days of the dry dock for tying slaves at night so that there would be available manpower for repairing ships the next day. I did not have to mention it, but it is important that past examples of "man's inhumanity to man" (e.g., the Holocaust) be remembered lest we fail to remind ourselves from time to time of humankind's particularly ignoble history.

In recent years, other plant layout types have come into common use. Figure 2-32 illustrates a group technology layout type that was first popularized in Britain and has been utilized extensively in the United States as well. Products are grouped by common process families, and then processes are developed to accommodate each family. What makes products common to a family is that they require a generally similar sequence of operations on similar equipment.

A particular layout type popularly adopted in the 1980s is a manufacturing cell. Figure 2-33 illustrates a "before" and "after" example of a three-step manufacturing process. Rather than moving all product between all three process departments, a sequence of required equipment is set up as an independent producing unit and is repeated as many times as necessary to provide the required output. Such an arrangement provides production planning with the means to easily adjust the level of output by selecting the appropriate number of cells to run at one time. If any one cell has problems, it does not interfere with the production of the other cells.

The manufacturing cell also typically reduces the material handling cost of product going through the process, because successive machines for a process are located in sequence, one next to another. A shortcoming of the manufacturing cell, however, is that if the inherent cycle times of equipment dedicated to a cell are not of similar magnitudes, considerable equipment disutility can result. The professional industrial engineer knows how to evaluate the strengths and weaknesses of competing processes, and to select the arrangement that best serves the processing of products to be produced. It is a matter of matching the process to the need.

Hybrids of generic types often represent the optimum layout configuration. Note that the group technology approach illustrated in Fig. 2-32 combined with the manufacturing cell approach illustrated in Fig. 2-33 results in group technology cells. Such cells, providing processing capability for a family of similar products, may well represent the superior layout design.

FIGURE 2-32 Group technology layout.

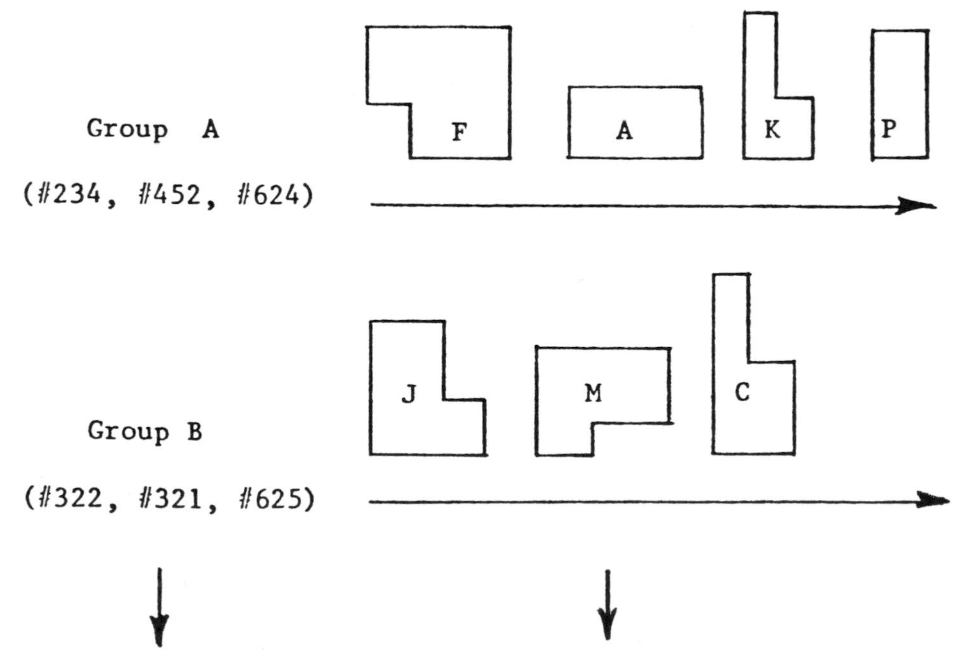

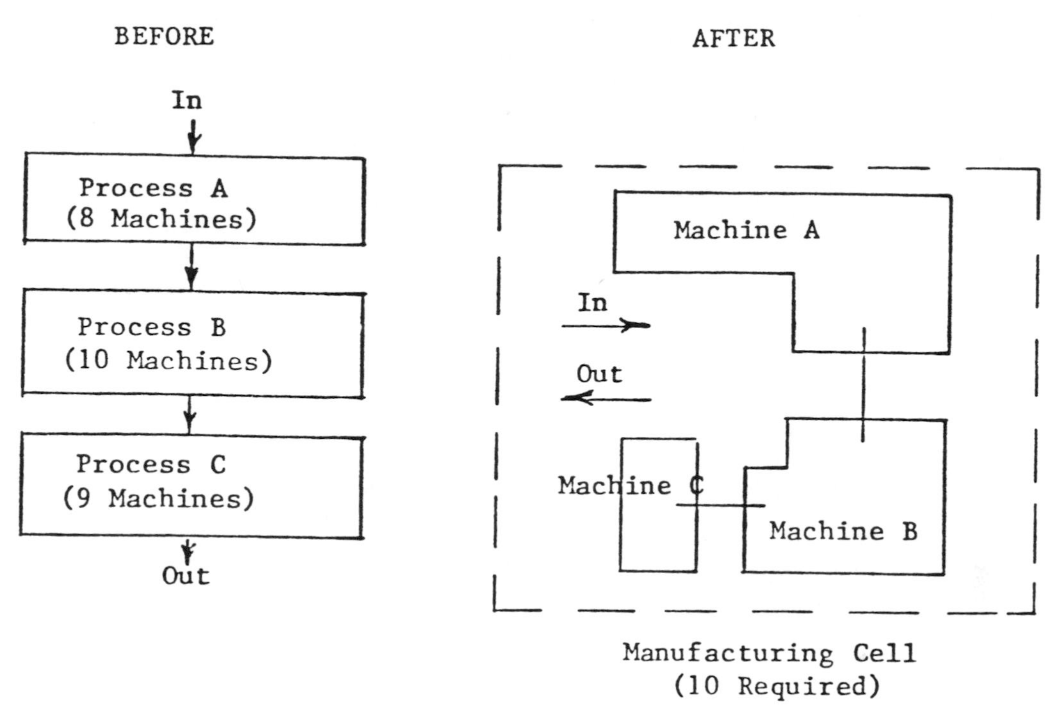

FIGURE 2-33 Manufacturing cell layout.

Figure 2-34 illustrates an example of flexible manufacturing. Utilizing transfer machines or automatic guided vehicles (AGVs) controlled by a computerized controller, product can be automatically routed to a workstation to meet specific routing needs.

Figure 2-35 illustrates a flexible manufacturing system (FMS) that is often referred to as a transfer line. A transfer line generally transfers the product directly (i.e., inline) to the next machine in sequence with fixed path mechanical handling. The boundary between a transfer line and a flexible manufacturing system is not well defined.

Equipment Calculations

The following example, employing some of the factors that need to be considered in determining the number of pieces of equipment required in a product type layout, is typical of calculations necessary in plant design. Figure 2-36 illustrates a process involving two sequential steps. Assume that the required output is 200 units per hour, overall process effective utilization is 85 percent because of expected labor pace and allowances with no relief labor, and standard times and scrap rates for steps 1 and 2 are as indicated in Fig. 2-36. How many step 1 and step 2 machines are needed? Make four assumptions in this example: (1) the

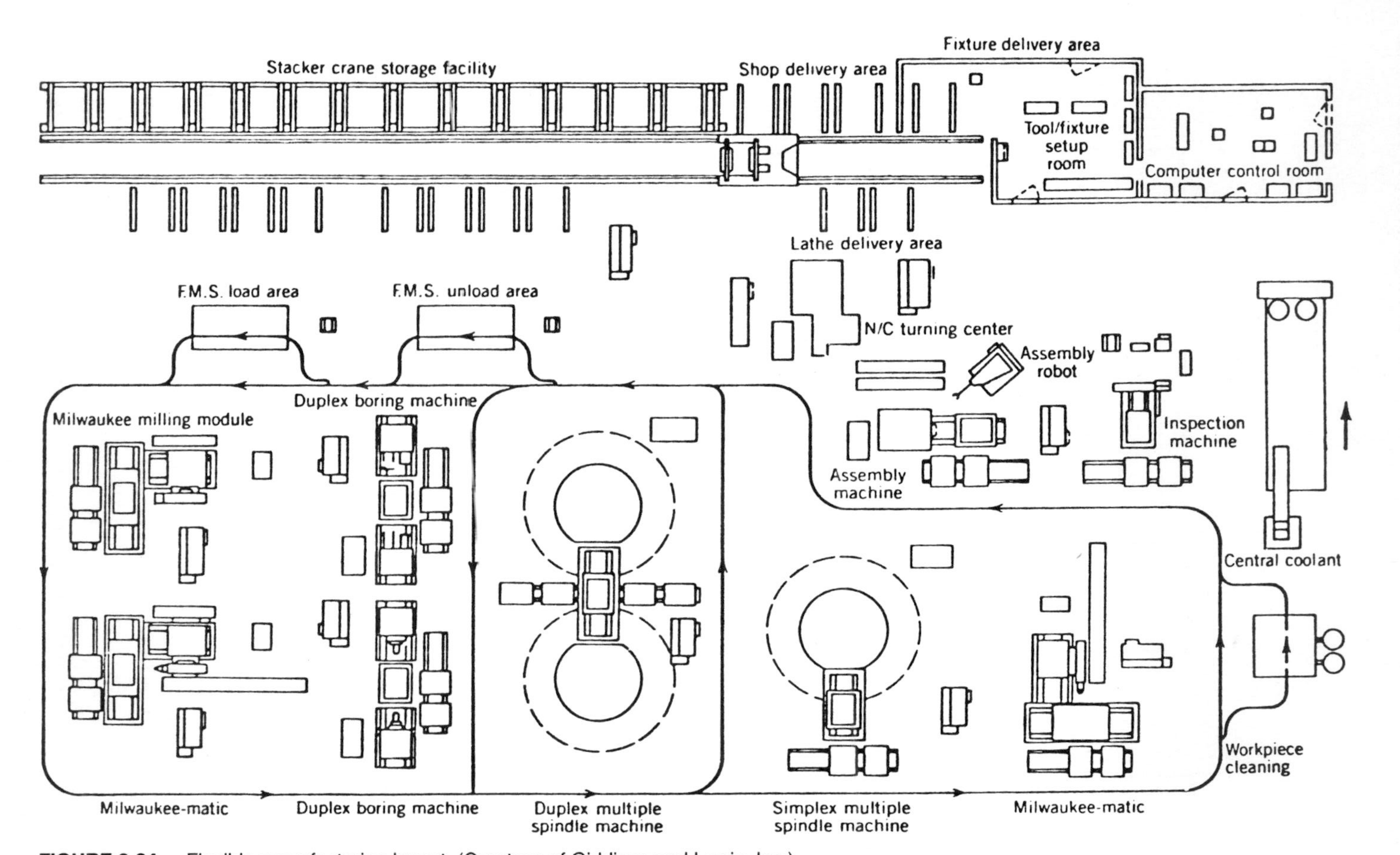

FIGURE 2-34 Flexible manufacturing layout. (Courtesy of Giddings and Lewis, Inc.)

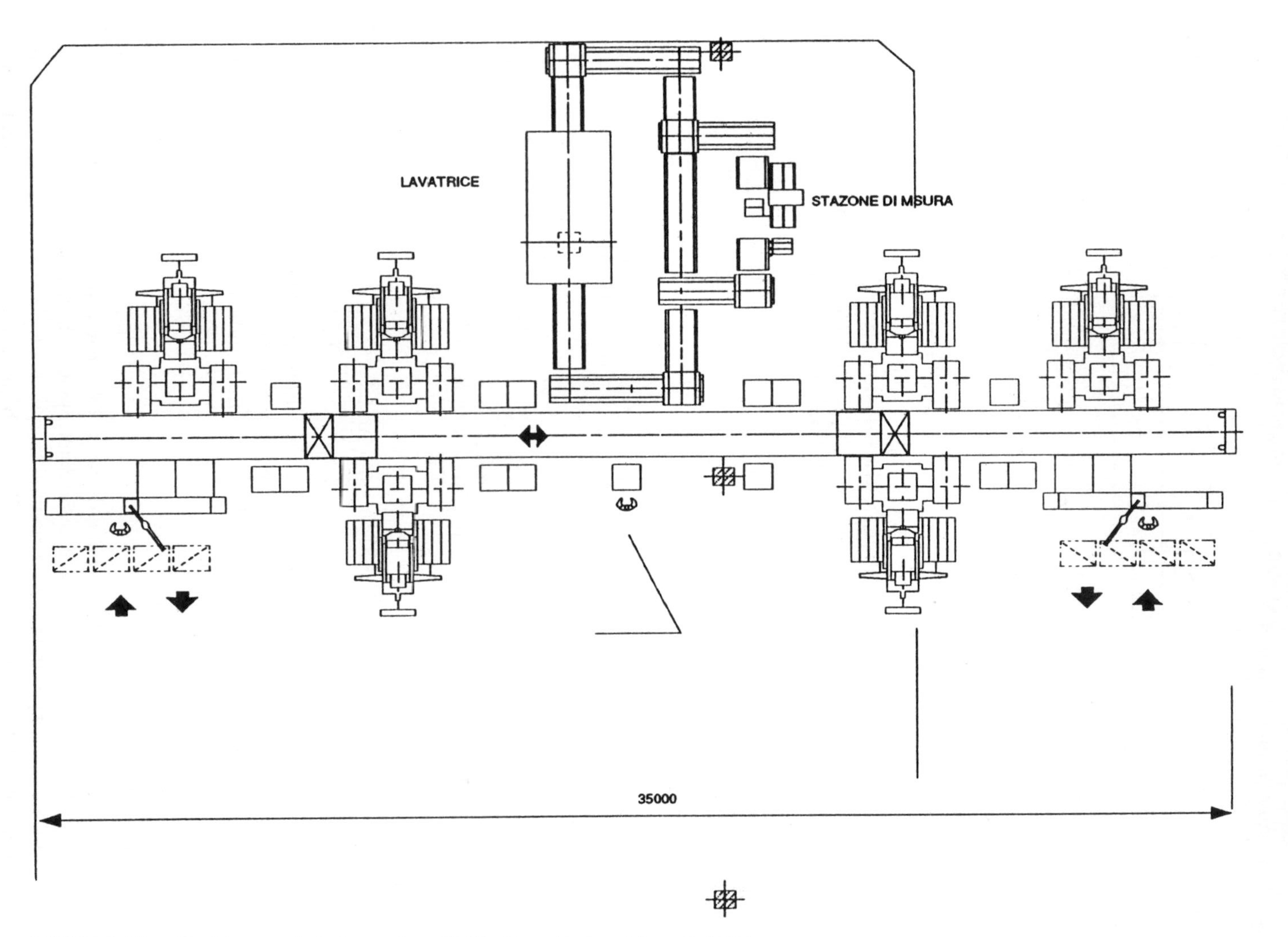

FIGURE 2-35 Flexible manufacturing system (FMS). (From Hodson [15, p. 7. 102].)

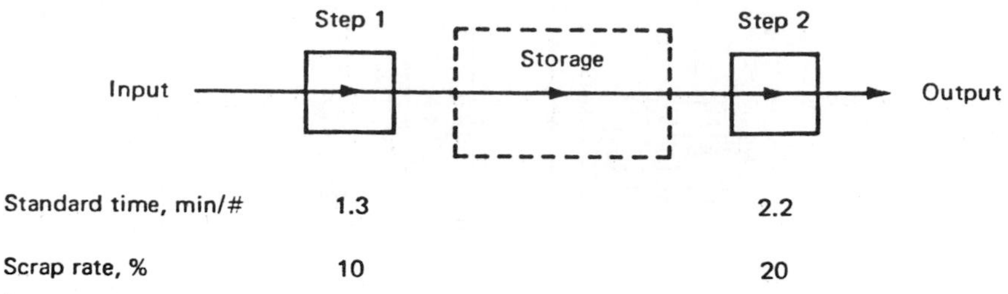

FIGURE 2-36 A two-step process.

process is such that when an employee is idle his machine is idle (i.e., no relief labor); (2) there is sufficient storage space between steps 1 and 2 such that the steps are independent with respect to blocking effects; (3) parts are tested for defectiveness after all processing is completed at each step; and (4) there is no reworking of defective parts. The calculation proceeds as follows:

Step 2

$$\text{Input} = \frac{200}{0.80} = 250$$

$$\text{Machines required} = \frac{250(2.2)}{60(0.85)} = 10.8 \cong 11$$

Step 1

$$\text{Input} = \frac{250}{0.90} \cong 278$$

$$\text{Machines required} = \frac{278(1.3)}{60(0.85)} = 7.1 \cong 7$$

This is a simplified example, because maintenance time on the machines and normal absentee effects of an employee may or may not have been included in the process efficiency, and so forth. The example is indicative, however, of calculations required in the design of a plant.

Activity Relationship Chart

As was mentioned earlier, a plant design is first macro in nature. One of the problems encountered at this stage is the determination of the relative spatial location of major areas of the plant. A method known as the "activity relationship chart," originally developed by Muther [35], can be used to record desired spatial relationships. Figure 2-37 illustrates a completed activity relationship chart for eight

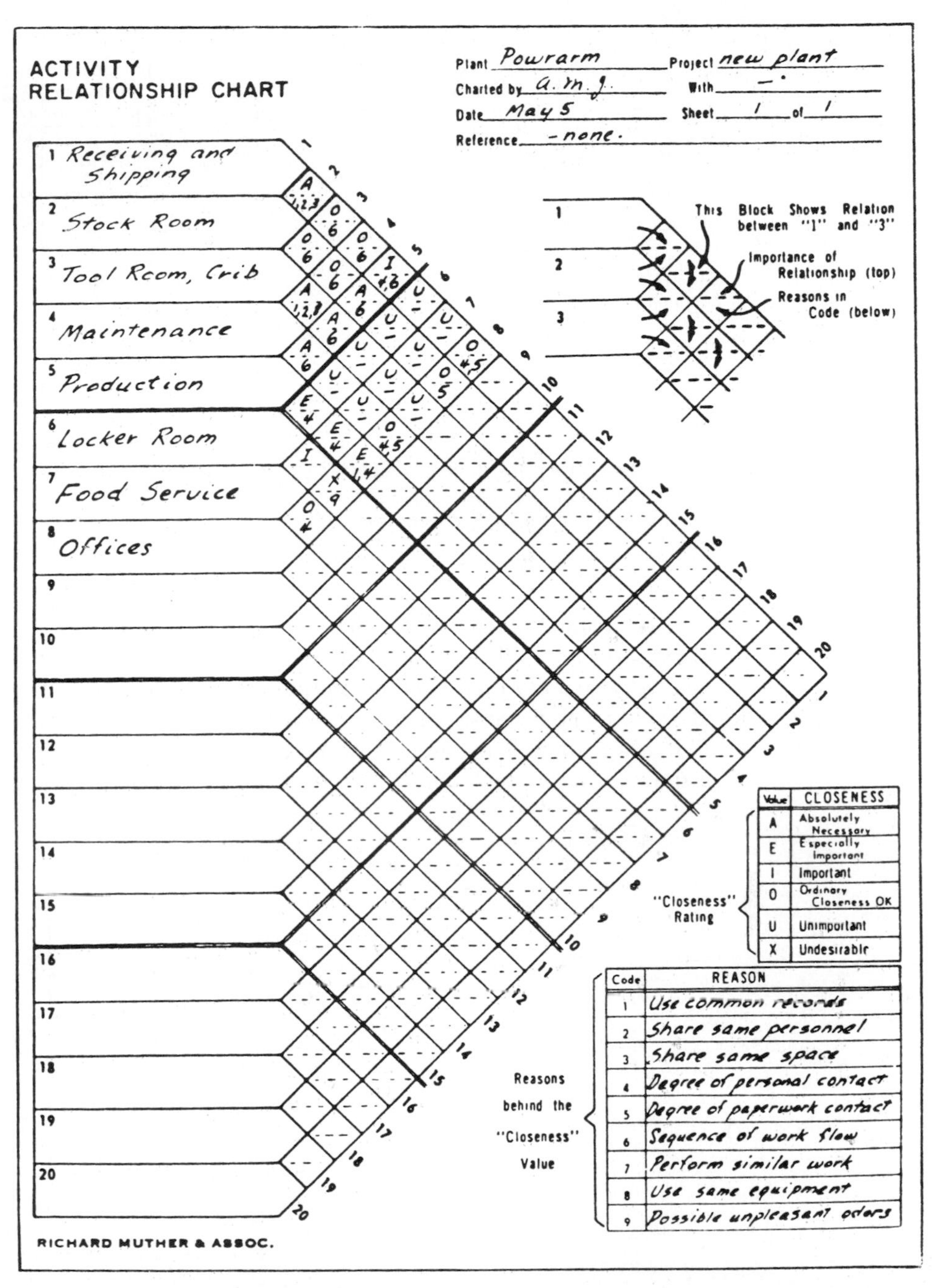

FIGURE 2-37 A typical activity relationship chart. (From James M. Apple: *Plant Layout and Materials Handling*, 3rd ed., p. 204. Copyright © 1977, John Wiley & Sons, New York.)

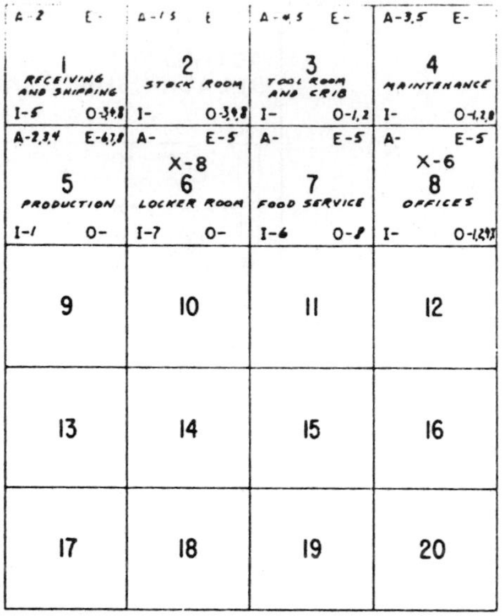

FIGURE 2-38 Unarranged activity areas. (From James M. Apple: *Plant Layout and Materials Handling*, 3rd ed., p. 208, John Wiley & Sons, Inc., New York.)

areas in a small plant. The relationships are transferred to equal-sized blocks, as shown in Fig. 2-38, and then cut apart and rearranged spatially in light of the desired proximity relationships, as illustrated in Fig. 2-39. The next step is usually to change the sizes of the areas to more appropriately represent their scaled individual sizes. This author prefers to use squares that conform in size to the estimated size of each department. These templates are arranged spatially, main-

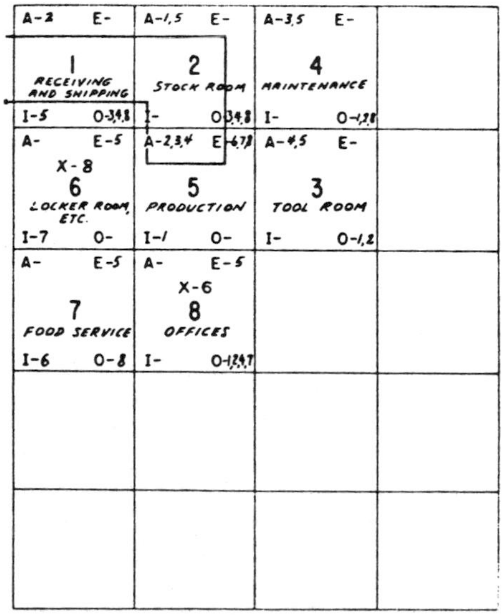

FIGURE 2-39 Arranged activity areas. (From James M. Apple: *Plant Layout and Materials Handling*, 3rd ed., p. 209. John Wiley & Sons, Inc., New York.)

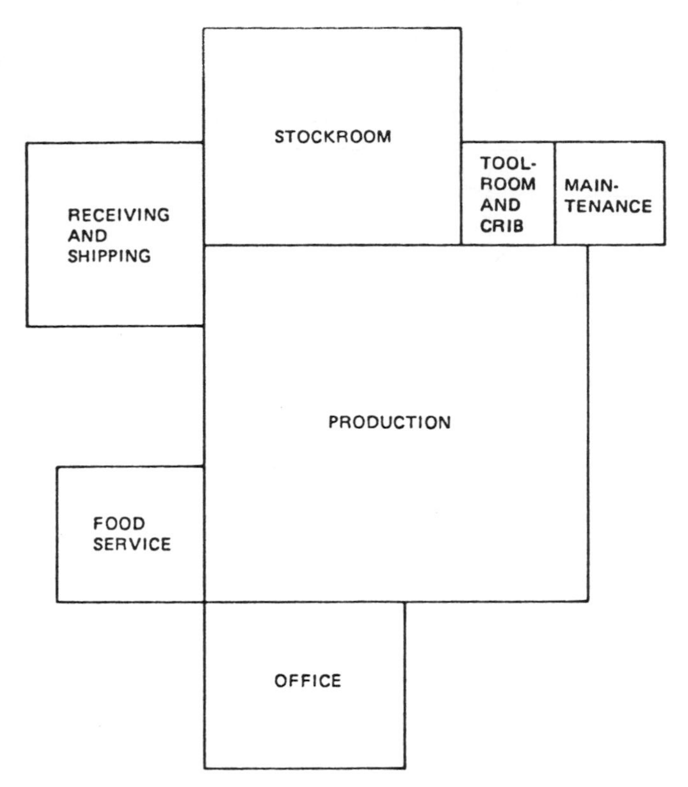

FIGURE 2-40 Arranged and scaled activity areas. (From James M. Apple: *Plant Layout and Materials Handling*, 2nd ed., p. 193. Copyright © 1963, The Ronald Press Company, New York.)

taining consistency with the previous step, as indicated in Fig. 2-40. Finally, the general shapes of the plant areas are devised by adjusting this arrangement to conform to a simple, typically rectangular, exterior shape, as indicated in Fig. 2-41. The final detailed design of the plant should conform generally to this design, as indicated in Fig. 2-42.

This approach differs from the approach typically taken by individuals unfamiliar with these methods. An intuitive solution to a plant layout problem often evolves by detailing elements at the beginning of a process and proceeding until the sum of the details produces a layout. Such an approach can result in a large number of arrangements, depending on the sequence of decisions that are made. One wrong decision in this sequence can produce an overall layout with considerable deficiencies. Methods such as the activity relationship chart produce generally consistent results with respect to desired overall characteristics of the plant. The larger is the scale of the mistake, the greater the cost effect. This is the primary reason for analyzing from macro to micro in plant layout work.

Plant layout analysis has been and still is relatively unquantifiable as compared to many other areas within industrial engineering. A plant layout is a capstone

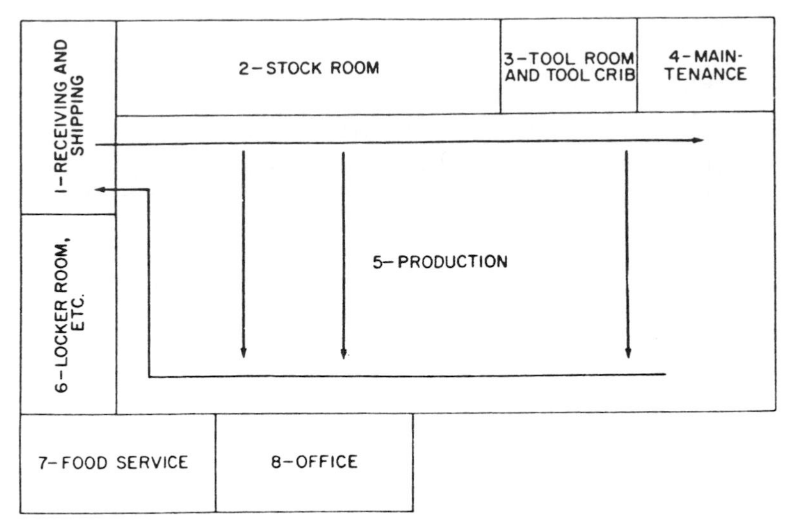

FIGURE 2-41 The final block layout. (From James M. Apple: *Plant Layout and Materials Handling*, 2d ed., p. 202. Copyright © 1963, The Ronald Press Company, New York.)

design effort in that it requires consideration of all of the design and control aspects of a plant. It is fairly common in industrial engineering departments in universities for the plant layout course to be the last course in the industrial engineering sequence related to design and control, and it is often the senior design course concerned with combining all of these techniques in the form of a term project. The number of factors to be considered makes quantification difficult. Table 2-3 is a list of marks of a good layout from Apple [2, pp.18–19]. Of the 35 desirable characteristics listed, 20 directly relate to material handling and are marked with an asterisk to draw attention to the importance material handling plays in plant layout. All characteristics are important, however, and are interrelated, yet many other important factors could have been mentioned, such as flexibility and density. In many respects, good plant layouts typically represent successful combinations of good layout principles.

The experienced plant layout engineer is constantly aware of a great many individual guiding principles of good plant layout practice. The engineer is also aware that many of these principles can be incompatible with one another. Compromise is an understood reality in plant layout.

One method of evaluating a plant layout is to make a list of its advantages and disadvantages. To be successful in doing this, one must be capable of assuming the roles, one by one, of all the functional managers in the plant affected by the layout and critically evaluating its effects on their activities. Only by this means is it possible to remove unnecessary and undesirable characteristics from a plant layout design. Ultimately, after much redesign, a point is typically reached where there is a minimum list of disadvantages. Removal of any of these, however, for the given design will produce an even greater disadvantage than the one being removed. At

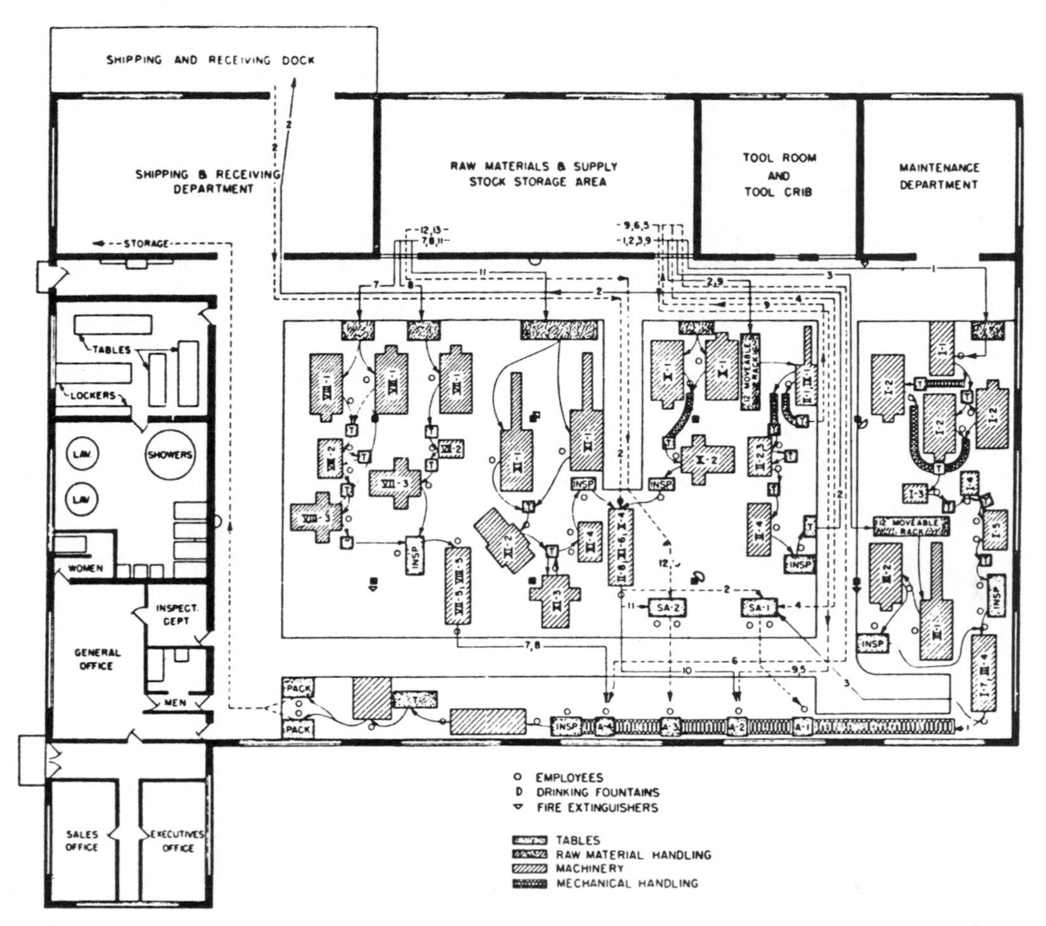

FIGURE 2-42 A partially detailed plant layout. (From James M. Apple: *Plant Layout and Materials Handling,* 2d ed., p. 353. Copyright © 1963, Ronald Press Company, New York.)

this point one probably has the best design for the general approach taken. Of course, this minimum list of disadvantages has to be valued along with the advantages the design offers in comparison to other totally different design approaches.

COMPUTERIZED PLANT DESIGN

Computerized approaches have been developed in recent years to assist in specific areas of plant design. A method known as CRAFT [4] was developed to determine desirable spatial arrangements of production departments in an attempt to minimize the total material handling cost incurred between all departments. An initial layout is given, and the cost per linear foot of material handled between each pair of departments is also given. The heuristic (i.e., the layout development logic embedded within the program) switches departments in the layout to effect a

TABLE 2-3 MARKS OF A GOOD PLANT LAYOUT

1.	Planned activity interrelationships
*2.	Planned material flow pattern
*3.	Straight-line flow
*4.	Minimum back-tracking
5.	Auxiliary flow lines
*6.	Straight aisles
*7.	Minimum handling between operations
*8.	Planned material handling methods
*9.	Minimum handling distances
*10.	Processing combined with material handling
*11.	Movement progresses from receiving towards shipping
*12.	First operations near receiving
*13.	Last operations near shipping
*14.	Point-of-use storage where appropriate
15.	Layout adaptable to changing conditions
16.	Planned for orderly expansion
17.	Minimum goods in process
18.	Minimum material in process
19.	Maximum use of all plant levels
20.	Adequate storage space
21.	Adequate spacing between facilities
22.	Building constructed around planned layout
*23.	Material delivered to employees and removed from work areas
24.	Minimum walking by production operators
25.	Proper location of production and employee service facilities
*26.	Mechanical handling installed where practicable
27.	Adequate employee service functions
28.	Planned control of noise, dirt, fumes, dust, humidity, etc.
29.	Maximum processing time to overall production time
*30.	Minimum manual handling
*31.	Minimum re-handling
32.	Partitions don't impede material flow
*33.	Minimum handling by direct labor
*34.	Planned scrap removal
*35.	Receiving and shipping in logical locations

Source: From James M. Apple, *Plant Layout and Materials Handling,* 3d ed, pp.18–19, John Wiley & Sons, New York, 1977.

reduction in total material handling cost. It iteratively (i.e., successively) attempts additional desirable switches until no more can be found. The last layout produced offers the lowest total material handling cost and is considered best in this respect. Figures 2-43 and 2-44 are initial and final layouts, respectively, for a sample problem using the CRAFT method.

A number of other computerized plant layout programs have been developed in the past [3, 22, 40]. Their use has been primarily academic, however, with limited use to date in actual practice, typically because of practical limitations of the software.

Use of quantitative methods in general has increased in facilities planning and design, however. Operations research techniques have made inroads in quantifying some aspects of this field.

CIMTECHNOLOGIES Corporation in Ames, Iowa, has developed two recent software products of use to the facilities planner: FactoryCAD [42] and FactoryFLOW [43]. FactoryCAD is a CADD software extension to AutoCAD, developed specifically for assisting the industrial engineer in performing industrial facility design. FactoryFLOW permits the determination of material handling costs associated with a particular design. Together they are a resource for developing and representing cost-effective industrial facility designs. Figure 2-45 illustrates a FactoryCAD layout with the material handling flows indicated. The thicknesses of the flow lines represent the material handling costs associated with each flow. The flow representations were developed from the FactoryFLOW program. A typical material handling report from FactoryFLOW is shown in Fig. 2-46.

Location pattern

	1	2	3	4	5	6	7	8	9	10	11	12	13	14	15	16	17	18	19	20	21	22	23	24
1	A	A	A	A	F	F	F	F	F	I	I	I	I	I	I	I	I	R	R	R	R	S	S	S
2	A			A	F				F	I							I	R			R	S		S
3	A			A	F				F	I	I	I	I	I	I	I	I	R			R	S		S
4	A	A	A	A	F				F	J	J	J	K	K	L	L	L	R			R	S		S
5	B	B	B	B	F				F	J		J	K	K	L		L	R	R	R	R	S	S	S
6	B			B	F				F	J	J	J	K	K	L	L	L	T	T	T	T	T	T	T
7	B			B	F				F	M	M	M	M	O	O	P	P	T						T
8	B	B	B	B	F	F	F	F	F	M	M	M	M	O	O	P	P	T						T
9	C	C	C	G	G	G	G	G	G	N	N	N	N	O	O	P	P	T						T
10	C		C	G					G	N	N	N	N	O	O	P	P	T						T
11	C		C	G					G	Q	Q	Q	Q	Q	Q	Q	Q	T						T
12	C	C	C	G					G	Q							Q	T						T
13	D	D	D	G					G	Q	Q	Q	Q	Q	Q	Q	Q	T	T	T	T	T	T	T
14	D		D	G	G	G	G	G	G	H	H	H	H	H	H	H	H	U	U	U	U	U	U	U
15	D		D	H	H	H	H	H	H	H							H	U						U
16	D	D	D	H	H												H	U	U	U	U	U	U	U
17	E	E	E	E	H												H	V	V	V	V	W	W	W
18	E			E	H												H	V			V	W		W
19	E	E	E	E	H	H	H	H	H	H	H	H	H	H	H	H	H	V	V	V	V	W	W	W

Total cost 47.90 Est. cost reduction 0.0 Move A Move B Move C Iteration 0

FIGURE 2-43 An initial CRAFT area layout.

Location pattern

	1	2	3	4	5	6	7	8	9	10	11	12	13	14	15	16	17	18	19	20	21	22	23	24
1	A	A	A	A	F	F	F	F	F	Q	Q	J	J	J	H	H	H	H	H	H	H	C	C	C
2	A			A	F				F	Q	Q	J	J	J	H	H					H	C		C
3	A			A	F				F	Q	Q	Q	J	J	J	H	H	H			H	C		C
4	A	A	A	A	F				F	Q		Q	K	K	L	L	L	H			H	C	C	C
5	B	B	B	B	F				F	Q		Q	K	K	L		L	H	H	H	H	H	H	H
6	B			B	F				F	Q		Q	K	K	L	L	L	S	S	S	S	H		H
7	B			B	F				F	Q		Q	Q	N	V	V	V	S			S	H		H
8	B	B	B	B	F	F	F	F	F	Q	Q	Q	Q	N	V		V	S		S	S	H		H
9	T	T	T	T	P	P	O	O	O	O	O	N	N	N	V		V	S	S	S	H	H		H
10	T			T	P	P	P	O	G	O	O	N	N	N	V	V	V	H	H	H	H	H	H	H
11	T			T	P	P	P	G	G	W	W	W	W	W	H	H	H	H			H	H	M	M
12	T	T	T	T	T	T	T	G	G	W	W	W	W	H	H						H	M	M	M
13	D	D	D	T			T	G	G	I	I	I	I	I	H	H	H	H	H	H	H	M	M	M
14	D		D	T			T	G	G	G	G	G	G	I	I	I	I	I	I	I	H	R	R	R
15	D		D	T			T	G					G	G	I					I	R	R		R
16	D	D	D	T			T	G		G	G	G	G	G	I	I	I	I	I	I	R	R		R
17	T	T	T	T			T	G		G	U	U	U	U	U	U	U	E	E	E	E	R		R
18	T						T	G		G	U						U	E			E	R		R
19	T	T	T	T	T	T	T	G	G	G	U	U	U	U	U	U	U	E	E	E	E	R	R	R

Total cost 32.24 Est. cost reduction 0.01 Move A P Move B T Move C Iteration 15

FIGURE 2-44 A final CRAFT area layout.

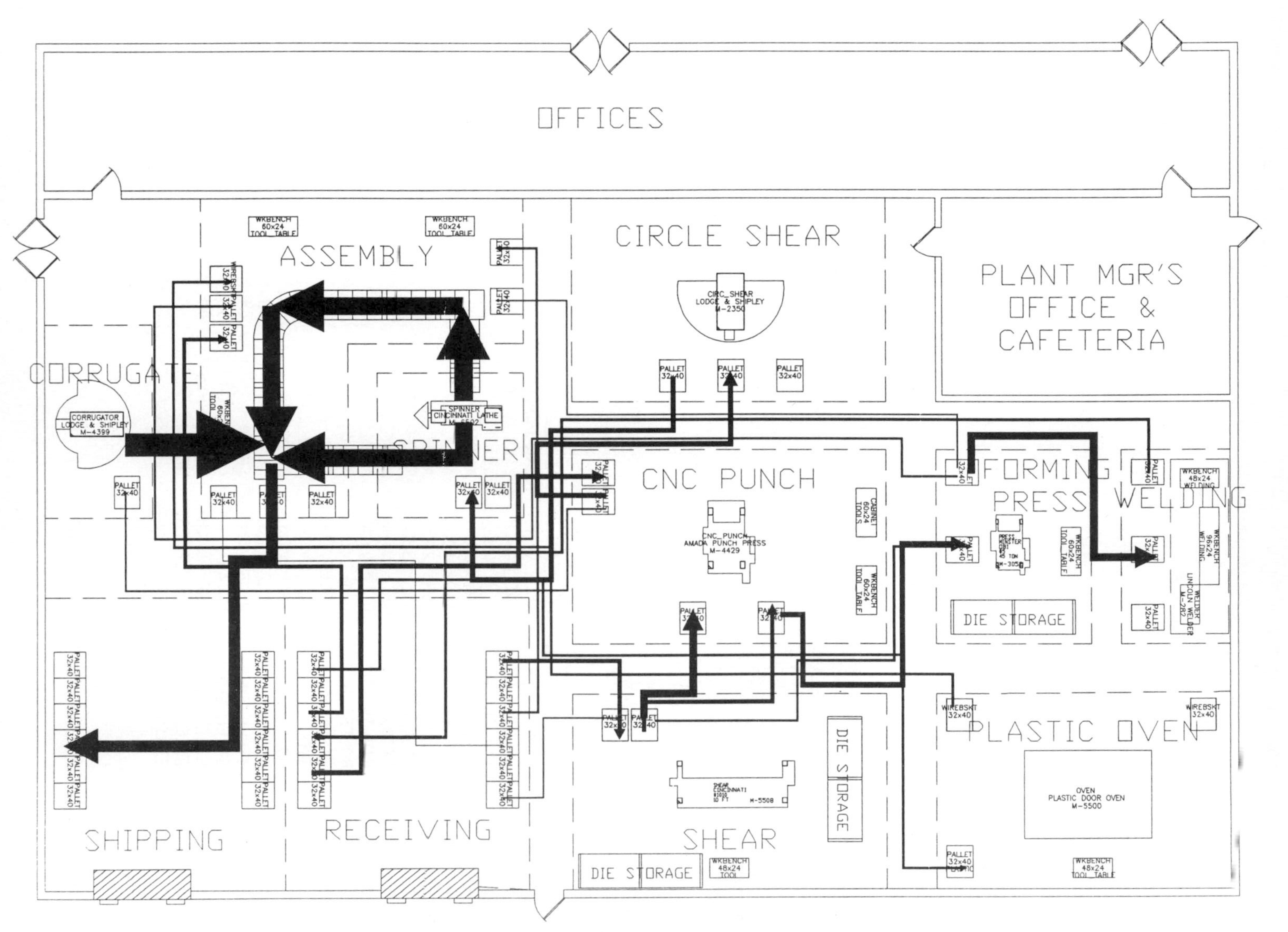

FIGURE 2-45 A FactoryCAD facility layout with flows. (From Sly: *FactoryCAD* [42].)

Advanced Material Handling Report

COMPANY NAME: ABC CORPORATION
PROJECT NAME: SAMPLE REPORT
DATE: 12/31/1992
TIME: 14:13

MATERIAL HANDLING REPORT

PRODUCT	PART	FROM	TO	TOTAL MOVE DISTANCE (FT.)	TOTAL COST ($/YEAR)
FEEDER	SHEET_STEEL	RECEIVING	CNC_PUNCH	134.85	164
FEEDER	SHEET_STEEL	CNC_PUNCH	CIRC_SHEAR	88.25	121
FEEDER	BOTTOM_BLANK	CIRC_SHEAR	SPINNER	113.10	146
FEEDER	FDR_BOTTOM	SPINNER	ASSY_100	0.00	215
FEEDER	SHEET_STEEL	RECEIVING	SHEAR	44.74	39
FEEDER	MIDSECTION	SHEAR	CNC_PUNCH	58.41	117
FEEDER	MIDSECTION	CNC_PUNCH	CORRUGATE	151.02	92
FEEDER	FINISHED_MID	CORRUGATE	ASSY_300	32.31	413
FEEDER	BOX_PLASTIC	RECEIVING	OVEN	233.66	260
FEEDER	PLASTIC_LIDS	OVEN	ASSY_200	287.73	315
FEEDER	SHEET_STEEL	RECEIVING	SHEAR	44.74	39
FEEDER	CONE_BLANK	SHEAR	FORM_PRESS	111.23	36
FEEDER	CONE_BLANK	FORM_PRESS	ASSY_100	142.95	71
FEEDER	SHEET_STEEL	RECEIVING	SHEAR	44.74	10
FEEDER	DIVIDER_BLANK	SHEAR	FORM_PRESS	111.23	73
FEEDER	DIVIDER_BLANK	FORM_PRESS	ASSY_200	285.26	221
FEEDER	SHEET_STEEL	RECEIVING	CNC_PUNCH	134.85	164
FEEDER	SHEET_STEEL	CNC_PUNCH	CIRC_SHEAR	88.25	121
FEEDER	TOP_BLANK	CIRC_SHEAR	SPINNER	113.10	146
FEEDER	FEEDER_TOP	SPINNER	ASSY_300	0.00	499
FEEDER	FEEDER	ASSY_100	ASSY_200	0.00	415
FEEDER	FEEDER	ASSY_200	ASSY_300	0.00	332
FEEDER	FEEDER	ASSY_300	SHIPPING	112.47	716
PURCHASED_PART	SKIDS	RECEIVING	ASSY_100	123.64	61
PURCHASED_PART	RIVETS	RECEIVING	ASSY_100	123.64	50
PURCHASED_PART	BOLTS	RECEIVING	ASSY_200	138.56	131
PURCHASED_PART	WASHERS	RECEIVING	ASSY_200	138.56	22
PURCHASED_PART	RODS	RECEIVING	ASSY_200	138.56	31
PURCHASED_PART	NUTS	RECEIVING	ASSY_200	138.56	53
PURCHASED_PART	DECALS	RECEIVING	ASSY_300	108.50	11
STAINLESS_FDR	STEEL_SHEET	RECEIVING	SHEAR	44.74	19
STAINLESS_FDR	STEEL_BLANK	SHEAR	CNC_PUNCH	36.03	140
STAINLESS_FDR	STEEL_BLANK	CNC_PUNCH	FORM_PRESS	90.11	247
STAINLESS_FDR	STEEL_BLANK	FORM_PRESS	WELD	78.93	322
STAINLESS_FDR	STAINLESS_FDR	WELD	SHIPPING	275.33	386
Grand total					**6,195**

FIGURE 2-46 A FactoryFLOW material handling report. (From Sly: *FactoryFLOW* [43].)

An article entitled A *Graphical Spreadsheet for Factory Layout* by Ron Gilbert [13], former president and CEO of CIMTECHNOLOGIES Corporation, describes FactoryFLOW and includes discussion of a case at John Deere (of Horican, Wisconsin) concerning use of the software [13].

Computer Simulation

Of all the quantitative techniques applied to facilities planning and design, and probably to industrial engineering as a whole, simulation has probably been the most productive to date. A plant layout offers only a static view of a producing system; if a photograph were made of a production process from a helicopter after the roof of the plant was removed, the photograph would be analogous to a layout, and would offer only a static view of where things were at a point in time.

Computer simulation makes it possible to evaluate the dynamic operational characteristics of a particular layout of a production process. Queuing theory analysis could also offer an analytical approach to gaining this dynamic appreciation. The complexities of modeling a real-world production process, however, typically make computer simulation of the system the better choice. The simulation program models analogous flows of material and assemblies in time, accumulations that develop in relation to processing times, handling and storage restrictions, and many other effects. The completed simulation provides a preview of how the total production process will operate over time. The typical result of simulating a production system, and adjusting it accordingly, is a considerable reduction of material in process and a system that generally performs as originally intended. Without a prior simulation, storage needs are difficult to estimate and consequently are often overestimated in an effort to play it safe.

Figure 2-47 illustrates a "before and after" design for part of a production process. As a result of simulating the dynamics of this process, the design was greatly simplified, resulting in less initial and operating costs and a simpler, more reliable process. Such a result is typical of simulating complex systems.

There are numerous computer simulation software packages on the market today, each with its unique strengths, weaknesses, costs, and benefits; GPSS-H by Wolverine [48], WITNESS by AT&T [47], SLAM and others by Pritsker & Associates (37), AutoSimulations [5], XCELL+ Factory Modeling System [50], ProModelPC [38], and SIMFACTORY [41], to name a few. Many are advertised in *Industrial Engineering* [18].

Material Handling

Material handling analysis is always a part of plant layout. Methods study, plant layout, and material handling are all part of the design of a production facility and can hardly be treated separately—in the final analysis it is all one design.

Material handling is a field in which there is a considerable body of information about equipment and a general inability to quantify problems; therefore, the use of experience seems to offer the greatest hope for solving material handling

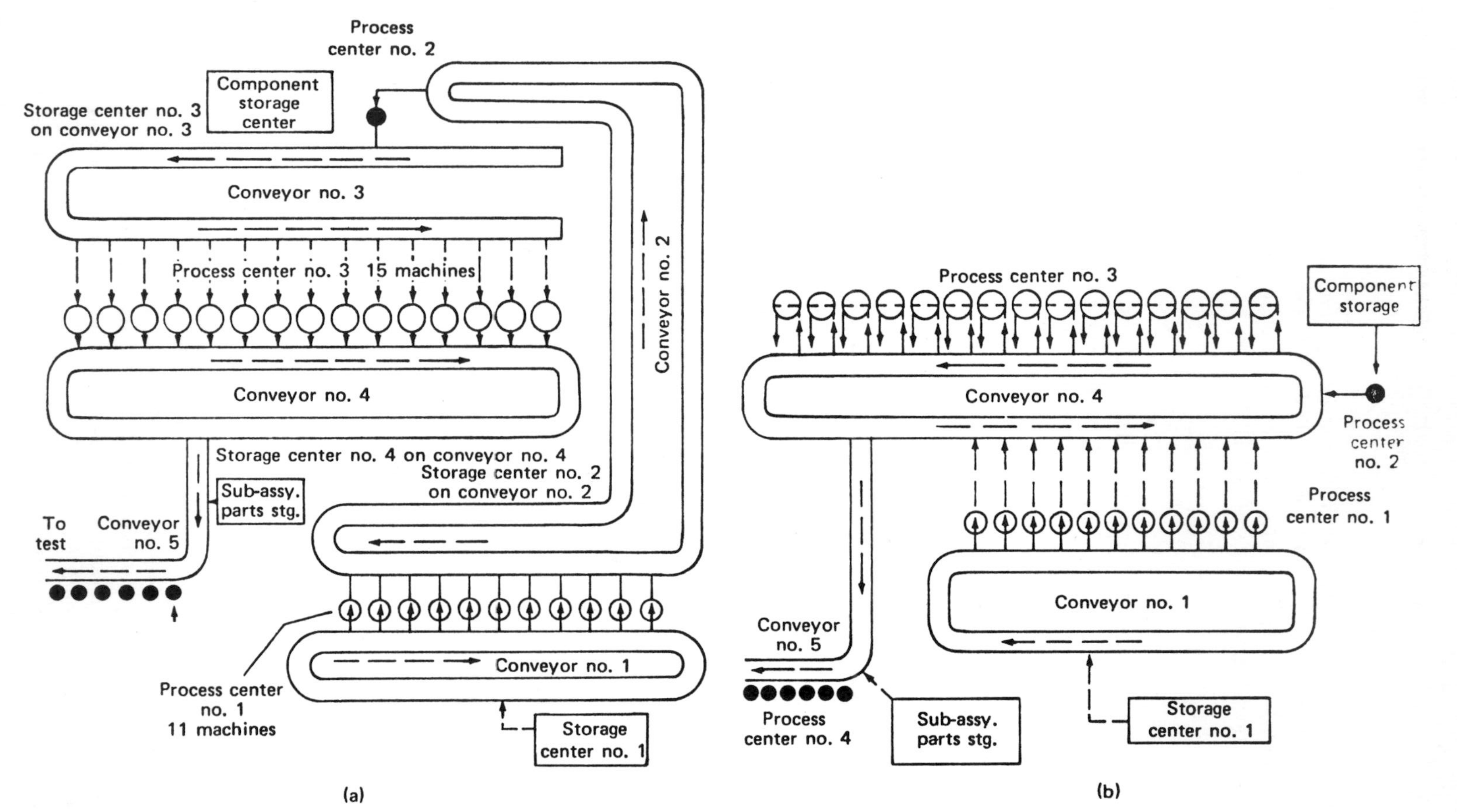

FIGURE 2-47 Conveyor designs: (a) before; (b) after. (From Hurley [16, p. 49].)

problems. There are more than 430 different types of material handling equipment, and each type is represented by 6 to 10 major manufacturers, each with a line of equipment for that type offering a range of capacities and options.

To be effective, a material handling consultant must be generally cognizant of the primary capabilities and limitations of this broad range of equipment. At the same time, the consultant must be cognizant of and effectively consider a broad range of guiding principles of material handling. Table 2-4 lists 20 primary princi-

TABLE 2-4 THE 20 PRINCIPLES OF MATERIAL HANDLING

1. *Orientation Principle:* Study the problem thoroughly prior to preliminary planning in order to identify existing methods and problems, physical and economic constraints, and to establish future requirements and goals.
2. *Planning Principle:* Establish a plan to include basic requirements, desirable options, and the consideration of contingencies for all material handling and storage activities.
3. *Systems Principle:* Integrate those handling and storage activities that are economically viable into a coordinated system of operations, including receiving, inspection, storage, production, assembly, packaging, warehousing, shipping, and transportation.
4. *Unit Load Principle:* Handle product in as large a unit load as is practical.
5. *Space Utilization Principle:* Make effective utilization of all cubic space.
6. *Standardization Principle:* Standardize handling methods and equipment wherever possible.
7. *Ergonomic Principle:* Recognize human capabilities and limitations by designing material handling equipment and procedures for effective interaction with the people using the system.
8. *Energy Principle:* Include energy consumption of the material handling systems and material handling procedures when making comparisons or preparing economic justifications.
9. *Ecology Principle:* Use material handling equipment and procedures that minimize adverse effects on the environment.
10. *Mechanization Principle:* Mechanize the handling process where feasible to increase efficiency and economy in the handling of materials.
11. *Flexibility Principle:* Use methods and equipment that can perform a variety of tasks under a variety of operating conditions.
12. *Simplification Principle:* Simplify handling by eliminating, reducing, or combining unnecessary movements and/or equipment.
13. *Gravity Principle:* Utilize gravity to move material wherever possible, while respecting limitations concerning safety, product damage, and loss.
14. *Safety Principle:* Provide safe material handling equipment and methods that follow existing safety codes and regulations in addition to accrued experience.
15. *Computerization Principle:* Consider computerization in material handling and storage systems, when circumstances warrant, for improved material and information control.
16. *System Flow Principle:* Integrate data flow with physical material flow in handling and storage.
17. *Layout Principle:* Prepare an operation sequence and equipment layout for all viable system solutions, then select the alternative system that best integrates efficiency and effectiveness.
18. *Cost Principle:* Compare the economic justification of alternative solutions in equipment and methods on the basis of economic effectiveness as measured by expense per unit handled.
19. *Maintenance Principle:* Prepare a plan for preventive maintenance and scheduled repairs on all material handling equipment.
20. *Obsolescence Principle:* Prepare a long-range and economically sound policy for replacement of obsolete equipment and methods with special consideration to after-tax lifecycle costs.

ples of material handling provided by the College Industry Council of the Material Handling Institute [25]. This list is a summary of earlier lists, provided by authors such as Apple [2], that were on the order of 5 to 10 times as extensive, and obviously more detailed. Each detailed principle has its value in a particular environment, but some of the more detailed principles are occasionally incompatible. One has to choose the principles that are applicable in a given work situation. Some principles become less applicable over time as new considerations come to light. For example, the trend today in a Just-in-Time environment is to have as little work in process as possible. An earlier material handling principle that stated "Handle as many pieces as practical in one unit" is no longer valid in a JIT work environment.

The solution to material handling problems, and taking advantage of material handling opportunities, in most plants typically involves understanding applicable material handling principles for a given application, and making the appropriate choices based on the guiding principles. Little, if any, mathematics is involved; it is simply a matter of understanding "what works when and why" based on the guiding principles relating to each situation.

The material handling engineer must select the best combination of equipment in light of a best combination of material handling principles for a particular situation. Typically, the best solution is a least-cost method of handling the required material. It is indeed rare when a more costly alternative is selected if all cost effects have been considered. Of course, if the application environment is unique, for example, defense, cost may be a far less critical issue, as compared to more important issues relating to winning a war.

ERGONOMICS

Almost all industrial engineering curricula contain at least one undergraduate course entitled ergonomics, human engineering, human factors, or human-machine systems. For the student wanting to specialize in ergonomics, curricula often permit electives in related courses taught in the psychology department as well. All these courses are concerned with understanding humans as a key element in a human-machine system. Industrial engineering systems are "people systems," and an understanding of the physical and mental capabilities and limitations of humans is necessary in designing these systems. Equally important is the fact that people systems work only if people let or make them; therefore, the body of knowledge concerned with human motivations, drives, and attitudes requires some measure of technical appreciation, if not thorough understanding, for an industrial engineer to be effective in developing designs and controls.

In designing a production system, the choice between human and machine for performing various functions has to be resolved. Figure 2-48 attempts to identify the functions that are performed best either by man or machine. There is little question that through mechanization and computerization humans are less involved in physical and mental drudgery. This is as it should be, if only from the point of view that the general physical and computational abilities of humans are

MAN VS. MACHINES

Man excels in	Machines excel in
Detection of certain forms of very low energy levels	Monitoring (both men and machines)
Sensitivity to an extremely wide variety of stimuli	Performing routine, repetitive, or very precise operations
Perceiving patterns and making generalizations about them	Responding very quickly to control signals
Detecting signals in high noise levels	Exerting great force, smoothly and with precision
Ability to store large amounts of information for long periods—and recalling relevant facts of appropriate moments	Storing and recalling large amounts of information in short time-periods
Ability to exercise judgment where events cannot be completely defined	Performing complex and rapid computation with high accuracy
Improvising and adopting flexible procedures	Sensitivity to stimuli beyond the range of human sensitivity (infrared, radio waves, etc.)
Ability to react to unexpected low-probability events	Doing many different things at one time
Applying originality in solving problems: i.e., alternate solutions	Deductive processes
Ability to profit from experience and alter course of action	Insensitivity to extraneous factors
Ability to perform fine manipulation, especially where misalignment appears unexpectedly	Ability to repeat operations very rapidly, continuously, and precisely the same way over a long period
Ability to continue to perform even when overloaded	Operating in environments which are hostile to man or beyond human tolerance
Ability to reason inductively	

FIGURE 2-48 Man versus machine. (From Wesley E. Woodson and Donald W. Conover: *Human Engineering Guide for Equipment Designers,* 2d ed., 1964, pp. 1–23. Reprinted by permission of University of California Press, Berkeley.)

poorer than those of machines, particularly if consistency and reliability are important factors. As a case in point, count the number of *F*'s in the following statement:

FINISHED FILES ARE THE
RESULT OF YEARS OF SCIENTIFIC
STUDY COMBINED WITH THE
EXPERIENCE OF MANY YEARS

If you counted three, go back and count again, and count a letter at a time if you have to until you find all six. Why was it hard to see the three *f*'s in the three *of* words? Simply because *of* is pronounced *ov* in English, and therefore, as far as your phonetically scanning brain is concerned, there is no *f* in *ov*.

Anyone who has previously been significantly involved in a production environment has probably had the experience of having a meeting interrupted while someone in the meeting went on to the production floor to count the number of products sitting on a bench or shelving unit, the number being the point of contention in the meeting. Assume for the moment that you have just counted the number of products on the bench and it is 23. It is suggested that you go immediately to the meeting and report that the number is 23. A momentary lack of confidence in the number that you just counted could cause you to count them again. What are you going to do if you count them again and the second count is 22? Now you have to count them a third time to resolve the conflict. It is suggested that if you count either 22 or 23 this time, you go immediately back to the meeting and report whichever number you just counted. Do not tell anyone in the meeting, however, that you once counted a different number, because that will raise concern about what the number really is. If another person goes out to count the products she or he may return with still another number (e.g., 24), and it only gets worse from there. Humans do not like to admit it, but they can have great difficulty doing simple tasks reliably and consistently.

One of the realities in quality control circles is the practical fallacy of assuming that 100 percent inspection actually works when performed by humans. Ask a class of students to count the number of *e*'s in this paragraph, collect the responses by secret ballot, and plot the results, and you will have an excellent example of a skewed normal distribution. It is skewed because it is easier to overlook an *e* than to add one.

The role of humans today is shifting toward employing their analytical decision-making ability and judiciously using their sensory abilities. Humans still do the coffee tasting and the bad-orange sorting, run the bridge crane, and decide when to shoot a burglar. Certain combinations of analytical and sensory ability will remain human functions for a long time to come.

In designing a workstation, the size and variability in size of those who will perform the job must be taken into consideration. Figure 2-49, as an example, illustrates recommended design dimensions based on anthropometric data. It should be noted, however, that only nominal dimensions are provided. Range values would greatly enhance these dimensions. There is a wealth of information in human engineering, and these examples are intended to offer some insight into the field. Quite often, in many practical situations, the desired anthropometric or other data are assumed to exist. The problem becomes one of finding the data and evaluating their applicability to the specific situation or environment.

It is not uncommon in military systems for the human links to be the most fragile system components when it comes to system performance and reliability. In the Hellfire missile system, for example, a U.S. air-to-ground missile system was well suited to destroying enemy (at the time, Soviet) ground armor such as a tank. In this system, the forward observer is potentially the weak link in the system. In early simulations of the system, technical details of the Soviet tank, the missile, the missile laser seeker, the helicopter for launching the missile, and the communication links between the forward observer and the helicopter were all

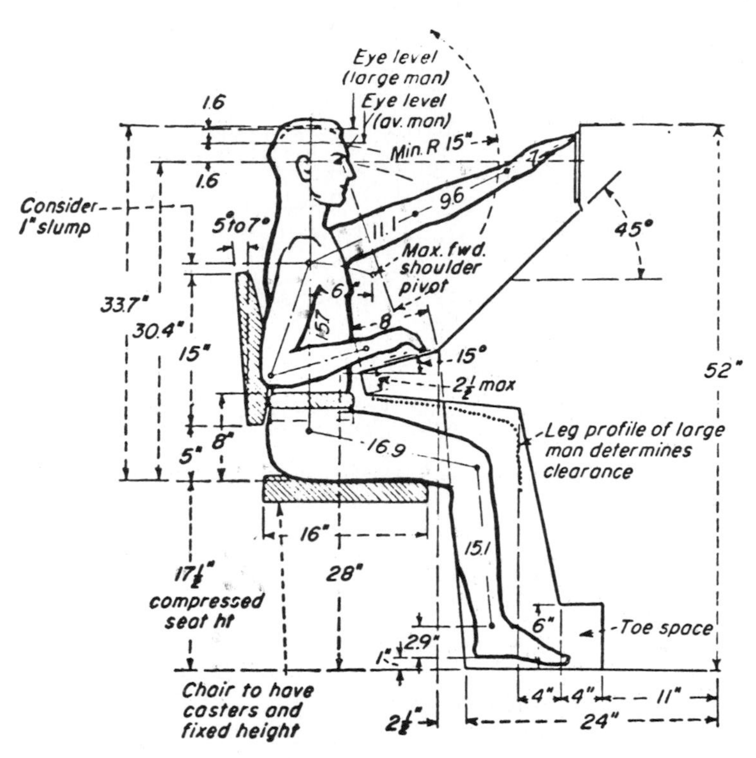

FIGURE 2-49 Recommended console dimensions.

well-known and reliable system components. The forward observer, on the other hand, is typically alone in a foxhole; he is often the first to observe the enemy coming; he communicates with his supporting helicopter under considerable duress; and at the designated time, he engages the enemy. He fires his laser rifle at the tank while the helicopter rises into position to release the missile, which will seek the laser energy on the tank and guide the missile to the tank.

All goes well if all goes well. But the forward observer knows that the Soviet tank has a laser energy sensor, and fire control and countermeasure systems. As soon as laser energy is sensed, the tank stops, the gun turret rotates and is set to a calculated range and azimuth, and if it gets a round off before the missile hits the tank, the forward observer will probably become a casualty. In addition, a countermeasure smoke canister may be released in the direction of the laser energy, in which case the laser energy lights up the smoke ball, and the missile is guided to the smoke ball instead of the tank. A key question in the development of one of the original Hellfire missile simulations was "How well does the forward observer aim his laser gun under such stressful conditions?" This is a key variable in determining the outcome of the simulation. However, data for this are not readily obtainable.

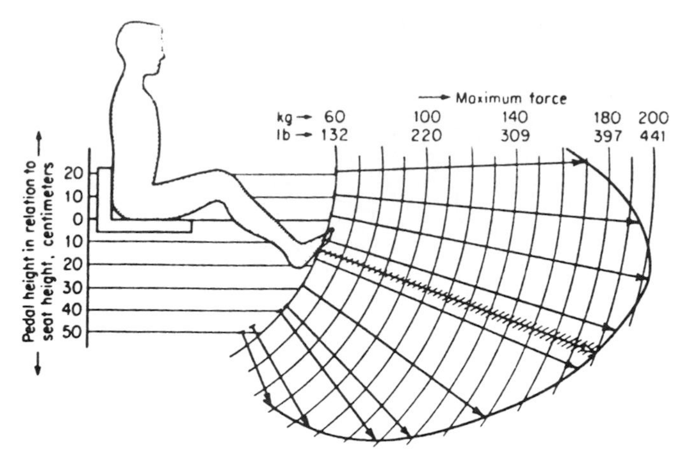

FIGURE 2-50 Maximum pedal thrust. (From McCormick [27, p. 436].)

Here is another example. A few years ago the *Vincennes*, a United States naval ship in the Persian Gulf with advanced radar systems, mistakenly shot down a commercial airliner. A likely cause of the system failure was "human error."

The physical abilities and limitations of humans represent a significant body of useful data in workstation design. Figure 2-50, for example, provides data on maximum pedal thrust as a function of the relative position of an operator and a pedal. It is typical of very useful, detailed, specific information in this field.

With respect to elements of design related to mental functions, Fig. 2-51 details generally preferred choices concerning direction of motion. Such a figure in most cases summarizes exhaustive tests to identify the preferred choice of alternatives under various conditions based on a broad audience of typical users. A wealth of human engineering data has evolved from military tests of human performance—for example, at Wright Patterson Air Force Base, Dayton, Ohio, since World War II. Also, the United States Air Force, under Project THEMIS, developed an ergonomics laboratory at Texas Technological University in Lubbock, Texas.

It is common knowledge that a person feels warmer as humidity increases. Figure 2-52 is a chart developed for determining effective temperature as a function of both dry- and wet-bulb temperatures. The temperature that an employee "feels" is the temperature that must be considered in accounting for temperature effects on human performance in situations where perceived rather than actual temperature is the controlling factor. Actual conditions obviously play their role as well.

When the attributes of a work environment are understood, it is possible to predict various effects concurrent with work performance as a function of the environment. Figure 2-53 illustrates the number of accidents that occurred as a function of temperature and age, based on a study of coal miners in England. The extent to which data of this type can be useful in a different occupation is questionable; the trends it suggests are likely to exist, however, in similar work environments.

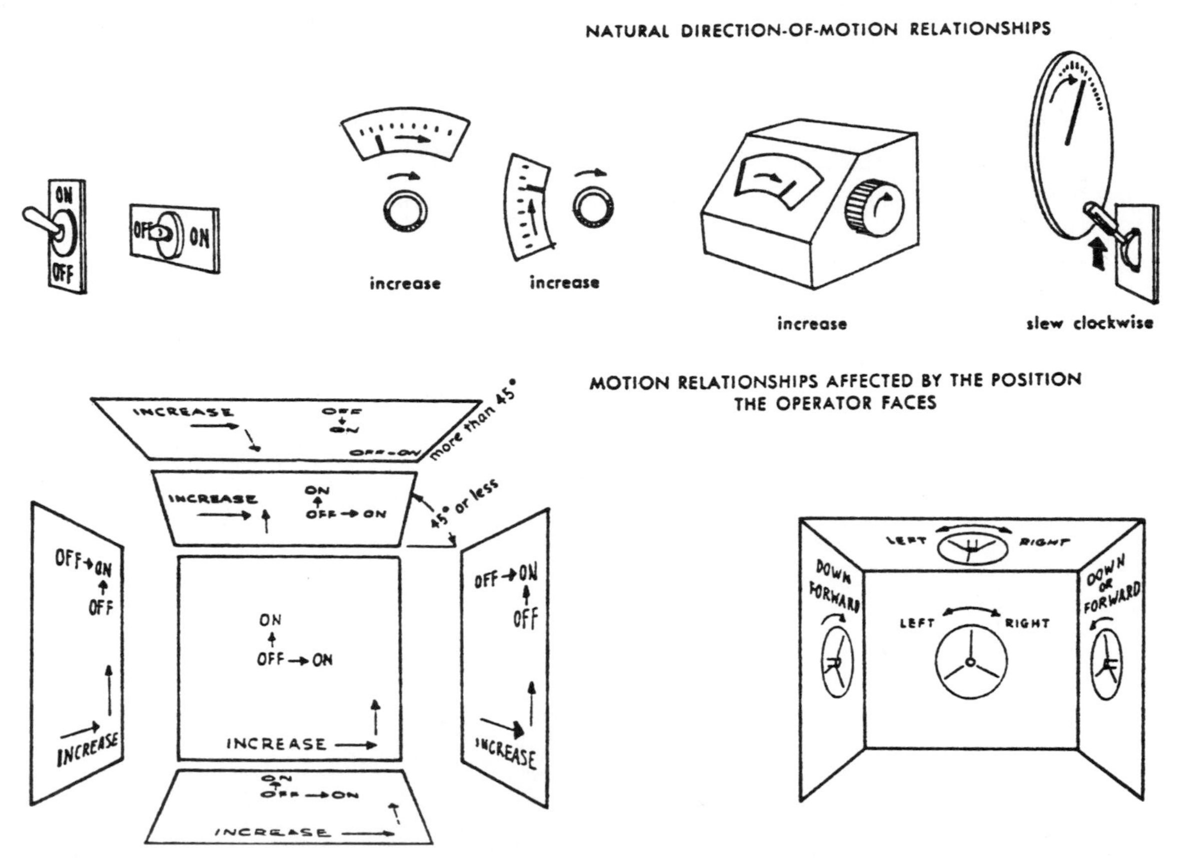

FIGURE 2-51 Motion relationships. (From Wesley E. Woodson and Donald W. Conover [49, pp. 2–112].)

In 1982, the International Union, United Automobile, Aerospace, and Agricultural Implement Workers of America, UAW, published an excellent 35-page brochure entitled *Strains & Sprains: A Worker's Guide to Job Design*, developed with the technical assistance of Don Chaffin and Tom Armstrong, both of the Center for Ergonomics at the University of Michigan. It is available as Publication #460, presently for $2.00 each, from the UAW Purchase and Supply Department, 8000 E. Jefferson, Detroit, Michigan, 48214. It contains well-illustrated descriptions of both bad and good examples of workplace design that impact the health of workers. Although photographs and examples are drawn from UAW work environments, the principles are generally applicable throughout industry. It is an excellent publication, highlighting the predominant ergonomics problems in industry, and it belongs on every professional's bookshelf.

The three major ergonomics injury categories covered in the brochure are (1) twisting wrists, (2) weary arms, and (3) aching back. In the first category, the tendon passing through what is called the carpal tunnel in the twisted or bent wrist

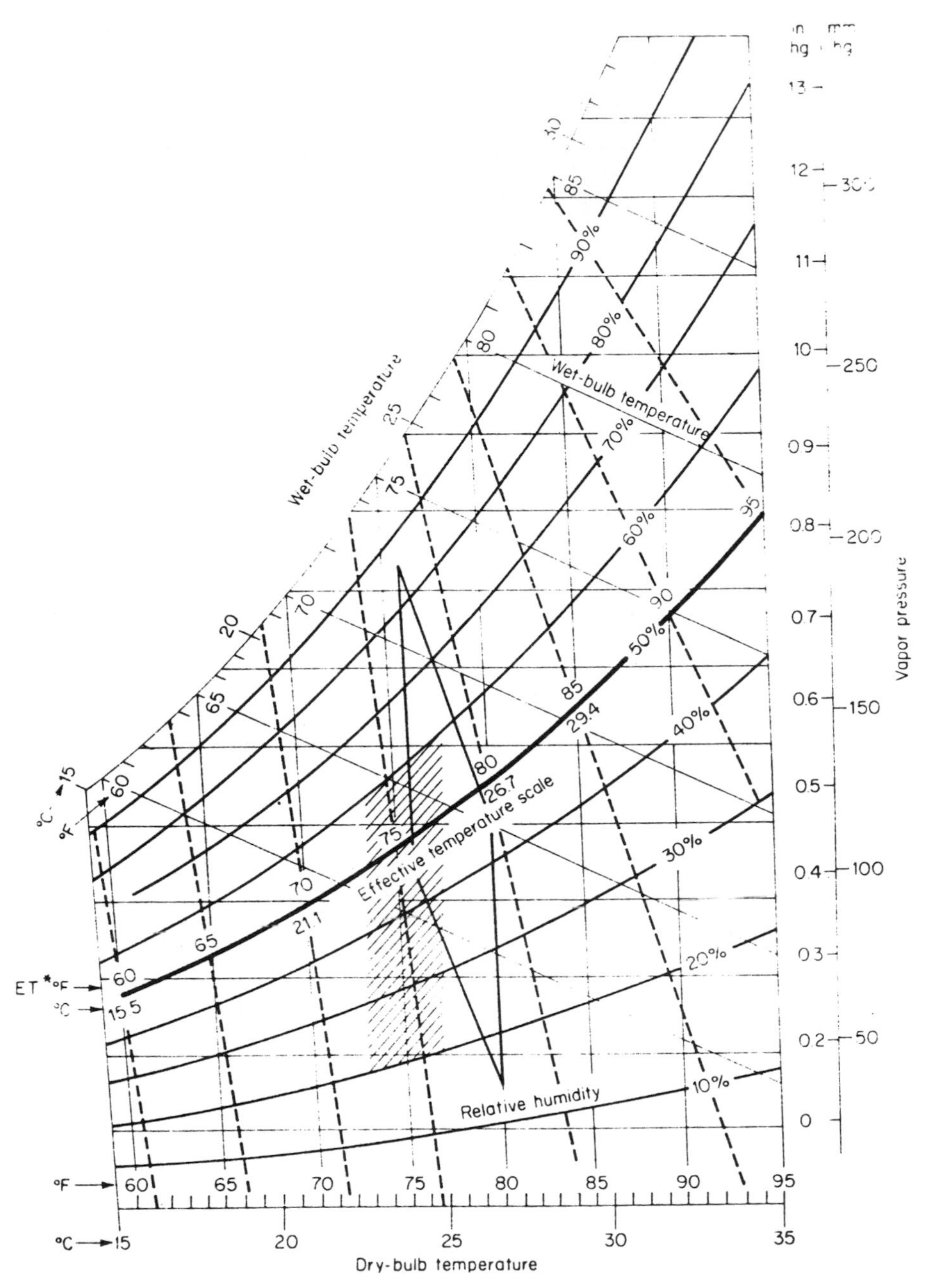

FIGURE 2-52 Scale of effective temperature. (From McCormick [28, p. 399].)

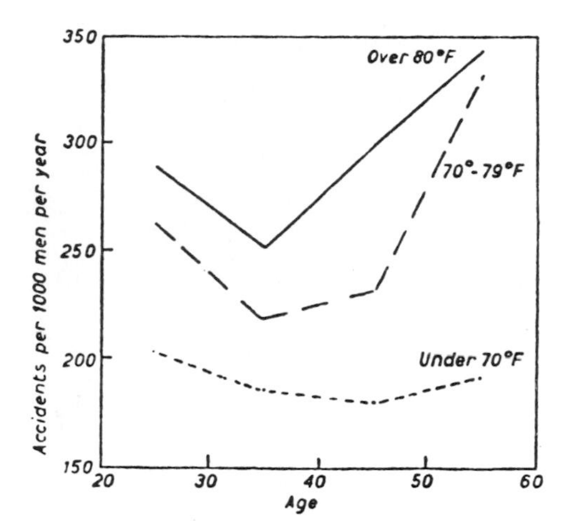

FIGURE 2-53 Accidents in relation to age and temperature. (From Murrell [33, p. 259].)

has to make a change in direction going to the fingers, and excessive and repeated pressure on the nerve can and does lead to partial or permanent disability of the affected wrist.

Powered screwdrivers and nutrunners create a torque in the wrist for those using them. This torquing of the wrist is a common cause of carpal tunnel syndrome. The screwdriver balance arm, shown in Fig. 2-54, is an example of an assembly device that absorbs the torque, essentially eliminating the torquing action of the wrist. Such devices, if employed universally, could tremendously reduce, if not eliminate, a major category of carpal tunnel syndrome cases in industry, yet relatively few are in general use today. The person displaying the screwdriver balance arm in Fig. 2-54 is not a production worker, and her loose-fitting blouse would be a potential safety hazard in such an environment. A production worker would wear clothing more suited for such an environment.

Figure 2-55 illustrates the scars from operations performed on an arm and wrists of a young woman, who as an admissions clerk at a major state university research hospital developed a severe case of carpal tunnel syndrome. The operation was performed on the tendons to reduce the pain in both arms. Her carpal tunnel syndrome condition was traced first to her computer terminal keyboard being too low and later to the keyboard being too high. In both instances, the forearms were not level with the wrists. This young woman will probably suffer from carpal tunnel syndrome for the rest of her life. If this can happen at a university teaching and research hospital it should come as no surprise that carpal tunnel syndrome is a major health problem in industry. More importantly, it is time to recognize carpal tunnel syndrome conditions when they exist and eliminate them, to prevent people from having to suffer needlessly for the rest of their lives.

The following examples are all from the *Strains and Sprains* publication. Figure 2-56 is an illustration of the twisted or bent wrist. Tendonitis is another

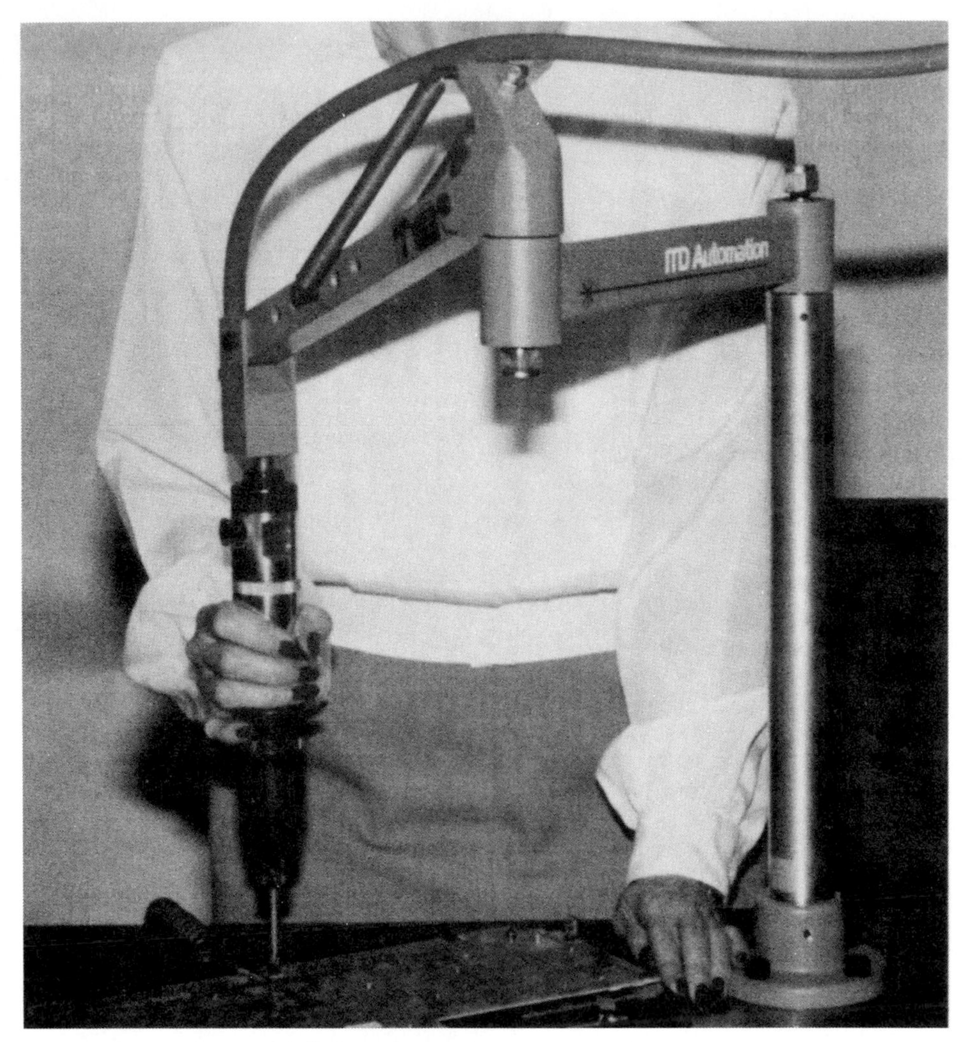

FIGURE 2-54 Screwdriver balance arm. (From ITD Automation [17].)

injury to tendons that become inflamed with repeated twisting actions, as indicated in Fig. 2-57. White finger, as indicated in Fig. 2-58, causes blood vessels to spasm and collapse. Such effects can be improved through improved tool design, through the use of vibration-absorbing materials, and obviously by eliminating continuous exposure of a single worker by rotating such work among a number of workers, limiting their individual exposures to tolerable exposure times. Of course, a better solution would be the elimination of the need for manual use of the vibrating tool in the first place.

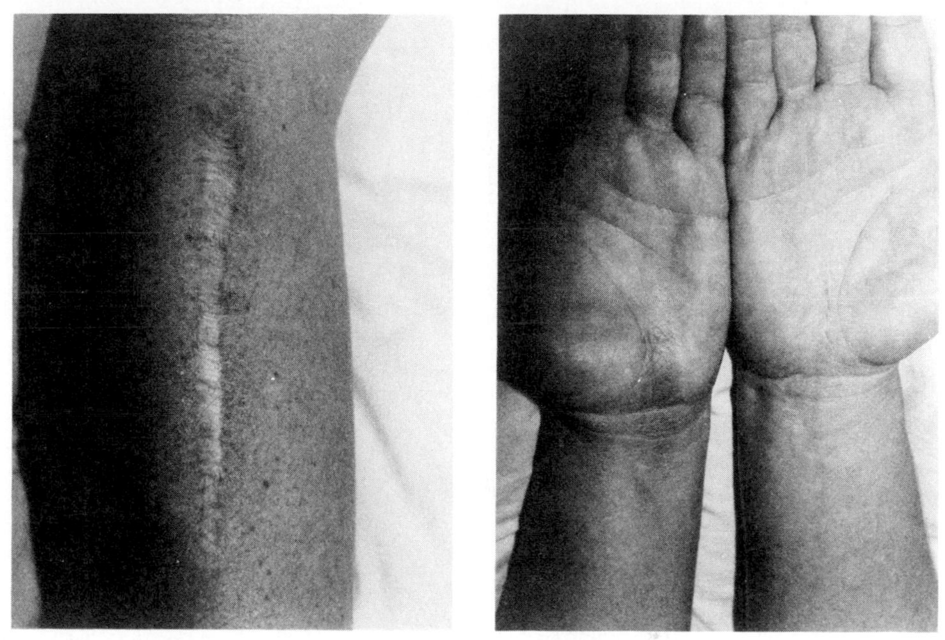

FIGURE 2-55 Carpal tunnel syndrome operation scars: (a) forearm; (b) wrists.

FIGURE 2-56 Bent wrist. (From Strains & Sprains [44, p. 8].)

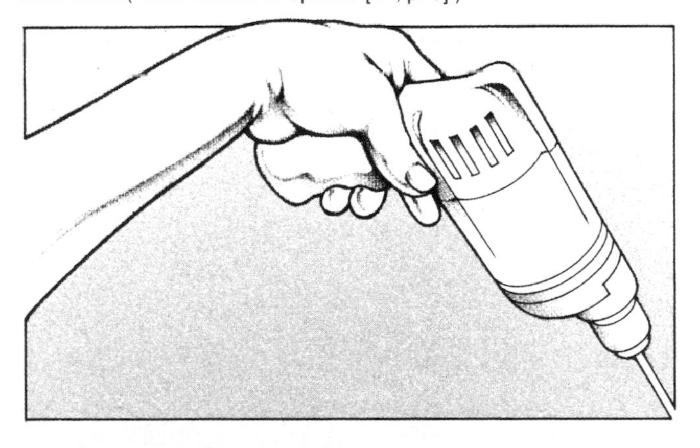

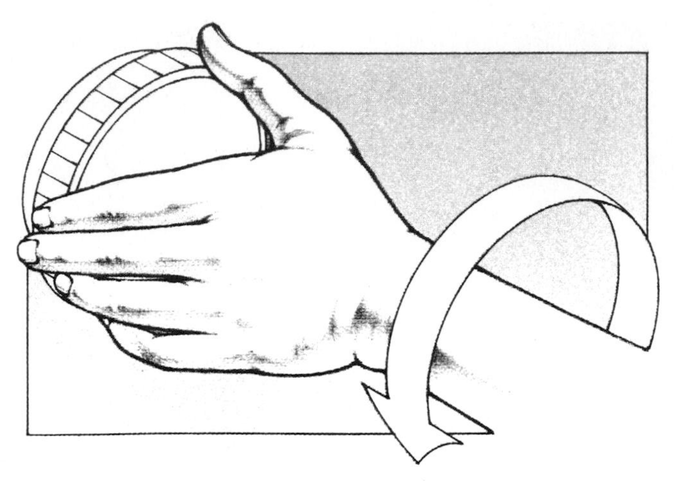

FIGURE 2-57 Tendonitis. (From Strains & Sprains [44, p. 9].)

FIGURE 2-58 White finger. (From Strains & Sprains [44, p. 10].)

Often it is not simply an example of poor tool design, but a question of tool selection in relation to the work, as indicated in Fig. 2-59. It should be noted that although the example in the lower left corner of Fig. 2-59 identifies the pistol grip tool as the better tool choice, a higher placement of the work would be preferred to limit the back bending observable in this figure.

In the weary arms category, Fig. 2-60 offers an excellent example of bad and good design. You might wonder why an equipment designer would not know to use the good design in preference to the bad design. The answer to that question may simply be that the equipment designer did not have a course in ergonomics, and may not have been made aware of this UAW brochure, for example. As a matter of fact, very few mechanical engineers—the type of engineer most likely to be the designer of this equipment—receive education in ergonomics, even though ergonomics is taught in practically every industrial engineering academic department in the United States. It is simply because ergonomics is not typically taught in the mechanical engineering department in a university and therefore is probably not a required course for mechanical engineers. It may not even be available as an elective within a mechanical engineering curriculum at some universities. Undergraduate engineers may only realize after graduation that there was an ergonomics course that they should have taken dealing with these issues. There are education turf battles that limit engineers from getting the optimum curriculum. This is an issue that needs to be more effectively addressed by the engineering education community in the United States.

FIGURE 2-59 Tool selection. (From Strains & Sprains [44, p. 13].)

FIGURE 2-60 Weary arms. (From Strains & Sprains [44, p. 18].)

The third category, aching back, is illustrated in Fig. 2-61. In American industry, people reach down into bins sitting on the floor; it is a predominant fact of life. Why? Because industrial managers have not taken the leadership necessary over the years to make the necessary investment in stands and other devices to present the work to the worker at an appropriate height. Has the investment in a

FIGURE 2-61 Aching back. (From Strains & Sprains [44, p. 25].)

simple stand at a workstation not been made because the cost of the stand is not cost justified? Not really. The answer is simply a lack of industrial leadership. In years past, the Du Pont Corporation distinguished itself in its high level of investment in safety equipment at its many plants. Did this represent a major cost drain for Du Pont? On the contrary, the Du Pont safety investments were a bargain, when the cost effects of such investments are fully evaluated, irrespective of the additional savings in human life and suffering that came as an extra. A Du Pont employee is safer at work than at home, because of the leadership provided by Du Pont management over the years.

Typical of ergonomics analysis that is taking place in industry is that described in the following article [39, pp. 181–194]. The case lacks a figure for viewing the machine involved, but it discusses the problem in sufficient detail to describe the effectiveness of the analysis.

Making a Difference at Johnson & Johnson: Some Ergonomic Intervention Case Studies

by Arthur R. Longmate and Thomas J. Hayes

A large bandage-making machine combines several wide layers of material and cuts the final product into the final shape. Each component is fed into the machine from separate raw material feed rolls, each of which must be replenished approximately six times per day. While no medical incidents had been reported on this job, the machine operator had submitted a suggestion that the unwind stand for the largest raw material roll be modified to reduce the risk of back strain. This requirement was based upon his feeling that he had experienced several close calls where he felt that he was placing "too much" strain on his back.

The employee's supervisor requested a review of the situation by the staff ergonomics engineer. The job elements required to handle the material rolls were analyzed utilizing the National Institute for Occupational Safety and Health (NIOSH) Work Practices Guide for Manual Lifting. The results verify that a problem exists, particularly when positioning a new roll of material into the unwind stand. The extreme horizontal (H) distance is required for two reasons:

- The roll is 18″ in diameter and is lifted using a mandrel inserted through the core. This requires the person to lift in the most stressful manner with the axis of the roll parallel to the front of the body.
- The unwind stand base was built with a piece of angle iron running across the floor right where the person should be standing when positioning the roll. This foot interference substantially increased the required H distance.

Largely based upon input from the machine operator, the unwind stand and mandrel were redesigned such that the roll could be positioned into the unwind stand from the side rather than the end. This required that the mandrel be redesigned and permanently attached to the unwind stand. Elimination of the need to lift the mandrel along with the roll reduced the amount of weight lifted by nineteen pounds (approximately 35 percent of the original lifting requirement). Also, the redesign of the unwind stand allowed the operator to lift the roll positioned with the flat side of the roll against the body. This engineering change reduced the required H distance to ten inches [as indicated in Fig. 2-62]. The allowable weight limit was increased from 25 to 51 pounds, which resulted in a much greater percentage of the weaker population being capable of safely performing this job element. An additional safety benefit was the elimination of two severe finger pinch points where the mandrel slid into slots in the original unwind stand.

A key factor in "selling" this engineering change to management was the use of the NIOSH Work Practices Guide for Manual Lifting to demonstrate the reduction in physical stress gained through the proposed modification. Perhaps the greatest use of the NIOSH Guide and other models is the ability to demonstrate relative improvements which can be achieved through equipment redesign before the change is actually implemented.

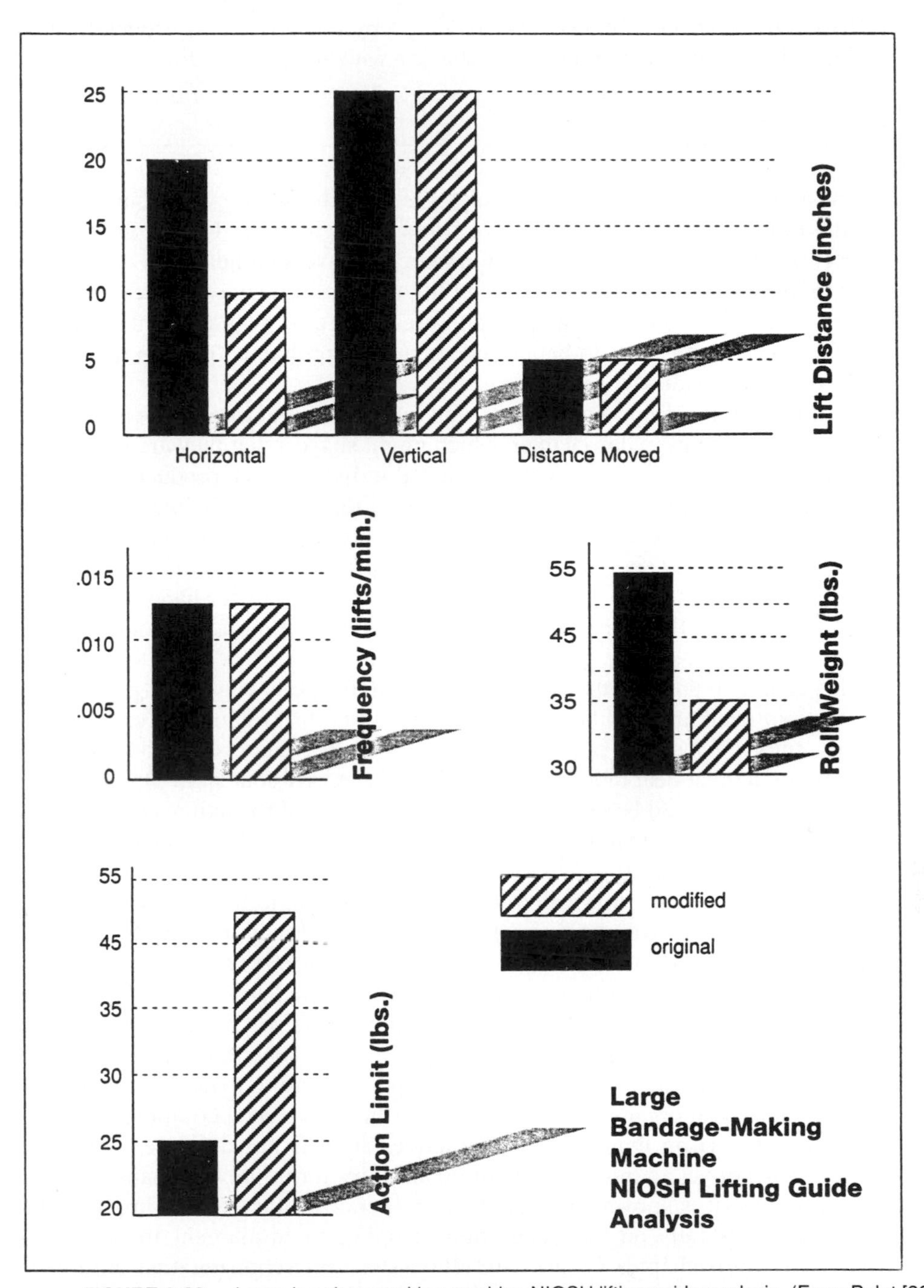

FIGURE 2-62 Large bandage-making machine NIOSH lifting guide analysis. (From Pulat [39].)

This modification was favorably received by the operators, especially two women who are now able to rotate into this job without apparent difficulty.

The above described ergonomic analysis is typical of good work being done in the field of ergonomics in progressive firms throughout the United States. The hope is that industrial managers of the future will have a higher sense of commitment to improving the quality of working life than unfortunately has been the case in far too many instances in the past and present. Tools available today make it possible for managements to better understand the problems they are facing and their potential remedies.

Methods engineering, plant layout, and human engineering are treated under separate subheadings in this chapter, which might suggest that they are three separate areas of study. In reality, what is designed in industry is a production system, and it draws from all these areas simultaneously; they are highly interrelated. It is also incorrect to speak of material handling and plant layout as if they were separable; generally, they are not. Even the words "production system" are meant by the author to possess a broad meaning, and include systems for handling laundry in a hospital as well as for processing tax forms in an Internal Revenue Service regional center.

This chapter represents the heart of traditional industrial engineering. The techniques employed are, for the most part, conceptual or graphical. Control techniques, such as inventory control or quality control, to be discussed in the next chapter, are a great deal more analytical in character. Of course, in the final analysis, techniques should be judged by what they produce. Production system design has been a major contributor in the development of the standard of living enjoyed in the United States today.

In most recent years, however, the standard of living in the United States has dropped, at least partially because of a failure of American industry to maintain our industrial leadership. Industry needs new blood that is ready, willing, and able, as most industrial engineering graduates are, to make the hard choices that need to be made to regain American industrial leadership in the world. More will be said on this topic in the last chapter. A student reading this chapter might assume that because the principles of good manufacturing practice are so logical and constitute nothing more than organized common sense, they have been completely implemented in American manufacturing practice. Such is not the case. American industry has not yet addressed much of opportunity for improving manufacturing operations. What is known about productive operations far exceeds implementation to date. It falls on the generations coming to implement the concepts expressed in this chapter. Figure 2-63 illustrates nine representative failures to implement good practice in industry. Much of the failure to implement good practices can best be explained in many instances as simply ignorance about what constitutes good manufacturing practice. Their solution is essentially a training issue.

FIGURE 2-63 Violations of good manufacturing practice.

In many other instances, the condition can best be explained as simply a lack of managerial discipline. Conditions exist because managements that know better have not chosen to require that good manufacturing practices be employed. Draw your own conclusions in considering the following questions relating to Fig. 2-63.

In (a), does it make sense to carry round pipe on a pallet? In (b), should material be stacked on a pallet this high? In (c), should mixed scrap be placed in a single container, requiring separation into steel, copper, aluminum, etc., as an additional required later step? In (d), should the punch press operator place the scrap on the floor, which requires that it be picked up off the floor later, or should the operator place it directly into a scrap container instead? In (e), should the press operator be stacking completed parts on the bench, all of which will have to be picked up again in order to place them into a container before moving them to the next operation, or should he simply drop them into a container for moving them to the next operation? In (f), is the operator employing good industrial seating that minimizes his fatigue, or is the chair instead both contributing to fatigue while being unsafe? In (g), should the workers be using a rule to measure inches from foot markers that were installed when this building was built probably a century ago, or should someone by now have marked inches between the foot markers so that using a rule to measure between foot markers can be eliminated? In (h), can the employee see the part she is sandblasting because of her height or does she need a platform so that she can see what she is doing? Finally, in (i), can anyone produce anything of quality, cost-effectively, on this bench? It is not a question of "Do we have the budget to fix this problem?"; the question is "Have we done the training necessary to make employees and management aware of what constitutes good manufacturing practice, and then have we exercised the industrial leadership (i.e., discipline) to see that such practices are employed?" The answer far too often in American manufacturing is emphatically *no*!

There is little question that some quantitative techniques, particularly simulation and statistics, have had an impact on production system design. These disciplines have had a considerable upgrading effect on industrial engineering practice. The developing analytical frontier in quantification for facilities planning is discussed in *Facility Layout and Location* by Francis and White [12] and *Facilities Planning* by Tompkins and White [46].

As will be discussed in more detail in later chapters, it is likely that the considerable volume of work being performed today by industrial engineers using traditional methods will be performed in the future by industrial engineering technicians skilled in modern techniques. This will come about when the system of higher education in the United States senses and responds to the ever-increasing need for industrial engineering technicians.

SUMMARY

This chapter began with a discussion about Arts-Way Manufacturing, a small but very capable farm machinery manufacturer in Armstrong, Iowa, that represents the backbone of our industrial economy. It is the kind of aggressive and unencumbered company from which desperately needed new jobs are created in our national industrial economy.

It is in such firms that analysis, reflection, and implementation of the concepts of this chapter will most readily lead to significant improvements in the physical aspects of production systems, such as workstation design, material handling, and facility design. Larger firms will likely eventually follow their lead; if they do not, they may well disappear as familiar corporate names, as many already have. The motion economy principles, for example, that were brought to light by Barnes a half-century ago are just as applicable today as they were then.

When a firm then adds concurrent engineering to its management practices it progresses a step closer to being what it needs to be. If its products and processes are also suitable to a JIT manufacturing approach, to be discussed in the next chapter, it may well move forward another giant step, to a lower-cost operation with increased and continuous quality improvement, with increased responsiveness in bringing new products to market and meeting customer needs.

Those firms that reinvest a modest amount of their returns into improving their manufacturing engineering efforts in the form of manufacturability, producibility, or design for assembly (DFA) have the opportunity of reaping additional rewards, as has Matsushita corporation, for example, over the years.

Those managements that wish to really know how they are doing by tracking labor performance, which translates directly to plant accomplishment, will utilize work measurement and labor reporting. If they do not, they will essentially be taking what they can get from their direct labor personnel, in the honored nonfunctional behaviorist tradition, to be discussed in Chapter 4. So much depends on what direct labor accomplishes that the author cannot understand why plant management would elect the latter.

The trend today toward JIT has caused many manufacturing firms to question their past process layouts; getting past vested interests concerning past decisions is always a potential problem. Manufacturing cells often work exceptionally well for JIT operations. The author's consulting firm is now implementing a JIT installation in a plant in New Jersey; this is opening up considerable manufacturing space for future use, and greatly reducing rejects and work in process while eliminating much of the previous MRP tracking that went with the previous high levels of work in process. Combined with going to a second-shift operation, in the future the same plant will be producing four times its present plant accomplishment without any additional "brick and mortar." The plant is also embracing a continuous productivity and quality improvement renaissance as part of the same project. It is the kind of renaissance numerous American industrial plants need to go through if they hope to compete internationally in the twenty-first century.

Ergonomics typically represents a fruitful opportunity for productivity improvement and a significant improvement in quality of working life for many direct labor personnel. Managements need to be trained in recognizing the opportunity for improvement in this area; such things as raised elbows, workers standing on their tiptoes to perform work, extended arms, fingers and hands wrapped in white surgical tape, and bent backs are all clues that should raise flags for those trained to see what needs to be corrected. If you have not had the training you will not see it, and fatigue, reduced performance, and pain and suffering will be common elements of daily production activities for hundreds, if not thousands, of direct and indirect labor personnel for which management is responsible.

Surely these principles are not new. A student might ask "Haven't these things already been fixed in industry?" The honest and sad truth is that often they have not been fixed. The opportunity has been there and is still there. Go back and look at Fig. 2-63. Those conditions exist today; any management worth its salt should not allow conditions like these to exist in their plant. To do so is an overt abdication of their assigned responsibility as managers.

REFERENCES

1 Apple, James M.: *Plant Layout and Materials Handling*, 2nd ed., The Ronald Press Company, New York, 1963.
2 Apple, James M.: *Plant Layout and Materials Handling*, 3rd ed., John Wiley and Sons, Inc., New York, 1977.
3 Apple, James M., and M. P. Deisenroth: "A Computerized Plant Layout Analysis and Evaluation Technique (PLANET)," in *Proceedings of the 23rd Annual Conference and Convention*, American Institute of Industrial Engineers, Anaheim, CA, 1972, pp. 121–127.
4 Armour, G. C., and E. S. Buffa: "A Heuristic Algorithm and Simulation Approach to Relative Location of Facilities," *Management Science*, vol. 9, no.1, 1963, pp. 294–309.
5 AutoSimulations, Inc., P.O. Box 307, 522 West 100 North, Bountiful, UT 84010.
6 Barnes, Ralph M.: *Motion and Time Study: Design and Measurement of Work*, 6th ed., John Wiley and Sons, Inc., New York, 1968.
7 Barnes, Ralph M.: *Motion and Time Study: Design and Measurement of Work*, 7th ed., John Wiley and Sons, Inc., New York, 1980.
8 Deming, W. Edwards: *Quality, Productivity and Competitive Position*, Massachusetts Institute of Technology, Center for Advanced Engineering Study, Cambridge, MA, 1982.
9 Deming, W. Edwards: *Out of the Crisis*, Massachusetts Institute of Technology, Center for Advanced Engineering Study, Cambridge, MA, 1986.
10 Divilbiss Electronics, 1912 Robert Drive, Champaign, IL, 61821.
11 Floyd, Bryan: "CAD/CAE/CAM for the Automotive Industry," Intergraph Corporation, February 25, 1991.
12 Francis, Richard, and John White: *Facility Layout and Location*, Prentice-Hall, Inc., Englewood Cliffs, NJ, 1974.
13 Gilbert, Ron: "A Graphical Spreadsheet for Factory Layout," *CADENCE*, April 1990, pp. 51–53.
14 Hillkirk, John: "Huge audit may attack fed waste," *USA Today*, February 23, 1993, p. 1A.
15 Hodson, William K. (ed.): *Maynards Industrial Engineering Handbook*, 4th ed., McGraw-Hill Book Company, New York, 1992.
16 Hurley, O. R.: "Simulation Finds the One Best Conveyor Layout," *Modern Materials Handling*, vol. 18, no. 10, 1963, pp. 47–49.
17 ITD Automation, INDRESCO, Inc., 1765 Thunderbird, Troy, MI 48084.
18 *Industrial Engineering*, Institute of Industrial Engineers, Norcross, GA 30092.
19 Koenig, Daniel T.: *Manufacturing Engineering*, Hemisphere Publishing Corporation, New York, 1987.
20 Krick, Edward V.: *Methods Engineering*, John Wiley and Sons, Inc., New York, 1962.
21 Lee, Kok-Meng, Georgia Institute of Technology, The Woodruff School of Mechanical Engineering, Atlanta, GA 30332-0405.
22 Lee, R. C., and J. M. Moore: "CORELAP—Computerized Relationship Layout Planning," *Journal of Industrial Engineering*, vol. 18, no. 3, 1967, pp. 195–200.
23 Ludema, Kenneth C., et al. *Manufacturing Engineering: Economics and Processes*, Prentice-Hall, Inc., Englewood Cliffs, NJ, 1987.
24 *Material Handling Engineering*, Penton Publishing, Inc., 1100 Superior Avenue, Cleveland, OH 44114-2543.

25 Material Handling Institute, Inc., 8720 Red Oak Blvd., Suite 201, Charlotte, NC 28217.
26 Maynard, H. B. (ed.): *Industrial Engineering Handbook*, 2nd ed., McGraw-Hill Book Company, New York, 1963.
27 McCormick, Ernest J.: *Human Factors Engineering*, 3d ed., McGraw-Hill Book Company, New York, 1970.
28 McCormick, Ernest J.: *Human Factors Engineering*, 5th ed., McGraw-Hill Book Company, New York, 1982.
29 *Modern Materials Handling*, The Cahners Publishing Co., 275 Washington Street, Newton, MA 02158-1630.
30 MTM Associates for Standards and Research, 16-01 Broadway, Fair Lawn, NJ 07410.
31 Mundel, Marvin E.: *Motion and Time Study: Improving Productivity*, 6th ed., Prentice-Hall, Inc., Englewood Cliffs, NJ, 1985.
32 Mundel, Marvin E.: "Now Is the Time to Speak Out in Defense of Time Standards," *Industrial Engineering*, vol. 24, no. 9, 1992.
33 Murrell, K. F. H.: *Ergonomics*, Chapman and Hall, London, England, 1969.
34 Muther, Richard: *Practical Plant Layout*, McGraw-Hill Book Company, New York, 1955.
35 Muther, Richard: *Systematic Layout Planning*, Industrial Education Institute, Boston, MA, 1961.
36 Niebel, Benjamin W.: *Motion and Time Study*, 8th ed., Richard D. Irwin, Inc., Homewood, IL, 1988.
37 Pritsker & Associates, Inc., P.O. Box 2413B, West Lafayette, IN 47906.
38 ProModelPC, ProModel Corporation, 1875 South State, Orem, UT 84058.
39 Pulat, Mustafa, and David Alexanders (eds.): *Industrial Ergonomics: Case Studies*, Industrial Engineering and Management Press, Norcross, GA, 1991.
40 Seehof, J. M., and W. O. Evans: "Automated Layout Design Program," *Journal of Industrial Engineering*, vol. 18, no. 12, 1967, pp. 690–695.
41 SIMFACTORY II.5, CACI Products Co., 3344 N Torrey Pines Ct., La Jolla, CA 92037.
42 Sly, David P: *FactoryCAD*, Release 2.01, CIMTECHNOLOGIES Corp., Ames, IA, 1991.
43 Sly, David P: *FactoryFLOW*, Release 2.00, CIMTECHNOLOGIES Corp., Ames, IA, 1991.
44 *Strains & Sprains*, United Auto Workers, Publication #460, 3rd printing, 8000 E. Jefferson, Detroit, MI 48214.
45 Tanner, John P.: "Product Manufacturability," *Automation*, vol. 36, May–September 1989, pp. 5–9.
46 Tompkins, James A. and John A. White: *Facilities Planning*, John Wiley & Sons, New York, 1984.
47 WITNESS, AT&T ISTEL, 25800 Science Park Drive, Beachwood, OH 44122-9662.
48 *GPSS-H*, WOLVERINE, 4115 Annandale Road, Annandale, VA 22003-2500
49 Woodson, Wesley E., and Donald W. Conover: *Human Engineering Guide for Equipment Designers*, 2nd ed., University of California Press, Berkeley, CA, 1964.
50 XCELL+ Factory Modeling System, The Scientific Press, Inc., 651 Gateway Boulevard, South San Francisco, CA 94080-7014.

REVIEW QUESTIONS AND PROBLEMS

1 What is the meaning of the acronym CAD/CAE/CAM? Does the term "electronic drafting" generally describe it?
2 What is a time standard?
3 Differentiate between the purpose of an operation process chart and the purpose of a flow process chart.

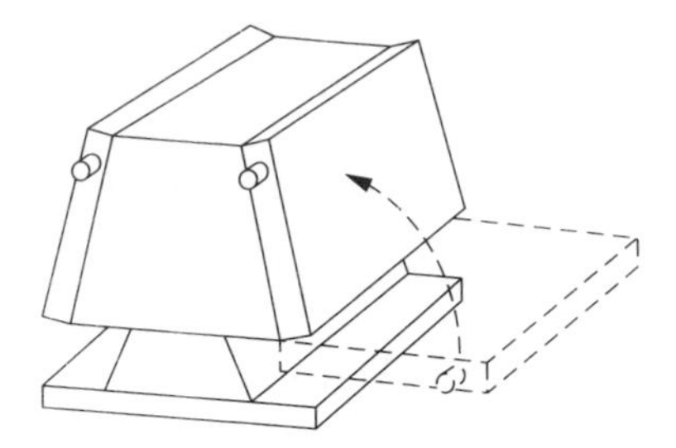

4 Rickmonald's Burger Company is introducing a new double-decker broasted bread sandwich that requires toasting three slices of bread per sandwich on a special gas flame toaster. The toaster toasts one side of each of two slices of bread at one time. As corporate industrial engineer, you have been asked to design an improved method for all 1000 toaster operators throughout the corporation so that they will toast the three slices of bread on both sides in a minimum cycle time. Each order is treated as a single intermittent order. Specific work elements are as follows: (1) Place a slice of bread in one side of the toaster, 4 seconds; (2) toast one side of a slice of bread, 24 seconds; (3) turn a slice of bread, 2 seconds; and (4) remove a slice of bread from the toaster, 4 seconds. Note that elements 1, 3, and 4 require the simultaneous use of both hands. Develop a man-machine chart to determine minimum cycle time. Employ a "left side of toaster," "right side of toaster," multiple activity chart similar to Fig. 2-6, with a vertical center scale marked in seconds up to 150 seconds. Does the new method provide improved product quality?

5 What is the relationship between the principles of motion economy and intraoperation handling?

6 If an operator sprays paint on an object moving past him on a paint spray conveyor on its way to its next operation, what flow process chart symbol might best represent the paint spray operation?

7 Design a fixture to hold the bolts in Fig. 2-9 so that both hands can be productive while performing the assembly.

8 After performing some assembly, an operator sets an assembly aside on a storage shelf next to her bench to let the adhesive dry for a minimum of one hour. She performs other work and later continues work on the assembly. Should the assembly waiting on the shelf be designated as a delay or storage on a flow process chart? When the operator takes the item off the shelf to complete the assembly after the adhesive has dried, should the continuation of the assembly be considered as continuation of the previous operation or a new operation?

9 What aspect of an operation might suggest that an alternate sitting and standing work chair would be most suitable for that operation?

10 Why is it important in bench assembly workstation design to limit reaches to a motion classification of level three or less?

11 What relationship is there between motion classification and fatigue in bench assembly?

12 How does operations analysis differ from workstation analysis and design?

13 What advantages does concurrent engineering offer as compared to value engineering as a design improvement approach?

14 How does concurrent engineering minimize bruised egos, and why is that important?

15 What does product design have to do with quality control?

16 You have been asked to select a robot to inspect the inner surface of 6-ft-diameter, 4-ft-long steel cylinders open at both ends and standing on end in a typical assembly plant. The inspection device must remain a distance of between one and two inches from the inner surface while the device scans the surface. Of the four basic robot configurations provided in Fig. 2-14, which would you consider the likely best candidate, and how might you consider mounting the robot?

17 What unique capability does the robot in Fig. 2-17 have that previous simpler robots have not had?

18 What is a performance rating factor?

19 If the rating factors in Fig. 2-18 had been 1.00, 0.90, 0.85, and 0.90 for elements 1 through 4, respectively, what standard time would be calculated?

20 What is meant in time-study if a time standard is referred to as being "too tight?"

21 What is a "guaranteed time standard"?

22 What is the obvious reason unions attempt to establish a maximum time period that a temporary time standard can be kept in effect?

23 What is a simo chart and why is it employed?

24 What role does videotaping of time-studies, as compared to direct observation, have in defending the validity of developed time standards?

25 How many minutes does it take to reach an object 20 inches away in a location that may vary slightly from cycle to cycle according to the MTM predetermined time system?

26 Using Eq. (2-1), calculate the number of observations that would be required if the percent occurrence of an activity of interest was estimated to be 40 percent, and a relative accuracy of ± 5 percent was desired at a confidence level of 95 percent (i.e., $K_\alpha \cong 2$)?

27 Determine n for problem 26 by using Table 2-2.

28 Determine n for problem 26 by using the nomograph in Fig. 2-26.

29 What is a "rated observation" in work sampling?

30 If direct labor in a plant constitutes only 7 percent of the manufactured cost of the product, does their performance really matter?

31 What is standard data?

32 A plant produces 12 products labeled A through L. Management is considering manufacturing the 12 products on three group technology production lines. The products utilize the following equipment in sequence:

Product	Equipment sequence
A	A, B, F, I
B	B, E, F, I
C	B, C, D, G
D	B, D, F, I
E	B, H, I, J
F	A, C, D, G
G	B, F, H, J
H	A, B, D, I
I	A, B, C, D
J	A, B, D, G
K	E, F, H
L	A, D, F, I

Which products would you recommend allocating to each of the three production lines, and which equipment would be associated with each line? Sufficient capacity exists on each machine.

33 Your boss, Margaret Johnson, the manufacturing engineering manager, is considering utilizing manufacturing cells as a new process design for the manufacture of a major component for the AK27 product, a small electronic product. The AK27 process involves three sequential, scrapless process steps with equipment costs and unit standard times as follows: step A, \$100,000 ea., 3.2 min/#; step B, \$150,000 ea., 1.6 min/#; and step C, \$10,000, 5.1 min/#. Ms. Johnson's calculations indicate that 12 step C machines should provide sufficient capacity to meet projected product requirements. She has suggested the following manufacturing cell layout and has asked you to review it and provide her with any recommended design (i.e., layout) improvements.

In – A – B – C – Out

How many cells do you recommend, and what equipment is associated with each cell?

34 What is meant by the statement "Plant design is first macro, then micro, and ultimately macro"?

35 Assume that a three-step process possesses the following characteristics:

	Step 1	Step 2	Step 3
Standard time, min/#	2.0	3.0	4.0
Scrap rate, %	10	15	10

If 100 good units are needed per hour from the process, effective utilization of equipment is 80 percent, and the same assumptions apply as in Fig. 2-36, how many step 1, step 2, and step 3 machines would be required?

36 A single product plant is to operate an estimated 250 work days per year, employing an eight-hour work day on a one-shift basis. Production requirement is 60,000 units per year. Process equipment estimates are to be made on the assumption of a uniform production rate throughout the year and a 5 percent loss of product following manufacture due primarily to deterioration and pilferage in final storage prior to shipment. Process information is as follows:

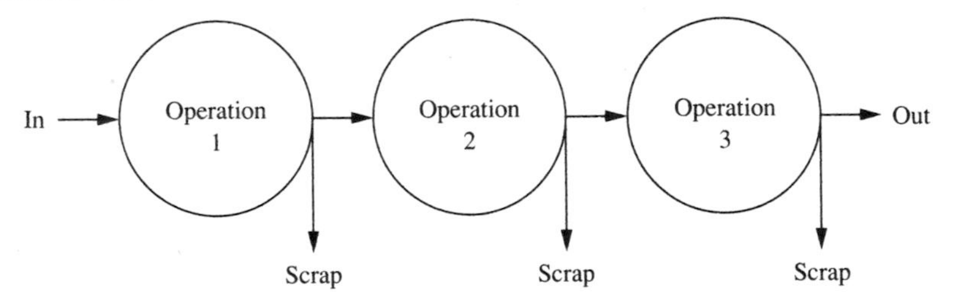

	Operation 1	Operation 2	Operation 3
Standard time per unit (min.)	5.2	7.5	11.4
Scrap rate for operation	3%	10%	15%
Estimated equipment utilization	91%	87%	89%

It is anticipated that operators for operations 1, 2, and 3 will have overall performances against standard of 100 percent. Estimated disutility of equipment will match the summation of estimated personal, fatigue, and unavoidable delay allowances employed in the establishment of standard times for these operations.

Operation inspection requirements are included within the normal times for each operation. Inspection occurs at the end of each operation, and the quality outcome is undetectable during the cycle; therefore, good or defective units take the same production time in any of the three operations. Defective units are not reworkable and are discarded when determined to be defective.

(a) How many operation 1, 2, and 3 equipment units are needed?
(b) How many units per hour must enter operation 1 during a normal work shift?
(c) How many good units per hour leave operation 3 during a normal work shift?

37 What is the purpose of employing the activity relationship chart approach in plant layout?

38 You have been asked to review a block layout for a plant design and have been provided with Figs. 2-64, 2-65, and 2-66. Figure 2-64 would indicate that the Muther (34) Activity Relationship Chart technique was employed in recording activity relationships. Figure 2-65 is an intermediate step in employing this approach. Figure 2-66 is the final block layout provided.

(a) Does Fig. 2-65 represent the best spatial arrangement of activities based on Fig. 2-64 information? If not, how could Fig. 2-65 be improved?
(b) Does Fig. 2-66 appear to have resulted from employing the Activity Relationship Chart technique, as indicated in Figs. 2-64 and 2-65? If not, why not?
(c) Does Fig. 2-66 fail to attain high-level relationships indicated in Fig. 2-64? If so, which ones?
(d) In Fig. 2-65, activity area 4, "Finished Goods" has "A-2,6" in the upper left corner of its block. What does "A-2,6" mean, and why is it recorded on the block?
(e) After making any improvement(s) to Fig. 2-65, as indicated in (a) above, prepare a block layout in Fig. 2-67 employing information from Figs. 2-64 and 2-65.

FIGURE 2-64

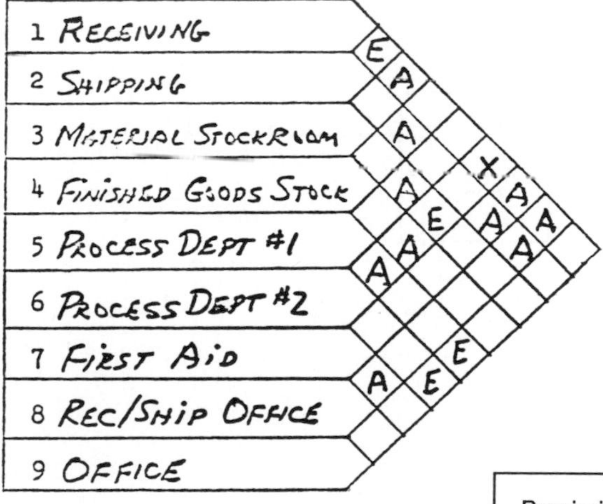

Proximity Level
A - Absolutely Necessary
E - Especially Important
U - Unimportant (Not recorded)
X - Undesirable

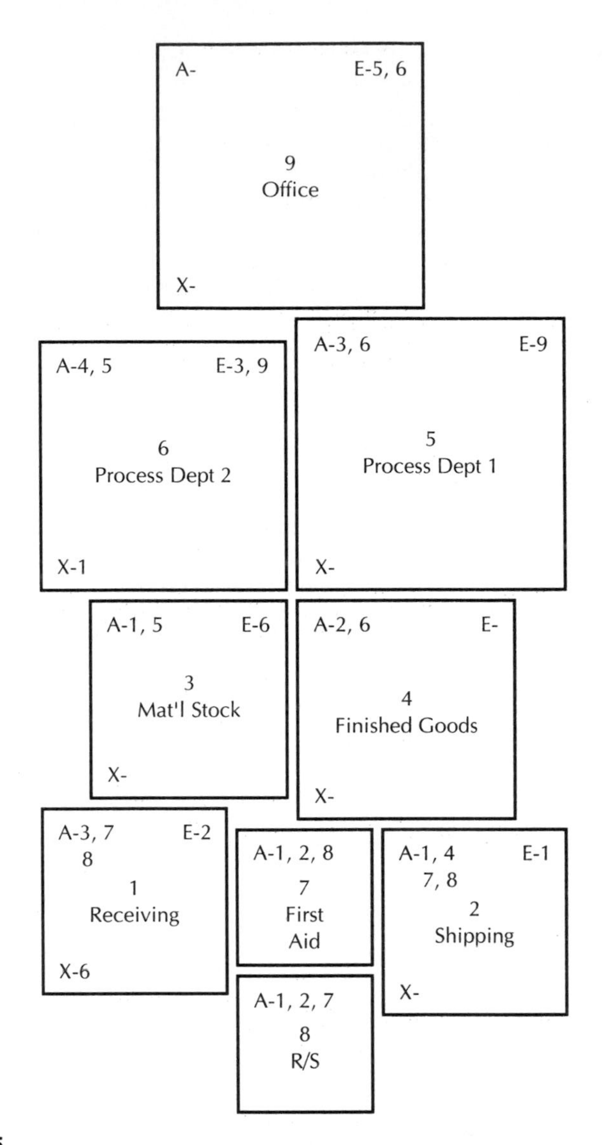
A-
E-5, 6
9
Office
X-
A-4, 5
E-3, 9
6
Process Dept 2
X-1
A-3, 6
E-9
5
Process Dept 1
X-
A-1, 5
E-6
3
Mat'l Stock
X-
A-2, 6
E-
4
Finished Goods
X-
A-3, 7
8
E-2
1
Receiving
X-6
A-1, 2, 8
7
First
Aid
A-1, 4
7, 8
E-1
2
Shipping
X-
A-1, 2, 7
8
R/S

FIGURE 2-65

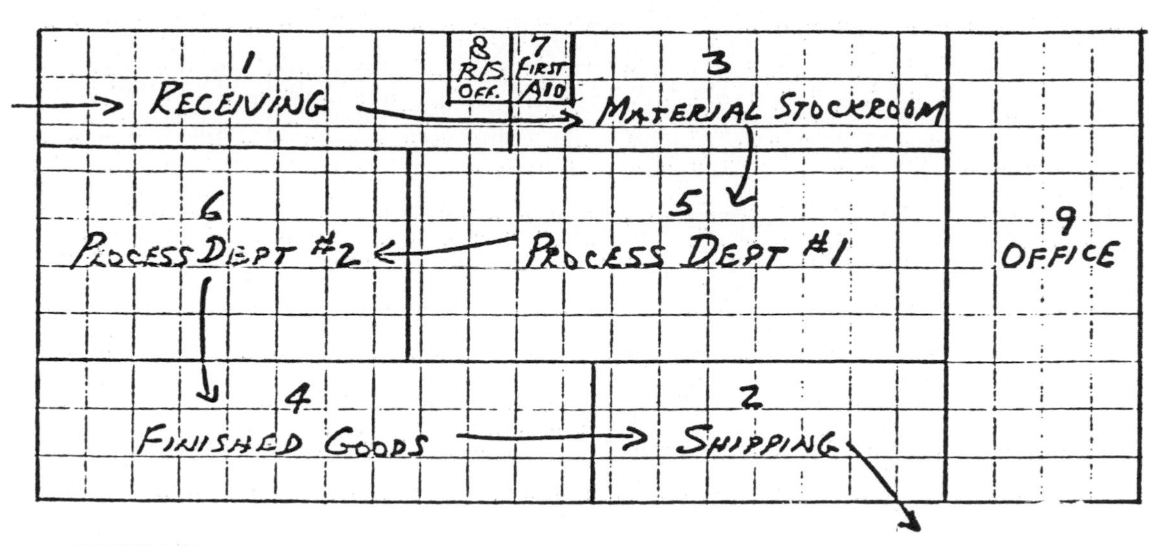

FIGURE 2-66

39 The following Fig. 2-68 is an organization chart for office personnel of the Super Duper Corporation. The following is a description of existing work relationships between office personnel:

1 The president has a full-time personal secretary who supervises a three-person secretarial pool consisting of one stenographer and two typists. The secretarial pool is utilized by the engineering and sales departments.

FIGURE 2-67

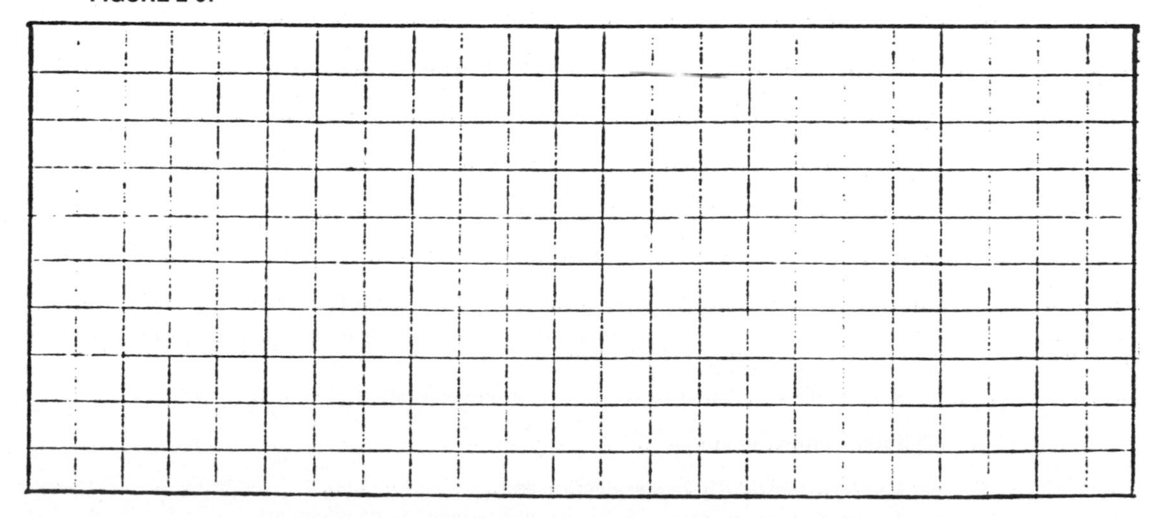

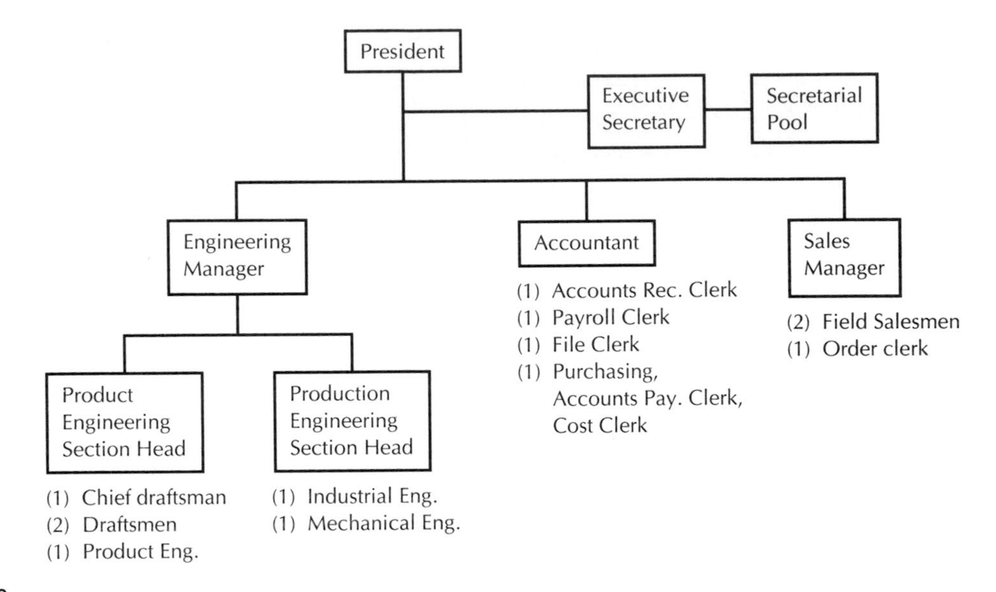

FIGURE 2-68

2 The engineering manager requires one-third of a stenographer and one-half of a typist. The product engineering section head and production engineering section head have a combined requirement of one-third of a typist. The engineering manager is in consultation with customer and supplier personnel, up to four people at a time, eight hours per week.
3 The product engineering section head spends about eight hours a week in consultation with some combination of the president, engineering manager, sales manager, two field salesmen, and up to three customer personnel dealing with confidential matters.
4 The production engineering section head and her two engineers spend 90 percent of their time in the production area. Production personnel come to the production engineering staff for various reasons about 20 times per day. Product and production engineering share technical records.
5 All accounting personnel utilize common records. Paperwork input to the accounting section is from outside sources for accounts payable, from the order clerk for accounts receivable, and from production for cost accounting and payroll.
6 The sales manager has frequent meetings with the president concerning present and future work.
7 The accounting section requires no secretarial service from the secretarial pool whereas the sales department requires two-thirds of a stenographer and three-fourths of a typist.
8 One-fourth of a typist is required as a receptionist.

Additional information:

1 Restrooms are in the production area.
2 Present office space is $60' \times 70'$, and appears to be adequate.

Measurement of the office indicates the following present space use:

	Square feet
Accounting	1100
Production engineering	760
Product engineering	380
Secretarial pool	300
Lobby	220
Receptionist	120
President's office	170
Engineering manager's office	120
Sales	880
Executive secretary	150

The president has asked you to re-layout the office for improved effectiveness in its existing space. Develop an activity relationship chart solution to this design problem.

40 What particular quality of a plant layout design is typically better understood as a result of digital simulation of the design?

41 As a transfer student from New Mexico State University to the University of Florida, you note that bending aluminum on a hand bender at your part-time job in a non-air-conditioned aluminum chair factory seems more demanding than when you did the same job back in Las Cruces. In New Mexico a typical day had a temperature of 75°F with a relative humidity of 50 percent. The humidity in Gainesville, Florida, today is 100 percent. What effective temperature in New Mexico would be equivalent to the temperature you feel today in Florida?

42 What is carpal tunnel syndrome, and how can it be avoided?

43 How can the placement of material for employees be generally improved to reduce back problems in industry?

44 To what extent have ergonomics problems already been solved in industry?

45 In observing direct labor employees in a plant, what three primary undesirable conditions described in the *Strains and Sprains* brochure [44] should be noted as indications of ergonomic opportunities for improvement?

LABORATORY EXERCISES 1–8: The Toy Train Factory

It is assumed that this project will be solved by laboratory teams of three to five students per team.

This project involves the design of an industrial plant for the manufacture of toy trains. Fig. 2-69 illustrates the product design for the toy train.

The plant will constitute a pilot operation for testing the market for this product. For that reason, the processing equipment (saws, sanders, nailers, drill, etc.) should constitute low-investment, general-purpose production equipment. If the product demand proves to be sufficient, the company may design and construct a much larger and more machine-intensive (i.e., automated) process for making this product. This limitation also prevents teams from proposing that one large machine called the "Kawasaki Toy Train Machine" be purchased and placed in the middle of the production area, which would be fed by raw material at one end and

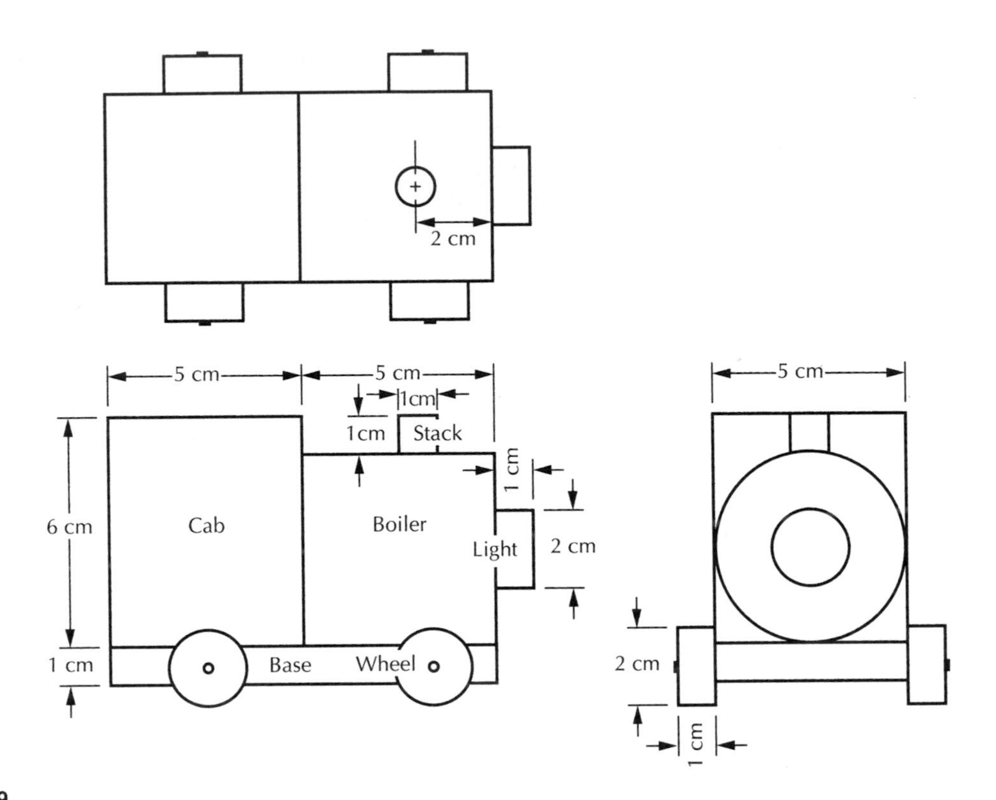

FIGURE 2-69

would have finished trains coming out the other end. Subcontracting the design and installation of the process to a Japanese, German, Swiss, or any other potential subcontractor is not allowed.

The equipment to be employed in this pilot operation should probably be available from numerous familiar retail sources (e.g., Sears, Black & Decker, Home Depot, etc.). Less familiar, but still readily available, storage and material handling equipment, such as roller conveyors, belt conveyors, monorail conveyors, tote boxes, and carts, are described in numerous industrial catalogs in general circulation.

The following are names, addresses, and telephone numbers of a few such equipment suppliers; catalogs are obtainable from these or other suppliers on request.

Global, 1070 Northbrook Pkwy., Dept. 77, Suwanee, GA 30174
Dozier Equipment International, Sidco Industrial Park, Box No. 11036, Nashville, TN 37222-0336
Arrow Star, Inc., 6087 Buford Hwy., Dept 87, Norcross, GA 30071

Two other sources of material handling information and material handling sources are *Modern Materials Handling* [29] and *Material Handling Engineering* [24]. Copies of their annual handbooks may be available from their publishers as well.

In developing this plant design you may choose to employ low investment tooling, such as holding fixtures, turntables, and so on.

In developing the design, technical questions will arise such as "How long does it take for glue to dry?" Your instructor may choose to serve as a source of common information for uniformly resolving such questions, or he or she may simply choose to let you make your best decision based on available reference sources, and state your assumption. Using hot glue, for example, a glued joint may set sufficiently in as few as three seconds to maintain position of the adjoining parts removed from a holding fixture. It can be assumed that sufficient glue is utilized in joining parts to eliminate the need for filling cracks in the body assembly.

Other technical questions, such as "How far must a paint dryer be from a paint booth for safety purposes?" will likely come to light in developing the design for this plant. Your instructor will advise you as to how such questions are to be handled in performing the project.

The incoming wood supplies are as follows:

5 cm dia rod in 2-m lengths
2 cm dia rod in 2-m lengths
1 cm dia rod in 2-m lengths
1 cm × 5 cm rectangular rod in 2-m lengths
5 cm × 6 cm rectangular rod in 2-m lengths

The wood material is called "dimensional stock." That means that it is supplied to the plant by a wood processor that both cuts and sands the wood to your specifications. The specification is such that the wood surface supplied is ready for finishing (i.e., painting). However, every saw cut made to the wood leaves a cut surface that must be sanded to meet this surface preparation standard.

You may assume that the wood surface must first be "sealed" employing a wood sealer that is brushed or sprayed on the exposed wood surfaces. A single coat of paint over the dried sealing coat is sufficient for meeting the required final surface specification. The body of the train will be painted black, and the wheels are to be painted white.

Assume that wheels must be drilled before being nailed to ensure that the wheels will turn when being used.

Sales are 30,000 per month (181 trains/hour). This hourly output rate is based on the assumed number of holiday and vacation days in a year, and an assumed one-shift operation (i.e., eight hours/day, five days/week). The plant is to be designed assuming the following scrap rates:

1 15 percent scrap loss when the glued body is inspected as an assembly
2 10 percent scrap loss when the wheels are nailed to the body assembly

These scrap rates are high but prototype efforts to reduce these rates have been unsuccessful to date. If an inspection is failed, all material associated with the inspected assembly or assemblies is to be discarded (i.e., all product-associated parts and assemblies).

All production times are to be estimated. The typical estimation process is to have one member of a design team pantomime the operational task (e.g., sawing parts) while another team member times the operation with a stopwatch. It is recommended that a number of cycles be timed (e.g., 10), and that an average value for one cycle be determined from the resulting total time for the group of cycles. It is suggested that the person performing the pantomimed work hold something of equivalent size and weight in his hand as a substitute for the product to provide comparable work conditions. The pace of the pantomimed worker should be what one would expect of a typical worker who performs the task over an entire shift.

A one-day supply of wood material for the above five raw material pieces required in the manufacture of each train is received each day from your wood supplier's plant, which is located next door to your plant. Each Wednesday afternoon the previous week's production (i.e., Wednesday through Tuesday) is loaded onto your truck and transported to a distribution warehouse. All other materials (nails, glue, saw blades, etc.) are delivered in monthly quantities with a one month supply on hand on the date of delivery (i.e., one-month safety stock).

With respect to personnel, all functions should be performed by plant personnel except sales (which is through an agent) and bookkeeping; both of these are done by people who come in one day per month. The sales agent and bookkeeper come in on different days and utilize one drawer each in the same desk, and each has two drawers in one five-drawer filing cabinet.

Trains are put in individual boxes. The boxes are then placed in corrugated shipping cartons containing 120 trains per carton.

Design a plant (i.e., develop a detailed plant layout) and determine space and personnel requirements for this plant.

At the conclusion of the project, each team will make a presentation describing the primary features of their layout and present supporting figures and calculations to support their design. Their presentation should describe the sequential flow and processing of materials from receiving to shipping for the final product.

Each team should indicate the total square feet of their plant design and the number of direct and indirect workers needed for this plant.

Figure 2-70 provides $\frac{1}{8}'' = 1'$ templates that may be used in the design effort. Use as many of these templates as you wish, and make any additional templates your team needs.

The following is a recommended sequence of toy train project steps for progressing through this project. These fourteen steps are divided into eight laboratory sessions. Your instructor may choose to alter this assignment of project steps to laboratory assignments.

Step	Description	Laboratory number
1.	Product analysis	1
2.	Operation process chart	1
3.	Workstation layouts	1
4.	Estimate operaton times	2
5.	Estimate equipment required	3

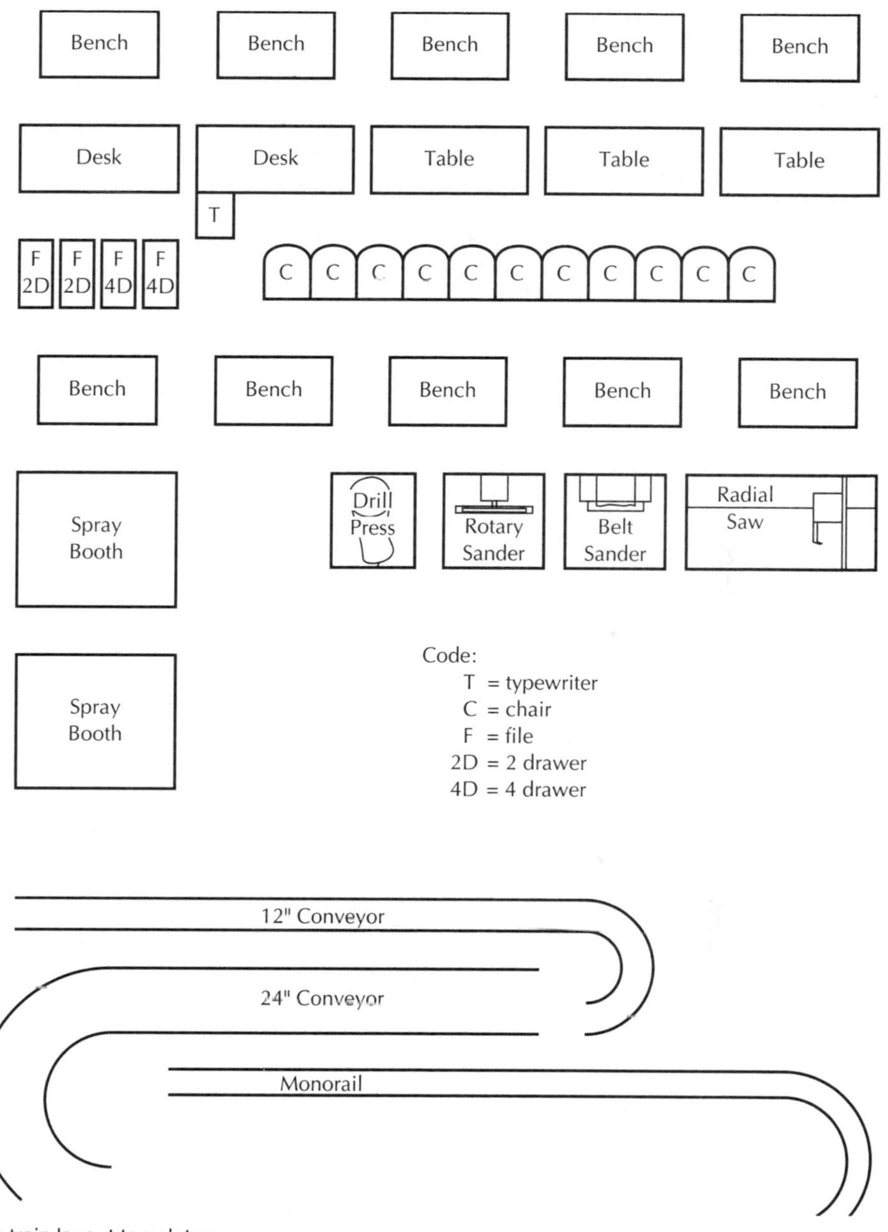

FIGURE 2-70 Toy train layout templates.

6. Estimate direct manpower	3
7. Estimate indirect manpower	3
8. Determine organization chart	3
9. Perform storage analysis	4
10. Estimate activity areas	4

11.	Perform activity relationship analysis	4
12.	Perform material handling system analysis	5
13.	Develop both gross square feet (GSF) and net square feet (NSF) block layouts	5
14.	Develop a detailed plant layout	6
15.	Provide a project presentation	7
16.	Provide a second project presentation (optional)	8

Toy Train Project Steps

Step 1 involves analyzing the product design (i.e., Fig. 2-69) until all members of the team understand the parts of the train and how they are assembled into a final product. The team must determine, for example, the piece part design of the stack to accommodate its attachment to the boiler.

Step 2 requires preparation of an operation process chart for this product process, similar to that indicated in Fig. 2-3. The operations indicated by circles represent the operations to be timed in step 4.

Step 3 involves developing "plan" (i.e., top) view sketches of each workstation, similar to that indicated in the top of Fig. 2-12. A gridded paper ($\frac{1}{4}''$) may be utilized for this purpose. Not every circle on the operation process chart will require a different workstation sketch because more than one operation can be performed at the same machine (e.g., numerous cutting operations may be performed at the same generic cutting workstation).

Step 4 involves pantomiming the operation, as mentioned above, and timing the pantomimed operation. An allowance should be added to the "normal time" (e.g., 15 percent) for the operation to produce a "standard time" for the operation, if the machine will sit idle when a worker is idle, during the eight-hour work shift.

Step 5 involves estimating the amount of equipment required as described for the problem in Fig. 2-36. Equipment obviously comes in integer values. However, employing overlapping break periods, overlapping lunch periods, and overlapping shifts (i.e., by adjusting start and end of shift times for different workers) it is possible to extend the availability of one unit of equipment well beyond what first appears to be integer value availabilities. At some point, rounding up to the next integer value is in order, and depends greatly on the relative cost of the equipment.

Step 6 involves calculating the number of direct labor personnel required. If equipment calculations indicated that 1.5 machines for both process steps A and B are required, how many direct labor personnel are required? Obviously one employee would run a process A machine all of the time, and another employee would run a process B machine all of the time. A third employee, however, would probably run an additional process A machine in the morning and another process B machine in the afternoon. Manpower, therefore, is calculated with real, not integer, process requirement values.

Step 7 is concerned with estimating indirect manpower requirements. After all direct labor functions are estimated, consideration should be given to supporting functions and the number of personnel required to perform these support functions (maintenance, stocking, loading/unloading of trucks, material handling, sweeping, scrap removal, janitorial functions, etc.).

Step 8 simply involves preparation of an organization chart to illustrate reporting relationships of all workers in the plant.

Step 9 involves estimation of the plant space required to store materials. Material storage space must be provided for raw material, work-in-process materials, and finished product. Space should also be provided for supplies and scrap as well.

Step 10 involves estimation of all of the activity spaces throughout the plant. The first task in this step is to list each of the different activity spaces that would be required in the plant. Each activity space would then be estimated in terms of size for that specific activity. An additional space, often called "circulation space," is reserved for aisles. Circulation space can be assumed to include both aisles and outer wall space. A typical estimate for circulation space is 20 percent of the sum of activity space.

Step 11 involves use of the Activity Relationship Chart technique, as described earlier in this text. Figures 2-37 through 2-42 illustrate the technique for use in developing a plant layout.

Step 12 involves analysis of the material handling requirements and the selection of storage and material handling equipment for this plant.

Step 13 involves development of a gross square footage (GSF) block layout of the plant employing estimated gross square footage requirements. Gross square footage is estimated by summing the activity space requirements for an area, and then adjusting this sum with a circulation adjustment factor (e.g., multiplying by 1.2 for a 20 percent circulation factor). Placing aisles in this block layout will then culminate in a net square footage (NSF) block layout for the remaining activity use space.

Step 14 involves placing individual equipment templates into a net square footage block layout to create a detailed layout of the plant, as illustrated in Fig. 2-42.

Step 15 involves making team presentations to other members of the laboratory group to acquire critical review of their plant layout design. It obviously provides team members with an opportunity to practice making a technical presentation, which they will probably be doing the rest of their professional lives.

Step 16 involves a second presentation opportunity after receiving feedback from the first presentation. This laboratory assignment provides the team with the experience of reconsidering its initial design and making adjustments, and another team opportunity to practice developing and making another technical presentation. This laboratory assignment step is considered optional.

3

PRODUCTION SYSTEMS CONTROL

Therefore, if any man objects to time study, the real objection is not that it makes him nervous. His real objection is that he does not want his employer to know how long it takes him to do his job.

F. W. Taylor

Methods engineering, material handling, plant layout, and human factors are all important considerations in developing a best design for a productive system. Once the system is installed, however, industrial engineering attention usually shifts to devising the best methods for operating the system. Inventory, production, and quality control all represent evolved combinations of qualitative and quantitative approaches for maximizing the economic utility of a productive system. Effective day-to-day management represents the third element of the triad of design, control, and management of productive systems.

In most instances, these control approaches attempt to minimize the unit cost associated with transforming incoming materials into a completed product in a timely manner at a desired quality level. One commonly recurring objective of industrial engineering in practice is cost reduction—producing something for less. Of course, there are numerous other goals as well.

Inventory control is concerned with effectively managing the considerable investment most corporations have had on-site in past years in terms of stored raw materials, work-in-process, and finished stock; it has traditionally represented a major cost of doing business in manufacturing. In the late 1970s, at a time when interest rates were exceptionally high (approaching as high as 20 percent), the combined costs of owning, storing, and handling materials on-site reached a cost peak and received considerable attention as a result.

As will be discussed in more detail shortly, the Japanese have also had unique manufacturing logistic characteristics: limited space and a dense population that greatly limits the desirability of storing inventory on-site. These two factors have in recent years led to aggressive attempts to limit the storage of inventory on-site. The Japanese have also discovered important benefits of ensuring that material flows through operations rather than sitting for long periods of time. As a result, most progressive firms today store a fraction of the inventories they stored on-site only a few years ago.

Production planning is concerned with determining what resources (e.g., materials, supplies, space, people, and equipment) must be available on-site over time to ensure that manufacturing goals are accomplished. Production control is concerned with the timely issuance of available on-site materials to the manufacturing process in such a way that materials are made available in a cost-effective manner where and when needed.

INVENTORY CONTROL

Introduction

Most productive systems contain inventories. In a manufacturing plant, inventories consist of basic materials, goods in process, and finished stock; in a hospital they may well consist of disposable hypodermic needles and aspirin tablets.

It costs money to maintain an inventory; therefore, inventories are inherently undesirable in the sense that storing them does not contribute to the direct transformation of materials and represents a non-value-added cost of doing business. The fundamental purpose of inventory is to provide an essential decoupling between unequal flow rates, as will be apparent from the following example.

Assume for the moment that you are in the bologna sandwich business. You run the business alone; it takes you five minutes to make and sell a bologna sandwich and you work eight hours a day. Each day you make and sell 96 bologna sandwiches, most of which are sold around lunchtime, with a few being sold in the morning and close to closing in the late afternoon. The refrigerator in which you store your sandwiches is a buffer between a constant—and limited—production rate and a variable demand rate. If you could sell bologna sandwiches at a constant rate, you would not need the refrigerator to store the completed sandwiches. If you had unlimited labor available, you could make all the sandwiches on demand and therefore would not need any inventory; in addition, the sandwiches would be fresher (i.e., a better-quality product). No inventory translates to no refrigerator, which means one less cost of doing business. That is what inventories are all about.

Economic Order Quantity (EOQ)

Around 1915, F. Harris developed what came to be known as the Wilson formula (Wilson publicized his work more than Harris did). This formula represents both the typical starting point for the development of inventory models, and an early lesson in the idea that "it pays to advertise."

Figure 3-1 illustrates this simple and highly idealized inventory model. Assume that at time zero exactly Q units of material are on hand and that as time progresses there is a constant demand or issuance of the material from stock. This decreases the quantity in stock in a linear fashion, as indicated in Fig. 3-1. Assume also that when the remaining stock reaches a level R at time t_1, an order is placed for Q units of material to be delivered to the plant in L days. L is the number of days that, based on the daily reduction rate, will result in the remaining R units of inventory being depleted at the same time that the Q units of material are received. Assume also that there is an annual requirement A, a unit material cost M, an inventory holding cost per unit per year H, an order preparation cost P, and an inventory level I.

This simple inventory model deals with three costs: material, holding, and order preparation. Holding costs are all costs associated with providing facilities for and maintaining an inventory. Order preparation costs are assumed constant and are the costs associated with placing and receiving an order of materials. If orders are placed very frequently and the annual requirement A is fixed, Q is small and therefore holding costs are small; however, the cost of placing numerous orders is high. If orders are placed infrequently, order placement costs for the year are low but a larger quantity of material Q must be ordered each time, resulting in high inventory holding costs. The Harris model can be employed to determine the optimum lot size

FIGURE 3-1 The Harris inventory model.

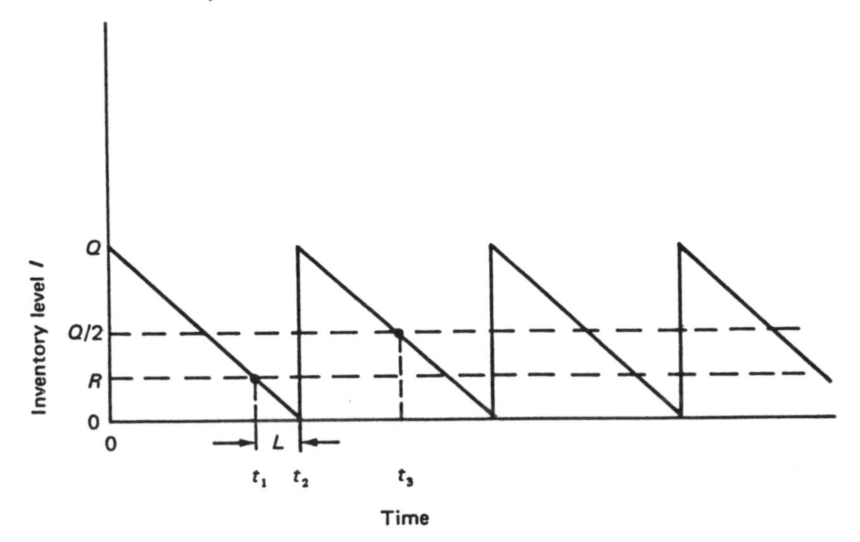

Q_0 to purchase so as to minimize the sum of all costs associated with order preparation and storage of the material for the year. Total annual costs are

$$C = \text{material cost} + \text{inventory holding cost} + \text{order preparation cost}$$

The annual inventory holding cost is the product of average inventory on hand, $Q/2$, and the annual unit holding cost H, because the inventory on hand is assumed to vary linearly between Q and 0, as indicated in Fig. 3-1 at time t_3. The annual order preparation cost is the product of the cost of placing an order, P, and the number of times orders are placed per year, A/Q. Therefore, total annual cost is

$$C = MA + H\frac{Q}{2} + P\frac{A}{Q} \tag{3-1}$$

To determine a minimum for C as a function of Q by employing differential calculus, one differentiates the function with respect to Q and sets it equal to zero. The value of Q needed to minimize C is then determined from this equation. Differentiating C with respect to Q and setting it equal to zero yields

$$\frac{dC}{dQ} = \frac{H}{2} - \frac{PA}{Q^2} = 0 \tag{3-2}$$

$$Q_0 = \sqrt{\frac{2PA}{H}} \tag{3-3}$$

Note that the term MA drops out in the differentiation step, Eq. (3-2). This indicates that the optimum order quantity is independent of material cost. An annual quantity A of material will be purchased during the year and will have the same cost regardless of how frequently material is purchased. Thus, the solution considers only the variable cost V during the year for both holding cost and order preparation costs.

Consider the following example:

$$A = 4{,}500 \text{ units}$$

$$H = \$2$$

$$P = \$20$$

$$M = \$10$$

From Eq. (3-3),

$$Q_0 = \sqrt{\frac{2(20)4500}{2}}$$
$$= 300$$

TABLE 3-1 TOTAL ANNUAL VARIABLE COSTS AS A FUNCTION OF LOT SIZE

Lot size Q	Inventory holding cost $HQ/2$	Order preparation costs PA/Q	Total annual variable cost V
0	$ 0	$ ∞	$ ∞
100	100	900	1000
200	200	450	650
300	300	300	600
400	400	225	625
500	500	180	680
600	600	150	750

Table 3-1 illustrates total annual variable cost V for different values of Q. It is apparent in reviewing the total annual variable cost column of Table 3-1 and Fig. 3-2 that total variable cost reaches a minimum when $Q = 300$. The quantity Q_0 is often referred to as the EOQ (economic order quantity).

The Harris model is a highly idealized model in the sense that a minimum number of cost factors are considered and all elements of the model are deterministic. In practice, at least three aspects of an inventory model would likely be probabilistic: (1) daily demand, (2) lead time, and (3) lost demand due to back orders. *Probabilistic* means, for example, that the number of units demanded each day

FIGURE 3-2 Annual variable costs as a function of lot size.

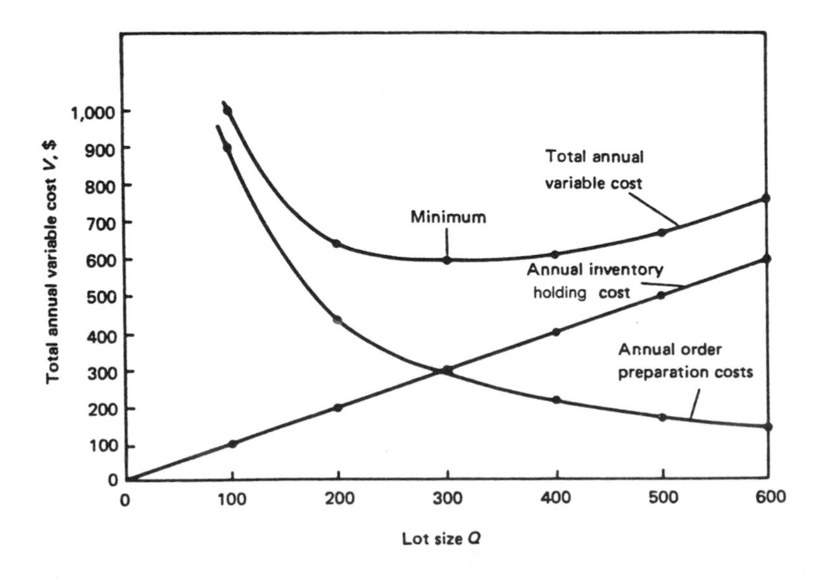

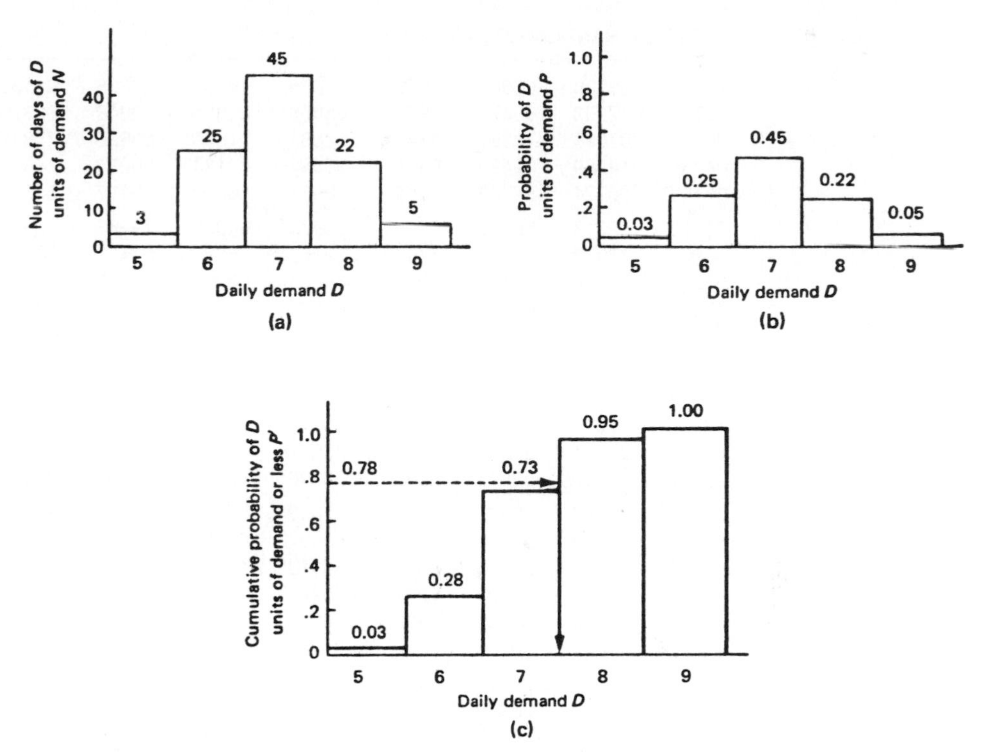

FIGURE 3-3 Probabilistic demand: (*a*) histogram; (*b*) density function, or distribution; (*c*) cumulative distribution.

would likely vary from day to day. If we were to collect data on the number of units demanded each day for 100 days, the histogram of Fig. 3-3*a* might represent our best estimate of typical demand.

One common technique for analyzing inventory systems containing stochastic elements is digital simulation. Note in Table 3-2 and Fig. 3-3 that the histogram data can be readily converted first into probabilities of daily demand, P , for varying levels of D and then into cumulative probabilities of demand, P', for levels of

TABLE 3-2 PROBABILISTIC DEMAND

Daily demand D	Number of days of D units of demand, N	Probability of a daily demand of D, P	**Cumulative probability of a demand of D or less, P'**
5	3	.03	.03
6	25	.25	.28
7	45	.45	.73
8	22	.22	.95
9	5	.05	1.00
	100	1.00	

TABLE 3-3 TABLE OF RANDOM DIGITS

78466	83326	96589	88727	72655	49682	82338	28583	01522	11248
78722	47603	03477	29528	63956	01255	29840	32370	18032	82051
06401	87397	72898	32441	88861	71803	55626	77847	29925	76106
04754	14489	39420	94211	58042	43184	60977	74801	05931	73822
97118	06774	87743	60156	38037	16201	35137	54513	68023	34380
71923	49313	59713	95710	05975	64982	79253	93876	33707	84956
78870	77328	09637	67080	49168	75290	50175	34312	82593	76606
61208	17172	33187	92523	69895	28284	77956	45877	08044	58292
05033	24214	74232	33769	06304	54676	70026	41957	40112	66451
95983	13391	30369	51035	17042	11729	88647	70541	36026	23113
19946	55448	75049	24541	43007	11975	31797	05373	45893	25665
03580	67206	09635	84612	62611	86724	77411	99415	58901	86160
56823	49819	20283	22272	00114	92007	24369	00543	05417	92251
87633	31761	99865	31488	49947	06060	32083	47944	00449	06550
95152	10133	52693	22480	50336	49502	06296	76414	18358	05313
05639	24175	79438	92151	57602	03590	2546	54780	79098	73594
65927	55525	67270	22907	55097	63177	34119	94216	84861	10457
59005	29000	38395	80367	34112	41866	30170	84658	84441	03926
06626	42682	91522	45955	23263	09764	26824	82936	16813	13878
11306	02732	34189	04228	58541	72573	89071	58066	67159	29633
45143	56545	94617	42752	31209	14380	81477	36952	44934	97435
97612	87175	22613	84175	96413	83336	12408	89318	41713	90669
97035	62442	06940	45719	39918	60274	54353	54497	29789	82928
62498	00257	19179	06313	07900	46733	21413	63627	48734	92174
80306	19257	18690	54653	07263	19894	89909	76415	57246	02821
84114	84884	50129	68942	93264	72344	98794	16791	83861	32007
58437	88807	92141	88677	02864	02052	62843	21692	21373	29408
15702	53457	54258	47485	23399	71692	56806	70801	41548	94809
59966	41287	87001	26462	94000	28457	09469	80416	05897	87970
43641	05920	81346	02507	25349	93370	02064	62719	45740	62080

Source: The Rand Corporation, *A Million Random Digits with 100,000 Normal Deviates,* Santa Monica, CA, p. 180.

demand D or less. A very powerful technique in digital simulation, known as the *Monte Carlo technique,* utilizes an empirically derived cumulative density function, such as P', to generate a sequence of representative values for a random variable (e.g., a stochastic variable such as daily demand D).

The technique employs uniformly distributed (i.e., equally likely) numbers over some range. For the distribution given in Fig. 3-3*c*, entries are drawn in a consistent manner from a table of random numbers, such as Table 3-3, so as to correspond on a direct basis with the 0 to 1 range of P'. For this problem, entries were drawn from the first two columns of digits in Table 3-3 (e.g., 78), from top to bottom, and converted to decimals. Doing this allows the selected numbers to correspond with the 0 to 1 scale of Fig. 3-3*c*. The first 30 numbers drawn from Table 3-3 provide the sequence of demands, as indicated in Table 3-4, which employs the cumulative density function of Fig. 3-3*c*.

TABLE 3-4 A GENERATED SEQUENCE OF DAILY DEMANDS

Day	Number from table	Scaled number	Corresponding demand D	Day	Number from table	Scaled number	Corresponding demand D
1	78	0.78	8	16	05	0.05	6
2	78	0.78	8	17	65	0.65	7
3	06	0.06	6	18	59	0.59	7
4	04	0.04	6	19	06	0.06	6
5	97	0.97	9	20	11	0.11	6
6	71	0.71	7	21	45	0.45	7
7	78	0.78	8	22	97	0.97	9
8	61	0.61	7	23	97	0.97	9
9	05	0.05	6	24	62	0.62	7
10	95	0.95	8	25	80	0.80	8
11	19	0.19	6	26	84	0.84	8
12	03	0.03	5	27	58	0.58	7
13	56	0.56	7	28	15	0.15	6
14	87	0.87	8	29	59	0.59	7
15	95	0.95	8	30	43	0.43	7

The 30 generated values of demand are assembled in histogram form in Fig. 3-4. Note the similarity of this histogram to that of Fig. 3-3*a*. If histograms of larger and larger samples were plotted, it would be noted that as the sample increases in size, the resulting distribution of values would come closer and closer to matching Fig. 3-3*a*. What should be obvious at this point is that, as the sample increases in size, Monte Carlo sampling of an empirically derived distribution produces a sample whose distribution comes closer and closer to matching the distribution from which it was drawn. The extent of the lack of correspondence for smaller samples is known as *sampling error* and is of a statistically calculable magnitude.

FIGURE 3-4 Histogram of sampled values for demand.

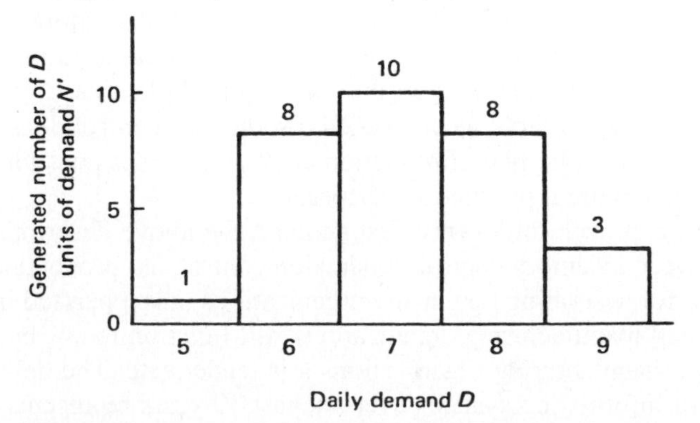

Similar sequences of values could be generated for the variables lead time and lost demand. A digital simulation containing these random variables along with other deterministic variables can simulate typical behavior of an inventory system over time. The accumulated results from such a simulation can then be evaluated to determine the desirability of the inventory policies employed. By the use of digital simulation, one can discover the performance behavior of a candidate inventory model before the model and its attendant policies are ever applied in an actual industrial environment. In this way, it is possible to know how the inventory control technique will work before it is applied, thereby minimizing the risks involved. The author, for example, simulated a multimillion-dollar computer-controlled stacker-crane-type storage system [5] to determine the performance capability of the system before it was installed. Because of the nonintuitive nature of the dynamic characteristics of queuing (i.e., flow) systems, analyses of this nature often yield invaluable information.

It should be understood that the Harris model and the few additional elements considered here to introduce the concept of a stochastic inventory model represent a very limited introduction to inventory theory and systems. Inventory control is really one element of a generally recognized larger area of control commonly referred to as production and inventory control.

PRODUCTION CONTROL

Production control and *production planning* are terms that are used somewhat interchangeably in some industries, whereas in other industries or companies they have distinctly different connotations. Generally speaking, production planning suggests at least a broader scope than production control. Most production control departments carry out a plant level staff function involving blue-collar workers concerned primarily with the day-to-day execution of production plans; production planning, on the other hand, is often done by white-collar production staff executives, engineers, or planners concerned with defining the overall production plan of a plant or a combination of plants, in order to be consistent with the overall short-term and long-term production goals of a company or corporation. In other words, the purpose of production planning is to determine what needs to be produced by specific time periods to accomplish established divisional or corporate goals. Production control is concerned with plant-level detailed planning and execution for today, tomorrow, and next month to ensure that required plant capabilities of the productive system are available when needed. Production control does the planning; production executes the plan. Production control makes it possible for production to perform as required; production performs.

As we approach the twenty-first century, the above description of the relationship between inventory control, production control and production planning may be adequate for describing how many organizations have operated in the past and how many small manufacturing organizations still function today, but for most medium and large manufacturing organizations it is inadequate. The development and integration of information systems over the past 30 years represents an ongoing developmental reality, as industries approach true CIM (computer integrated manufacturing). Whereas pencil and pad were adequate for production planning in the past, more and more the computer has become a means of utilizing common data from numerous sources within a manufacturing organization for the purpose of integrating manufacturing information and management systems for effective management decision making.

Production Planning

MPC (manufacturing, planning, and control) represents terminology that better describes how many medium and large manufacturing organizations function today in developing and sharing data concerning their numerous functions; the result is a single, common, computerized database of manufacturing data available to all aspects of the organization. Figure 3-5 describes the integrative nature of functional data in a manufacturing organization today.

Figure 3-6 describes the sequence of database elements that must be dealt with in the production and distribution of a typical product. It also suggests the types of techniques that are necessary in controlling the many sequential steps in this pro-

FIGURE 3-5 MPC as a central focus for CIM. (From Vollman et al. [16, p. 4].)

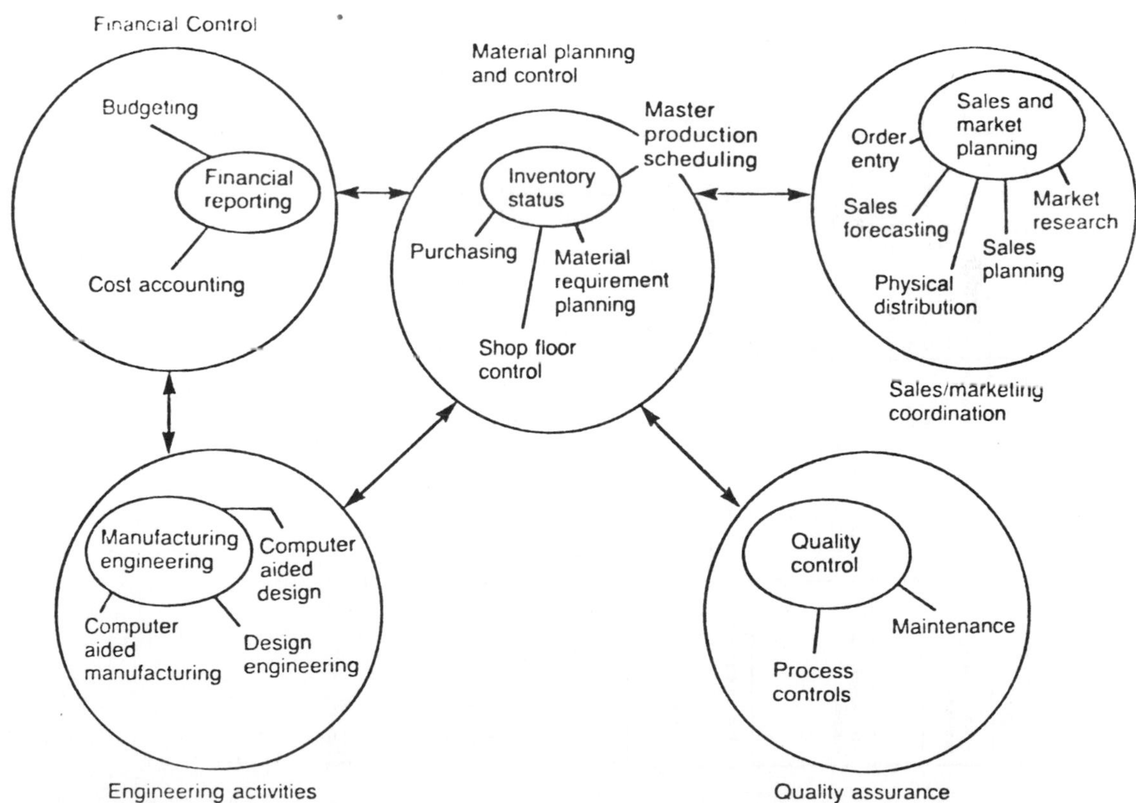

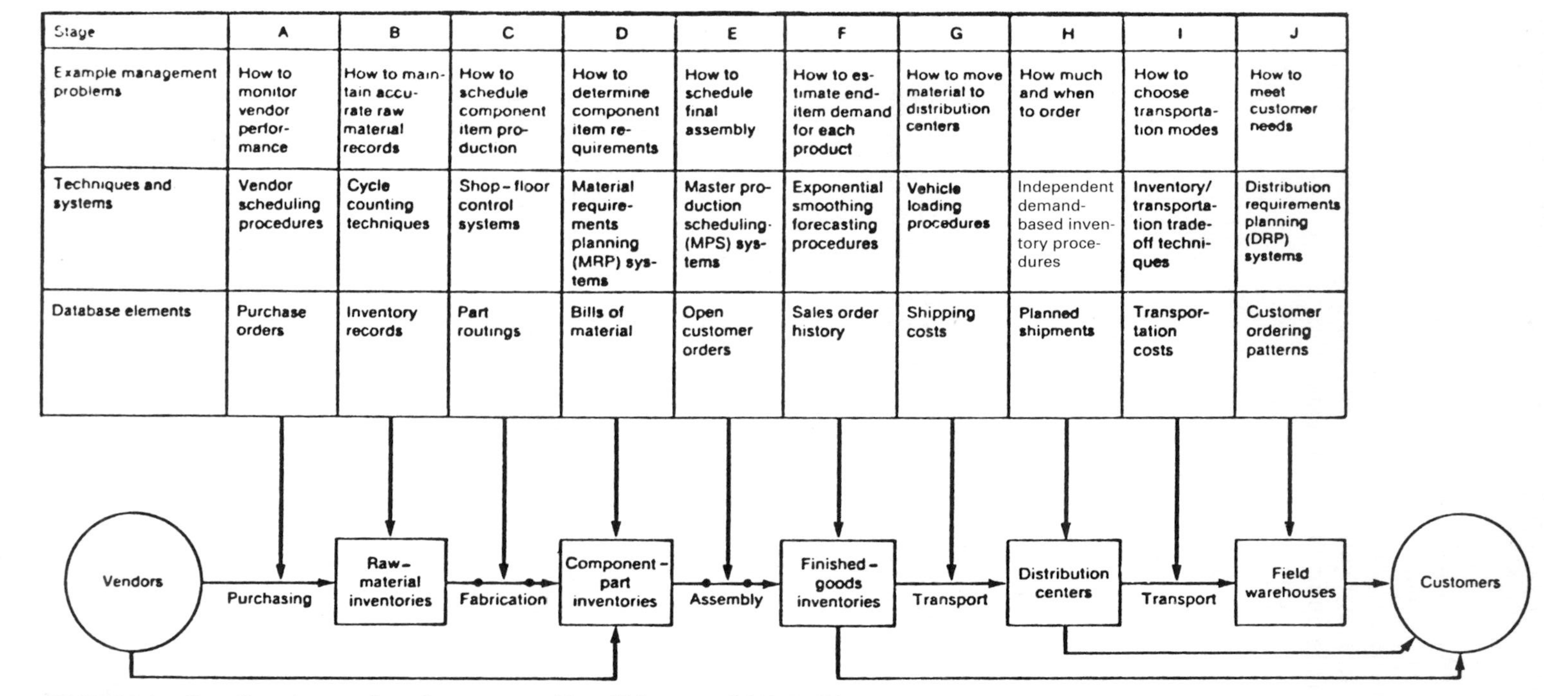

Stage	A	B	C	D	E	F	G	H	I	J
Example management problems	How to monitor vendor performance	How to maintain accurate raw material records	How to schedule component item production	How to determine component item requirements	How to schedule final assembly	How to estimate end-item demand for each product	How to move material to distribution centers	How much and when to order	How to choose transportation modes	How to meet customer needs
Techniques and systems	Vendor scheduling procedures	Cycle counting techniques	Shop-floor control systems	Material requirements planning (MRP) systems	Master production scheduling (MPS) systems	Exponential smoothing forecasting procedures	Vehicle loading procedures	Independent demand-based inventory procedures	Inventory/transportation trade-off techniques	Distribution requirements planning (DRP) systems
Database elements	Purchase orders	Inventory records	Part routings	Bills of material	Open customer orders	Sales order history	Shipping costs	Planned shipments	Transportation costs	Customer ordering patterns

FIGURE 3-6 Data flows in manufacturing systems. (From Vollman et al. [16, p. 8].)

cess. As indicated, each stage of the production and distribution system presents its own unique management problems that must be resolved at each stage in the overall process. As a product moves forward in the process, its accumulated information and needs must be known, available, and accommodated in order for downstream operational needs to be served.

Bar Coding

It should be obvious that the information flows described above represent an enormous amount of information. In years past, entry of information was typically performed by data entry clerks. One source [6, p. 12] indicates, "It is common for a key-entry operator to make one character error in every 300 characters entered." A typical page in a book contains on the order of 2500 characters. Therefore, in a typical page of material entered by a data entry clerk, there would be 8 errors; in a 400-page book, 3200 errors. The often hand-generated data sheet that served as the source of data to the data entry clerk was typically even more flawed and therefore generated even more errors. Thus it is understandable that the first bar code, invented by Joe Woodland and Benny Silver in 1949 (which, incidentally, was circular in design and involved a series of concentric circles), started a revolution. There are now more than 50 bar-code systems in use, some of the more extensive being UPC, Code 39, Code 128, and LOGMARS (employed by the Department of Defense). Harmon and Adams [6, p. 12] state that "the reliability of [bar-coded] data is at least 50 times better and can be more than a billion times better than a manual system."

It is ironic that there are still those today who, despite appearing to need it the most, have not yet fully embraced the full power of bar-code technology. The author recently observed, in a client supermarket chain, stock clerks who used portable bar-code pen scanners to enter replenishment orders for stock shelves. These clerks moved through the supermarket aisles, trying to place their orders, before the required early-morning deadline, for replenishment delivery that evening. Their responses to questions revealed that although store items were read by bar-code scanners at the time of sale, the information was not used to generate replenishment orders. The stock clerks were visually evaluating the quantity of product on the shelves and inputting the quantity of cases or items for the replenishment order. The scanning done by the stock clerks ensured only that the proper stock item number was entered. The clerks noted that the store invariably ran out of some items on the shelves during the day, whereas other items continued to be overstocked.

Humans are not known for effectively making a large number of repetitive decisions, especially when these decisions are made in haste. Software that combines such information as the number of products sold on the previous day and seasonally adjusted daily product forecasting, including special events such as football games, or holidays, would likely have produced far superior replenishment orders than those created by the stock clerks. It would not have surprised the author if the stock clerks and their managers were opposed to eliminating the task of inputting product replenishing quantities. It is the only interesting thing the stock clerks do at the end of a tiring and busy third-shift operation.

Employees in any industry generally resist any attempt to remove any aspect of "black magic" associated with their jobs (e.g., ordering just the right amount of replenishment stock, given the myriad of variables that affect the decision). It has a lot to do with their egos. "Ego maintenance," as it relates to worker and management behavior, will be mentioned elsewhere in this text. In the above stock clerk case, management had the technology but had not yet reached the point of effectively using it. Estimating replenishment stock as the deadline approached added some excitement to the clerks' tasks; without it, they would merely be stuffing products on shelves. Should employees be permitted to perform tasks that can be more effectively performed by a computer, just to protect their egos? The stockholder simply wants cost-effective performance. Progress almost never comes willingly or rapidly.

Material Requirements Planning (MRP)

Most manufacturing organizations today utilize a master production schedule that represents management's best guess as to the production requirements they anticipate in the future. Whereas production requirements in the short term represent firm orders, orders in the long-term portion of the schedule typically represent sales estimates. As indicated in Fig. 3-7, the *front-end* work involves developing the master production schedule estimates. As economic and other conditions change over time, all previous estimates are adjusted in light of present knowledge concerning future prospects and conditions. Updating of the master production schedule is obviously a never-ending and evolving process.

Material requirements planning (MRP) software packages are generally utilized today to "explode" (i.e., process the parts and assemblies hierarchical structure) product requirements, starting with required finished-product date commitments, to required dates by which all product material requirements must be purchased, received, subassembled, and final-assembled prior to shipment in order to meet delivery commitments. This phase is referred to as the *engine* part of the MPC process. The phase in which the product is being made on the production floor prior to shipment to the customer is referred to as the *back end* of the process.

Typically, a valid general assumption is that producing the product is relatively easy in the back-end phase if everything that was supposed to happen in the front-end and engine phases happened on a timely basis. Disruptions in manufacturing, however, are often caused by failure of the planning process to provide the required documentation, equipment, tools, materials, and human resources when they are needed. It is a lot like baking a cake: if all the ingredients specified are available, baking the cake is easy. If some ingredients are missing, substitutions or other adjustments are likely to take place and the quality of the final product will be in jeopardy. Most MPC systems attempt to minimize the occurrence of these problems over time.

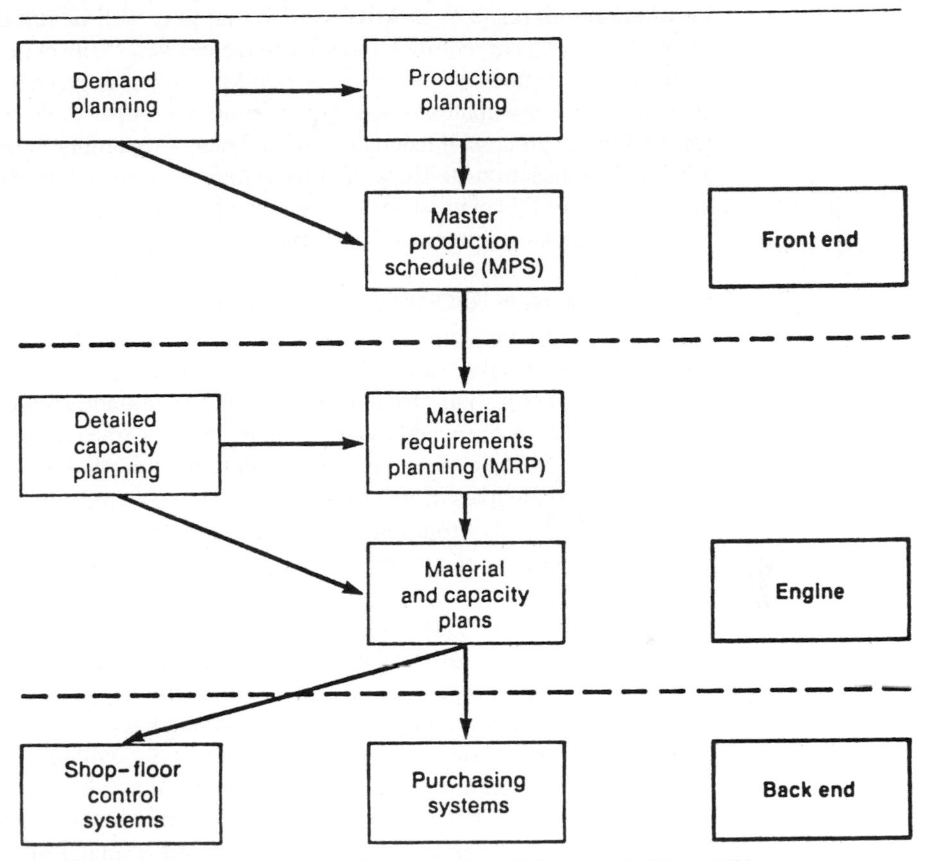

FIGURE 3-7 Manufacturing, planning, and control system. (From Vollman et al. [16, p. 16].)

Many firms today have either suffered through or have heard "horror stories" (many of which are true) about other firms' frustrating, disruptive, and expensive attempts to implement MRPII (i.e., Material Requirements Planning, Generation II) systems. However, there is typically no other rational choice than to make the transition. MRPII is a necessity for most manufacturing organizations in today's competitive world-class environment and must therefore be implemented if the organization is to remain competitive in the future. Thus, the objective should be to proceed carefully and to get the professional help necessary to maximize the success and minimize the pain of implementation. In an analogous way, many firms have suffered considerable cost, some to the point of extinction, when attempting to implement an appropriate and feasible CADD system for their operations. Nevertheless, for most firms in the design business today, CADD is a professional necessity. The only rational action is to attempt to ensure a successful transition to an appropriate CADD system for meeting present and future needs.

The adoption of a Just-in-Time philosophy, to be discussed in more detail shortly, will greatly simplify MRP requirements—so much so that a well-instituted JIT installation will result in an MRP system that may well be run on a PC, whereas in the past it may have required a medium-size or even mainframe computer system.

In the old sequential-department manufacturing approach, MRP transactions occurred when raw material was issued to the first department. Every time it moved forward in the process, it reached a higher level of assembly. It may have taken 20 to 100 MRP transactions to track the raw material usage and assembly of the finished product. Under this approach, there may well have been a ton of in-process material being tracked, so valuing in-process material was important and obviously time consuming.

In a typical plant that has transitioned from sequential-department manufacturing to JIT cellular manufacturing, raw material is often stored at the point of use. When the time comes to make a typical product, the material is used, and in a matter of minutes or hours a final product exits the JIT manufacturing cell. This product is shipped or put in the warehouse. At that point, a transaction made in the MRP system "back flushes" material requirements for the products produced (e.g., production of 100 bicycles of a particular model corresponds to using 100 handlebars, front tires, back tires, front fenders, rear fenders, etc.). The same transaction enters the finished products (say, the 100 bicycles) into the warehouse, with their labor added.

With JIT there are essentially only two categories of materials in the plant: raw material and finished goods. The raw material in a manufacturing cell, which will become a finished product some minutes or hours in the future, can be treated as raw material until it is reclassified as a finished product when it is completed, thereby eliminating one classification of material—work-in-process. Eliminating that classification eliminates the enormous non-value-added task of tracking it.

When backflushing is used to account for raw material usage by means of a manufacturing bill of material instead of a design bill of material, it is also possible to account for typical raw material losses such as scrap. Whereas a design bill of material indicates that one handlebar is needed on a bicycle, a manufacturing bill of material may indicate 1.03 handlebars are needed per bicycle. Thus, when 100 bicycles are to be made, 103 handle bars will be issued if 3 handlebars will typically not make it through the process to a final product. In addition, backflushing of 100 finished bicycles would involve assuming that 103 handlebars were consumed in making the 100 finished bicycles. In the bakery business, the term *baker's dozen* translates to 13.

Economic Production Quantity (EPQ)

Whereas the EOC, eq. (3-3), resulting from the Harris formulation attempts to compromise between the sum of costs associated with order preparation and inventory holding, production control often has an analogous sum-of-costs problem associated with machine lot sizes that determine machine setup cost and WIP (work-in-process) inventory holding cost.

Assume that a machine is used to make five different products and that a setup cost is involved every time the machine is set up to change from one product to another. Assume also that the longer the machine produces one product before shifting to another, the larger the inventory will be. The cost trade-off, then, is between setup cost, which does not produce units of product, and inventory holding cost.

Let t_P be the production period for a product, and let t_R be the time from starting production on the product until the machine is returned to production on that same product, as indicated in Fig. 3-8. Let D be the daily demand rate, P the daily production rate, A the annual requirement, S the setup cost, H the unit annual holding cost, V the total annual variable cost, and Q the production lot size.

Inventory reaches a peak value at time t_P of $t_P(P - D)$. Average inventory is therefore equal to $t_P(P - D)/2$. The quantity of material produced is $Q = t_P P$; therefore, $t_P = Q/P$. Substituting for t_P, average inventory is then

$$\frac{Q(P-D)}{2P} = \frac{Q}{2} - \frac{QD}{2P} = \left(1 - \frac{D}{P}\right)\frac{Q}{2} \tag{3-4}$$

and consequently,

$$V = S\frac{A}{Q} + H\left(1 - \frac{D}{P}\right)\frac{Q}{2} \tag{3-5}$$

Differentiating V with respect to Q and setting the result equal to zero provides the basis for determining the value of Q that results in a minimum annual variable cost for production setups and the holding of inventory.

$$\frac{dV}{dQ} = -S\frac{A}{Q^2} + \frac{H}{2}\left(1 - \frac{D}{P}\right) = 0 \tag{3-6}$$

$$Q_0 = \sqrt{\frac{2SA}{H(1 - D/P)}} \tag{3-7}$$

Optimum lot size is Q_0 (the economic production quantity, EPQ), and the optimum number of production runs per year for the product is $N_0 = A/Q_0$.

FIGURE 3-8 Production lot size model.

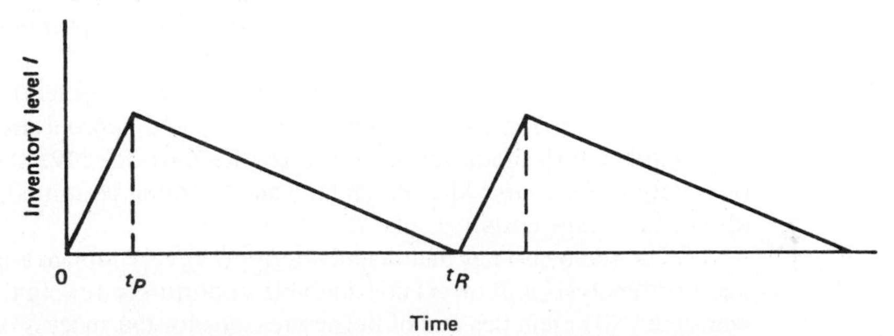

It should be noted that EOQ and EPQ are important today not so much as a means to setting operating policy but as a means to gaining a basic understanding of underlying relationships between certain inventory control costs. As will be noted in the discussion of more modern inventory management methods today, there are other issues and cost effects that are equally and often more important than those considered in these simple formulations. It seems most appropriate, however, to start an analysis of inventory control with an understanding of these basic cost relationships.

Just-in-Time (JIT)

Local conditions affect behavior. To understand the density of Japanese population, envision half of the U.S. population living in Montana. Space is premium in Japan. In addition, as with many Eastern cultures, the Japanese are a frugal people. Last but not least, Japan has limited natural resources (e.g., a major source of steel used in Japan for manufactured products is crushed used cars from the United States and other Western countries). Under such local conditions, the Japanese have always viewed the dedication of space to the storage of materials as undesirable. Instead of utilizing valued space for inventory, they much prefer to import required natural resources such as steel, immediately transform them into final products, and then, without delay, ship them to international customers, thereby tying up funds and space as little as possible while employing high material turnover rates. This also permits them to concentrate their labor efforts on value-added material transformation rather than non-value-added activities such as support storage activities. This difference in local conditions and culture partially explains their preconditioned interest in a JIT (Just-in-Time) type of production management philosophy. Under a JIT philosophy, inventories are kept to a minimum and material and value-added labor are expeditiously combined to meet final product requirements.

The EPQ formula derived in Eq. (3-7) describes the optimum relationship between the variables contained in its formulation. Any desire to operate under a JIT philosophy should preferably not violate these basic underlying relationships. Let us examine these variables from the perspective of a manufacturing manager who wishes to minimize the cost of present and future operations:

A The annual requirement is a given.

D The daily demand is a fraction of the annual requirement based on the number of production days to be worked in a year and is therefore a given.

P The daily production rate is relatively given by optimum process quantities for the specific process in use.

H The inventory holding cost per unit per year is essentially given if appropriate storage facilities have been put in place and appropriate material handling equipment and methods are being utilized for cost-effective storage and handling of materials. Of course, the less storage and handling permitted, the lower the handling and storage costs.

S The setup cost has traditionally been given very limited engineering and management analysis and offers considerable opportunity for significant reduction, if warranted. (Therein lies one of the key reasons for the success of JIT.)

Kanban In 1955, only 10 years after the end of World War II, Toyota Corporation management were making their best efforts to "raise per capita income" within the Toyota Corporation [12, p. 77]. Taiichi Ohno, with the support of board chairman Toyoda Eiji, initiated what is now called the *kanban system*. The initial kanban system's basic thesis, as stated by Ohno, was "What you need, only in the quantity you need, when you need it ... and as inexpensively as you can" [12, p. 96]. Note the hint of the inescapable triad: quality, cost, and delivery. Note also that there is no hesitancy to specifically state "as inexpensively as you can." There is a misconception in the United States that *quality* is the goal; whereas, in Japan, *quality, cost, and schedule* are all of primary importance.

For JIT to work without violating the underlying economic logic of the EPQ formulation, it became necessary to reduce setup costs as much as possible. More will be said about reducing setup costs once the kanban system has been described.

Kanban means "card" in Japanese, or more literally "visible record." In the kanban system, the card in the holder on the front of a production tote box (which designates the part number to be placed in that container) serves as authorization to convey or produce more of that specific product. The true kanban system is a two-card system: one card is the "conveyance" card, and the other is the "production" card. Figure 3-9 describes the original Toyota two-card Kanban system.

The following steps are identified in Fig. 3-9, starting with the point labeled "1. Start here":

1 When an additional tote of material is needed at the drilling work center, the conveyance kanban (C) is removed from a tote on the "in" side of stock point M and placed in the kanban collection box; the tote is placed in the drilling work center.

2 When a material mover (e.g., forklift driver) notes that there is one or more conveyance kanbans at the in side of stock point M, he or she returns the conveyance kanban and an empty tote from the drilling work center to the "out" side of the supplying workstation indicated on the conveyance kanban, in this case the milling work center.

3 When the conveyance kanban is received at the out side of stock point L, it replaces the production kanban in the card holder of a finished tote of the material that is indicated on the conveyance kanban, providing authorization for the tote to be moved to the drilling work center.

4 The production kanban removed from the tote in Step 3 is placed in the kanban collection box on the out side of stock point L, serving as authorization to make another tote of the material to replace the tote now going to the drilling work center.

5 The operator of the milling work center occasionally transfers kanbans found in the kanban collection box on the out side of stock point L to his or her work center dispatch box, retaining FIFO (i.e., first-in-first-out) order unless instructed otherwise.

6 The operator of the milling work center completes the next order by filling a tote box that has the production kanban in the dispatch box placed in the card holder of the completed tote. The tote is placed in the out side of stock point L for future delivery to the drilling work center when authorized.

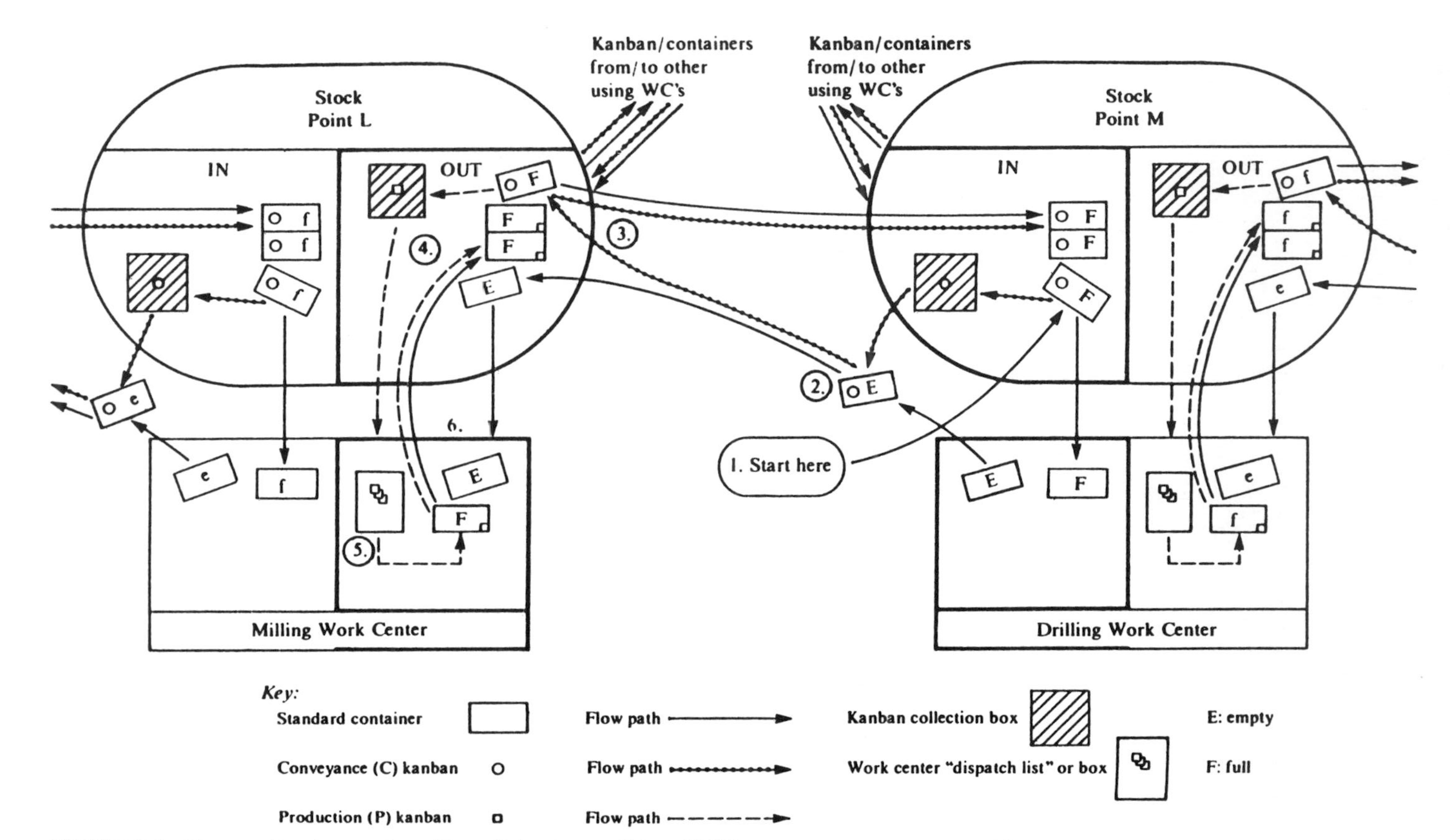

FIGURE 3-9 Two-card kanban system. (From Schonberger [13, p. 222].)

Note that there are two kanban loops going on, simultaneously connecting each pair of server and served equipment (i.e., milling work center and drilling work center): the conveyance kanban that authorizes the movement of product from stock point L to stock point M, and the production kanban that authorizes the production of more parts at the milling work center based on a replenishment authorization (i.e., conveyance kanban from the drilling work center). The kanban quantity provided in each loop (e.g., two, three, four, or five) determines the amount of "buffer" on the in and out sides of each stock point. This required quantity is a function of the longest downtime that the supplying equipment (i.e., the milling work center) is typically expected to experience, such that if delayed longer the machine being fed (i.e., the drilling work center) could not remain operational.

Few companies use the original two-card Toyota kanban system described above. Many, including Kawasaki [13, p. 231], use a one-card system. The one card is the conveyance card for authorization of material conveyance to the next machine; part production at a workstation is done on a scheduled, typically constant, daily basis.

One of the simplest versions of the kanban system utilizes parts containers, with permanently attached kanban, serving each supplied part between server and served machines. The container provides authorization for both conveyance and production. Control of the number of containers issued is necessary to control in-process inventory levels in the system.

One of the JIT philosophies practiced by the Japanese is to test required resource levels. For example, if a container that holds 30 parts is typically received back at the assembly line at a time when an on-line part container still has 20 parts remaining, the container size may be made smaller so that it holds only 20 parts. When that container is returned to the line from fabrication, there should still be a sufficient number of items in the on-line container to meet production needs. In an analogous way, if one worker on a ten-person assembly line is absent, the line supervisor may elect to have the nine-person crew perform for the day without adding a substitute for the absent worker. At the end of the day, if the nine-person crew performed the required work without difficulty and was able to keep up with assembly line requirements, the crew size may then be changed to nine in the future. Such practices are but one typical way Japanese management methods have greatly increased productivity of Japanese industrial plants.

In contrast, unions in the United States, believing that they were creating jobs, were successful in many instances in maintaining constant crew sizes long after technology improvements made these crew sizes unnecessary (e.g., firemen on diesel locomotives, three-person cockpit crews on airliners when two were adequate). In the short term, the unions were creating jobs, but unfortunately they were diluting productivity per worker as well. Over the long term, they were weakening the competitiveness of their firms in the international markets they served. Whole U.S. industries have been lost to our global competitors as a result of this backward, counterproductive long-term thinking.

The above kanban discussion was offered to suggest that JIT need not represent a theoretical departure from classical inventory control theory. Instead, JIT philosophy draws attention to the importance of setup time as the major management-controllable determinant of cost-effective production lot quantities.

Most American managers who visited Japan and saw JIT in action thought the Japanese were primarily interested in reducing inventory holding costs. The Japanese explained, however, that they have become more interested in the quality effects than the inventory holding cost effects of JIT. The Japanese have learned that the primary benefit of a JIT philosophy is the immediate quality feedback it offers. A JIT operating philosophy immediately exposes quality problems and demands effective correction. It provides the discipline necessary if a plant wishes to embrace a continuous and long-term improvement philosophy.

In typical U.S. manufacturing plants of the past, in which setup times were long and therefore relatively high in cost, economic production quantities were high as a result (e.g., 1000 units). If 1000 units were made but, when needed, were found to be defective, the tendency was to attempt to use the parts because of the investment in material and labor in the lot, as well as the immediate need for their use. A very familiar organizational activity in traditional U.S. manufacturing plants was typically called the material review board (MRB). Defective material was reviewed by this committee and often received a waiver based on a committee decision that the parts were "good enough."

The JIT philosophy provides much more immediate feedback concerning defective material; if the whole lot of 10 pieces is defective, all 10 can be disposed of immediately and good parts can immediately be made to replace them. If defective parts will immediately be discovered as defective (which would likely hold up the assembly line), workers are much more likely to make certain that the parts being made are not defective. In a JIT environment, defective parts are discovered almost immediately after being made.

The discovery and removal of defective work-in-process material is also consistent with the Japanese management philosophy of striving to produce only good material, discovering and disposing of defective material as soon as possible, and immediately correcting the underlying source of defective material.

Contrast the above philosophy with the following recollection. The author knows a former professor who was once employed in an American automobile factory. This professor spoke of an occasion in which master brake cylinders on an assembly line were suspected of being defective. While an order was placed for replacement master cylinders, the assembly line continued with the installation of defective master cylinders. When the replacement cylinders arrived some days later, a single mechanic spent days or weeks in the parking lot replacing defective master cylinders. In such an environment, it can be very tempting to try to prove that the suspected master brake cylinders are really "good enough." The phrase *good enough* typically means that something less than desirable has been produced. The person who asks, "Is it good enough?" is usually seeking a way to get rid of something without accepting personal responsibility for it.

Figure 3-10 is a list of concepts offered by JIT consultant Jack Harrison, of Hands-on-JIT Company in Orlando, Florida.

Key to understanding the effect that a JIT philosophy has on reducing both in-process inventories and cycle time is consideration of the flow of material between

1. Lot size reduction
2. Flat bills of material
3. Setup time reduction
4. Lead time reduction
5. On-the-line stocking
6. Schedule commitment: linearity
7. Level master schedule
8. Mixed model production
9. "Pull" vs. "push"
10. Management by sight
11. Opportunity signals (lights)
12. Stop the line: fix the problem
13. Visible data: quantity and quality
14. Housekeeping
15. Focused factories
16. "Product" type layout
17. Small machines, multiple copies
18. Operation checklists
19. Group technology
20. Manufacturing cells
21. Small group improvement activities (SGIA)
22. Discretionary SGIA. budget
23. Under-capacity scheduling
24. Multifunctional workers
25. Total plant involvement
26. KISS (Keep it simple, stupid)
27. Resource restriction
    Inventory
    Space
    Capacity (equipment, people)
28. Total preventative maintenance
29. Simultaneous inspection
30. Total quality control
    Quality at the source
31. Statistical process control
32. "Failsafe" the process
33. Standardize parts and processes
34. Value engineering
35. Minimum specifications
36. Minimum suppliers
37. Supplier development
38. JIT supplier deliveries
39. Macro-measurement

FIGURE 3-10 JIT Concepts. (From Harrison [7].)

machines A and B in Fig. 3-11. Cycle time is the time it takes to get from any designated starting point to any designated finish point in any sequential process. Assume in this example that cycle time is defined as the time to get from machine A to machine B, involving a typical interoperational handling between two machines in a traditional plant.

Assume that the machine B operation takes eight hours to perform and that there is one daily shift of operation. Once a day, the machine B operator takes the nearest box, containing one item, off the conveyor. About the same time, the machine A operator completes an item and puts a box on the vacated space created at that end of the conveyor. How long will it take for an item produced by machine A to get to machine B? There are 10 boxes in the queue, so the answer is 10 days. If there are 9 machines that the product has to go through in being processed, each with the same amount of buffer stock between operations, then the boxes will spend $8 \times 10 = 80$ days in queues after leaving the first machine and arriving at the last machine, exclusive of machine-related processing time. That is why one can directly relate the amount of inventory to the cycle time (i.e., "inventory = cycle time"). If the buffer conveyor held only five boxes, the time in queues between leaving the first machine and arriving at the last machine would be only 40 days, half as much as before.

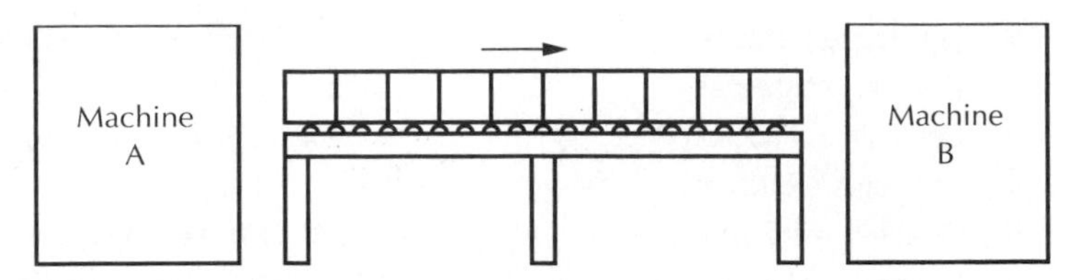

FIGURE 3-11 Typical interoperation inventory.

Therefore, buffer quantities and buffer pass-through times are directly related. If you wish to reduce the time that it takes for material to go through the system, limit the amount of work-in-process on the production floor; the less the better.

In Fig. 3-12, the water in the lake is analogous to inventory in a plant. The rocks in the lake are analogous to problems contained in a company's operating procedures, operations, and philosophy; the water (i.e., the inventory) is hiding the problems. Forcing operation under a JIT philosophy, as Harrison says at least 10 times per day, will expose the rocks (i.e., the problems). Thus—and this is the underlying key point in operating under a JIT philosophy—JIT (i.e., reduction of inventory, or lowering the water) forces continuous improvement. The more a company operates under a JIT philosophy, the more continuous improvement the company gets. Every time a rock appears it must be removed (i.e., the problem must be fixed), and the water level can be lowered again. But when the water level lowers, another rock will appear and the improvement process continues. The Japanese wanted to have less inventory, and what they discovered was that they got "rocks" that drove continuous improvement.

FIGURE 3-12 The lake–inventory analogy. (From Harrison [7].)

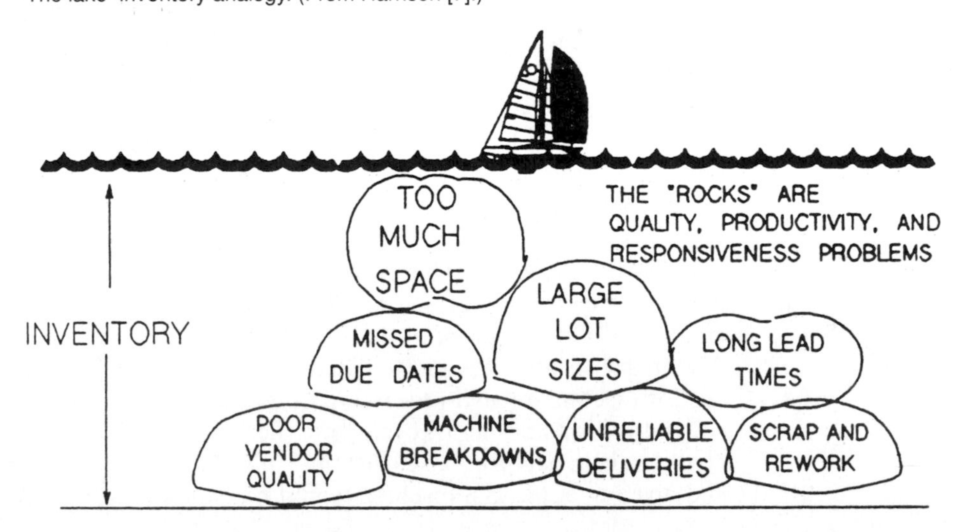

It should be noted that the caption for Fig. 3-11 called the figure a "typical inter-operation inventory." Far more typical in many plants, however, would be a pallet or pallets of material between two machines, often containing hundreds if not thousands of items.

It was indicated earlier in this chapter that MRP explodes product requirements (i.e., it calculates when material must be ordered and then received and assembled to meet delivery dates). In doing so, MRP utilizes traditional purchasing lead times.

Suppose a purchasing agent for a motorcycle manufacturer calls up his or her favorite seat manufacturer and says, "I would like to buy 12,000 more XY45 seats from you. Can you deliver them in 180 days, as you have done in the past?" What do you think the supplier's answer will be? "You betcha." Perhaps the seats could be supplied in 30 days, but why not take the full 180 days? Maybe some other jobs will be received in a day or two, in which case the additional time will be necessary; later deliveries are always safer than earlier deliveries. The supplier may well wait five months and then start making the needed seats.

If two people agree to meet in two weeks to continue a discussion and they need to spend one day in preparation for the meeting, which day would it be? It would almost certainly be the day before the next meeting. Have students ever waited until the day before a test to study? No contest! The point to this discussion is that traditional lead times are "self-fulfilling prophecies"; they become true if allowed to do so.

Suppose the question to the seat supplier is changed to "I have a problem. I would like to buy 12,000 seats from you at our previous price and terms, but I have to have them in 90 days. Can you help me, or would it be a problem for you?" There is an excellent chance that the supplier's response will be something like "It will be tough, but we want to serve a great customer like you; we will do it somehow." The supplier then waits two months and makes the seats. The motorcycle manufacturer can continue to ask for shorter and shorter lead times in an effort to determine the real lead time limit to the supplier's production system; subsequent orders can be made based on that time, plus maybe a little more, just in case. In time, however, the manufacturer might establish a long-term partner relationship with the supplier and help him or her to acquire the building down the street so that ultimately seats can be supplied on an hourly basis by forklift. That is the ultimate "win-win" solution for both parties—a mature, long-term, strategically and mutually sound partner relationship.

Figure 3-13 is a list of key JIT natural work team (NWT) techniques. Harrison of Hands-On-JIT believes strongly in an NWT approach to implementing these techniques. The combination of JIT NWT techniques, traditional industrial engineering approaches (i.e., ergonomics, workstation design, methods engineering, material handling, plant layout), and total quality management (i.e., a participative and continuous improvement team approach) represents a powerful world-class manufacturing philosophy for the future.

Single Minute Exchange of Dies (SMED) Another Toyota engineering manager, Shigeo Shingo [15], attacked the setup cost part of the problem with his single minute exchange of dies (SMED) system, which dramatically reduced setup costs, and provided the opportunity to operate cost effectively with much smaller

Cycle Time/Inventory Reduction
Inventory = cycle time. Cycle time is equal to what we schedule it to be (i.e., it's self-fulfilling). Continuously push to reduce WIP Inventory and thereby reduce cycle time. Note: Inventory is the enemy of quality!

Space Reduction
Squeeze out the waste. "Hand off" to the next operator. Minimize the number of racks, carts, storage bins. Move operations next to one another.

Setup/Lot Size Reduction
Purpose: Make only what we need, only when we need it. First objective: Single minute exchange of dies (SMED). Do as much setup preparation as possible while the machine is still operating. Standardize. Use quick clamping devices. Simplify/remove adjustments.

Schedule Commitment, by Shift
"The shift ends when the schedule is done." Your "customer" is the next operation. On-time delivery impacts all shops and the customer. Use schedule misses as another method to identify "rocks" that need to be eliminated.

"Pull" product where possible
Only make/move a product when your "customer" needs it. Signal this need through the use of a kanban (a visual signal—card, container, etc.). Do not violate the allowable kanban quantity at any location. Continuously strive to reduce the size and number of kanbans in the system.

Visible Data in the Area
Get all problems and ideas posted in the area. Utilize flip charts. Use the team measurements and goals as the stimulator. Get and record commitments, with a name and date, to fix problems and incorporate ideas.

Housekeeping
"Cleanliness is next to godliness." Remove all unnecessary material, tooling, scrap, trash, etc., from the work area. Keep benchtops and machines free of dirt and oil. "A place for everything, and everything in its place." A clean area improves quality and safety and helps early detection of problems.

Manufacturing Cells
Arrange equipment in such a way that product "flows" from process to process, with quick setups and minimal lot sizes. Build in flexibility for additional or reduced manning depending on the demand. Set work height for stand-up operation. Use high stools.

Cross Training
Flexibility is a key ingredient to becoming world class. Cross training allows us to effectively make what the customer wants, regardless of vacations, line mix changes, absenteeism, etc.

Sequential Inspection
Each operator verifies the quality from his or her supplier. Defects are immediately returned for rework. All defects are posted on the "problem/idea" chart for team failsafe efforts. Operator checklists are updated as required.

Failsafe
To failsafe an operation means to make the problem "impossible to reoccur." It may be as simple as color-coding tools, fixtures, parts bins, etc., or as complex as redesigning holding fixtures or even the part itself so that it physically cannot be manufactured incorrectly. Attempt to failsafe every source of a defect.

Do-See
If a proposed change (1) does not endanger anyone's safety, (2) can be corrected, in case of total failure, such that the customer will not be impacted, and (3) has had input from all involved—Just do it! Do not over-study simple changes. Try it and see if it works.

Continuous Questioning
To be world class means to rethink the entire way we have been doing business. Challenge everything. Why? Why not? What value does it add?

FIGURE 3-13 Key JIT NWT techniques. (Adapted from Harrison [7].)

tion lot quantities. Cost-effective operation with small production lot quantities provides the underlying basis for making JIT economically justifiable from an inventory cost perspective. In light of the local conditions in Japan, it became economically possible to operate with very little inventory if the process was effectively operated employing total quality control concepts and if management employed both a disciplined production and inventory control system and a management information system.

To demonstrate the opportunity for setup time reduction, consider a traditional punch-press die-change methodology in a traditional American plant. The setup worker might loosen 10 bolts and extract them from the press bed and ram, often using a box wrench and considerable excess threaded length on the bolts. After doing the unbolting, he would find and bring a die cart to the press and likely hand-crank the cart to raise the top of the cart to match the height of the press bed. He would then slide the die set onto the cart and push the cart to a die storage area. He might then hand-crank the cart to match the height of a storage shelf and push the die set onto the shelf. He would likely then identify the job to be run next and would locate the die set to replace the one he just stored. After locating the replacement die set, he would then hand-crank the die cart to accommodate the height of the storage shelf for the replacement die set, slide it onto the cart, and then push the cart to the punch press. He would then likely hand-crank the cart to raise or lower the cart top to match the height of the press bed. He would then push the die set onto the press bed. It is likely that different-length bolts would be necessary for this die set, so he might well look for and select bolts that will accommodate this die set. After sliding the die set front and back and left and right to line up the die set with the bolt holes, he would then likely hand-thread eight to ten bolts and tighten them with a box wrench, thus completing the setup. All of this may include a break time or lunch and some idle conversation time as well. Such a setup can and often does take two to three hours to perform.

Consider the Japanese alternative method. First, because this punch press utilizes only five different die sets, roller conveyor sections that come from the back of the press to the front have been purchased and welded to the press to store these five die sets at the press. As a die set is pulled from the back of the press, its replacement die set can be pushed in place from the front of the press. If all die sets were designed to utilize the same bolts, with common tightening heights, guide plates for locating the die set left to right, and a removable stop to determine the die set position front to back, locating these bolts on the press bed is easy and quick. With proper design, one loosening turn of each of four bolt heads with a nut runner (or, if possible, four toggle clamps attached to the press bed) and four bolts with removal C-washers, for attachment to the ram, would complete the setup. Under these assumptions, the setup may well be accomplished in one to five minutes.

Figure 3-14 illustrates the possibility for setup time reductions when engineering analysis is applied to setup times. Figure 3-15 illustrates numerous screw improvements that can be employed to reduce setup times.

Refer to the EPQ formula, Eq. (3-7), and consider setup times of 3 hours (180 minutes) and 5 minutes previously discussed. Consider the effect of this difference

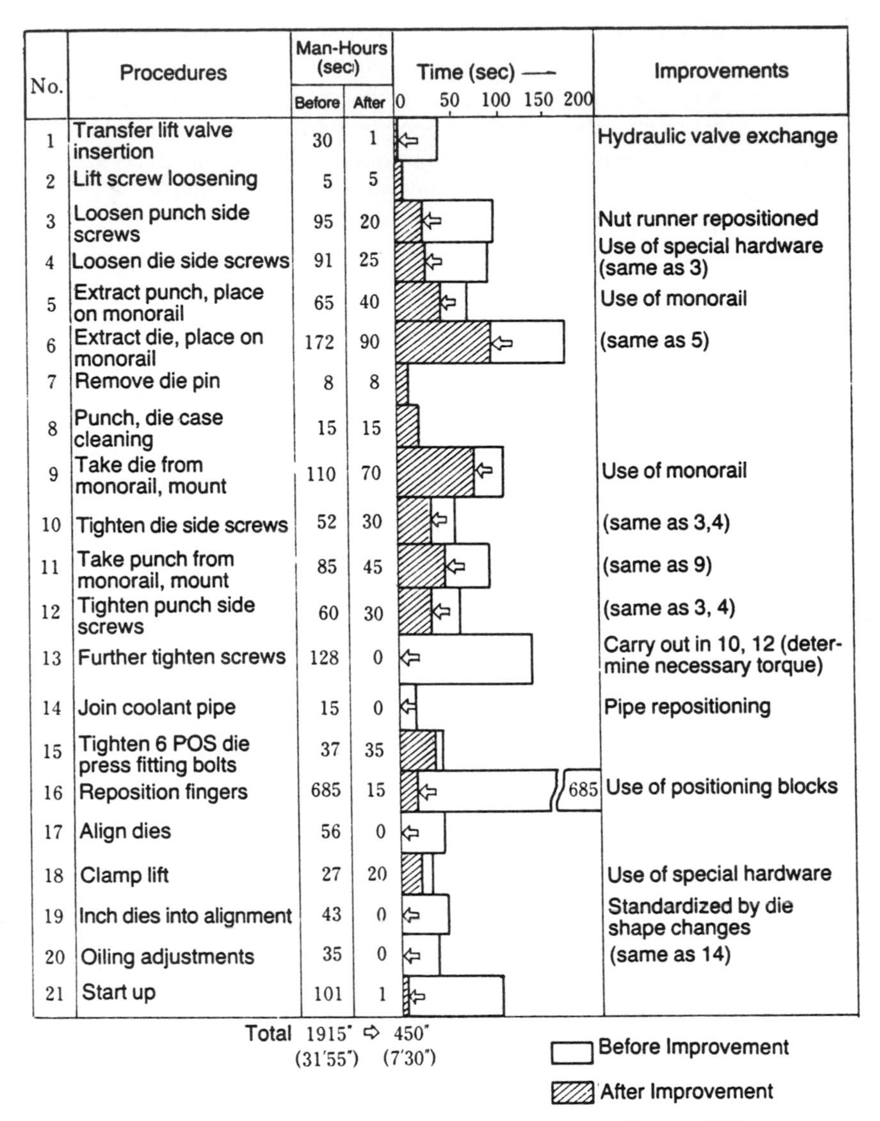

No.	Procedures	Man-Hours (sec) Before	Man-Hours (sec) After	Improvements
1	Transfer lift valve insertion	30	1	Hydraulic valve exchange
2	Lift screw loosening	5	5	
3	Loosen punch side screws	95	20	Nut runner repositioned
4	Loosen die side screws	91	25	Use of special hardware (same as 3)
5	Extract punch, place on monorail	65	40	Use of monorail
6	Extract die, place on monorail	172	90	(same as 5)
7	Remove die pin	8	8	
8	Punch, die case cleaning	15	15	
9	Take die from monorail, mount	110	70	Use of monorail
10	Tighten die side screws	52	30	(same as 3,4)
11	Take punch from monorail, mount	85	45	(same as 9)
12	Tighten punch side screws	60	30	(same as 3, 4)
13	Further tighten screws	128	0	Carry out in 10, 12 (determine necessary torque)
14	Join coolant pipe	15	0	Pipe repositioning
15	Tighten 6 POS die press fitting bolts	37	35	
16	Reposition fingers	685	15	Use of positioning blocks
17	Align dies	56	0	
18	Clamp lift	27	20	Use of special hardware
19	Inch dies into alignment	43	0	Standardized by die shape changes
20	Oiling adjustments	35	0	(same as 14)
21	Start up	101	1	

Total 1915" ⇨ 450" (31'55") (7'30")

FIGURE 3-14 Setup time reduction for a cold forging machine. (From Shingo [15, p. 171].)

on economic production quantities (EPQs). Whereas a setup time of 3 hours may indicate an economic production quantity of 1000 units, a setup time of 5 minutes may indicate a much smaller lot size (e.g., 5 units). Smaller lot sizes create much less work-in-process. Thus, reducing setup times is much of what makes JIT possible from an economic inventory control perspective. The primary benefit of the small lots is the instant quality feedback that such an arrangement provides.

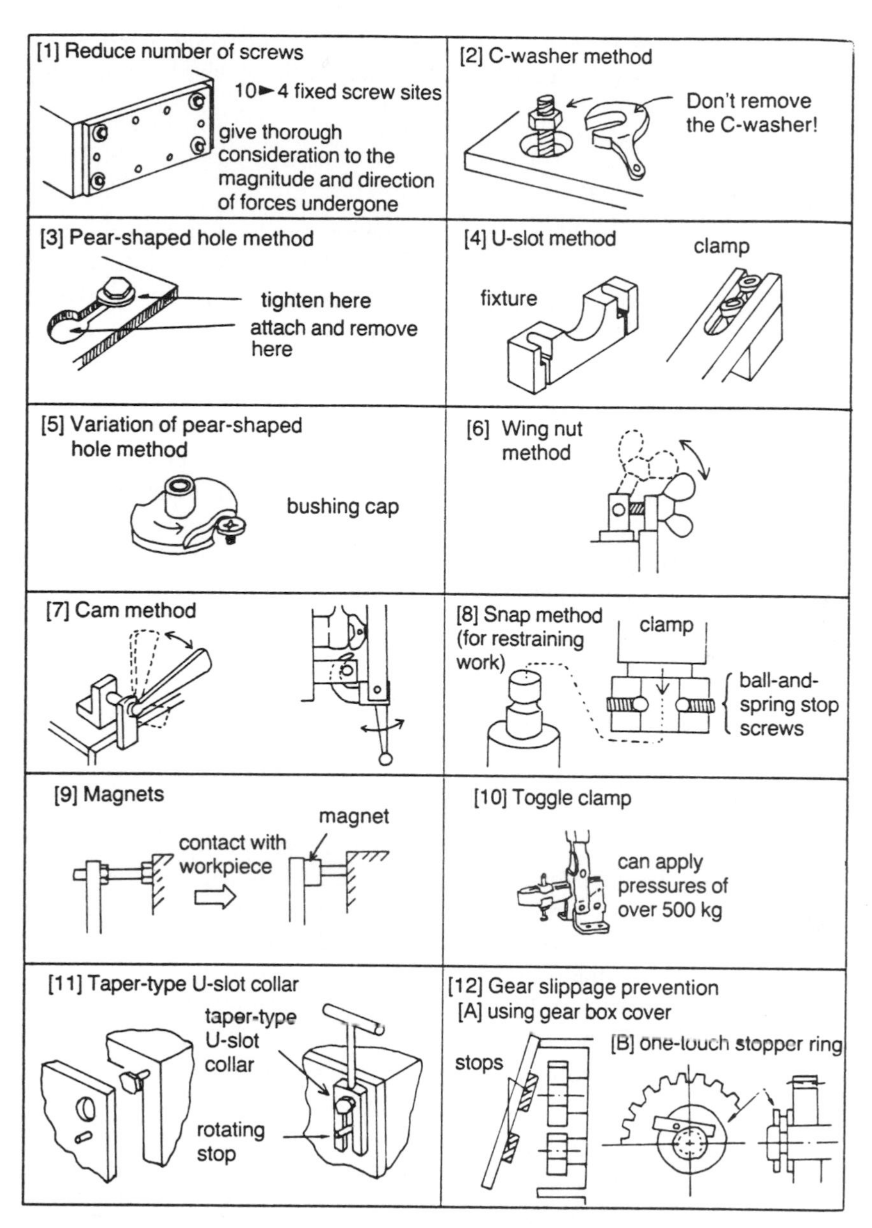

FIGURE 3-15 Screw improvement examples. (From Shingo [15, p. 196].)

Setup time analysis has traditionally been neglected in American industry, and large economic lot sizes and their consequent large in-process inventories were the result. Much opportunity exists for reducing setup time, lot sizes, and in-process inventories and for improving product quality by means of moving to a JIT manufacturing philosophy. Many of the more progressive American manufacturing plants have or are making the shift to a JIT philosophy. JIT does not apply to all production organizations; however, many of those who have not adopted JIT will have difficulty competing in the future.

Assembly Line Balancing

Another problem often dealt with in production control is that of assembly line balancing. The most common method in practice is to first attempt to distribute work elements to stations along an assembly line in such a manner that each station has approximately the same sum of elemental times (i.e., operation times). After the line has run for a short period of time, it usually becomes apparent that one employee on the line is having to work especially hard to keep up with the line or that work is beginning to collect at that workstation. The industrial engineer or time-study analyst noting this condition would then remove some elements of work from this station, if justified, and add them to another station along the line that appears to be low on total elemental time. This juggling of time elements on an assembly line might result in one employee installing the master brake cylinder on an automobile, for example, and then setting a muffler clamp on an automobile frame, because the muffler installer further down the line needs time relief and the master brake cylinder installer has excess time available.

In some cases, the number of elemental times is rather large and their respective precedence relationships (i.e., required task sequence) are sufficiently complex that some organized procedure is needed for determining an optimum assignment of elements to workstations. In the following example a heuristic first proposed by Moodie and Young [11] is used. *Heuristic* means that the approach produces good results but does not necessarily guarantee the best or optimum result. (An *algorithm,* on the other hand, is a procedure that by definition guarantees an optimum solution). This particular heuristic, however, does provide a test for determining whether an optimum solution has been produced, and it is rare for the heuristic to produce a solution that is less than optimum.

Figure 3-16 illustrates the precedence relationships and elemental times for 13 elements in a process. Assume that there is 8000 min. of production time available (i.e., approximately three weeks on a one-shift basis) for producing 1000 units of a product. A desired cycle time would therefore be:

$$\text{Cycle time} = 8000 \text{ min.}/1000 \text{ units} = 8 \text{ min./unit}$$

Therefore, an 8-min. cycle time will be assumed. If it is later determined that the 8-min. cycle time solution is not optimum, a 7- or 9-min. cycle time might be employed on a second attempt to find optimum balancing of the line.

The following terminology will be employed in this example:

$$\begin{aligned} \text{Cycle time} &= C \\ \text{Station number} &= K \\ \text{Element number} &= i \end{aligned}$$

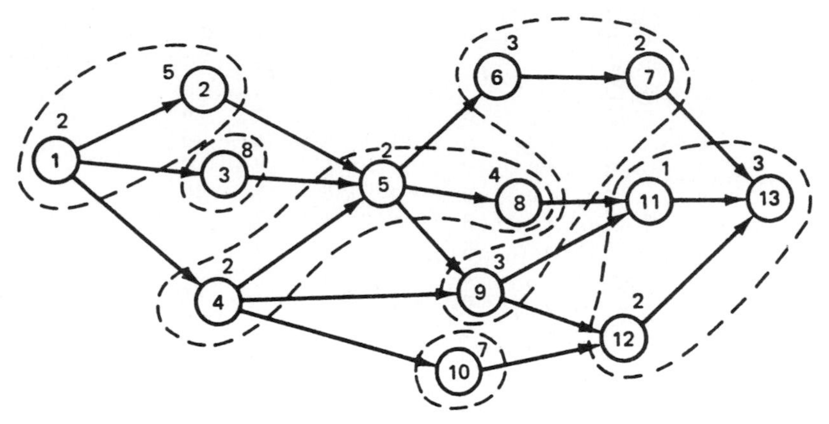

FIGURE 3-16 Precedence chart for elemental times.

$$\text{Elemental time} = E_i$$
$$\text{Station time} = S_K$$
$$\text{Balance delay} = C–S_K$$

In Fig. 3-16, the number within a circle is the element number i and the number above the circle is the elemental time E_i. The arrows in the figure show which elements must be completed before an element can be initiated. For example, element 5 cannot begin until elements 2, 3, and 4 are completed; elements 6, 8, and 9 cannot begin until element 5 has been completed.

This information (along with other marks that will be explained below) is shown in Table 3-5. The number of columns of the P matrix is determined by the largest number of elements preceding any element in the precedence diagram. In this example, elements 2, 3, and 4 precede element 5; therefore, three columns are required for the P matrix. In a like manner, the maximum number of elements following any element in Fig. 3-16 is three; therefore, three columns are needed in the F matrix.

Note that the sum of all 13 elemental times in Table 3-5 is 44 min. With a desired cycle time of 8 min., an optimal allocation of elements would result in six workstations, with 4 min. of balance delay left in the system (i.e., $6 \times 8 = 48$; $48 - 44 = 4$). Any feasible grouping of elements into six workstations for the purpose of this problem is therefore assumed to represent an optimal solution. The heuristic [11] is as follows:

1 Note the rows of the predecessor matrix **P** that contain all zeros [that have not previously been assigned], and assign the largest element possible [time] indicated by these rows if more than one exists. Check the element number to indicate that it has been assigned.

2 Note the element number in the row of the follower matrix F that corresponds to the assigned element and go to the row of matrix P indicated by this number, and replace the assigned element's identification number with a zero.

TABLE 3-5 LINE BALANCING DATA

i	E_i	P^*			$F^\dagger$		
1✔	2	0	0	0	2	3	4
2✔	5	~~1~~ 0	0	0	5	0	0
3	8	~~1~~ 0	0	0	5	0	0
4	2	~~1~~ 0	0	0	5	9	10
5	2	~~2~~ 0	3	4	6	8	9
6	3	5	0	0	7	0	0
7	2	6	0	0	13	0	0
8	4	5	0	0	11	0	0
9	3	4	5	0	11	12	0
10	7	4	0	0	12	0	0
11	1	8	9	0	13	0	0
12	2	9	10	0	13	0	0
13	3	7	11	12	0	0	0
	44						

*The numbers in the rows of matrix P are the element numbers directly preceding element i for that row.

†The numbers in the rows of matrix F are the element numbers directly following element i for that row.

3 Continue assigning elements to each station, following the restriction that

$$\text{Maximum } E_i(K) <= S_i <= C$$

where S_i is the sum of element times for the ith station.

Proceed until the P matrix contains all zeros.

In this example, the heuristic is applied to Table 3-5 to produce the station assignments indicated in Table 3-6. When all nonzero element numbers in a row of the matrix P have been crossed out and replaced with zeroes, that element is available as an assignment to a station, because all of its predecessor elements have been completed. The matrix F tells you which predecessor to remove. The crossouts in Table 3-5 appear as they would at the time when element assignments for station number 1 have been completed.

When all assignments have been made, all numbers other than zeros in the P matrix will have been crossed out and replaced by zeros. The station assignments of Table 3-6 are indicated by the dotted-line zones of Fig. 3-16.

Note, for example, that element 2 was selected to be added to element 1 to form station 1. This is because element 1's time of 2 min. left 6 min. of assignable time remaining and because element 2 had the largest possible assignable time available, 5 minutes. If the final sum of all the balance delays in this example had been 8 min. or greater, it would not be apparent that the solution obtained was an optimum solution, because a better assignment of elements to stations might have resulted in one less station.

It should also be noted that the heuristic as stated results in a well-defined sequence of assignments that can easily be computer programmed. A computer pro-

TABLE 3-6 LINE BALANCING STATION—SUMMARY

K	i	E_i	S_k	$C - S_k$
1	1	2		
	2	5	7	1
2	3	8	8	0
3	4	2		
	5	2		
	8	4	8	0
4	10	7	7	1
5	6	3		
	9	3		
	7	2	8	0
6	12	2		
	11	1		
	13	3	6	2
				4

gram could readily be developed to make station assignments for a large number of elements and precedence relationships. This example is designed to demonstrate how a simple heuristic can be devised to assist in making element assignments to workstations, including a means of checking the feasibility of using a smaller number of stations with a different assignment. Simple heuristics such as this one represent the "tricks of the trade" in applying industrial engineering on a day-to-day basis.

Note in Step 1 of the heuristic the explicit instruction to assign "the largest element [time] possible." This is one of the key aspects of the overall heuristic. Consider an analogy. Assume that a New England farmer has cleared a field and has created a pile of rocks of varying sizes. The land is rolling pasture land, and the farmer wishes to build a stone wall, with the top of the wall level, as indicated in Fig. 3-17. What rocks should he select as he builds the wall in order to maintain a level wall? If he always selects the "largest rock possible" as he is building the wall, he will save as many smaller rocks as possible; when he needs smaller rocks to produce his level wall, he will have maximized his opportunity to do so. If he indiscriminately selects rocks of varying size, a small rock to maintain a level wall may not be available when needed. Heuristics contain explicitly rational elements to force the applicator to adopt optimal rules in performing the assignment task. Chapter 7 contains numerous examples of the use of heuristics and algorithms to produce successful results.

Project Management: CPM and PERT

Another category of techniques in the general area of production planning and control are commonly referred to as *project management techniques*. The two most

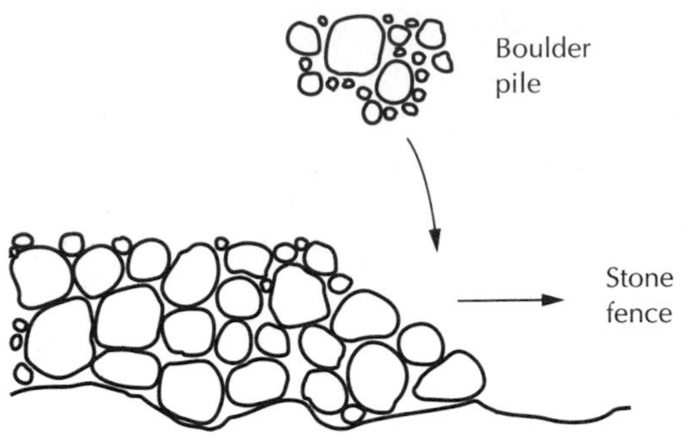

FIGURE 3-17 The level stone fence analogy.

common of these are CPM (critical path method) and PERT (project evaluation and review technique). CPM was developed in industry at the same time that PERT was being developed in the U.S. Navy, specifically in the Polaris submarine program. Papers describing both techniques were first published in 1959 [10, 17].

Assume that a tank is to be designed and assembled in the field and that Table 3-7 represents the project engineer's best estimate of the time required for each task and the key precedence relationships for each task. As indicated in Table 3-7, field assembly of the tank takes 15 days (activity K) but cannot begin until the tank is transported to the site (activity H) and tank site preparation (activity I) has been completed. Figure 3-18 is a basic CPM diagram for the tank project, indicating the precedence relationships by use of connected arrows. Note that arbitrary but unique members are often assigned to nodes.

TABLE 3-7 TANK INSTALLATION PROJECT DATA

Activity symbol	Activity	Time, days	Precedence
A	Design tank	30	
B	Design controls	10	A
C	Prepare parts list for tank	5	A
D	Prepare parts list for controls	3	B
E	Procure tank parts	40	C
F	Procure control parts	60	D
G	Partially fabricate tank	5	E
H	Transport tank to site	7	G
I	Prepare site	10	B
J	Transport controls to site	3	F
K	Field assembly of tank	15	H, I
L	Field assembly of controls	2	J
M	Install controls to tank	4	K, L
N	Inspect and test tank	3	K
O	Inspect and test controls	2	M
P	Document and clean up	1	N, O

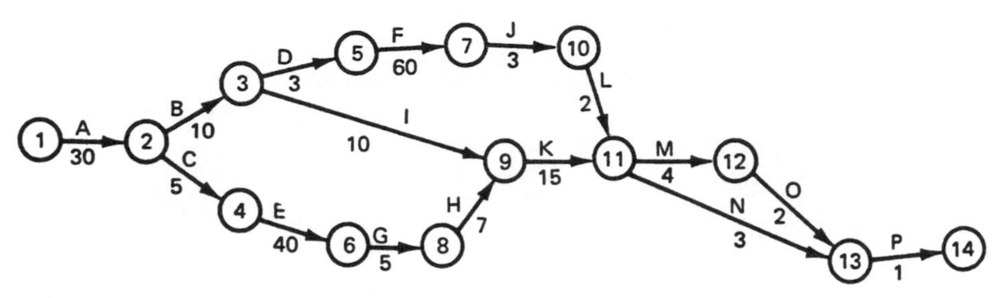

FIGURE 3-18 A basic CPM diagram for the tank project.

One of the primary reasons for performing a CPM analysis of a project is to determine the *critical path,* which is the sequence of project tasks, through the network, that determines the minimum completion time for the project. The minimum time for the tank project is found to be 115 days, which is the sum of task times for tasks A, B, D, F, J, L, M, O, and P. No other sequence of tasks in this network requires a longer time. If the project engineer is concerned with completing the project as soon as possible, which is usually the case, particular attention must be paid to tasks on the critical path. Essentially, a delay in any of these tasks will represent a similar delay in the total completion time of the project.

For a large-scale project, such as the research and development of the first Polaris submarine or the building of a sports stadium, only a small fraction of the tasks to be performed are on the critical path. Identification of and attention to critical path tasks often result in a dramatic improvement in total project completion time; the Polaris submarine project was completed 18 months ahead of its initial schedule, which was previously unheard of for this type of project.

In analyzing the network for Fig. 3-18, it would be common practice to record earliest start (ES), earliest completion (EC), latest start (LS), and latest completion (LC) times on the network, as indicated in Figure 3-19. As mentioned previously, both i and j are arbitrary, yet unique, node numbers used to numerically identify the beginning and ending points for each task. The letters X and t represent a task designation symbol and a corresponding task time, respectively. The CPM network for the tank project including these data values is shown as Fig. 3-20.

To determine earliest start and earliest completion times, one starts at the beginning of the network, adding the task time for an activity to the earliest start time to

FIGURE 3-19 Standard CPM network symbol.

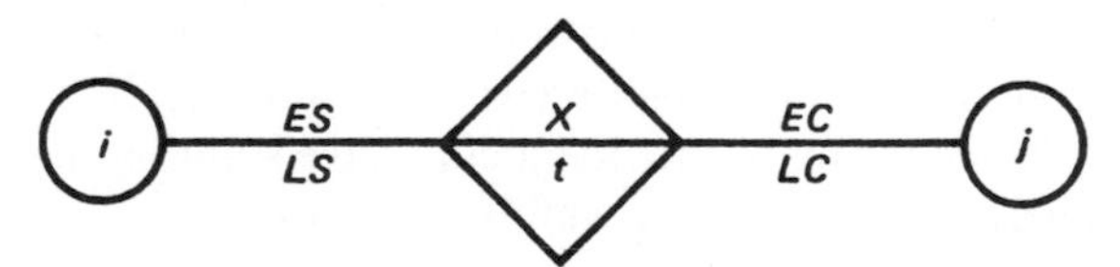

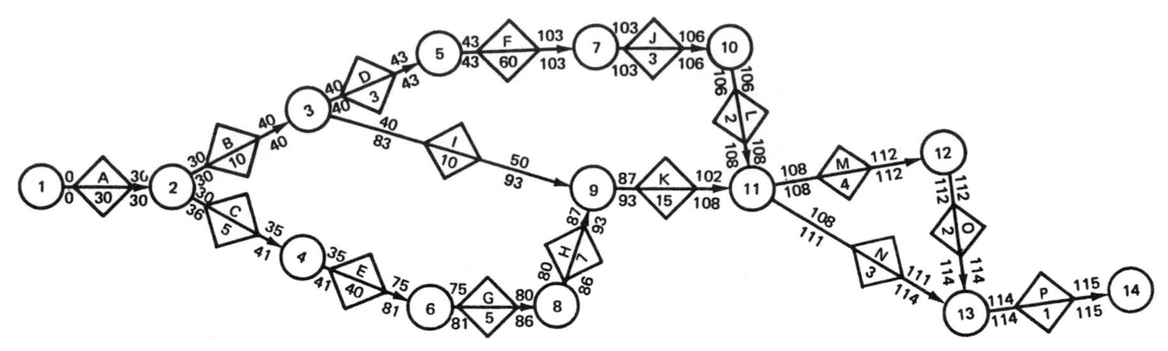

FIGURE 3-20 CPM diagram for the tank project

provide the earliest completion time. If two tasks precede a third task, the later completion time determines the earliest start time for the third task (e.g., task K).

In a similar manner, latest completion and latest start times are determined by starting at the end of the network and working backward. If more than one task, for example, M and N, follow a particular task (e.g., task K), the smallest latest start time of the two following tasks (108 for M versus 111 for N) determines the latest completion time of the preceding task (e.g., *LC* for task K is 108).

The earliest and latest start and completion times recorded on Fig. 3-20 provide useful information to the project engineer. Slack time is calculated as follows:

$$\text{Slack time} = LS - ES \quad \text{or} \quad LC - EC$$

For task J, for example, slack time is zero. This says that task J is on the critical path and, if delayed, will delay the total project. Slack for task N, however, is 3 days. This indicates that if task N is delayed 3 days, the total project would not be delayed, as noted by analyzing the later tasks (i.e., N, O, and P) in Fig. 3-20.

The assignment of node numbers *i* and *j* mentioned earlier facilitates tabular analysis of a CPM network on a computer, as indicated in Table 3-8. The graphical networks of Figs. 3-18 and 3-20 are essentially superfluous if a computer solution is employed. It has been reported by practitioners, however, that for relatively small networks (i.e., less than 100 activities), a time-scaled network, as indicated in Fig. 3-21, is often quite adequate for graphically displaying and subsequently controlling a project, without the need for a computerized solution. This is an example of a situation in which the use of a computer may not provide a sufficient return to justify the inconvenience that inputting data may involve. It is not uncommon today to observe someone trying to provide a computer solution to a problem for which a noncomputerized technique may well be superior.

In CPM, all tasks or activities are assumed to be deterministic with respect to both the sequence of tasks and their duration. PERT was originally developed with the hope of specifying tasks in a simple yet reasonably accurate stochastic manner. Three times were estimated for each task: the optimistic time *A*, the most likely time

TABLE 3-8 TABULAR ANALYSIS OF TANK PROJECT

Activity	From i	To j	t	Forward pass ES	Forward pass EC	Backward pass LS	Backward pass LC	Slack
A	1	2	30	0	30	0	30	0
B	2	3	10	30	40	30	40	0
C	2	4	5	30	35	36	41	6
D	3	5	3	40	43	40	43	0
E	4	6	40	35	75	41	81	6
F	5	7	60	43	103	43	103	0
G	6	8	5	75	80	81	86	6
H	8	9	7	80	87	86	93	6
I	3	9	10	40	50	83	93	43
J	7	10	3	103	106	103	106	0
K	9	11	15	87	102	93	108	6
L	10	11	2	106	108	106	108	0
M	11	12	4	108	112	108	112	0
N	11	13	3	108	111	111	114	3
O	12	13	2	112	114	112	114	0
P	13	14	1	114	115	114	115	0

FIGURE 3-21 Time-scaled CPM network for the tank project.

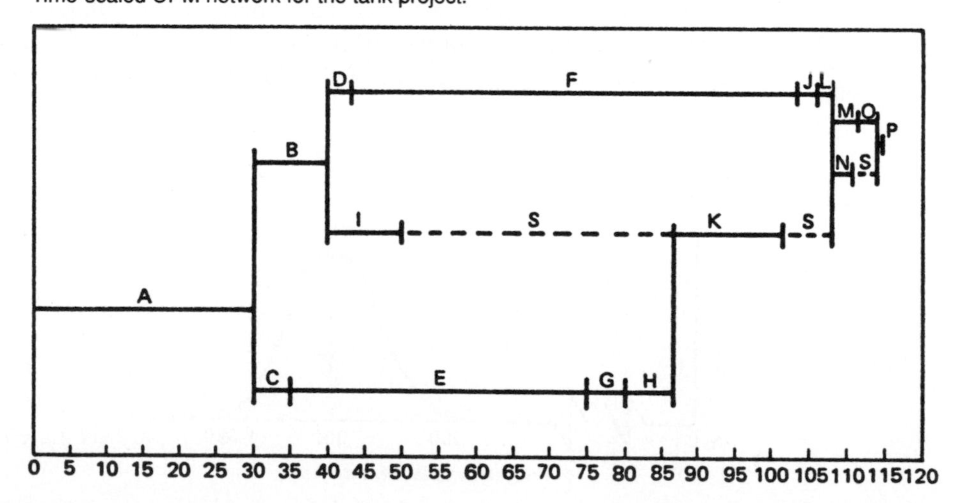

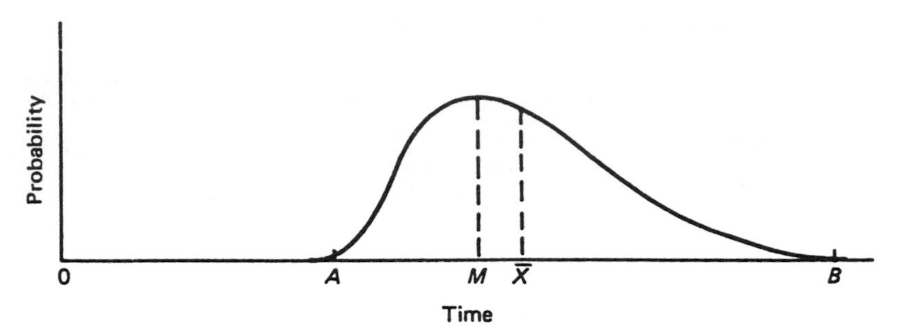

FIGURE 3-22 A beta distribution in PERT.

M, and the pessimistic time B. These three times were then used to specify a beta distribution with a mean $\overline{X}$ and a variance S^2 estimated as follows:

$$\overline{X} = \frac{1}{6}(A + 4M + B)$$

$$S^2 = \left[\frac{1}{6}(B - A)\right]^2$$

Figure 3-22 illustrates a typical beta distribution employing the A, B, and M estimates. Times A and B were to be estimated on the assumption that more extreme values than A or B at either end would be expected to occur only one percent of the time.

By having all tasks specified as density functions, it was possible to produce a density function for the overall project completion time, as exemplified in Fig. 3-23. This density indicates that it is 99 percent likely that the project would be completed in more than 280 days and 99 percent likely that the project would take no more than 320 days, or 98 percent likely that it would take between 280 and 320 days, with 300 days being most likely. With CPM, the estimate of completion time

FIGURE 3-23 Density function for project completion time.

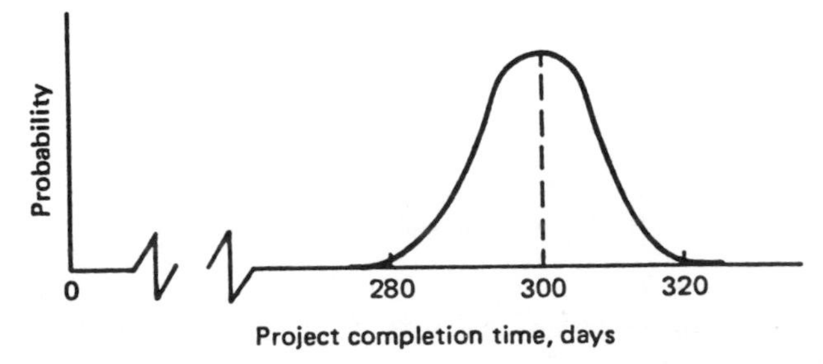

would simply be 300 days, assuming the constant task times selected in CPM summed to the sum of $\overline{X}_i$ for the same activities in PERT.

The three-estimates concept of PERT seems to have essentially disappeared in practice. Most practitioners seem to have adopted the single-estimate approach of CPM. Two possible reasons for this trend are (1) primary practical interest has been and still is in providing a "best estimate" of actual completion time, and (2) the beta distribution assumption may not have been well founded, introducing considerable error in some applications and causing the general applicability of the beta distribution assumption to become suspect.

There is little question today of the need for employing some project management technique, similar to CPM or PERT, to control the complex scheduling of limited resources for key activities. These techniques have provided a measure of control that makes timely completion of complex projects possible, predictable, and even expected, barring any unusual, unforeseen, or unpredictable factors. Bad weather, for example, is predictable, and it can be included as "lost days due to bad weather" based on accumulated data in project planning; a hurricane or earthquake at a specific location and time, however, is not predictable.

Production planning and control are primarily concerned with specifying who will do what when in a production activity. Another area of control—quality control—is primarily concerned with establishing procedures to ensure that a prespecified quality level for the finished product is achieved.

QUALITY CONTROL

In the days of Taylor, every plant, with rare exceptions, had an inspection department. The function of the department was to see that unacceptable material was removed throughout the process of making a product, from incoming material to outgoing finished product. In those early days, statistical sampling was essentially unknown in manufacturing. Consequently, key points were often selected throughout the process at which 100 percent inspection was performed. Because employees were often paid on the basis of the number of good units produced and because production managers were evaluated on the number of units shipped, inspection department employees and production managers traditionally endured often unfriendly if not hostile work relationships.

Quality control as a subdiscipline of industrial engineering has two basic components. One component involves the technical use of applied statistics and is typically referred to as either statistical quality control (SQC) or statistical process control (SPC). The other component involves the development and day-to-day administration of a programmatic and managerial basis for accomplishing quality control objectives in an organization. As was mentioned earlier in this text, no single course in a typical industrial engineering curriculum is more important than a course in probability and statistics. This is particularly true in the area of statistical quality control. Statistical quality control, in terms of its technical content, is simply the application of statistics and sampling theory in an industrial or productive environment. As important as probability and statistics are to the education of an industrial engineer, no attempt will be made to describe the contents of such a course in this text. It is assumed here that a statistics course is a prerequisite for the study of statistical quality control.

Statistical Process Control: $\overline{X}$ and R charts

Probably the most commonly known technique, and arguably the oldest, in statistical quality control (SQC) is the use of the $\overline{X}$ and R chart. It was originally developed by Shewhart [14] in the 1920s. This chart is employed when the goal is to maintain statistical control of some single variable of interest (e.g., a dimension for a part or assembly).

Figure 3-24 describes a cutting operation on a lathe, with dimension A of interest. This operation could be performed, for example, in preparation for cutting threads for mating two pipes in an oil field environment (e.g., in west Texas). The field measurements, to the nearest hundredth of an inch, for the first 48 pipes cut are provided in Table 3-9.

Note that the measurements have been divided into samples of size four. As is explained elsewhere in this text, the reason for doing so is to take advantage of the central limit theorem, which basically says that when you plot the means of samples from an unknown but essentially unimodal distribution of individual values, the means will be relatively normally distributed—that is, a Gaussian distribution. (It is not true, as one Ph.D. mathematics student reportedly stated during a Ph.D. oral examination upon failing to remember the name *Gauss,* that the name of the original developer of the normal distribution was Sir George Normal. It was such a totally wrong, yet innovative and humorous response, however, that he was reported to have survived the day.)

In reviewing the calculations for the upper and lower control limits for both the mean and range in Table 3-9, it will be noted that the products of the factor values and the calculated sample range can be used to determine these limits. The values for these factors, for various sample sizes, are given in Table 3-10.

FIGURE 3-24 Final outer diameter A in a screw machine.

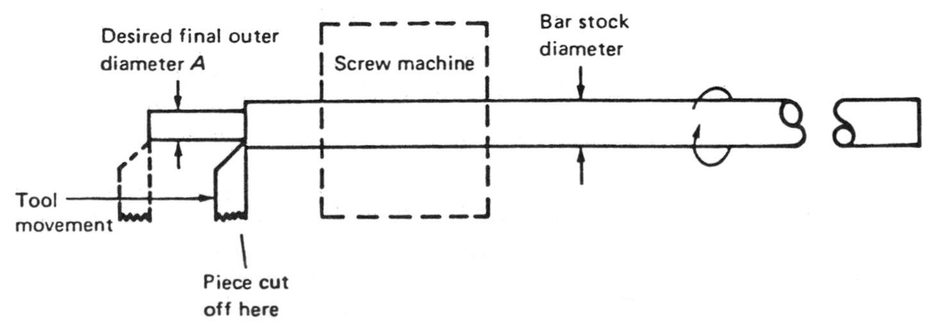

TABLE 3-9 $\overline{X}$ AND R DATA AND CALCULATIONS FOR DIMENSION A

	Sample number											
	1	2	3	4	5	6	7	8	9	10	11	12
1	3.50	3.60	3.60	3.60	3.50	3.70	3.70	4.00	4.50	3.50	3.70	3.70
2	3.50	3.50	3.70	3.70	3.50	3.80	4.00	4.40	4.60	3.60	3.70	3.60
3	3.60	3.50	3.50	3.60	3.60	3.90	4.20	4.50	4.20	3.60	3.60	3.80
4	3.50	3.60	3.60	3.60	3.60	3.90	4.10	4.20	4.10	3.50	3.60	3.80
Total	14.1	14.3	14.4	14.5	14.2	15.3	16.0	17.1	17.4	14.2	14.6	14.9
$\overline{X}$	3.525	3.55	3.60	3.625	3.55	3.825	4.0	4.275	4.35	3.55	3.65	3.725
High value	3.60	3.60	3.60	3.60	3.60	3.60	3.60	3.60	3.60	3.60	3.60	3.50
Low value	3.50	3.50	3.50	3.50	3.50	3.50	3.50	3.50	3.50	3.50	3.50	3.50
Range	0.10	0.10	0.20	0.10	0.10	0.20	0.50	0.50	0.50	0.10	0.10	0.20

$$\overline{\overline{X}} = \sum \overline{X}_i / n = 45.225/12 = 3.769$$

$$\overline{R} = \sum R_{i/}n = 2.7/12 = 0.225$$

$$UCL\overline{X} = \overline{\overline{X}} + A_2\overline{R} = 3.769 + 0.73(0.225) = 3.93$$

$$LCD\overline{X} = \overline{\overline{X}} - A_2\overline{R} = 3.769 - 0.73(0.225) = 3.60$$

$$UCL\overline{R} = D_4\overline{R} = 2.28(0.225) = 0.51$$

$$LCL\overline{R} = D_3\overline{R} = 0(0.225) = 0$$

It is apparent in analyzing plotted means for values of diameter A for the first 48 pieces cut (see Fig. 3-25) that this process appears to be "out of control," which means that far too many (i.e., more than 1 in 100) sample mean values fall outside the control limits. This suggests that there is an underlying *assignable cause* for the displayed degree of sample mean variability. The suspicion here is that something is wrong with how this operation is being performed. The underlying ability of the cutting operation to maintain diameter value dimensions is far superior to that reported in the sample mean values. Numerous questions concerning the process must be considered. For example, is the chuck on the lathe firmly holding the piece

TABLE 3-10 FACTORS FOR DETERMINING UPPER AND LOWER CONTROL LIMITS FOR $\overline{X}$ AND R CHARTS FROM $\overline{R}$

Sample size	A_2	D_3	D_4
2	1.88	0	3.27
3	1.02	0	2.57
4	0.73	0	2.28
5	0.58	0	2.11
6	0.48	0	2.00

[a] From "ASTM Manual" [1, p. 134].

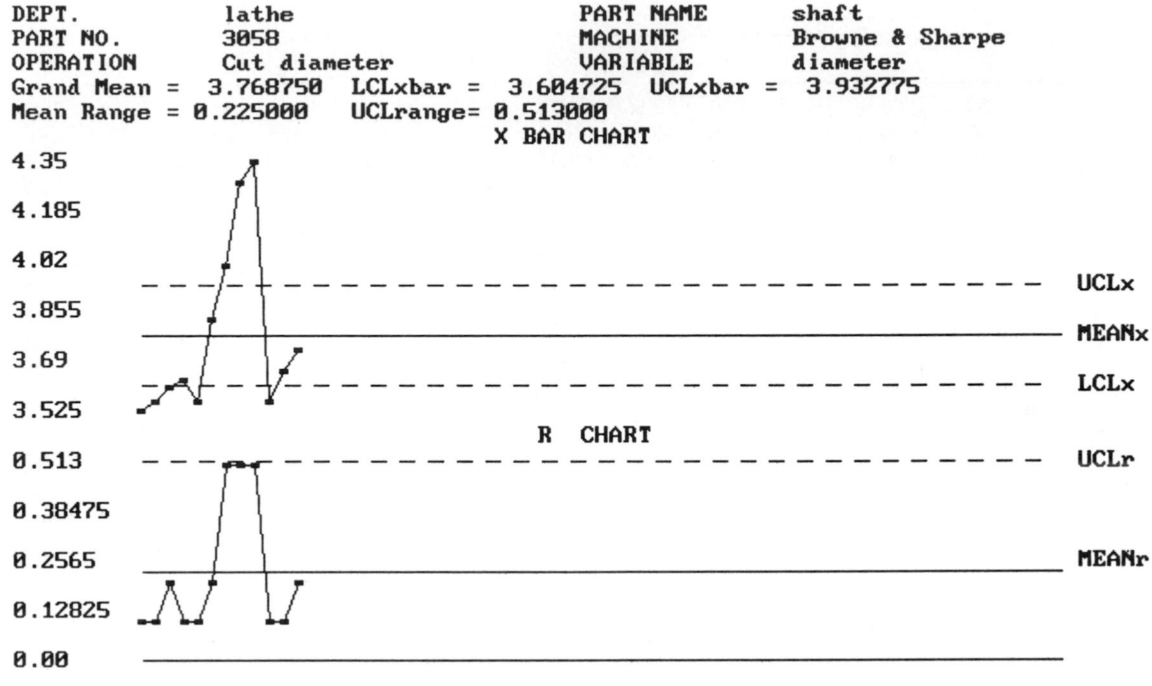

FIGURE 3-25 Plotted means for values of diameter *A*.

in the chuck? Is the tool dull, thereby causing an uneven cut that disturbs the diameter for some portion of the pipe? Is there material inconsistency along the length of the pipe?

Consider another example, shown in Table 3-11, concerning the pitch diameter of threads on aircraft fittings. Figure 3-26 illustrates values for individual members of sampled diameters in relation to the part nominal dimension and upper and lower tolerance limits. Of the 20 samples of 5 readings each (i.e., 100 individual diameters measured), only 1 diameter in sample 8 (i.e., 0.4023 in.), fell outside the tolerance limits. Is the process in control? That depends on what *control* means.

Figure 3-27 illustrates $\overline{X}$ and R control charts for the 20 samples. The control chart limits, upper and lower, are statistically determined such that there only should be 1 chance in 100 that a sample mean will fall outside these limits. Note that three sample means are outside the limits for $\overline{X}$ and that one sample range is outside the limits for R, which suggests that there are assignable causes for having this many samples fall outside the limits. In other words, something is probably going on that isn't supposed to. In quality control language, the process is not in control. Recall that 99 of the 100 parts measured fell within the tolerance limits (i.e., they are acceptable). Simply stated, the range of acceptability of the part dimension, in terms

TABLE 3-11 THREAD PITCH DIAMETER OF AIRCRAFT FITTINGS

Measurements of pitch diameter of threads on aircraft fittings. Values are expressed in units of 0.0001 inch in excess of 0.4000 in. Dimension is specified as 0.4037 ± 0.0013 in.

Sample number	Measurement on each item of five items per hour					Average $\bar{X}$	Range R
1	36	35	34	33	32	34.0	4
2	31	31	34	32	30	31.6	4
3	30	30	32	30	32	30.8	2
4	32	33	33	32	35	33.0	3
5	32	34	37	37	35	35.0	5
6	32	32	31	33	33	32.2	2
7	33	33	36	32	31	33.0	5
8	23	33	36	35	36	32.6	13
9	43	36	35	24	31	33.8	19
10	36	35	36	41	41	37.8	6
11	34	38	35	34	38	35.8	4
12	36	38	39	39	40	38.4	4
13	36	40	35	26	33	34.0	14
14	36	35	37	34	33	35.0	4
15	30	37	33	34	35	33.8	7
16	28	31	33	33	33	31.6	5
17	33	30	34	33	35	33.0	5
18	27	28	29	27	30	28.2	3
19	35	36	29	27	32	31.8	9
20	33	35	35	39	36	35.6	6
Totals	...	...	...	...	...	671.0	124

FIGURE 3-26 Individual measurements of diameter. (From Grant and Leavenworth [4, p.10].)

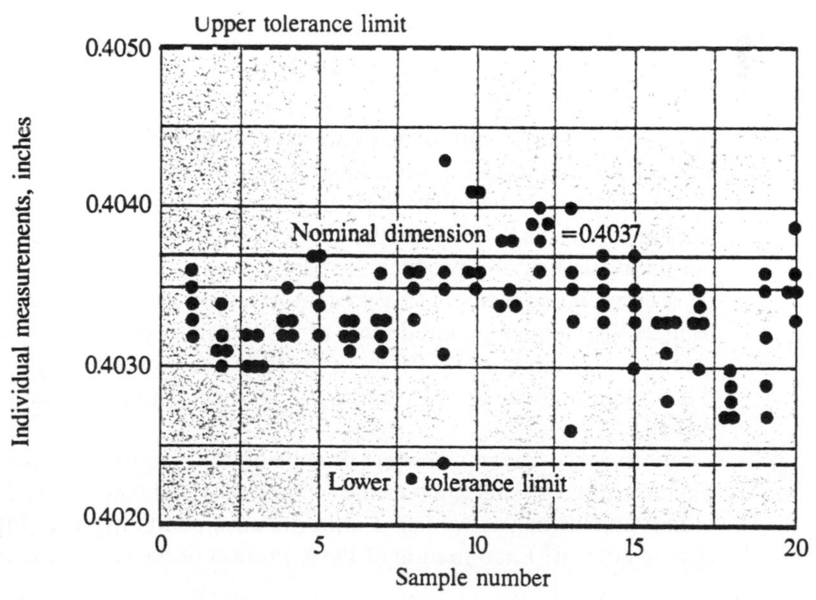

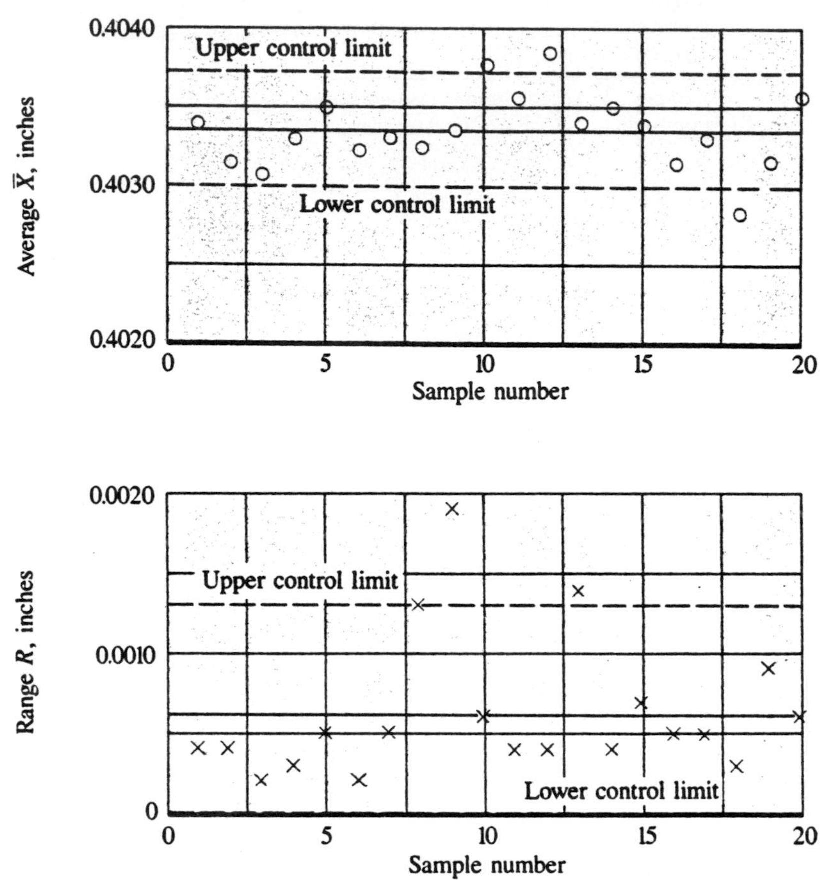

FIGURE 3-27 $\overline{X}$ and R control charts for pitch diameter. (From Grant and Leavenworth [4, p. 11].)

of its use, is sufficiently wide that even though the process is "out of control," it is still capable of producing acceptable parts (99 out of 100 times). With better control of the process—remember, there is reason to believe that something may be producing undue variability in the process—this process would have an increased ability to successfully produce the part.

A common adjustment to a process based on the information from such charts is to alter the frequency and degree of centering or positioning of the process (e.g., resetting it as close to 0.4037 in. as possible, as often as needed, if it is drifting from the center position; or setting it low within the acceptable range as the tool wears down).

Table 3-12 concerns another dimension that requires control: the distance from the back of a rheostat knob to the far side of a pinhole. Figure 3-28 illustrates individual dimensions as well as $\overline{X}$ and R control charts for this dimension. Is this process in control? Examination of the $\overline{X}$ portion of the chart indicates that the process

TABLE 3-12 MEASUREMENTS OF DISTANCE FROM BACK OF RHEOSTAT KNOB TO FAR SIDE OF PINHOLE

Measurements of distance from back of rheostat knob to far side of pinhole. Values are expressed in units of 0.001 in. Dimension is specified as 0.140 ± 0.003 in.

Sample number	Measurement of each item of five items per hour					Average $\bar{X}$	Range R
1	140	143	137	134	135	137.8	9
2	138	143	143	145	146	143.0	8
3	139	133	147	148	139	141.2	15
4	143	141	137	138	140	139.8	6
5	142	142	145	135	136	140.0	10
6	136	144	143	136	137	139.2	8
7	142	147	137	142	138	141.2	10
8	143	137	145	137	138	140.0	8
9	141	142	147	140	140	142.0	7
10	142	137	145	140	132	139.2	13
11	137	147	142	137	135	139.6	12
12	137	146	142	142	140	141.4	9
13	142	142	139	141	142	141.2	3
14	137	145	144	137	140	140.6	8
15	144	142	143	135	144	141.6	9
16	140	132	144	145	141	140.4	13
17	137	137	142	143	141	140.0	6
18	137	142	142	145	143	141.8	8
19	142	142	143	140	135	140.4	8
20	136	142	140	139	137	138.8	6
21	142	144	140	138	143	141.4	6
22	139	146	143	140	139	141.4	7
23	140	145	142	139	137	140.6	8
24	134	147	143	141	142	141.4	13
25	138	145	141	137	141	140.4	8
26	140	145	143	144	138	142.0	7
27	145	145	137	138	140	141.0	8
Totals	...	...	...	...	...	3,797.4	233

is fairly well centered and well within range limits; therefore, the process is under control and centered. Does the process produce acceptable parts? The answer to this question appears to be "Sometimes." As shown in the individual measurements chart, a large fraction of parts fell outside the tolerance limits. It appears that the process lacks the necessary inherent capability to produce parts within specified tolerances. A first question to raise might be "Are the tolerance limits of ±0.003 in. realistic?"

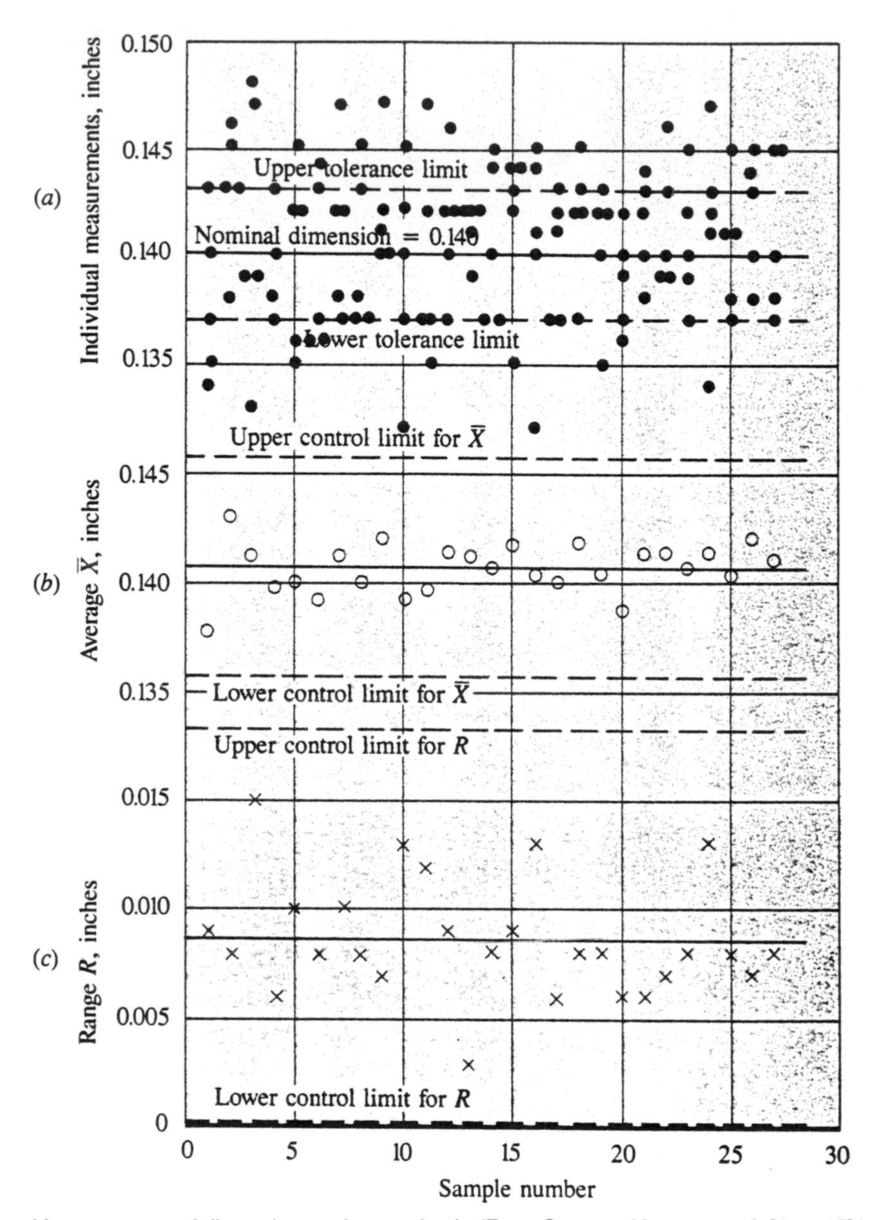

FIGURE 3-28 Measurements of dimension on rheostat knob. (From Grant and Leavenworth [4, p. 15].)

In far too many instances, tolerances are created by individuals who lack knowledge of process capabilities and who, in their ignorance, set tolerance limits that cause functionally acceptable parts to be labeled as unacceptable when they are, in fact, adequate for the intended use. Grant and Leavenworth [4] indicated that on further engineering analysis, tolerances for the rheostat knob distance were increased to +0.010 in., − 0.015 in. From Fig. 3-28*a* it is apparent that all of the individual measurements would have been acceptable at these revised tolerance limits. It should therefore be apparent that part of the job of quality control and production is to ensure an acceptable matching of machines to parts. This matching should be based on the inherent abilities of the machines to maintain dimensions relative to tolerable dimensional variability requirements of specific parts. Ideally, when the right machines are making the right parts and are "controlled" in a quality control sense, the result will be good parts coming off the machines.

When a machine simply lacks the ability to produce the desired percentage of good parts, one alternative is to simply sort good parts from bad parts as they leave the process until an improved capability becomes possible in the future. Redesign is often another alternative. Outsourcing to someone else who has better equipment is still another alternative.

The reason for plotting means of samples rather than individual values goes back to the desirable properties of the central limit theorem, discussed briefly in Chapter 2. It suffices to say that the reliable determination of control limits depends on the assumption that the means of samples are distributed normally regardless of the distribution of the individual values from which the samples were drawn.

$\overline{X}$ and R charts have been employed in statistical quality control since Shewhart [14] brought forth these concepts in the early 1920s. Over a period of half a century and on into the 1980s, statistical quality control grew to include a considerable array of statistical techniques for handling quality problems concerning manufactured products. Finding the source of quality problems was a part of the theory, but it was typically assumed that such problems in the manufacturing process could be resolved once they were discovered. The use of $\overline{X}$ and R charts, and other related statistical approaches, for finding the causes of quality problems on the production floor was unfortunately a case of "too little, too late."

There is an important point to be made here. Far more important than the most sophisticated mathematical formulations one can conceive or calculations one can perform—feats that students have been known to become intimately associated with in their engineering educational pursuits—are the guiding principles, concepts, or assumptions that lie at the heart of the practice of a profession. Quality control as practiced from the 1920s to 1980 was highly statistically based, at least in theory if not in practice. In hindsight today it is clear that although a great deal was accomplished, so much more might have been accomplished with a better understanding of the true underlying source of quality problems in industry. A review of the evolving definitions of quality offers some understanding of this gradual change in the approach to meeting more successfully the challenge of improving quality in the future.

Definitions of Quality

The general public often thinks of quality as "excellence" or "goodness." Juran [8] defined quality in 1964 as "fitness for use," which later evolved to "conformance to specifications." In 1979, Crosby [2] defined quality as "conformance to requirements" because specifications must truly reflect customer needs. In 1983, ASQC (American Society for Quality Control) [18, p. 4] defined quality as "the totality of features and characteristics of a product or service that bear on its ability to satisfy given needs." In 1986, Deming [3, p. 5], stated, "Quality should be aimed at the needs of the consumer, present and future."

Design of Experiments/Taguchi Methods

In 1986, Taguchi [9] gave the interesting definition of quality as "the loss a product causes to society after being shipped, other than any losses caused by its intrinsic functions." Reminiscent of "fitness for use," Taguchi's definition relates to long-term survivability of a product when used in its intended environment. In a sense, we have come "full circle" back to Juran's definition, with one major adjustment in perspective: focusing on unfitness for use rather than fitness for use. It has become readily apparent today that it is the customer who passes judgment on quality and that it is therefore his or her real or perceived satisfaction with a product or service that must be the primary focus. An associate of mine, when asked about the required upkeep and maintenance (one important dimension of quality) of her Toyota, responded, "I put a quart of oil in it every two to three years whether it needs it or not." She was indirectly and humorously stating quality from Taguchi's perspective. What consumers want, though they might be reluctant to admit it, is a product that will perform successfully when used and abused over long periods of use. The focus of quality control in this context then really has to address the question "How do we design and manufacture products that can continuously survive human abuse?"

The main point here is that the initial product design should take into account use and abuse by the customer and the inherent limitations and capabilities of the manufacturing process. Using highly complex statistical approaches to "fix" the product on the manufacturing floor after it has been designed is simply a case of "too little, too late," or "closing the barn door after the horse has left."

In industrial engineering environments, mathematics is typically secondary—what is fundamentally important is identifying the relevance of underlying primary guiding principles. In most instances, their consequences are known, at least in a qualitative sense. Techniques may be helpful in identifying the underlying guiding principles or essential in performing a detailed solution, and in that respect, they are important. But college students spend so much time acquiring useful techniques that they erroneously come to believe that employing the techniques constitutes the entire practice of industrial engineering. Good practice is 80 percent identifying underlying guiding principles relating to a specific situation and 20 percent employing techniques.

Quality control today is focused less on the product and more on the process. Control of the process will ensure optimization of product quality within the inherent limitations of the process. Selection of the process, however, can be fundamentally far more important than control of the process. There is a realization today that 90 percent of quality problems have as their true underlying source the design of the product. Therefore, if you want to maximize product quality, process

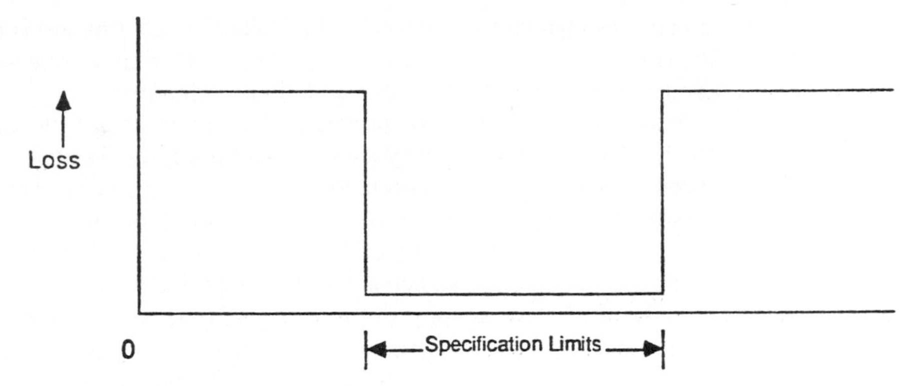

FIGURE 3-29 Traditional quality/loss function. (From Lochner and Matar [9, p. 14].)

selection must be an integral part of product design. That is why concurrent engineering, discussed in Chapter 2, is so important today as an underlying aspect of modern quality control.

Lochner and Matar [9] describe a traditional loss function (see Fig. 3-29) and an alternative loss function proposed by Taguchi (see Fig. 3-30). Having part dimensions fall within specified limits, as indicated in Fig. 3-29, assumes a finite limit between "good" and "bad" parts. However, Taguchi argues that in the machining of a part, the closer the dimensions of successive parts are to the target value, the better the performance of parts produced by that process, as indicated by the quadratic quality/loss function in Fig. 3-30. Lochner and Matar [9, p. 13] describe a Ford Motor Company example in which the same transmissions were made both at a Ford plant in the United States and on contract at a plant in Japan. All parts for the transmissions at both locations met all applicable specifications, but the Japanese transmissions performed better, based on a summarization of subsequent warranty data. Under examination it was found that the Japanese parts were closer to target

FIGURE 3-30 Quadratic (Taguchi) quality/loss function. (From Lochner and Matar [9, p. 15].)

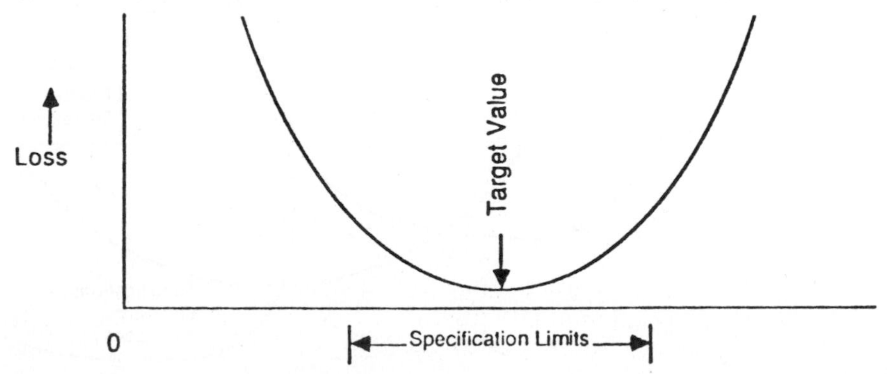

dimensions than those produced in the United States. Anyone familiar with the consequences of tolerance buildup in a product such as a transmission can readily appreciate the importance of this underlying principle.

Figure 3-31 illustrates the production/quality system cycle, a "birth" to "death" holistic approach to quality control proposed by Taguchi. This approach is becoming a major part of current thinking in today's quality control circles. Taguchi uses the term *off-line quality system* to refer to all of those functions prior to production that affect the quality of the product. As mentioned earlier, these functions are the likely source of 90 percent of quality problems in traditional quality control.

In Taguchi's terminology, the *on-line quality system* involves manufacture of the product, warehousing, delivery, and both after-sale service and all other contact with the customer. Upon analysis of Fig. 3-31, it becomes obvious that attempting to control quality by fixing quality problems as they become known in the factory can be only partially successful in providing what the customer perceives as a quality product.

When all the product and process design is complete and the product is manufactured, shipped, stored, and finally sold, it takes only one "surly and uncoopera-

FIGURE 3-31 Production/quality system cycle. (From Lochner and Matar [9, p. 16].)

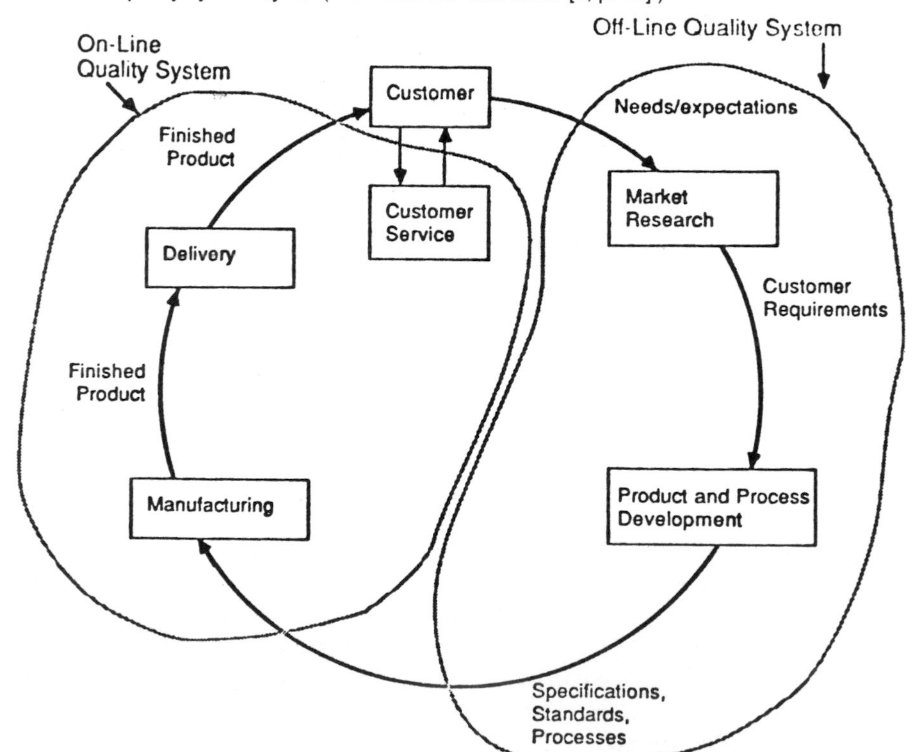

tive" service representative who, say, does not hear the rattle in the back of the car, to lead the customer to the final perception that the product is a "lemon." Tightening a bolt on the license plate, for example, might make a considerable difference in how that customer perceives the product.

Figure 3-32 describes Taguchi's overall approach to quality control. Note in Fig. 3-32 that process design is separate from product design. In a practical sense, there will probably always be some functional separation between process design (i.e., factory people) and product design (i.e., engineering people). Concurrent engineering (i.e., team design) can greatly eliminate this barrier.

Note also that Taguchi's methodology has been developing during a period in which concurrent engineering has also been developing as an organizational concept. One would hope that, to the extent possible, both product design and process design will be simultaneously performed as a team effort, with the goal of developing the very best combined product and process designs. More will be said about this in Chapter 10.

The preponderant technical approach (i.e., statistical approach) that Taguchi uses is called *design of experiments.* Design of experiments is by no means a new approach. As used in the United States for many years, it has been primarily employed as a means for determining machine settings or levels of ingredients for attaining optimum average product performance. Taguchi has been employing it for determining robust product designs, as well.

The concept of a "robust" product design, so fundamental to Taguchi's approach, is described in the following quotes from Lochner and Matar:

> The goal is to set controllable factors at levels which make the product robust with respect to noise factors. Examples of robust design include an automobile part which can withstand shock and vibration (external noise), or a food product with a long shelf life (internal noise), or a replacement part that will fit properly (unit-to-unit noise). [9, p. 18]

> Dr. Taguchi uses experimental designs primarily as a tool to make products more *robust*—to make them less sensitive to noise factors. That is, he views experimental design as a tool for reducing the effects of variation on product and process quality characteristics. [9, p. 20]

As an example of a design-of-experiments study, Lochner and Matar [9, pp. 54–61] describe a study, performed at the Charter Manufacturing Company, that concerned three ingredients—cleaner, copper, and zinc—used in zinc plating. The recurring problem was that the plating thickness was occasionally too thin. The typical action taken to resolve the problem was to increase the levels of ingredients in the standard recipe for the plating process. The result, however, was increased plating costs, with a continuing problem of inadequate plating thickness.

A technical understanding of this case would require a level of educational preparation in statistics beyond the scope of this text, so only limited technical results will be included in this discussion.

An eight-run (i.e., 2^3), two-level design-of-experiments study was designed and performed. Two levels of each of the three ingredients were employed. Existing

OFF-LINE QUALITY CONTROL	Stage 1: PRODUCT DESIGN	**Concerns:** 1. Identify customer needs and expectations 2. Design a product to meet customer needs and expectations 3. Design a product which can be consistently and economically manufactured	**QA Steps:** 1. System Design 2. Parameter Design 3. Tolerance Design
	Stage 2: PROCESS DESIGN	**Concerns:** 1. Develop clear and adequate specification standards, procedures and equipment for manufacture	**QA Steps:** 1. System Design 2. Parameter Design 3. Tolerance Design
ON-LINE QUALITY CONTROL	Stage 1: PRODUCTION	**Concerns:** 1. Manufacture products within specifications established during product design using procedures developed during Process Design	**Form 1:** Process Diagnosis and Adjustment **Form 2:** Prediction and Correction **Form 3:** Measurement and Action
	Stage 2: CUSTOMER RELATIONS	**Concerns:** 1. Provide service to customers and use information on field problems to improve product and manufacturing process designs	**Actions:** 1. Repair, replacement or refund 2. Feed back information on field problems 3. Change product and process specifications/design

FIGURE 3-32 Taguchi's quality system. (From Lochner and Matar [9, p. 17].)

TABLE 3-13 VERIFICATION RUN DATA

Run number	Average thickness
1	0.000593
2	0.000656
3	0.000685
4	0.000700
5	0.000083
6	0.000088
7	0.000593
8	0.000578

Source: Lochner and Matar [9, p. 61].

recipe levels were utilized to represent high levels, and lower levels of ingredients were employed for all three ingredients to represent the other (lower) levels. The conclusion of the study was to recommend lower levels of both cleaner and copper, and existing (higher) levels of zinc. Verification runs were done, and the resulting run-average thicknesses are indicated in Table 3-13 and plotted in Fig. 3-33.

A final report by the engineer at the plant indicated that the new settings resulted in "substantial cost savings." Of additional interest, however, were verification runs 5 and 6. Investigation revealed that an automatic method of tumbling had been employed only on those two runs. It was recommended that until this effect could be further explored, the use of automatic tumbling should be discontinued.

FIGURE 3-33 Time plot of verification runs for mechanical plating equipment. (From Lochner and Matar [9, p. 61].)

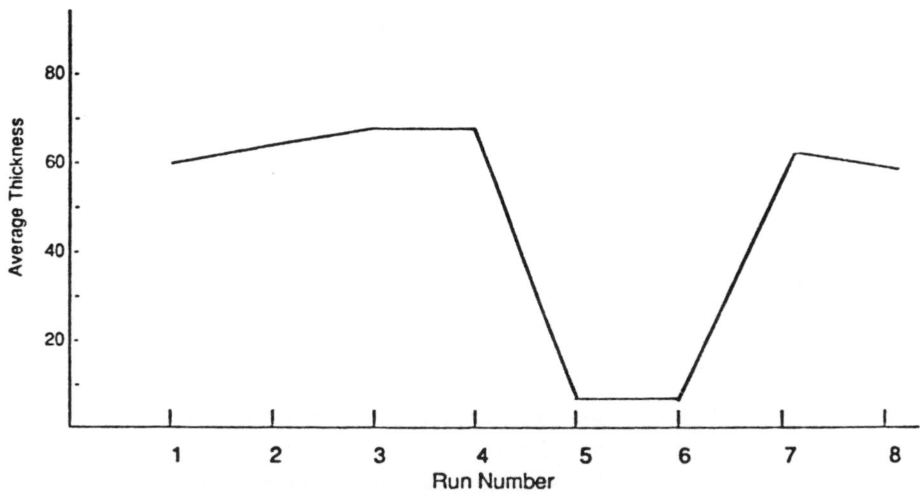

Sampling Plans

For both cost and reliability reasons, statistical sampling has been commonly employed for evaluating the quality of incoming materials. There has been considerable development over the years, particularly in the military, with respect to sampling plans. Table 3-14 demonstrates the use of a typical double sampling plan. It must first be assumed that of all sampling plans, this particular one is most appropriate for the sampling task to be performed. Assume that a batch of 1000 units is to be sampled to determine whether they meet the acceptable quality level (AQL) of 4.0, which means that if no more than 4 percent of the material is defective, the batch is to be considered acceptable. Assume that 1st and 2nd sample sizes of 50 each are to be taken. Examination of Table 3-14 indicates that for sample size code J and an AQL of 4, the accept-reject values for a first sample of size 50 are 3 and 7, respectively. If inspection of the random sample of 50 parts from the batch produces 3 or fewer defects, the batch is accepted; if 7 or more defects are found the batch is rejected; and if 4 to 6 defects are found, the sample size should be increased by inspecting 50 more parts. Based on the total of 100 parts inspected, if 8 or fewer defects are found the batch is accepted; if 9 or more defects are found, the batch is rejected. This demonstrates how a typical double sampling plan is employed in practice.

The emphasis in statistical quality control or quality assurance is not only to evaluate the quality of the manufactured product but also to effectively employ these data to search out the root cause of unacceptable product and correct such undesirable conditions where feasible. Whereas inspection merely sorts good parts from bad, quality control is concerned with identifying and correcting the source of unacceptable products.

Japanese managers have introduced many changes in the practice of quality control in recent years. The statistical sampling plans so commonly used in the United States for many years (e.g., MIL-STD-105D) are being discarded in many instances. The Japanese philosophy rejects the use of 4 percent or even 1 percent, for example, as acceptable quality levels (AQLs). The Japanese often prefer automatic (i.e., "go–no go" machine-processed) 100 percent inspection as an integral step during an operation to ensure that poor-quality parts are not allowed to enter an assembly process and contaminate the assemblies of which they become a part. When 1 percent rejection levels are reached, the Japanese then proceed to parts per million (PPM) as the appropriate dimension for measuring continuous quality improvement. Some plants in Japan and the United States are now measuring defects in the parts per billion (PPB) range.

So much emphasis has been placed on quality in the United States in recent years that it is now difficult to distinguish between quality management and management in most firms. In fact, quality has become such a common, revered, and almost singular goal of companies that "quality" in some instances appears to have become an end in itself. At that point, the trend toward quality control as *the* "silver bullet" has probably gone too far. In the final analysis, to enjoy long-term success, a company cannot have just one silver bullet; rather, it must have at least three silver bullets, thereby simultaneously producing (1) the right (i.e., of proper quality) product

TABLE 3-14 MASTER TABLE FOR NORMAL INSPECTION (DOUBLE SAMPLING)

Table *O* Master table for normal inspection (double sampling)—MIL-STD-105D (ABC standard)

Sample size code letter	Sample	Sample size	Cumulative sample size	Acceptable Quality Levels (normal inspection)																									
				0.010	0.015	0.025	0.040	0.065	0.10	0.15	0.25	0.40	0.65	1.0	1.5	2.5	4.0	6.5	10	15	25	40	65	100	150	250	400	650	1,000
				Ac Re	Ac Re	Ac Re	Ac Re	Ac Re	Ac Re	Ac Re	Ac Re	Ac Re	Ac Re	Ac Re	Ac Re	Ac Re	Ac Re	Ac Re	Ac Re	Ac Re	Ac Re	Ac Re	Ac Re	Ac Re	Ac Re	Ac Re	Ac Re	Ac Re	Ac Re
A				↓	↓	↓	↓	↓	↓	↓	↓	↓	↓	↓	↓	↓	↓	†	↓	↓	†	†	†	†	†	†	†	†	†
B	First	2	2	↓	↓	↓	↓	↓	↓	↓	↓	↓	↓	↓	↓	↓	†	↑	↓	0 2	0 3	1 4	2 5	3 7	5 9	7 11	11 16	17 22	25 31
	Second	2	4	↓	↓	↓	↓	↓	↓	↓	↓	↓	↓	↓	↓	↓		↑	↓	1 2	3 4	4 5	6 7	8 9	12 13	18 19	26 27	37 38	56 57
C	First	3	3	↓	↓	↓	↓	↓	↓	↓	↓	↓	↓	↓	↓	†	↑	↓	0 2	0 3	1 4	2 5	3 7	5 9	7 11	11 16	17 22	25 31	↑
	Second	3	6	↓	↓	↓	↓	↓	↓	↓	↓	↓	↓	↓	↓		↑	↓	1 2	3 4	4 5	6 7	8 9	12 13	18 19	26 27	37 38	56 57	↑
D	First	5	5	↓	↓	↓	↓	↓	↓	↓	↓	↓	↓	↓	†	↑	↓	0 2	0 3	1 4	2 5	3 7	5 9	7 11	11 16	17 22	25 31	↑	↑
	Second	5	10	↓	↓	↓	↓	↓	↓	↓	↓	↓	↓	↓		↑	↓	1 2	3 4	4 5	6 7	8 9	12 13	18 19	26 27	37 38	56 57	↑	↑
E	First	8	8	↓	↓	↓	↓	↓	↓	↓	↓	↓	↓	†	↑	↓	0 2	0 3	1 4	2 5	3 7	5 9	7 11	11 16	17 22	25 31	↑	↑	↑
	Second	8	16	↓	↓	↓	↓	↓	↓	↓	↓	↓	↓		↑	↓	1 2	3 4	4 5	6 7	8 9	12 13	18 19	26 27	37 38	56 57	↑	↑	↑
F	First	13	13	↓	↓	↓	↓	↓	↓	↓	↓	↓	†	↑	↓	0 2	0 3	1 4	2 5	3 7	5 9	7 11	11 16	↑	↑	↑	↑	↑	↑
	Second	13	26	↓	↓	↓	↓	↓	↓	↓	↓	↓		↑	↓	1 2	3 4	4 5	6 7	8 9	12 13	18 19	26 27	↑	↑	↑	↑	↑	↑
G	First	20	20	↓	↓	↓	↓	↓	↓	↓	↓	†	↑	↓	0 2	0 3	1 4	2 5	3 7	5 9	7 11	11 16	↑	↑	↑	↑	↑	↑	↑
	Second	20	40	↓	↓	↓	↓	↓	↓	↓	↓		↑	↓	1 2	3 4	4 5	6 7	8 9	12 13	18 19	26 27	↑	↑	↑	↑	↑	↑	↑
H	First	32	32	↓	↓	↓	↓	↓	↓	↓	†	↑	↓	0 2	0 3	1 4	2 5	3 7	5 9	7 11	11 16	↑	↑	↑	↑	↑	↑	↑	↑
	Second	32	64	↓	↓	↓	↓	↓	↓	↓		↑	↓	1 2	3 4	4 5	6 7	8 9	12 13	18 19	26 27	↑	↑	↑	↑	↑	↑	↑	↑
J	First	50	50	↓	↓	↓	↓	↓	↓	†	↑	↓	0 2	0 3	1 4	2 5	3 7	5 9	7 11	11 16	↑	↑	↑	↑	↑	↑	↑	↑	↑
	Second	50	100	↓	↓	↓	↓	↓	↓		↑	↓	1 2	3 4	4 5	6 7	8 9	12 13	18 19	26 27	↑	↑	↑	↑	↑	↑	↑	↑	↑
K	First	80	80	↓	↓	↓	↓	↓	†	↑	↓	0 2	0 3	1 4	2 5	3 7	5 9	7 11	11 16	↑	↑	↑	↑	↑	↑	↑	↑	↑	↑
	Second	80	160	↓	↓	↓	↓	↓		↑	↓	1 2	3 4	4 5	6 7	8 9	12 13	18 19	26 27	↑	↑	↑	↑	↑	↑	↑	↑	↑	↑
L	First	125	125	↓	↓	↓	↓	†	↑	↓	0 2	0 3	1 4	2 5	3 7	5 9	7 11	11 16	↑	↑	↑	↑	↑	↑	↑	↑	↑	↑	↑
	Second	125	250	↓	↓	↓	↓		↑	↓	1 2	3 4	4 5	6 7	8 9	12 13	18 19	26 27	↑	↑	↑	↑	↑	↑	↑	↑	↑	↑	↑
M	First	200	200	↓	↓	↓	†	↑	↓	0 2	0 3	1 4	2 5	3 7	5 9	7 11	11 16	↑	↑	↑	↑	↑	↑	↑	↑	↑	↑	↑	↑
	Second	200	400	↓	↓	↓		↑	↓	1 2	3 4	4 5	6 7	8 9	12 13	18 19	26 27	↑	↑	↑	↑	↑	↑	↑	↑	↑	↑	↑	↑
N	First	315	315	↓	↓	†	↑	↓	0 2	0 3	1 4	2 5	3 7	5 9	7 11	11 16	↑	↑	↑	↑	↑	↑	↑	↑	↑	↑	↑	↑	↑
	Second	315	630	↓	↓		↑	↓	1 2	3 4	4 5	6 7	8 9	12 13	18 19	26 27	↑	↑	↑	↑	↑	↑	↑	↑	↑	↑	↑	↑	↑
P	First	500	500	↓	†	↑	↓	0 2	0 3	1 4	2 5	3 7	5 9	7 11	11 16	↑	↑	↑	↑	↑	↑	↑	↑	↑	↑	↑	↑	↑	↑
	Second	500	1,000	↓		↑	↓	1 2	3 4	4 5	6 7	8 9	12 13	18 19	26 27	↑	↑	↑	↑	↑	↑	↑	↑	↑	↑	↑	↑	↑	↑
Q	First	800	800	†	↑	↓	0 2	0 3	1 4	2 5	3 7	5 9	7 11	11 16	↑	↑	↑	↑	↑	↑	↑	↑	↑	↑	↑	↑	↑	↑	↑
	Second	800	1,600		↑	↓	1 2	3 4	4 5	6 7	8 9	12 13	18 19	26 27	↑	↑	↑	↑	↑	↑	↑	↑	↑	↑	↑	↑	↑	↑	↑
R	First	1,250	1,250	↑	↑	0 2	0 3	1 4	2 5	3 7	5 9	7 11	11 16	↑	↑	↑	↑	↑	↑	↑	↑	↑	↑	↑	↑	↑	↑	↑	↑
	Second	1,250	2,500	↑	↑	1 2	3 4	4 5	6 7	8 9	12 13	18 19	26 27	↑	↑	↑	↑	↑	↑	↑	↑	↑	↑	↑	↑	↑	↑	↑	↑

↓ = use first sampling plan below arrow. If sample size equals or exceeds lot or batch size, do 100% inspection.

↑ = use first sampling plan above arrow.

Ac = acceptance number.

Re = rejection number.

†Use corresponding single sampling plan (or alternatively, use double sampling plan below, where available).

Source: MIL-STD-IO5 D (ABC Standard).

at (2) the right time (i.e., on the right schedule), and at (3) the right price (i.e., at low relative cost). More will be said about quality control, as an integral part of management philosophy, in Chapter 5.

Lochner and Matar [9, pp. 215–216] offer their thoughts as to the improvements needed in the United States concerning the application of quality control.

Engineering Design and Quality Improvement

Robert H. Lochner and Joseph E. Matar

The Quality profession has come a long way since the days when the Quality Control Department was headed by a "Chief Inspector." But changes in Western perception of quality have not come fast enough. In the 1950s Japanese engineers and managers traveled to the United States to learn how to improve the quality of manufactured products. Today American engineers and managers are visiting Japan to learn the secrets of Japanese quality. But there really are no "secrets." There are just some basic principles which must be followed and tools which must be learned, such as:

- Management cannot be left to the managers. Everyone in the company is a manager of one or more processes, and those closest to a process should participate in its management.
- Random variability is present in all processes. Engineering design and control methods which fail to take randomness in measurements into account lead to out-of-specification products and high production costs.
- Markets cannot be held through use of the latest technology alone. Today customers want proven, reliable, low-cost products designed to meet their needs. Companies must become champions of their customers.
- Experimentation belongs on the manufacturing floor as well as in the research lab. It is naive to assume that the production environment can be exactly reproduced in the lab or that production people cannot be taught methods of experimentation.
- Quality by inspection is no longer a competitive option. Use of Statistical Process Control as the primary method of assuring quality will delay but not prevent a company's demise. The only way to increase market share and profits is by designing quality into all products and processes.

The United States used to loan money to the world. Now it is the largest debtor nation in the world. Today the per capita GNP and wages paid to U.S. workers lag significantly behind those in Japan. In 1982 the United States machine tool business was the biggest in the world. By 1988 the U.S. had sunk to fifth. Many people blame Japan for these dramatic changes. Complaints about the "playing field not being level" are frequently voiced. But on closer scrutiny the problems can be found a little closer to home. The message is clear—many products produced in Western countries cannot compete in world markets because they do not meet customer expectations for quality, cost, and performance. And they will not meet those expectations if Western countries continue to depend on quality control departments for product and process quality. It is time for action:

- Management must reevaluate the way in which they manage. They must provide ongoing leadership in the quality improvement effort. A major obstacle to quality in most companies is a rigid top-down management system which does not encourage process improvement. Dr. Deming's *Out of the Crisis* is required reading.
- Processes must be standardized, evaluated for capability to meet requirements, and brought under statistical control. Statistically designed experiments must be used to improve processes.

- Everyone must be trained in team building, a structured approach to process improvement, and tools of data analysis. Employees must become involved in managing and improving processes.

An active quality program focused on designing quality and performance into products and processes is essential for companies which hope to survive into the twenty-first century.

SUMMARY

Inventory, production, and quality control are all part of the production system control responsibility. First a plant is designed, as discussed in Chapter 2, and then procedures are devised for operating and controlling the operation in an optimal manner. Inventory control, production control, work measurement, and quality control all originated back in the 1920s or earlier. EOQ and EPQ were early mathematical formulations that considered some of the important costs associated with inventory, but less mathematically reducible aspects, such as immediacy of quality feedback effects, were not represented in these early formulations. The use of JIT in recent times has demonstrated that these earlier formulations were therefore significantly incomplete and misleading.

Most manufacturing firms learned to incorporate MRP systems to keep track of lead times and assembly builds. With the introduction of JIT systems, so much less inventory is work-in-process and is on the production floor for such a short time that, as a component of overall inventories, work-in-process inventory has typically been greatly reduced or even no longer exists as a material category.

The development of the kanban system by Toyota and the concentration of effort on the reduction of setup time by such innovative thinkers as Shigeo Shingo have greatly enhanced the benefits of adopting a JIT philosophy. Reducing the setup time is now seen as a productivity opportunity, whereas in the past it was typically ignored—something we wanted to do but never got around to doing.

Quality control and quality assurance promised but often failed to deliver in identifying quality problems at the source. It was never a production responsibility, and so, as a staff function, it rarely got the resources it needed or the attention it deserved for doing the job right. Making quality a production responsibility has focused increased attention on quality needs, and making the production team take responsibility for identifying and resolving quality issues has greatly enhanced the team's effectiveness. The quality control department, a technical staff resource, can now be effectively employed as the technical means of assisting production productivity and of helping quality teams employ the technology needed to resolve problems (i.e., opportunities).

Taguchi techniques, a technical area that for years in the United States has been called *design-of-experiments,* is a fairly technically demanding area that needs to be applied to production problems by those with sufficient statistical training. A partnership is needed between those in production who have the problems and those skilled in statistical analysis (e.g., Taguchi methods) to effectively capitalize on these opportunities.

In recent years, some mathematical techniques have been developed that concentrate, in particular, on identifying optimal procedures. Operations research, the topic of Chapter 7, can be thought of as a subdiscipline concerned primarily with the application of mathematical optimization procedures and approaches to productive systems.

REFERENCES

1 "ASTM Manual on Presentation of Data and Control Chart Analysis," STP15D, 1976, p. 134.

2 Crosby, Philip B.: *Quality Is Free: The Art of Making Quality Certain,* McGraw-Hill Book Company, New York, 1979.

3 Deming, W. Edwards: *Out of the Crisis,* Massachusetts Institute of Technology, Cambridge, MA, 1986.

4 Grant, Eugene, and Richard Leavenworth: *Statistical Quality Control,* 6th ed., McGraw-Hill Book Company, New York, 1988.

5 "Hallmark's Brainy Warehouse," *Fortune,* pp. 91–93, August, 1973.

6 Harmon, Craig K., and Russ Adams: *Reading Between the Lines: An Introduction to Bar Code Technology,* Helmers Publishing, Inc., Peterborough, NH, 1989.

7 Harrison, Jack: Hands-on-JIT Company, P.O. Box 26, Gotha, FL, 32734.

8 Juran, J. M.: *Juran on Planning for Quality,* The Free Press, New York, 1988.

9 Lochner, Robert H., and Joseph E. Matar: *Designing for Quality: An Introduction to the Best of Taguchi and Western Methods of Statistical Experimental Design,* Quality Resources, New York, 1990.

10 Malcolm, D. G., J. H. Rosenbloom, C. E. Clark, and W. Fazar: "Applications of a Technique for R&D Program Evaluation (PERT)," *Operations Research,* vol. 7, no. 5, pp. 646–669, 1959.

11 Moodie, C. L., and H. H. Young: "A Heuristic Method of Assembly Line Balancing for Assumptions of Constant or Variable Work Elemental Times," *Journal of Industrial Engineering,* vol. 16, no. 1, January–February, 1965.

12 Ohno, Taiichi: *Workplace Management,* Productivity Press, Cambridge, MA, 1988.

13 Schonberger, Richard J.: *Japanese Manufacturing Techniques,* The Free Press, New York, 1982.

14 Shewhart, Walter A.: *Economic Control of Quality of Manufactured Product,* D. Van Nostrand Co., New York, 1931.

15 Shingo, Shigeo: *A Revolution in Manufacturing: The SMED System,* Productivity Press, Cambridge, MA, 1985.

16 Vollman, Thomas E., et al., *Manufacturing Planning and Control Systems,* 2d ed, Irwin, Homewood, IL, 1988.

17 Walker, M. R., and J. Sayer: "Project Planning and Scheduling," Report no. 6959, E. I. DuPont de Nemours, Inc., Wilmington, DE, March, 1959.

18 American Society for Quality Control: *Glossary and tables for statistical quality control,* Milwaukee, WI: ASQC, 1983.

REVIEW QUESTIONS AND PROBLEMS

1 On average, do progressive plants today carry more or less inventory than they did 20 years ago?

2 What factors unique to Japan make the storing of materials there particularly unattractive?

3 What is the fundamental underlying purpose of maintaining an inventory?

4 **a** Using the Harris model, determine the EOQ for the following data:

$$A = 10{,}000 \text{ units}$$
$$H = \$5$$
$$P = \$10$$
$$M = \$4$$

b What is the optimum number of purchases per year, N_0?

c Plot a figure similar to Fig. 3-2 for the above data.

5 Assuming that the following data represent the number of days of lead time L that occurred on the last 50 orders that were placed, and using the fifth and sixth columns of Table 3-3 from the top down, develop a sequence of 10 typical lead times using the Monte Carlo technique.

Days of lead time L	Frequency
3	20
4	17
5	6
6	4
7	3
	50

6 Is production control responsible for producing the required quantities of products indicated on the company production plan?

7 Is production responsible for ensuring that the required equipment to meet forecasted production demands is available when it must be utilized to meet the company production plan?

8 Determine the optimum lot size to run on a machine for the following data:

$$A = 1000 \text{ units}$$
$$S = \$200$$
$$H = \$4$$
$$D = 10 \text{ units/day}$$
$$P = 50 \text{ units/day}$$

9 Who invented the bar code?

10 Should employees be permitted to perform tasks that can be more effectively performed by a computer, so that they will have something important to do (thereby protecting their egos)?

11 Explain the difference between the front end and the back end of an MPC (manufacturing, planning, and control) process? What is the engine of this process?

12 Why is backflushing with a manufacturing bill of material superior to backflushing with a product engineering bill of material?

13 "What you need, only in the quantity you need, when you need it … and as inexpensively as you can" is a famous quote in JIT manufacturing. Who made it, and what does it mean?

14 What is the primary benefit typically derived from adoption of a JIT production philosophy?

15 Explain "inventory = cycle time."

16 In the lake–inventory analogy, why is it important to lower the lake and expose the rocks? What happens if the lake is not lowered (i.e., inventory levels are allowed to stay the same)?

17 Assume that a revision in the process shown in Fig. 3-16 results in a doubling of the element time for element 10 from 7 to 14 min. Can a cycle time of 8 min. as found to be optimum for this process, still be employed in the future? If so, how?

18 Explain the phrase "Lead times are self-fulfilling prophecies."

19 Assume that 10,000 min. of production time is available to produce 1000 units on a production line. The precedence relationships and times for the 16 production elements are as indicated in Fig. 3-34. Determine station assignments of elements that will result in an optimum number of stations on the production line.

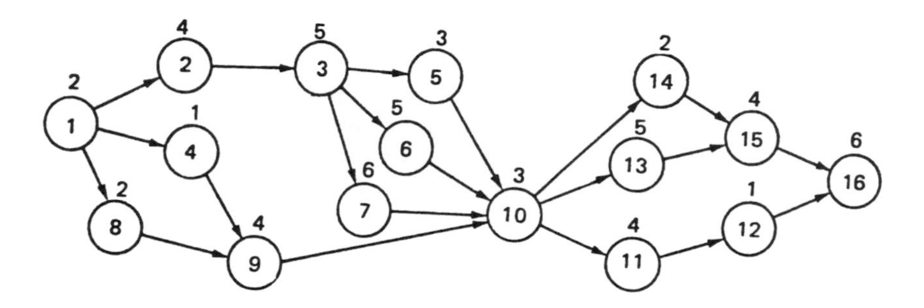

FIGURE 3-34

20 **a** The activities and their times and precedence relationships are indicated below for building and testing a boat.

Activity symbol	Activity	Time, days	Precedence
A	Construct hull	10	
B	Construct cross beams	4	A
C	Attach transom	4	A
D	Lay the deck	5	B, C
E	Install the mast	2	B
F	Paint the boat	4	D
G	Install hardware	5	F
H	Install the engine	2	F
I	Test for leaks	2	H
J	Trial run	1	E, G, I

Which activities are on the critical path?

b If construction of cross beams is delayed a week, what effect will it have on the overall completion time?

21 Develop your definition of "quality," and then explain it.

22 What does Taguchi mean by a "robust design"? How can the design-of-experiments approach help in achieving robust designs?

23 What is an AQL? What do PPM and PPB mean? Compare the newer PPM and PPB measures of quality with the previous AQL measure; has the emphasis shifted toward producing more acceptable parts and assemblies today?

LABORATORY EXERCISE 9: The JIT House Project*

This project involves the in-plant production of houses, utilizing three different production and marketing assumption sets: cases A, B, and C, as follows:

Case A: A house is produced by means of a traditional production approach, and customers can order one of two house colors. The color ordered applies to both the roof and the base of the house.

Case B: A house is produced by means of a JIT approach to production. The colors are the same as in case A above.

Case C: The available roof and base colors are increased to seven colors each. The number of color combinations is increased to 49 (i.e., $7 \times 7 = 49$). This case employs a JIT approach to production.

Figures 3-35 and 3-36 are the product designs for the base and roof of the house. The single solid lines are the outer dimensions of the cut card stock. Double solid lines are cut lines, and quality control specifications require that cuts remain within these lines. A dotted line is a fold line. The quality control specifications require that folds be made within 1⁄16 in. of these lines.

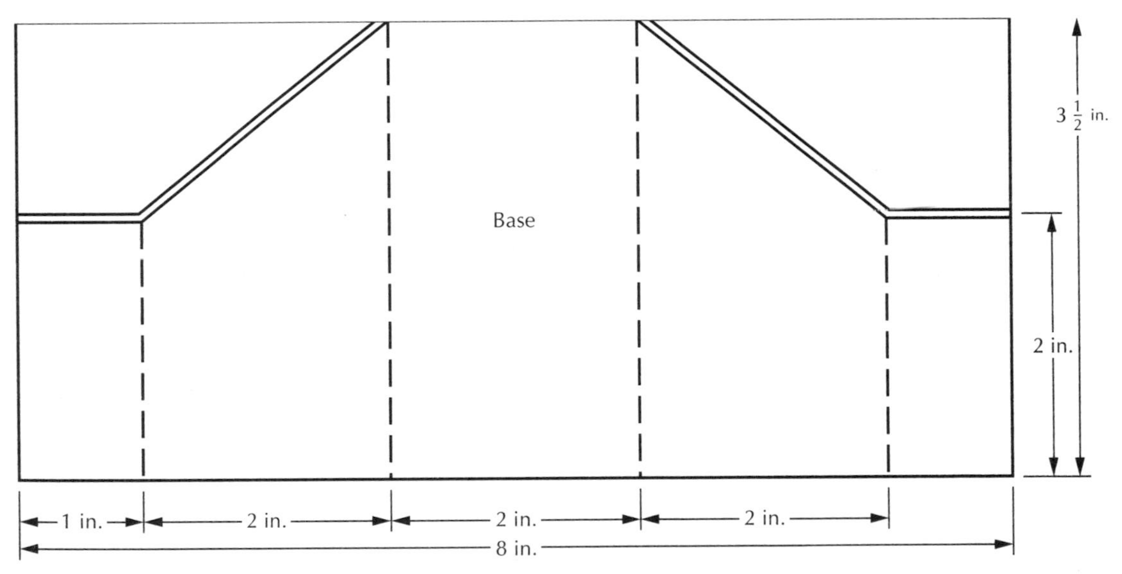

FIGURE 3-35 The house base product design.

*This exercise is modeled after a similar case developed by Jack Harrison, Hands-on-JIT, PO Box 26, Gotha, Florida, 32734.

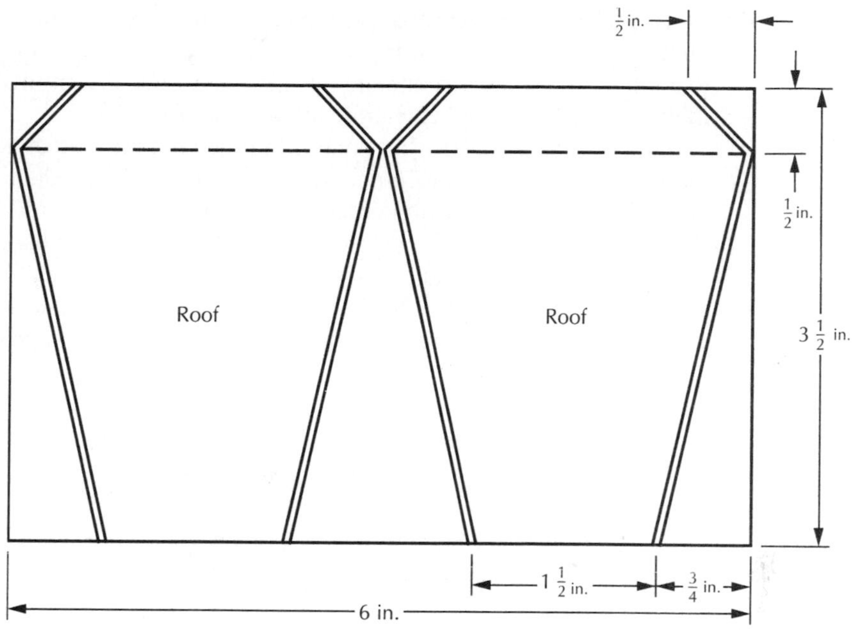

FIGURE 3-36 The house roof product design.

Case A: traditional manufacturing (two colors)

The laboratory instructor should select one student each to fill the following roles for this case:

1 Base cutter
2 Base assembler
3 Roof cutter
4 Final assembler
5 Warehouse manager
6 Customer A
7 Customer B
8 Sales manager
9 Production supervisor
10 Production planner
11 Material handler
12 Quality control inspector
13 A case timer

This case is initiated on the assumption that the sales manager has received a sales order from both customers A and B. Customer A has ordered a green house, and customer B has ordered a gray house.

The sales orders were transmitted to the production planner to plan for the production of the two houses. Because of the setup times for shifting from one color to another in the plant, economic lot sizes have been determined for each operation in the plant. Assume all economic lot sizes are four.

The production planner has prepared production orders that he or she can issue as required to initiate various production operations. The production planner has prepared four sets of production orders, with each production order on a 3×5 card, for each of the following four types of orders: (1) base cut, (2) base fold and assemble, (3) roof cut and fold, and (4) final assemble. That is, there are four card types for each color. The card color should match its stated color. Three sets of each card type and color should be prepared. A sample production order appears as follows:

PRODUCTION ORDER
Base cut
Green
Lot size: 4

The warehouse manager utilizes an inventory record (see Fig. 3-37) to keep track of his or her inventory.

The production planner hands production orders one at a time to the material handler, who moves all material between the warehouse and production operations

INVENTORY RECORD

Base Cut-Out			Base Assembly			Roof		
In	Out	Balance	In	Out	Balance	In	Out	Balance

FIGURE 3-37 Inventory record.

and between production operations. The warehouse manager cannot issue material to the material handler without first examining the production order that the handler brings.

The instructor should select the first five players in this case—the four direct employees and the warehouse manager—from fairly distributed locations in the room in which this case is being performed.

The tools issued for each of the three cases are (1) scissors for the base cutter, (2) tape for the base assembler, (3) scissors for the roof cutter and folder, and (4) tape for the final assembler.

When the case timer begins the case, the production planner should give an order to the material handler, who then goes to the warehouse to get the material indicated in the production order and delivers it to the worker location for that material. As workers complete their required production tasks, they should inform their production supervisor that they have completed a production order. The production supervisor should then inform the production planner that the work has been completed, and the planner then tells the material handler to move the completed order to the warehouse, or to the customer in the case of the final assembly. A maximum of one additional work order can be issued and given to a production worker by the material handler in addition to the work order that the worker is presently attempting to complete.

The material handler takes his or her orders from the production planner. Cut bases, folded and assembled bases, and cut and folded roofs are all inventoried in the warehouse following their completion. They are issued to the final assembly operation by means of a final assembly work order, following withdrawal of the assemblies (base and roof) from the warehouse by the material handler.

During the production period for the houses, if either of the two customers feel that they have been waiting long enough for receipt of their ordered houses, they should contact the sales manager with whom they placed their orders and ask him or her to expedite their order. The sales manager may then choose to contact the production supervisor and inquire as to expected completion of the houses. The production supervisor may then choose to exert some influence in an attempt to expedite completion of the houses.

Throughout the production period, the quality control inspector should monitor the quality of the product, remove defective material, and pass information as to noncompliance of quality standards to the production supervisor, so that he or she can direct employees to improve their quality performance.

When both customers have received their houses, the case timer should note the elapsed time for case A. Information concerning case A should be recorded in Fig. 3-38, the case summary.

Case B: JIT manufacturing (two colors)

The first four players in the previous case should continue in their role—the four direct employees. The instructor should ask the four direct employees to relocate

Case summary			
	Case A	**Case B**	**Case C**
Production time (two houses)			
Inventory[1]			
Employees/space[2]			
Product choices			

[1]Remaining house pieces or assemblies in warehouse or at workstations.
[2]Count one for each plant person involved in this production effort.

FIGURE 3-38 Case Summary.

together, so that the base cutter can pass his or her completed product to the base assembler and so that the base assembler and the roof cutter can both pass their completed work directly to the final assembler. One green card and one gray card should be placed, turned over, on the table between the base cutter and the base assembler to act as a kanban between them. The base cutter's instruction is to cut a base and place it on an empty kanban of the same color when one becomes empty. Two green cards and two gray cards should be set next to the final assembler. One green card should be marked "base" and the other "roof." Two gray cards should be marked "base" and "roof" as well. Instructions for the base assembler and the roof assembler are to produce a finished assembly (roof or base) of the same color as the empty kanban and to place it on the final assembler's kanban when a space becomes empty.

The material handler's job now involves keeping sufficient raw material at each production location initiating a base or roof. Therefore, prior to timing the case, the material handler should place an adequate supply of raw material at each workstation; this is called *point-of-use storage,* and in practice the material may actually be delivered to the workstation by the material supplier (e.g., seats are delivered at Kawasaki by the seat supplier [13]).

When the case timer indicates the start time, the base and roof cutters start their work, and two finished houses should eventually be delivered to the customers by the material handler. When customer A and customer B receive their respective houses, the elapsed time should be noted by the case timer. The resulting case B information should be recorded in the case summary.

Case C: JIT manufacturing (49 colors)

In this case, the sales manager should ask each of five customers what base and roof colors they would like on their individual houses. There are seven different colors to choose from for both the base and the roof. The sales manager should list the houses ordered on a piece of paper (the base color on the left and roof color on the right). Once the five houses are listed, the sales manager should provide a duplicate copy to the final assembler and then tear the original sheet of paper in half, handing the base color list to the base cutter and the roof color list to the roof cutter. Both cutters should produce product in the sequence indicated on their respective color lists.

The base and roof workstations should be provided raw material prior to initiating timing of the case, as was done in case B. When the two first houses ordered are completed, they should be delivered to the customers by the material handler, and the elapsed time should be recorded. The information for case C should be recorded in the case summary.

Discuss the results. How much elapsed time would it take to deliver the first two houses ordered, assuming the availability of 49 color combinations and employing the traditional production approach? How much in-process inventory would be on hand at the point in time when the two houses are delivered?

4

MANAGEMENT

We have met the enemy, and he is us.

Pogo (Walt Kelly)

Design, control, and management represent an essential triad of interests in any comprehensive effort to improve operations. Chapters 2 and 3 dealt with the design and control of productive systems. This chapter focuses on the management of productive systems. For any operation to be truly effective, all of the following must simultaneously be in place: (1) a physical production system with the capability of being fully productive (production system design); (2) optimal procedures for providing operational control of the physical production system (production system control); and (3) a management system fully capable of providing effective management of the physical, control, human, and all other resources of the operation (production systems management).

Industrial engineers have traditionally designed production systems and the controls for those systems, but they have not typically borne responsibility for the design of management systems. Management theory has always resided in the domain of management, not engineering. However, for two compelling reasons, industrial engineers have become increasingly involved in management in recent years.

The first reason is that, although the management system is but one of three legs of the triad, it is typically the single greatest determinant of the success or failure of a production system on a day-to-day basis. The best-designed physical system with the best-designed control systems will not live up to its potential if the employees are out on strike or, as is far more often the case, are sufficiently at odds with or unmotivated by management that they choose not to produce at potential plant capacity levels. Restriction of output by employees is an age-old tradition in the United States; it is what Taylor called "soldiering," and it has not fundamentally changed much in the century since he first discussed it.

The second reason is that there is fairly universal agreement that *management* is the most apparent problem with industry in the United States today. It is difficult to avoid this conclusion when yet another Japanese company leases one of our empty industrial buildings, hires our employees, purchases our materials, purchases the same equipment that our plants purchase, sells to us the products we produce, and makes a profit. What has changed in the plant? The answer is simple and inescapable: the management. In these circumstances, the frequently mentioned excuses—undereducated labor, lazy employees, environmental costs, taxes, government interference, product dumping, safety legislation (i.e., OSHA), environmental regulations, and product liability—do not preclude success.

It is time to admit that of late the Japanese have been managing better than us. Nowhere in this text will it be proposed that Americans surrender to Japanese management. And I hope Americans will never hear the English equivalent of the Japanese word *karoshi*—which means "death from overwork." But if they know something we do not, why not listen and learn? Patriotism is a desirable attribute for any citizen, but not when it produces an attitude that inhibits learning.

Industrial engineers are becoming increasingly more involved in assisting in the development of effective management systems to ensure that the production systems and controls they design will be implemented in a management environment that will permit such systems to be employed successfully on a day-to-day basis.

MANAGEMENT BASED ON BEHAVIORIST THEORY

A new era in management was initiated with the publication in 1954 of Abraham H. Maslow's text *Motivation and Personality* [13]. People were seen to be far more complex in terms of their needs, and therefore their motivations, than had earlier been fully appreciated. Prior to Maslow, managers typically assumed that a $0.05 per hour increase in pay could solve just about any employee problem; they often appeared surprised when it did not. Maslow's ultimate goal appears to have been self-actualization of the masses (or in this context employees). The following quote describes his belief:

> Scientists [employees] are motivated, like all other members of the human species, by species-wide needs for food, etc.; by needs for safety, protection, and care; by needs for gregariousness and for affection-and-love relations; by needs for respect, standing, and status, with consequent self-respect; and a need for self-actualization or self-fulfillment of the idiosyncratic and species-wide potentialities of the individual person. These are the needs that are best known to psychologists for the simple reason that their frustration produces psychopathology. [13, p. 2]

Being an industrial engineer, I may not fully comprehend what "psychopathology" means, but I believe I know it when I see it. The mental image from the evening news a few years ago of an automobile assembly employee walking out of his plant with "fire in his eyes" and declaring his sincerest wish that his company would "go under" might be something that qualifies. The employee appeared to have no immediate concern about the obvious fact that if his company "goes under," he is out of a job. In far too many instances, management and labor seem to have honed a remarkable inability to work together effectively. The development of the management–labor relationship is a key management responsibility. Too often, it has evolved into a "my lawyer against your lawyer" relationship that offers no chance for long-term cooperation between the two parties.

Maslow's original text discussed what he referred to as the *hierarchy of needs*. The five categories of needs, from bottom to top of the hierarchy, were

1 Physiological needs
2 Safety or security needs
3 Belongingness and love needs
4 Esteem needs
5 Self-actualization

A need, when satisfied (e.g., food, water, sex, warmth, self-esteem, or love), no longer represented an immediate need, so other, less recently met needs would assume a higher position on the ordered list of unmet needs. The behaviorist theory that developed from Maslow's research then treats human motivation as the singular determinant of productive accomplishment.

Rensis Likert, as Director of the Institute for Social Research at the University of Michigan, summarized institute research efforts involving numerous business organizations in his 1961 text *New Patterns of Management* [11]. His *principle of supportive relationships* was offered as a general principle that "high-producing managers seem to be using" [11, p. 102]:

> The leadership and other processes of the organization must be such as to ensure a maximum probability that in all interactions and all relationships with the organization each member will, in the light of his background, values, and expectations, view the experience as supportive and one which builds and maintains his sense of personal worth and importance. [11, p. 103]

This principle obviously builds on the theory offered by Maslow in highlighting the need for management to be sensitive to employees' personal needs. As Maslow clearly indicated, personal needs go well beyond pay.

In 1959, Frederick Herzberg, Bernard Mausner, and Barbara Snyderman published *The Motivation to Work* [5], which dealt with the relationship between job attitudes and productivity. Figure 4-1 indicates results of their study concerning satisfiers and dissatisfiers [6]. The numbers relating to each factor in Fig. 4-1 identify the number of times that factor was significantly mentioned in interviews that were performed during the study. Satisfiers are on the right of the chart, and dissatisfiers are on the left. Note, for example, that company policy and administration was identified 31 times and that in 28 of those instances, it was a dissatisfier (i.e., as a job attitude factor, company policy and administration tended to aggravate people and rarely served to motivate them). This result is quite consistent with Robert

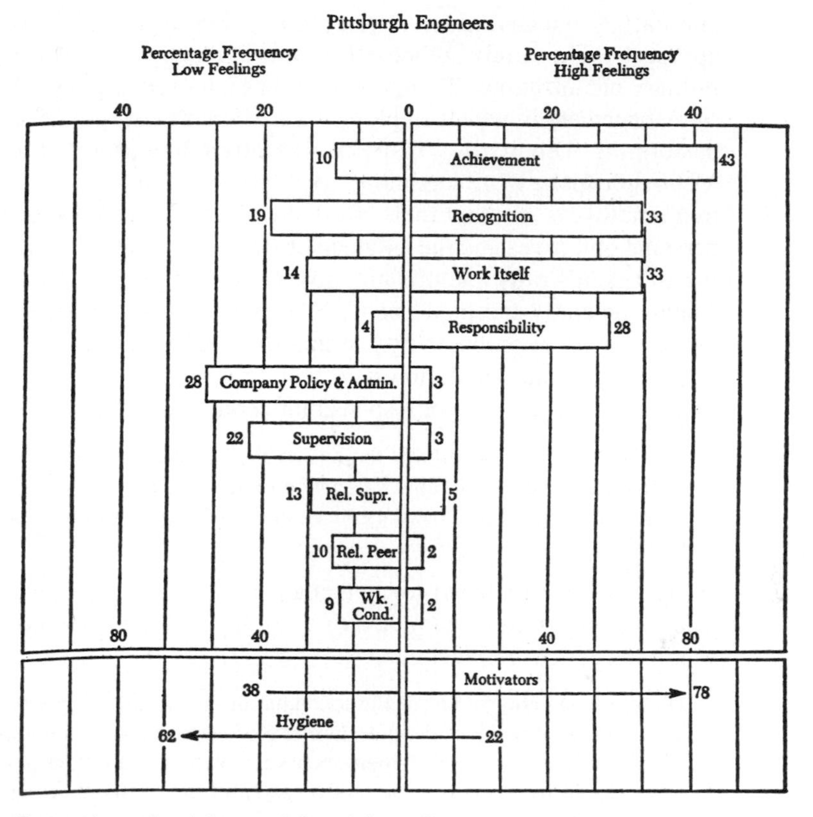

FIGURE 4-1 Comparison of satisfiers and dissatisfiers. (From Herzberg [6, p. 97].)

Townsend's recommendation in *Up the Organization* [25] to eliminate personnel departments. He prefers having personnel coordinators, assigned to each functional area and under control of the respective functional managers, to deal on a local basis with personnel matters. This arrangement eliminates the extensive and insensitive bureaucracy.

One of the hypotheses that Herzberg, Mausner, and Snyderman formed, based on their results, was that any single job attitude factor tends to serve predominantly as either a satisfier or a dissatisfier in their dual-factor theory, with some exceptions. The *motivation-hygiene theory* they developed equates job satisfaction factors to motivation factors, and dissatisfaction factors to hygiene or maintenance factors. Their theory suggests that dissatisfiers must be overcome before any efforts concerned with meeting motivation factor needs can be effective (i.e., "Make me safe, warm, and secure before you talk to me about motivation"). Herzberg expanded on the earlier study in *Work and the Nature of Man* [6].

In 1960, Douglas McGregor, probably the most influential behaviorist to date, published *The Human Side of Enterprise* [15], in which he compared two different approaches to management: Theory X and Theory Y. The epitome of the Theory X approach is the purely authoritative form of management typically associated with military organizations. Theory Y describes a looser and much less oppressive form of management in which employees are offered much greater freedom of action in performing their work. Theory Y was offered as a means to self-actualization. McGregor spoke of "integration," appropriate compromises, as a means of permitting employees to adapt their efforts to achieve a closer alignment between their personal objectives and the assumed or stated corporate objectives—a concept similar to Likert's principle of supportive relationships [11]. Human relations professionals generally accepted Theory Y because it closely matched their beliefs—that the attainment of productivity is primarily a question of employee motivation.

McGregor cautioned, however, that Theory Y was not to be interpreted as an abdication of management responsibility concerning the workplace:

> We have now discovered that there is no answer in the simple removal of control... that abdication is not a workable alternative to authoritarianism.... We recognize today that "industrial democracy" cannot consist in permitting everyone to decide everything. [15, p. 46]

There is a common belief today that a participative style of management will typically be more effective in motivating employees than an authoritative approach, as suggested earlier by McGregor:

> Direction and control are of limited value in motivating people whose important needs are social and egoistic. People, deprived of opportunities to satisfy at work the needs which are now important to them, behave exactly as we might predict—with indolence, passivity, unwillingness to accept responsibility, resistance to change, willingness to follow the demagogue, unreasonable demands for economic benefits. [15, p. 42]

In 1974, a monograph by Mitchell Fein entitled *Motivation for Work* [2], published by the Institute of Industrial Engineers, offered an alternative perspective on what motivates industrial employees. Whereas Maslow, Herzberg, and McGregor (the behaviorists) seem to offer self-actualization at work as a universal desire and goal of all employees at all levels, Fein questions its universality. Do all employees search for fulfillment in their work? Fein states:

> What I have tried to show by various citations and experiences is that the concepts of McGregor and Herzberg regarding workers' needs to find fulfillment through their work is sound *only for those workers who choose to find fulfillment through their work.* In my opinion, this includes about 15% to 20% of the blue-collar work force. These behaviorists' concepts have little meaning for the others. Contrary to their proposals, the majority of workers seek fulfillment outside of their work. [2, p. 39]

Fein also questioned another relatively universal and cherished assumption of the behaviorists' theory concerning job enrichment—that the opportunity for significant enrichment generally exists:

> Managers know from experience that job enrichment is possible only for some jobs. Few jobs can be altered to include the vertical enlargement responsibilities suggested by Herzberg, who also recognizes this limitation in his KITA [7, p. 62] article when he says: "Not all jobs can be enriched, nor do all jobs need to be enriched." [2, p. 16]

I can identify with Fein's perspective, having asked numerous blue-collar workers in the past their feelings about doing work that I perceived to be monotonous, only to learn that they liked their work. What seems to be important to most employees is job security, peer contact and acceptance, reasonable working conditions, safety, and pay, in roughly that order. It might be noted that many of the jobs people do—such as filling cavities in teeth (dentists), selling homes (real estate agents), spraying for bugs (exterminators), or teaching in elementary school (elementary school teachers)—can be very monotonous as well but are never mentioned in the management literature. Somehow teeth, houses, bugs, and grammar school students are assumed to be inherently more interesting than machines. Is that true? Has anyone tested that hypothesis?

A seminar on Herzberg's theories is probably best received by an audience of manufacturing managers, plant managers, or CEOs who identify with the self-actualization and job enrichment aspects of *their* work. It is a perfect audience for his concepts—a room full of "workaholics" who search for fulfillment in their work. The same concepts, when presented to Harold, a 52-year-old drill press operator who works to "put bread on the table" and finds personal fulfillment after work in a barber-shop quartet, may not be as applicable. How do you enrich a drill press job? There is a considerable amount of work to be performed in industry that does not require imagination, just a pair of hands. Those who perform such work do not necessarily dislike it; more often than not it is a white-collar worker who seems to have a problem with it (i.e., the work is not something he or she would want to do).

In 1981, William Ouchi published his best-selling book *Theory Z* [20], with the appealing subtitle *How American Business Can Meet the Japanese Challenge*. Ouchi offered his considerable insights as to differences in industrial culture between Japan and the United States, as indicated in the following comparative list [20, pp. 48–49]:

Japanese organizations	American organizations
Lifetime employment	Short-term employment
Slow evaluation and promotion	Rapid evaluation and promotion
Nonspecialized career paths	Specialized career paths
Implicit control mechanisms	Explicit control mechanisms
Collective decision making	Individual decision making
Collective responsibility	Individual responsibility
Wholistic concern	Segmented concern

Ouchi describes an experiment he performed in which he listed seven characteristics of the Japanese type of manager, without identifying that they were Japanese, and then asked a number of American executives to list American firms that had managers with those characteristics. Ouchi found that

> Surprisingly managers named the same companies repeatedly: IBM, Procter & Gamble, Hewlett-Packard, Eastman Kodak, the U. S. Military. These organizations, all commonly thought to be among the best managed in the world, were identified by our respondents as having the same characteristics as Japanese companies! [20, p. 57]

This raises a question as to whether Japanese-style management characteristics are as culturally based as we might have previously assumed. Ouchi generally refers to companies that "somehow manage to develop talent and are known for doing so" [20, p. 58] as *type Z* firms.

In describing McGregor's Theory X and Theory Y approaches, Ouchi offers the following description:

> A Theory X manager assumes that people are fundamentally lazy, irresponsible, and need constantly to be watched.... A Theory Y manager assumes that people are fundamentally hard-working, responsible, and need only to be supported and encouraged. [20, p. 58]

Some of my consulting firm personnel had occasion to follow aircraft mechanics around on one assignment, at the request of an airline management, to determine if the company was experiencing a slowdown. The company's labor-reporting system was unable to determine if a slowdown was occurring, and they obviously could not ask the union.

One mechanic selected to be followed was performing a task of greasing wing flaps. First, he stood in line waiting to use the microfiche machine because there were too few available. The mechanic went to the tool crib and stood in line because there were not enough tool crib employees; he was finally issued a grease gun. He noted that it did not have any grease in it, so he went three hangars away to find a grease barrel and proceeded to fill the grease gun. He then walked back to the other hangar but could not reach the flaps because there were no wing stands, so he went looking for a ladder (ladders were in short supply as well). To grease the flaps, he also had to search for an air hose, which he then had to beg to borrow because they were in short supply as well. Three hours after starting, he completed the task, which with proper technology, procedures, and tools should have been performed in half an hour. His annual pay, including fringes, was in excess of $50,000, and there were over 800 mechanics like him at this facility.

A day or two later, as I entered the deep-carpet conference room at the airline's headquarters, the opening management conversation had moved in the direction of suggesting that the airline had unmotivated, lazy mechanics. I felt compelled to relate the above story and to indicate that if I were an FAA-approved mechanic at their facility and had to do the same job in the same circumstances, I too would likely become an "unmotivated, lazy mechanic." At that point I mentioned the Pogo quote at the beginning of this chapter, which may have damaged some egos, and then got on with business.

One of the most frequently repeated sayings on production floors across this nation is "If they don't care, why should I?" In too many instances today, top managers are so busy designing their "golden parachutes" that they are often not dealing effectively with problems on the production floor.

Ouchi's text describes his proposed steps for becoming a type Z firm. Central to his approach is Japanese-style participative management and the utilization of ad hoc teams to solve specific problems that develop. Probably his greatest contribution has been getting managers to go off by themselves to a retreat in order to determine who they are and to explicitly define what their corporate goals and objectives are

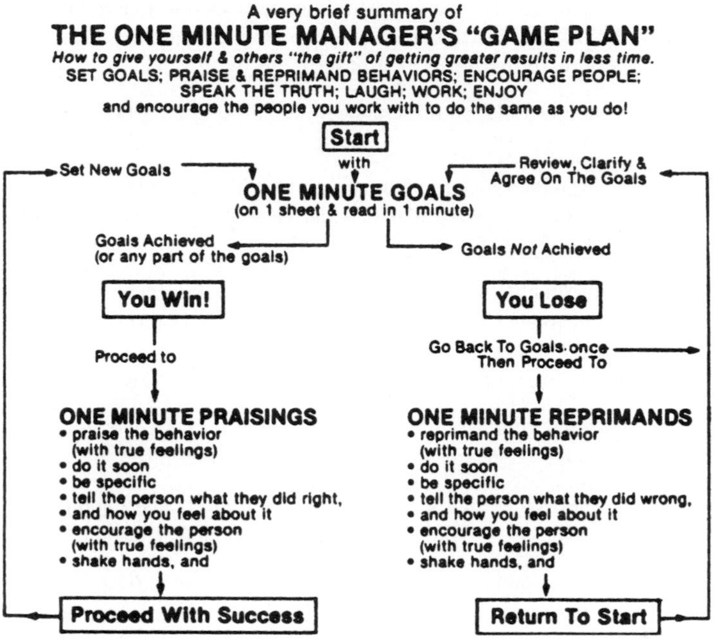

A very brief summary of
THE ONE MINUTE MANAGER'S "GAME PLAN"
How to give yourself & others "the gift" of getting greater results in less time.
SET GOALS; PRAISE & REPRIMAND BEHAVIORS; ENCOURAGE PEOPLE;
SPEAK THE TRUTH; LAUGH; WORK; ENJOY
and encourage the people you work with to do the same as you do!

Start
with
Set New Goals
Review, Clarify & Agree On The Goals
ONE MINUTE GOALS
(on 1 sheet & read in 1 minute)
Goals Achieved
(or any part of the goals)
Goals *Not* Achieved
You Win!
You Lose
Proceed to
Go Back To Goals once
Then Proceed To

ONE MINUTE PRAISINGS
- praise the behavior (with true feelings)
- do it soon
- be specific
- tell the person what they did right,
- and how you feel about it
- encourage the person (with true feelings)
- shake hands, and

ONE MINUTE REPRIMANDS
- reprimand the behavior (with true feelings)
- do it soon
- be specific
- tell the person what they did wrong,
- and how you feel about it
- encourage the person (with true feelings)
- shake hands, and

Proceed With Success
Return To Start

FIGURE 4-2 The One Minute Manager's Game Plan. (From Blanchard and Johnson [1, p. 101].)

and what their relationships are to one another, their employees, their suppliers, their clients, their community, etc.

Kenneth Blanchard and Spencer Johnson, in their 1983 text *The One Minute Manager* [1], suggested that neither managers primarily interested in results (i.e., "autocratic") nor managers primarily interested in people (i.e., "democratic") were the answer. They indirectly proposed a balanced approach by stating, "Effective managers manage themselves and the people they work with so that both the organization and the people profit from their presence" [1, p. 15]. Much of their book deals with the role of the manager as a teacher of employee or manager behavior. Their "one minute goal setting," "one minute praising," and "one minute reprimand," are all aimed at providing psychologically effective ways of motivating people through positive reinforcement and information feedback.

Blanchard and Johnson offer the One Minute Manager's "Game Plan" as shown in Fig. 4-2.

In 1981, Richard Pascale and Anthony Athos published *The Art of Japanese Management* [21]. Much of their text praises the Matsushita Corporation, the largest electronics company in the world. Familiar Matsushita product lines are Panasonic, Quasar, Technics, and National. The following quotes indicate Pascale and Athos' concern about American management and, in contrast, praise for Matsushita's management system:

> In this book we will argue that a major reason for the superiority of the Japanese is their managerial skill. [21, p. 21]

> [Matsushita and Takahashi] exhibit an approach to management that involves its getting into the factory, into the field, and out with the customers. Matsushita executives are noted for spending less time in their offices. [21, p. 46]

> Praise and positive reinforcement are an important part of the Matsushita philosophy. [21, p. 55]

> Matsushita, interweaving human values with hard-edged efficiency, has created an organization of astonishing resilience and vitality. Perhaps Matsushita is like an army—in the best sense. [21, p. 57]

> The prime qualification of a Japanese leader is his acceptance by the group, and only part of that acceptance is founded on his professional merits. The group's harmony and spirit are the main concern.... In the West, work group leaders tend to emphasize task and often neglect group maintenance activities. [21, p. 126]

> It has been accepted only slowly by many American executives that part of our economic problems are the result of inadequate management. [21, p. 200]

> There has been an ineffective grabbing and gimmickry and quick fixes as we [American management] have felt the pinch, and that tendency has hurt us. [21, p. 201]

It can be noted from the Pascale and Athos text that Matsushita relies heavily on its production engineering department. The suspicion is that not only does Matsushita have excellent management but that they have also developed and supported one of the premiere industrial engineering and manufacturing engineering capabilities in the world, and that is one of the keys to their considerable success:

> Matsushita's success has been won in a large part through the use of managerial tools that we think of as "invented" in the West. In fact, Matsushita has often beaten us at our own game. [21, p. 29]

> At the heart of Matsushita's followership strategy is production engineering. [21, p. 31]

> Matsushita has consistently invested five percent of sales in R&D, of which the major portion goes to Production Engineering. [21, p. 31]

I know of no firm in the United States that makes a comparable commitment to manufacturing or industrial engineering.

The behaviorist approach offers little empathy for control systems. In contrast, Pascale and Athos state:

> There are a great number of Japanese firms that are remarkably close to the American model—Matsushita, YKK, Sony, and Honda, to name a few. Their control systems are very tight and they are more bottom line oriented. [21, p. 42]

> A highly focused managerial information system [at Matsushita] reports on a dozen key indicators of divisional performance on a monthly basis. [21, p. 39]

Another key factor in the Matsushita management system is implied in the following quote from an employee in a plant owned by a Japanese company:

> This was a sloppy, losing operation before they acquired it. When they came in, they said they wanted to turn a profit within three years, and—by God—we made it.

The quote hints that one important ingredient of the Matsushita management system is discipline. Of course, discipline is a necessary precondition for the success of any management system. The fact that the employee said, "we made it" is highly significant. In far too many American plants, the focus is on *they* instead of *we*, and therein lies the difference.

Thomas Peters and Robert Waterman, Jr., published *In Search of Excellence* [23], a much-acclaimed best-seller, in 1982. True to the behaviorist tradition, they indicate in their first chapter, "A theme we shall return to continually in the book, is that it is *attention to employees,* not work conditions per se, that has the dominant impact on productivity" [23, p. 6]. Of course, attention to employees is essential, but there are a lot of other things that must also be "made right" if the true potential for productivity is to be achieved. A motivated workforce is one essential factor, but it is only one of many. Production system design, control, and management must all be represented if potential productivity is to be achieved.

Peters and Waterman offer some very worthy advice to those in American management responsible for corporate vision and product conception. Many of their concepts are very compatible with Japanese management concepts regarding employees—such as developing a respectful, open, and honest relationship; utilizing positive reinforcement; and creating shared values. They also hint as to the complexity of the labor–management relationship by mentioning a dilemma—that most employees simultaneously want self-determination and security, yet the perceived reality in our culture is that security is gained through conformity, not self-determination.

Peters and Waterman display distaste for what they label "The Rational Model" [23, p. 29]. I understand their distaste for a management approach that embodies operationally detached senior managers who "live and die" by what their financial staff's spreadsheets tell them. When the lawyers, financial staff members, and accountants acquire and exercise direction of a manufacturing company, without adequate corporate vision and knowledge of products and plant operations, the company is almost certainly in trouble. If their taking control simply killed the company it probably would not be so bad. Unfortunately, the greater likelihood is that, without vision and operational knowledge, they will simply direct it poorly and simply cause it to hemorrhage in an internationally noncompetitive way for some predictably limited amount of time.

H. Ross Perot founded Electronic Data Systems (EDS), which was one of the early stars in the computer applications business. EDS was later acquired by General Motors (GM), but the "honeymoon" ended with Perot leaving GM when the GM board agreed to pay him double the value of his GM stock. Perot's very revealing *Fortune* article entitled "How I Would Turn Around GM," written some time after his departure from GM, offers his advice about letting accountants make product decisions:

> The historic power struggle between the financial staff and car builders will not be tolerated. Financial people will be responsible for maintaining accounting information. People who know how to build cars and serve customers will make the product decisions. Accountants will not sap the productivity of car builders. [22, p. 44]

Peters and Waterman clearly dislike large technical organizations:

> In company after company, we found ten-person skunk works that were regularly more innovative than fully equipped R&D and engineering groups with casts of hundreds. [23, p. 112]

It is impossible to defend large, ineffective, and inefficient organizations. But are they saying that a "skunk works" of 10 people can replace a cast of hundreds? It is an appealing and romantic concept, but on reflection it seems rather unlikely and obviously simplistic, in light of the complexities of modern products, processes, and techniques. Consider the following Peters and Waterman statement:

> At a $5 billion survey company, for example, three of the last five new-product introductions have come from a classic skunk works. It consists at any one time of eight to ten people, and is located in a dingy second-floor loft six miles from the corporate headquarters. The technical genius is a fellow whose highest degree is a high-school equivalency diploma earned in the Army in Korea (although the company has literally thousands of Ph.D. scientists and engineers on its payroll). One of the other members of the group was arrested for sneaking into a manufacturing facility to which he had no pass and swiping some material needed to get on with an experiment. [23, p. 211]

This quote raises a number of questions. Do scientists work better in dingy lofts? Are people with high-school equivalency diplomas, on average, more likely to offer greater technical contributions to a firm than Ph.D.s? Is stealing material a policy one wishes to embrace? It is psychologically appealing to root for the "little guy" or the "underdog," but you wouldn't want to bet the company on it. It would make sense, of course, to support the technical genius with the high-school equivalency diploma mentioned above and to reevaluate the large organization unable to match his accomplishments in order to understand why and to take appropriate action. Managers are supposed to do that.

Peters and Waterman state, "The old rationality is, in our opinion, a direct descendent of Frederick Taylor's school of scientific management and has ceased to be a useful discipline" [23, p. 42]. Industrial engineering is a direct descendent and extension of Taylor's school. Does that mean that industrial engineering is no longer a useful discipline? What Peters and Waterman never seem to recognize is that, sooner or later, a product must be manufactured on the production floor. When it is manufactured, it is important that good production system design, control, and management be in place to provide the means for effective and efficient manufacture. Highly motivated production workers are an essential component to that accomplishment, *but* they will never be an adequate substitute for industrial, manufacturing, and quality engineers in providing the production system designs and controls necessary for an effective and efficient operation. The most effective result will likely come from the effective utilization of industrial, manufacturing, and quality engineers in conjunction with the sum of contributions from all employees of the firm. There is no question that the direct labor employee has a great deal to offer, as do the engineers. It is the combination that is so powerful; they are not mutually exclusive resources.

There is much merit to behaviorist philosophy. Theory Y, as an overall management philosophy, appears to be far superior to the theory X authoritarian approach of the past. But management has both a need and a responsibility to know what production levels to expect from a production system and to see that that level of production is accomplished under normal conditions. The underlying assumption often implied in behaviorist management philosophy is "If you are nice to them, they will be nice to you." Such a philosophy has its limitations, given human nature. It also places the aspiring manager in an often unjustified catch-22 situation: "I have been as nice as I know how to be, and it isn't working; what do I do now? If I try to be nicer, it probably won't work, and if I am less nice, I will surely be accused of not being nice." It can be a "no win" situation for the manager. He or she may be a potentially excellent manager but may become demoralized, thinking that he or she has failed when, in fact, the underlying management philosophy has failed.

In any worker–manager relationship, there has to be an underlying understanding of what the worker, working for the manager, is expected to produce. Labor performance reporting simply makes that expectation, that understanding, explicit. If the manager does not know how much employees are supposed to produce, how could they be expected to respect their manager? A management system based on hoping that employees will produce if the manager is nice enough is doomed to failure. The last 30 years of behaviorist management philosophy have more than demonstrated this notion.

A CRITIQUE OF BEHAVIORIST PHILOSOPHY

In 1980, Dr. James A. Lee, a sociologist at Ohio University, published an important and atypical management text entitled *The Gold and the Garbage in Management Theories and Prescriptions* [10]. The following quotes suggest why it is atypical:

> This book is for anyone who would rather see the pros *and* cons rather than just the pros of management theories, principles and techniques. [10, p. ix]
>
> An inventory of the effectiveness of the applications of the sciences in the management of organizations is overdue. [10, p. 5]
>
> Who in management has not read of program announcements at TRW (Thompson-Ramo-Woolridge) Systems (career development program), Volvo (job enrichment), Union Carbide (Organizational Development Department and Management by Objectives), and the Non-Linear Systems "experiment"? Who has read follow-up accounts of these efforts with enough data for careful analyses of the results? Almost no one. And yet some of these programs were abandoned completely. One almost bankrupted the corporation, and others fell far short of achieving stated goals. [10, p. 6]
>
> No scientific discipline can afford to ignore the opportunity to learn from its critics and its failures—especially from its failures—least of all fledgling sciences such as psychology, sociology, and anthropology, or if one prefers—the behavioral sciences. [10, p. 6]

Dr. Lee reviewed the behavioral science literature concerning the leading management theories of the day and then offered a critical assessment of each of the theories presented. It is a timely and exceptional work, and I thank Marvin Mundel, Ph.D., for drawing my attention to it.

The first theory Lee reviewed was Maslow's hierarchy of needs:

> It is not likely that many human relations courses for supervisors over the last 20 years excluded any mention of Maslow's need hierarchy theory. In most such courses, it is unlikely that the presentation was billed as untested "theory."... One reason for the lack of supportive evidence is simply that there isn't much. In Maslow's own words, as late as 1970, "[the theory] seems, for most people to have a direct personal, subjective plausibility. And yet it still lacks experimental verification and support. I have not yet been able to think of a good way to put it to the test" [13, p. xii]. This does not mean that there have not been a few attempts to validate his theory.... What this does mean, given the disappointing results of these studies, is simply that any present evaluation of Maslow's theory must necessarily be based mostly upon logic, common sense, personal observation, and experience. [10, p. 63]

Concerning Maslow's theories, Lee concludes that

> Maslow's need hierarchy, an up-to-date "dynamicized" version of Aristotle's hierarchy of the soul, which included a "nutritive" (physiological) soul at the lower end and the noetic (self-actualization) soul at the top [16, p. 596], remains unverified, yet personally plausible to many. . . . In summary, it can be said that science seems to have no quarrel with Maslow's physiological need level. This seems to be the most personally plausible of the five levels, because, although everyone does not necessarily know if he self-actualizes, he knows that he eats, wants sex and goes to the bathroom. Maslow's highest need level—self-actualization—seems to be not much of a problem, either, but mainly because most people are somewhat unsure of what it means. In between these, his model appears sufficiently ambiguous and therefore difficult to refute or support, thus permitting individual interpretations with sufficiently wide margins to allow everyone to be right. [10, p. 72]

In his conclusions regarding Herzberg's [6, p. 75] dual-factor theory, i.e., maintenance and motivation, Lee finds that

> The evidence to date eliminates Herzberg's theory as a general or universal theory of work motivation. Far too many findings fail to support the Dual-Factor model. [10, p. 101]

Since the 1930s, there have been various "path-goal" theories concerning motivation. Most have taken a hedonistic philosophical view that behavior accommodates a search for pleasure and avoidance of pain. One such theory, the Vroom path–goal model, was described in Vroom's text *Work and Motivation* [26, pp. 14–28], published in 1964. Lee's observation is that

> Until the early 1960s it was evidently widely accepted that there was a simple and direct relationship between job satisfaction and performance. Furthermore, job satisfaction levels were generally thought to *cause* job performance levels, although there appears to have been almost no evidence to support this. Some observers, not to mention a number of researchers and theorists, were caught in one of the oldest traps in scientific interpretation—that since two things (apparently) happen together, one caused the other. [10, p. 104]

Lee then offers the following quote from George C. Homans [8, p. 40], whom he references extensively throughout his text: "There is no [proven] general relationship between productivity and satisfaction." [10, p. 105]

Lee goes on to state that

> The original impetus pressing for research evidence that job satisfaction caused variations in job performance was likely the Hawthorne studies. The popular conclusion, although inaccurate, perpetuated by the inordinate publicity of these studies, was that a change in mental attitude brought about by a friendly relationship with supervisors caused job satisfaction which, in turn, caused the reported improved job performance. The satisfaction-causes-performance notion was perpetuated for years in the face of (1) continuously recorded low or negative correlations between the two variables, (2) employees' and unions' continued demands, not for "tender loving care," but for wage and fringe benefits, and (3) failures in human relations training of supervisors to replicate the alleged advantages of the Hawthorne experimental supervisory style. [10, p. 105]

Considerable group theory has been developed in the past 40 years. The theory is likely responsible for a common belief today that "group work" is superior to "individual work." But is there evidence to support such a conclusion? In the chapter entitled "Work Group Theory," Lee writes that

> It does not follow automatically that happy, cohesive groups always perform well. Groups do not necessarily solve problems better nor do they make better decisions than do individuals. Although non-groups are not studied as such by behavioral scientists, we can suspect that they exist here and there in well-performing organizational units. [10, p. 191]

We should not always assume that groups will be happy.

If behavioral scientists are *scientists,* shouldn't they have studied non-groups as well as groups to offer a comparison of effects? Have behaviorists studied incentive systems over the past 40 years, or were incentive systems counter to their theory and therefore essentially dismissed out of hand?

The most common basis of paying for a worker's labor has been and still is the purchase of a worker's time. The worker suffers the inconvenience of being at work rather than elsewhere and is rewarded for the inconvenience. There need not be any direct relationship, however, between the purchase of a worker's time and his or her productivity.

High on the list of worker interests is job security. A common belief of adult life is that one should work, whether remuneratively or not, and a job provides one with the means to acquire the necessities of life. It provides a means to play the expected societal role of breadwinner, and it is a source of additional discretionary family income of spouses (of either sex). Behavioralists such as Herzberg [6], McGregor [15], and Maslow [13] believed the key to productivity was essentially motivation, of the nonremunerative variety. In dealing with professional people, they are probably at least partially right. In dealing with a worker putting nut B on bolt C, however, they may well be wrong. Fein takes the following view:

> We know from first-hand experience that job security is not enough; witness the low productivity in the civil service jobs. I believe that job security is an essential *precondition* to

> increasing the will to work. Without it employees resist management's attempts to raise productivity. But adding job security by itself will probably not appreciably change the productivity picture.
>
> In my opinion, the 85% of the non-involved employees need job security *plus* the incentive of a piece of the action to raise their productivity. These workers come to work to eat, to exchange their efforts and skills for what they can buy back outside of the work place. Their work has too little meaning to them for it to motivate them. Management must develop financial incentive plans which will appeal to these employees and encourage them to apply themselves more diligently to their work. [2, p. 2]

It has been well documented in numerous instances over the past century that if a plant on day work (i.e., paying for time) is converted to an incentive plant (i.e., paying for output), output and employee pay (assuming a 100 percent participation plan) increase by 25 to 35 percent. As fixed costs become a greater and greater percentage of the total manufacturing cost, such increases in total facility output represent significant decreases in unit manufactured cost. It is one of those rare "everybody wins" situations. The employee brings home more pay, and the owners and consumers share in the advantages of a lower unit cost. Why is it that plants in the United States have moved away from incentive plans, for other than top management, in the last 30 years? The answer is not obvious.

To press the point a bit further, what is particularly unfortunate is that the approach suggested above is not new and untried. Over 30 years ago, James F. Lincoln [12] initiated what he called *incentive management.* The abnormally high productivity of the Lincoln Electric Company has been well known and discussed for the past 30 years. Why does it stand alone? Why have so many other companies failed to adopt similar approaches?

Fein states, "These [Lincoln Electric] employees participate in their work to a far greater extent than visualized by psychologists in their writings. The key to this participation is that the employees *want* to do it; there is no holding back. A manager from the outside would drool on witnessing Lincoln employees at work" [2, p. 57].

Lee, in a chapter entitled "Theories of Economic Motivation," describes a number of successful incentive plan installations: Conley Mill [10, p. 134], Midland-Ross [10, p. 135], and Lincoln Electric [10, p. 137]. The success of such plans runs counter to behaviorist philosophy. Pay was not viewed by behaviorists as a significant motivator. The following quote from Leon Meggison shows the effectiveness of incentives as motivators:

> Studies conducted immediately after World War II in hundreds of plants in a great variety of industries showed that labor productivity under the straight hourly rate form of compensation seldom exceeded 60% of the performance obtained with good wage incentive methods. . . . One of the more comprehensive studies of the effects of incentive plans on productivity was conducted in 1959. The sample for this study included 29 industries and 305 plans, most of which had been installed during the 1950s; the study showed that productivity increased an average of 63 percent. . . . In summary, with a properly developed and administered wage incentive plan, the output per man-hour should increase 20 to 50 percent over what it was before the installation of the scheme. It has been proved that the

> output of the individual in any group of employees on incentives will distribute itself in a pattern which follows the normal distribution curve about the midpoint. This midpoint is generally found at about 130 percent of the standard, or at the 30 percent bonus level. [17, pp. 454–455]

Under behaviorist philosophy, employees were believed not to be particularly concerned with compensation as a motivator. As a result, both the continuation of old incentive plans and the implementation of new incentive plans dropped significantly as behaviorist philosophy became widely adopted. Is it possible that employees are more interested in compensation than the behaviorist philosophers assume, and that the resulting reduction of incentive programs has been a factor in the continuing deterioration in productivity in the United States, relative to competing economies, over the past 40 years? Was the behaviorist philosophy a step forward or a step backward with respect to real productivity, and more importantly, management practice?

There has been some renewed interest in incentives systems in recent years; much of this effort was spearheaded by authors such as Mitchell Fein [3]. A system that he developed and has received considerable attention is called gain sharing. Figure 4-3 illustrates a typical distribution of monthly earnings, employing a value-added version of a gain-sharing plan.

As Lee mentions in the introduction to his text, "The essential problem is simply that ignorance of these [behaviorist] theories does not prevent success in management" [10, p. 3]. Have "management theorists" been giving practicing managers

FIGURE 4-3 Bonus calculation for a value added gain sharing plan. From Hodson [14, p. 6.46].

Value Added		
Monthly Calculation		
1. Value of Production (Sales + Various adjustments)		$1,500,000
2. Less Outside Purchases		
Material and supplies	$760,000	
Other outside purchases and non-labor costs	350,000	(1,100,000)
3. Value Added		400,000
4. Allowed Employee Costs Base ratio × value added (0.43 × 400,000)		172,000
5. Less Actual Employee Costs		(152,000)
6. Employees' Share		20,000
7. Less Reserve for Deficit Months (33% × 20,000)		(6,000)
8. Bonus Distribution		13,400
9. Participating Payroll		90,000
10. Bonus Percentage = (13,400/90,000)		14.9%

bad advice, and have the managers been following it, for the past 40 years? And if the advice did not work, would a manager know it or admit it? We must not forget the influence of egos.

Concerning the ideas of Douglas McGregor [15], Lee writes that "theory Y has been common fare in management training programs since it was first introduced in his book in 1960. His major contribution, however, was not theory, as he called it, but philosophy" [10, p. 259]. McGregor's theory X assumptions, referred to in his text [15, pp. 33–34] as "The Traditional View of Direction and Control," are stated as follows:

> **1** The average human being has an inherent dislike of work and will avoid it if he can.
>
> **2** Because of this human characteristic of dislike of work, most people must be coerced, controlled, directed, threatened with punishment to get them to put forth adequate effort toward the achievement of organizational objectives.
>
> **3** The average human being prefers to be directed, wishes to avoid responsibility, has relatively little ambition, wants security above all.

Lee states:

> McGregor's arguments that these assumptions prevailed as the guiding view of employees by managers included no cited research into the matter at all. He *stated* that they were *implicit* in the literature and "in much current managerial policy and practice." [10, p. 259]

To his credit, McGregor reported changes in his thinking following a period in which he served as president of Antioch College. McGregor reported:

> I believed, for example, that a leader could operate successfully as a kind of adviser to his organization. I thought I could avoid being a "boss." Unconsciously, I suspect, I hoped to duck the unpleasant necessity of making difficult decisions, of taking the responsibility for one course of action among uncertain alternatives, of making mistakes and taking the consequences. I thought that maybe I could operate so that everyone would like me––that "good human relations" would eliminate all discord and disagreement.
>
> I could not have been more wrong. It took a couple of years, but I finally began to realize that a leader cannot avoid the exercise of authority any more than he can avoid responsibility for what happens to his organization. [15, p. 48]

Most revealing in Lee's book is Chapter 20, entitled "Case of a Company Attempting To Apply a Variety of Modern Human Relations Theories." The case, developed by Professor Erwin L. Malone and described in Lee's text [10, pp. 436–451], concerns Non-Linear Systems, Inc. (hereafter referred to as NLS), which was established in Del Mar, California, in 1952. Between 1952 and 1960, it enjoyed considerable success, growing from 5 employees to 430 during that period. It became one of seven firms sharing 95 percent of the market in digital electrical measuring instruments.

In 1960, Andrew F. Kay, NLS's president, initiated an ambitious adoption of behaviorist management throughout the firm. Employees were reorganized into three zones: (1) trustee management, (2) general management, and (3) departmental management. Zone 1 had four members, including Kay, for providing basic policy. Zone 2 was an eight-member executive council that included Kay. The council's purpose was to "establish operating policies; plan, coordinate, and control the business as a whole; and appraise results" [10, p. 439]. Kay, as a member of the council, was no longer president of NLS. Zone 3 consisted of 30 departments, of 3 to 12 employees each, including the department manager. Department managers were responsible for day-to-day management (i.e., methods, operations, procedures) of their departments.

A number of significant actions followed reorganization in 1960:

1 The executive council received numerous lectures from many of the leading behavioral theorists in the field of human and industrial relations, including James V. Clark, Richard E. Farson, Abraham H. Maslow, Vance Packard, Carl Rogers, Robert Tannenbaum, and Frances Torbert.

2 Every employee was interviewed by a retained Los Angeles testing service, and industrial relations counselors were retained to provide one-on-one counseling of employees.

3Time clocks were eliminated.

4 Lengthy on-the-job training was provided to each new employee.

5 "Douglas McGregor had stated that keeping performance records would vitiate the principle of self control, inasmuch as records could be used by management to control a subordinate and as a check on his performance. So far as production and personnel records were concerned, NLS became a company that operated essentially on a 'put nothing in writing' basis" [10, p. 441].

6 The inspection department was eliminated.

The following are quotes from a final section of the case entitled "Why Did the Experiment Fail?":

> In the period from 1953 to 1960 the company had experienced a phenomenal rate of growth. [10, p. 443]

> Sales volume did increase each year from 1960 to 1963, but the rate of increase diminished. Late in 1963 and again in 1965 sales dropped sharply. At these times orthodox management would have reduced plant employment in line with the incoming order rate, but such an action would have been in conflict with the goals of job and personal security. . . . Even when layoffs did occur, the higher-salaried engineering and research personnel were shifted to lower-rated production jobs without the usual downward salary adjustment, in the belief that this would keep them available when improved conditions warranted their services. [10, p. 443]

> The seven vice-presidents formerly had been vigorously active in the midst of daily problems and were more or less experts in their individual specialties. Under participative management these men were practically immobilized as "sideline consultants." [10, p. 444]

> Many [department managers] found it disquieting to have to seek aid from the Executive Council as a whole. And no matter how well experienced, young men in subordinate positions appear to want some measure of authoritative direction—someone else to help shoulder some of their problems. [10, p. 444]

> In 1963 the output from the factory was in the neighborhood of 30 percent more than it was in 1960. But the number of employees rose from 240 to 340, an increase of 42 percent over the same period. . . . There is no indication of NLS ever having increased plant efficiency during the entire run of the experiment. [10, p. 445]

> In 1965, after the experiment's end, salaried salesmen were replaced by commission sales agents. [10, p. 446]

> At the lower management level, department managers, many of them engineers and technicians, experienced a drop in morale during the course of the experiment. These technically skilled employees had often remained working at the plant after quitting time. Toward the end they departed promptly after putting in an 8-hour day, and their offices were deserted on Saturdays, Sundays, and holidays as never before. Despite "ideal" working conditions and high salaries, 13 of the 30 department managers who were with the company in 1962 were not with NLS in 1965. [10, p. 447]

> The seven subordinate executives [on the executive council] felt that participative management had restricted rather than widened their horizons. [10, p. 447]

> No thought may have been given to raising profits above the 1959–60 level reached prior to the experiment, but it hardly could have been expected that profits, which had increased yearly from 1953 to 1960, would begin in 1961 to spiral downward until the losses sustained compelled abandonment of the experiment in 1965. [10, p. 448]

> We have the much more recent remark by a behavioral scientist who had been employed on the NLS experiment since its inception: "I think we know that Human Relations don't have a lot to do with profit and productivity." [10, p. 449]

> The company probably avoided bankruptcy by abandoning the experiment and returning to orthodox methods of management; it is now operating with a small profit and expanding into other, but related, fields. [10, p. 451]

The case involving Non-Linear Systems, Inc.—probably the most extensive effort to embrace the totality of behaviorist management theory in the history of management—appears to have been a documented failure. Lee's text raises serious questions about the behaviorist theory of management as espoused and practiced in the United States for the last 40 years. Surely the theory is not totally wrong. The job that lies ahead is to distinguish between what is right and what is wrong about the behaviorist theory of management.

MANAGEMENT BASED ON INDUSTRIAL ENGINEERING

Frederick W. Taylor offered his industrial management concepts nearly a century ago. It should come as no surprise that what he espoused then needs to be adjusted to meet the societal conditions of today. His underlying concept, however—that the work to be performed on the production floor should be engineered and that production system design and controls should be put in place to permit workers to be effective and efficient in their work—is as valid today as it was when he conceived it.

An experienced industrial engineer can walk through a company that has not embraced industrial engineering in its operations and point out to management an almost unending list of obvious ways in which that company needs to improve. Some managements find such an experience to be upsetting; it often damages their egos. Management egos are usually very sensitive. That is a serious problem that will be discussed in more depth later in this text.

The Japanese management culture strongly embraces a participative and consensual type of employee relations at all levels in the firm. It is an interesting historical fact that efforts to restore the Japanese economy after the end of World War II by the GHQ (General MacArthur's general headquarters) resulted in the creation and passing of Japanese labor laws that made it almost impossible for a Japanese employer to fire an employee, resulting in the near "lifetime employment" relationship between many Japanese workers and many of the larger employers today. In addition, as a result of GHQ efforts, the *zaibatsu,* the prewar wealthy industrial class closely associated with the Japanese military before and during the war, saw their enormous power through interlocking holding companies taken away to produce greater democratization of their industrial economy. This resulted in a considerable expansion of the middle class in Japan. Over 90 percent of Japanese, when asked in polls, consider themselves members of the middle class [18, p. 135]. Japanese management's respect for people as the key resource in a firm is clear from the following quote from Akio Morita's 1986 book, *Made in Japan*:

> The most important mission for a Japanese manager is to develop a healthy relationship with his employees, to create a family-like feeling within the corporation, a feeling that employees and managers share the same fate. . . . But in the long run—and I emphasize this—no matter how good or successful you are or how clever or crafty, your business and its future are in the hands of the people you hire.

Morita's book is exceptional in detailing many of the weaknesses of American management today; it should be required reading for all American managers. One theme that he makes apparent is his respect for technical excellence and his reservations about letting lawyers, accountants, and business "professionals" make all the decisions at high levels based on what appears to best suit this quarter's profitability.

In the same year that *Made in Japan* was published, Eliyahu M. Goldratt and Jeff Cox [4] published *The Goal*. Part of the book's appeal is its novelistic style. The book concerns a plant manager trying desperately to improve plant operations. One major lesson in the book comes from recognizing that in any process there are "choke points" that determine overall plant capacity, and that unless those choke points receive adequate attention, they will restrict total plant output and productivity. Goldratt and Cox have drawn attention to the need to know where capacity restrictions exist in a plant and to deal with those restrictions, as needed, to improve overall plant capacity and productivity.

One dimension of the departure of the "behaviorist school" from the "Taylor school" (i.e, traditional industrial engineering) has been the reduced use of work standards and the labor performance reporting that engineered standards provided. There are two prerequisites to effective labor control by management: (1) management knows how long it takes to perform a task, and (2) the employee knows that management knows. If management does not know how much production is appropriate, blue-collar workers may give them "the quota," not to be confused with the quantity one would expect based on an engineered time standard. This quota is an amount of production that has historically been sufficient to keep management happy. The employees may think, "If management seems to be pleased with that much production, why should we kill ourselves to give them more? If they don't care, why should we?" Can the employees be blamed for not producing more? Is it not a failure of management, in not knowing the appropriate level of production and their relationship with their employees, that constitutes the underlying production system (i.e., management) failure?

I was once asked, as an additional increased-scope consulting effort, to look into a management suspicion concerning whether all seven mechanized floor sweepers in a plant were needed. A brief analysis determined that four sweepers would be more than adequate. According to behaviorist philosophy, Harold, one of the sweeper drivers, should have driven over to his supervisor's office and said, "Jack, you don't need seven sweepers, you only need four. I don't think you need me running this sweeper." In the typical industrial plant in the United States, do you expect Harold to say that? No; he just might get fired if he did. Harold may not be a rocket scientist, but he isn't a fool.

In another situation, there was a nine-person paint line crew. I realized that there were rarely more than five employees on the line at a time and recognized the need for industrial analysis. A subsequent engineering analysis indicated that only five people were needed and recrewing of six parallel production lines on four-shift operations (i.e., 24 hours per day, seven days a week) constituted an annual labor savings in excess of a third of a million dollars a year.

Consider another example involving a tipple crew (a tipple rotates a railroad car upside down to empty it). My associates and I had indicated a level of output that was clearly possible, but management declared that the estimate was well beyond reason based on past crew performance. The crew heard of the management response and then proceeded to outproduce the estimate, using the occasion to embarrass their management. The reaction of the crew speaks to their relationship with management.

In another plant, it was noted on a 20-minute first tour of a large electronics plant that four employees, on four identical benches, were all trimming leads on the back sides of printed circuit boards with hand-held diagonal cutters. At each bench, in the holder beside each employee, was a more efficient but heavier air-powered lead trimmer. I mentioned to my guide, an industrial engineer, that the plant did not seem to have effective labor control. She admitted that they had standards but that they "weren't any good," and she then asked how I knew. I told her that if the time standard for lead trimming had been based on the powered trimmers—a reasonable assumption because they were at each bench—it would have been impossible for the four employees to accomplish the required amount of lead trimming with their degraded, but more comfortable, method of employing hand-held diagonal cutters. The fact that such a situation could continue without management intervention indicated that the standards and the labor performance reporting system were not what they should have been. It also says something about the plant management's philosophy and dedication to productivity.

Note the contrast between American and Japanese management in two respects. In too many American plants, even with a supposed behaviorist management philosophy, management does not give the level of respect employees deserve, and managers often do not know how much their employees should produce. In the typical Japanese plant, respect is shown for and to employees, and management knows how much their employees should produce; the resulting Japanese accomplishment often greatly exceeds that of the comparable American plant. In Japan, workers are respected, and they produce; managers manage. It is really not all that complicated.

Assume for the moment that you are an alien and that you have landed your spaceship near an American industrial plant. You have altered your appearance to look human, and your plan is to enter the plant to observe, listen to those who work there, and, if necessary, engage them in conversation by posing as a visitor from another related company plant. Your mission is to determine how a typical American company competes with other national and international producers.

You enter the plant 30 minutes after the work shift started, but it does not appear that more than half of the employees are working. Those not working at workstations appear to be talking in groups of two to five employees. You also note, as you approach a production area some distance away, that it has a material take-away conveyor system and that there are only three assemblies on the 100-ft. conveyor. As you approach that section of the plant, some of the groups of people stop talking, go to their individual workstations, and start working and placing completed product on the conveyor. After standing and observing a machine for five to ten minutes, you note when you leave the area that the number of assemblies riding the conveyor has risen to 20. After walking another 100 ft. away from the area and pausing for five to ten minutes to observe a particular process, you look back at the conveyor and see only three assemblies riding the conveyor.

In walking through the plant office area, near a water cooler outside a conference room, you overhear the plant manager talking with the comptroller and the human resources director. The plant manager explains that "the production departments have been meeting their production goals [which are based on historical standards], but that is not good enough, because unless we do better, we will have to close the plant. Our costs are too high and we cannot compete at these cost levels." They then discuss the possibility of having a plant picnic to raise plant morale, and hopefully, the employees' motivation to work. (This is the "If you are nice to them, they will be nice to you" approach.)

You return to the plant and watch a typical employee running a punch press; you observe that the employee drops the scrap generated at the punch press onto the floor as he runs the press.

You then hear an alarm sound and note that a paint machine tender on the line has stopped a paint finishing line because corrugated box material has gotten into the paint rollers, preventing proper application of the paint. After removing soggy material from the rollers, the machine tender goes to a pile of folded corrugated box material, selects a piece, and puts it in the paint on the inside edge of the paint tank to keep waves of paint from going over the top of the tank. He then pushes a button, and the line begins again. Twenty minutes later, you note the same employee stopping the line again; this time he changes a paint filter on a paint feeder line that leads to his machine. After he changes the filter in the line, he turns on the finishing line again. It is apparent after a period of observation that the line is "down" more often than it is "up."

You also note that between the 4 ft. × 8 ft. sheets of material being painted are gaps of 6 to 10 ft. in length (i.e., the line is running at half of its potential capacity, as compared with having no gap). You ask the worker putting sheets on the line why there is an average gap of 8 ft. between sheets, and she tells you that her supervisor told her in "no uncertain terms" not to let any sheet get on top of another sheet and that she wants to make sure to avoid doing so.

You, the alien, may well have formed a preliminary conclusion that this plant is in trouble for the following reasons:

1 Only half of the employees are productive 30 minutes into the shift.

2 It appears that the production rate of the production area served by the conveyor belt varies considerably based on who is walking through the area at the time. The change in the number of assemblies—starting at a low of 3 before you passed by, peaking at 20 when you were at the line, and then returning to 3 soon after your departure—suggests a lack of consistent dedication to purpose.

3 If the employee running the punch press is dropping scrap onto the floor as he runs the machine, he will have to pick up the scrap on the floor after he completes the run and put it in a scrap container. This task may take as long as it takes to run the machine. It would be more efficient to place a scrap container at the press before he makes the run, so that he can place the scrap directly into the container. The fact that he is not using such a method indicates a lack of training concerning efficient methods of operation; it also shows that his supervisor lacks similar knowledge, concern for discipline, or interest in seeing that the most efficient method is employed.

4 It is apparent that the plant manager thinks that she should expect only what the historical standard says she should get. Item 1 above, however, suggests that she is not getting the production she should, so something appears to be wrong with the historical standards or the labor-reporting system. The plant manager is either unaware of, or does not wish to deal with, this problem.

5 When the employee at the paint machine puts corrugated material into the paint tank to prevent the paint from going over the edge of the tank, it should be apparent that the material will get soggy in time, thereby generating a future downtime in the paint rollers. Such circumstances are causing the numerous downtimes on the line, and the employee's actions say a lot about management of the line.

6 Shutting down the line to change the paint filter is presently required because there is only one paint filter line installed in the paint line that feeds the machine. If there were two filter lines in parallel, one line could be closed while the other is opened, and one filter could be changed while the other line feeds paint to the machine, thereby eliminating the downtime.

7 The fact that an employee is allowing an 8-ft. gap between sheets and is much more concerned about getting in trouble with her supervisor than about production speaks clearly about where the employees' interest must lie in this plant. It also demonstrates the relationship between management and employees in the plant.

These seven observations suggest that the line organization, plant maintenance, engineering, and therefore plant management are not performing well in this plant. If so, it is likely that the plant organization is also not performing other necessary but less visible functions and services properly; therefore, the assumption that the plant is in trouble is probably valid.

I sincerely wish that the above description were purely hypothetical, but such examples abound in existing American plants. For instance, in a matter of three hours of touring a facility, 70 photos were taken at my direction to demonstrate things the employees ought not to be doing in that facility (and even more important—things that management ought not to be letting them do). There is a "deadly disease" in American management today, characterized by managers who think, "Don't criticize employees; say positive things and demonstrate through positive examples instead"—permissive behavior is alive and well in industry. Refer to Fig. 2-63i and reread the caption. The slides were taken to develop awareness of the need for improvement and were shown as part of an in-plant seminar presentation. Most of the improvements would involve no, little, or only modest investment by management; a large majority are training issues.

Consider the confusion that would be added to the above example if the alien were also to view some combination of nonsensical union rules. The adversarial relationship that exists in unionized plants between management and labor could only be born out of tradition; no reasonable person would intentionally create such an unproductive work environment. A discussion of management is incomplete without some discussion of unionism in the United States.

There is a generally held belief that unions came into existence because they were the only means whereby workers could make a rather thoughtless management give serious consideration to their grievances. In the early days of the labor movement, there were valid grievances that begged for solution. Management's insensitivity to the issue of safety alone offered more than enough justification for the initiation of unions. At present, however, most corporations (though certainly not all) work hard at being thoughtful and responsible; they also would prefer not to have to deal with unions on a day-to-day basis. In a typical unionized company, it is possible by an imaginative combination of union rules for a union to drive productivity to essentially zero. Unions sometimes do this to punish management; unfortunately, they punish themselves and everyone else at the same time. One of the most frustrating aspects of unionism to a sincere management is the "manufactured" union incident, designed to convince union membership that their monthly union dues are really buying something.

There has been a considerable reduction in unionism in recent years in the United States. Only 16 percent of the workforce is now unionized. However, that doesn't mean that unions will disappear. When management manages poorly enough, an environment that fosters unionization is created. If the mismanagement continues, the workers will likely become unionized whether management likes it or not. A union will probably not make the plant more competitive or create a nicer work environment, but it is the price management pays for not managing well. Most companies have learned this lesson over the years. It seems only reasonable to assume, however, that there will always be a fresh supply of uninformed or inherently bad managers who will see to it, through their actions, that new unions will continue to form.

It seems reasonable that the space traveler in the above example might simply get in his or her spaceship and leave; he or she may have seen enough. Obviously, a world society that has been producing products as long as our world cannot be very advanced if the organization in a plant has progressed only to the level described above. The reality in the United States today is that if the plant described above is not typical, it is at least common. Typical U. S. trade magazines dedicate much article space to such topics as AI (artificial intelligence), CIM (computer-integrated manufacturing), and other similar "blue sky" topics when the typical plant in the United States could often be far better served with solutions that are simpler, more effective, and far less costly and esoteric. More will be said about this idea in Chapter 10.

WHAT AMERICAN INDUSTRY NEEDS TO DO

The intended point of the above plant discussions is that while management discussions continue ad infinitum on whether to use one theory or another, there are real problems in production plants today that urgently need to be addressed. I had the experience recently of being in a plant on the day that it was acquired by a French firm. Management did not know what their company name was that day, but they did know the names of their new owners, and learning to speak French went to the head of their list of good ideas. American managers need to learn to manage better so that their firms will not be acquired by foreigners.

There are those who believe that we do not need to maintain an industrial economy—that we should compete as an information society instead. *Megatrends 2000* [19], so well received just a few years past, suggested such an approach to our national dilemma. I do not subscribe to this theory; a successful manufacturing base is essential to a competitive national economy.

A. A. Imberman, in a study concerning job enrichment, surveyed workers for "suggestions they might have to make their job situations happier, more productive, and more satisfying." He described the results as follows:

> The sharpest complaints that the assembly line worker had about his job were (1) supervisors who didn't know how to supervise, were abusive, unfeeling, dictatorial, and unhelpful; and (2) poor management that was responsible for the poor planning of material, shortage of parts, ill-maintained machinery, inefficient working conditions (lights, drinking water, etc.), unbalanced inventories (trips to the tool crib wasted 15 minutes, when the foreman could maintain a small inventory of tools in the department), wrong size parts, poorly machined parts that didn't fit, ignored safety regulations, arbitrary management policies, and disregard of employee sentiments. [9, pp. 31–32]

There is little question that, prior to Maslow, Theory X prevailed and that it did not work well in normal work environments. There are exceptions, of course—such as guarding inmates on a road gang or leading an infantry charge. Brainstorming or taking a vote just does not seem appropriate in those environments.

The breadth of needs that employees possess and that management needs to address is certainly one of the lessons learned from the behaviorist movement. It also seems reasonable to assume that security and maintenance needs are prerequisite to motivational needs, as proposed by Herzberg. However, it appears to be true, as Lee [10] indicates, that much of what has been sold as "theory" by behaviorists is untested and advocative. Much of it cannot be proven or disproven. Nevertheless, there is much in behaviorist theory that makes intuitive sense and should therefore be embraced, whether it can be proven or not. A blend of behaviorist and industrial engineering approaches would encourage use of the following approaches today:

1 Eliminating the adversarial relationship between management and labor.

2 Attempting to align corporate goals to match employee goals, and vice versa, to minimize the mismatch between goals, as suggested by Likert [11].

3 Recognizing that employees face legitimate dissatisfiers in their work, which should be identified and eliminated, or at least minimized, to the extent possible.

4 Recognizing that some employees are more motivatable at work, whereas others possess primary motivations outside of work, as Fein [2] suggested. However, management should attempt to accommodate all employees across the full range of "inside to outside" motivations.

5 Utilizing participative management. Certainly a participative environment provides all employees with the opportunity to maximize their contribution to the overall effort, and all other parties benefit as well.

6 Encouraging the use of consensus management, because decisions made with full participation and clarification of all interests and points of view typically produce better results. When decisions are made without the benefit of input, the result is usually a bad decision. Simply having the opportunity to express an opinion makes it easier for an employee to support whatever decision is finally adopted.

7 Using praise and positive reinforcement to provide a more productive environment for mutual cooperation and motivation.

8 Selecting managers who are respected by those they will manage and who will create harmony within the group. This attribute is as important as the managers' technical attainments.

9 Developing employee empowerment, responsibility, and accountability, to the extent possible, when effective and appropriate.

10 Utilizing either group participation or individual work, depending on the nature of the task. Group participation is certainly beneficial in developing solutions and arriving at decisions that enhance commitment and bind the group toward mutually developed group goals.

What does a typical American employee want? I have no proven theory to guide me here, but I have talked at length with many employees. With rare exception, American workers want the opportunity to go to work, be effective in performing the work that they do, and be praised for doing it. They want to go home at the end of the day with a feeling of accomplishment. They want to tell their spouse and their children and anyone else who will listen what they accomplished that day, feel tired, and feel good about it. The following story, which is pure tragedy, makes the point.

A shop manager in a shipyard noted that one of the shipyard workers within his organization seemed hard to find on occasion. When the manager mentioned this to the employee's first line supervisor, the supervisor indicated that the worker produced excellent work very expeditiously when asked to do so but was indeed difficult to find sometimes. The supervisor made a special effort to keep track of the worker and was surprised to note that the worker was spending half of his time in the shop restroom. The supervisor became concerned that maybe the worker had a drug or alcohol problem. After an unsuccessful attempt to determine why the worker was spending so much time in the restroom, the supervisor finally invited the employee into his office and simply asked him.

The worker was very proud of his family's tradition of being shipyard workers. He learned his skill well and practiced it well, but he had great difficulty sitting out in the open shop, trying to look busy, when there was no work to do. He worked very expeditiously when he had work, but that left a lot of time with no work to do. He simply found it difficult to work slowly, so he chose not to do so. Because of his pride in being a shipyard worker, he simply elected to sit in the restroom when he had no work to do, rather than sit exposed in the shop.

Consider what management has done to this proud shipyard worker. He sits in a dark restroom stall to protect his justifiable pride in what he does for a living. The behaviorist movement does not solve this problem. In fact, this situation is more likely to occur under a behaviorist management system as now practiced in the United States, because most behaviorist-based systems prevent work measurement, which provides the basis for labor reporting and personnel requirements analysis.

The point to be made here is that Theory Y is not the culprit. The culprit is a behaviorist perspective that precludes engineering analysis of what employees are doing at workstations, as well as other aspects of production systems. Part of the problem is the sole reliance on the operator as the "expert" concerning operations, without the use of methods engineering, work measurement, and labor reporting to evaluate labor improvements and performance over time.

Many past behaviorist writings treated industrial engineering as "the enemy," when in fact the two approaches are quite compatible. The point is that good behaviorist philosophy, blended with modern industrial engineering, represents a powerful and effective combination of theory for managing in the future.

SUMMARY

This chapter summarizes many recent leading management theories offered by such authors as Maslow, Likert, Herzberg, McGregor, Ouchi, Fein, Blanchard and Johnson, Pascale and Athos, Peters and Waterman, Perot, Morita, Goldratt, and Lee. This chapter is an attempt to describe the management approach that has lead to today's managerial climate and conditions.

Upon contrasting the managerial approaches of Japanese and American managers, it seems apparent that the Japanese prefer a more compassionate, intimate, consensual, mutually respectful, and participative relationship between workers and managers.

The behaviorist philosophy that has developed over the past 40 years generally assumes that if direct employees are sufficiently motivated by their managers, the workers will ensure that the proper production system will come into being and will be effectively utilized. Management need do little else besides ensuring that their employees are sufficiently motivated. This approach was taught in human resources (HR) courses before the behaviorist theory was developed and was therefore fully adopted by HR management. Facilitating emplyee motivation is an important management responsibility, and one means to attain it is to provide employees with the proper equipment, tools, procedures, training, and other resources they need to do their jobs in an effective and efficient manner. Industrial and manufacturing engineering analysis can assist in defining and meeting these needs. It is time to accept that a greater utilization of industrial and manufacturing engineering capabilities will provide a better approach.

Lee performed an invaluable service by writing his atypical management text [10], which attempted to offer an objective assessment of management theory of the past 40 years. His work is noteworthy in pointing out both the strengths and weaknesses of the research performed to date in the field. He very convincingly indicates that the preponderance of behaviorist theory of the past 40 years appears to be advocative in nature, and not well supported by research. It appears evident that very little technically based research has been conducted to support such theory. What research has been conducted has often, in the final analysis, been inconclusive or contrary to the expected results.

Lee showed that strict adherence to behaviorist theory, as exemplified by the case involving Non-Linear Systems, Inc., is clearly ineffective. The full-blown experiment, which utilized an orthodox behaviorist philosophy, was an obvious and well-documented failure. Hopefully, the extensive damage done by blind adherence to behaviorist philosophy will lead to a more suitable approach that takes into account more than just managerial kindness in the establishment of a productive work environment.

Management should provide conditions that will permit workers to achieve their true work potential. Workers want to attain this goal.

REFERENCES

1 Blanchard, Kenneth, and Spencer Johnson: *The One Minute Manager,* William Morrow, New York, 1983.

2 Fein, Mitchell: *Motivation for Work,* Monograph No. 4, Institute of Industrial Engineers, Atlanta, GA, 1974.

3 Fein, Mitchell: *Improshare: An Alternative to Traditional Managing,* Institute of Industrial Engineers, Norcross, GA, 1981.

4 Goldratt, Eliyahu M., and Jeff Cox: *The Goal: A Process of Ongoing Improvement,* Rev. ed., North River Press, Croton-on-Hudson, NY, 1986.

5 Herzberg, Frederick, Bernard Mausner, and Barbara Snyderman: *The Motivation to Work,* 2d ed., John Wiley and Sons, Inc., New York, 1959.

6 Herzberg, Frederick: *Work and the Nature of Man,* World Publishing Co., New York, 1966.

7 Herzberg, Frederick: "One More Time: How to Motivate Employees," *Harvard Business Review*, Jan–Feb, 1968, p. 62.
8 Homans, George C.: "Conversation—An Interview with George Homans," *Organizational Dynamics*, Autumn 1975.
9 Imberman, A. A.: "Assembly Line Workers Humbug Job Enrichment," *The Personnel Administrator*, March–April, 1973.
10 Lee, James A.: *The Gold and the Garbage in Management Theories and Prescriptions*, Ohio University Press, Athens, OH, 1980.
11 Likert, Rensis: *New Patterns of Management*, McGraw-Hill Book Co., New York, 1961.
12 Lincoln, James F.: *Incentive Management*, Lincoln Electric Company, Cleveland, 1951.
13 Maslow, Abraham H.: *Motivation and Personality*, Harper & Row, Publishers, Inc., New York, 1970.
14 Hodson, William K. (ed.): *Maynard's Industrial Engineering Handbook*, 4th ed. McGraw-Hill, Inc., New York, 1992.
15 McGregor, Douglas: *The Human Side of Enterprise*, McGraw-Hill Book Co., New York, 1960.
16 McKeon, Richard (ed.): *The Basic Works of Aristotle*, Random House, New York, 1941.
17 Meggison, Leon: *Personnel: A Behavioral Approach to Administration*, Richard D. Irwin, Inc., Homewood, IL, 1972.
18 Morita, Akio, with Edwin M. Reingold and Mitsuko Shimomura: *Made in Japan*, E. P. Dutton, New York, 1986.
19 Naisbitt, John: *Megatrends 2000: Ten New Directions for 1990*, Morrow Publishing, New York, 1990.
20 Ouchi, William G.: *Theory Z: How American Business Can Meet the Japanese Challenge*, Avon Books, New York, 1982.
21 Pascale, Richard, and Anthony Athos: *The Art of Japanese Management*, Simon and Schuster, New York, 1981.
22 Perot, H. Ross: "How I Would Turn Around GM," *Fortune*, February 15, 1988.
23 Peters, Thomas J., and Robert Waterman, Jr.: *In Search of Excellence: Lessons from America's Best Run Companies*, Harper & Row, New York, 1982.
24 Spock, Benjamin: *Dr. Spock's Baby and Child Care*, E. P. Dutton, New York, 1985.
25 Townsend, Robert: *Up the Organization*, Alfred A. Knopf, Inc., New York, 1970.
26 Vroom, Victor H.: *Work and Motivation*, John Wiley and Sons, Inc., New York, 1964.

REVIEW QUESTIONS AND PROBLEMS

1 What is the triad of interests so essential to effective operation? Which of the three legs of the triad is typically the greatest determinant of how well an operation performs on a day-to-day basis?
2 What are the five categories (i.e., levels) of needs in Maslow's hierarchy of needs?
3 Does learning about company policy from management motivate employees to perform more effectively? Would you recommend frequent discussions of company policy by management with employees?
4 What do Herzberg's self-actualization theory and "CEO workaholics" have in common?
5 Is there any evidence to suggest that successful managers from both Japanese and American firms have similar managerial traits?
6 Do Blanchard and Johnson prefer autocratic or democratic managers?
7 Based on quotes from Pascale and Athos, what emphasis does the Matsushita Corporation place on efficiency and production engineering?

8 What do managers' egos have to do with industrial engineering?
9 Do Japanese generally consider themselves to be lower, middle, or upper class in their own society?
10 What book should be required reading for American managers?
11 What main point does Goldratt make about choke points in a process?
12 What are two prerequisites to labor control?
13 Why is Lee's text, *The Gold and the Garbage in Management Theories and Prescriptions*, an atypical management text?
14 According to Lee, how much "supportive evidence" is there for Maslow's hierarchy of needs theory? Does Maslow suggest the level of testing of his theory?
15 What relationship is there between Maslow's hierarchy of needs theory and Aristotle's hierarchy of the soul?
16 A commonly held belief of behaviorists has been that job satisfaction leads to job performance. What does Lee state concerning the level of evidence in support of this hypothesis?
17 Does Fein believe that most workers find fulfillment at work?
18 How do unit overhead costs in a typical incentive plant compare with unit overhead costs in a daywork plant? What is a typical percent change in pay for an employee in an incentive plant compared with a daywork plant?
19 If employees take home more pay from an incentive plant, does the additional employee compensation cost take away from profit? Explain.
20 What did McGregor learn as a result of his experience as president of Antioch College?
21 Non-Linear Systems, Inc. embraced behaviorist theory. Describe the results of their experiment.
22 Is the cost of fixing many of American management's industrial problems simply infeasible?
23 Are behaviorist theory and industrial engineering theory incompatible?

5

TOTAL QUALITY MANAGEMENT

Total quality control, Japanese style, is a thought revolution in management.

Kaoru Ishikawa

As mentioned in Chapter 3, quality control has both technical and managerial components. The quality control material in Chapter 3 deals with statistical process control (SPC) as one of many control procedures; this chapter focuses on the management component of quality control.

Total quality management (TQM) is a management philosophy concerned with broad-based, continuous quality improvement throughout an organization. In its totality, it has more to do with management than quality. TQM is also difficult to define, which is typical of an emerging subdiscipline; it is what professionals say it is, and they have different opinions. The author has endeavored to review in this chapter the most common aspects of this subdiscipline.

"Quality" has been an increasingly popular stated objective of organizations during the past half-century. The following briefly describes the historical sequence of these changes in emphasis:

Inspection—sorting the good from the bad.

Quality control—continued inspection, but including recognition of the need to identify and eliminate poor quality at the source.

Statistical process control (SPC)—the application of statistics to quality control.

Quality assurance—quality control, but including (i.e., promising) sufficient process analysis and administration to ensure that the required customer quality requirements are met. It is more audit based than previous direct control had been.

Product assurance—quality assurance that recognizes the need for and extends the search for quality solutions in production to product design, which is the primary source of most quality problems.

Total quality control—product assurance, including a commitment to continuous improvement employing participative management, and making the line organization and suppliers primarily responsible for quality performance.

Total quality management—total quality control, including continuous review and assessment of all operational and management policies, procedures, and practices, to perfect quality performance in meeting customer needs.

Any practicing industrial engineer may well disagree with the above definitions. The definitions of these popular terms vary depending on who read which article or book at what time, and then what one believes the words and thoughts expressed might or should have meant. Inspection, quality control, quality assurance, product assurance, and some of the means to product assurance (concurrent or simultaneous engineering and producibility engineering, discussed in Chapter 2) are predominantly engineering management issues.

In 1961, A. V. Feigenbaum wrote *Total Quality Control* [7]. His book was typical of the period in predominantly addressing statistical quality control. The book, however, did address more of the management aspects of quality control than had previously been the case. The book title alone proposed a broader scope for quality control, beyond the purely technical materials that had traditionally been covered. Quality control had been an engineering topic, taught primarily in industrial engineering academic programs and practiced by engineers; it was not considered an element of the primary management system of a firm.

In 1965, Robert N. Lehrer wrote *The Management of Improvement* [13]. The primary thesis throughout his text was that attention needs to be focused on improvement in organizations. He proposed a Vice-President of Improvement position in typical organizations to ensure that the long-term strategic plans, as well as the short-term quality and productivity improvement programs, were being developed, nurtured, and managed. The book was ahead of its time; along with Feigenbaum's, it served as a catalyst in initiating the considerable improvement focus in the United States today. Of course, one of the primary motivators for many improvement efforts today is simply survival. There is a sense in today's highly competitive world economy that if a company does not learn and adopt improvements in quality and productivity, it may well not survive. Survival can be a powerful motivator.

In 1979, Philip B. Crosby wrote *Quality Is Free* [4]. He had earlier headed the quality control function in the International Telephone & Telegraph (ITT) Corporation, and had developed the "zero defects" program, which was adopted almost universally as a quality control philosophy throughout the Department of Defense as well as in a broad range of commercial corporations. His "quality is free" theme was somewhat analogous, in a sense, to an earlier theme espoused by the Du Pont Corporation, that safety not only pays its way but is an excellent investment. In the foreward material for his book, Crosby offers a quote from his previous chief executive officer at ITT, Harold Geneen: "Quality is not only right, it is free. And it is not only free, it is the most profitable product line we have" [4].

Typical of Crosby's compelling arguments concerning quality control was his description of the "factory within the factory." The factory within the factory was the typical 20 percent of total manpower, materials, and other resources involved in reworking defective product to make it acceptable product. His intuitively appealing concept was that it costs far less to do it right the first time, which is hard to dispute. To a large degree, it represents the cost of lack of discipline in the total process of manufacture (i.e., lack of discipline in the way products are conceived, designed, specified, made, tested, packaged, warehoused, etc.).

Crosby's texts, of which there are many, are not technical treatises as such; they are more managerial in nature. In some ways, they are somewhat reminiscent of Norman Vincent Peale's *The Power of Positive Thinking* [16]. Although this kind of material might not get the author an invitation as a speaker in an engineering graduate seminar series at a prestigious engineering school, it does offer the practical philosophy that can guide management in improving product quality while simultaneously reducing the cost of manufacture. It is often not necessary to be technically complicated to be highly valuable, and quite often the true assets are not technical, but simply well-packaged and appropriate common sense. (Frankly, all science is common sense if and when you finally understand it.) For a corporation, the bottom line is where it's at, and Crosby is bottom line stuff, assuming of course that the seminar participants listen and act effectively on what they have been taught. Like most prophets, Crosby can lead you to water but he cannot make you drink; that you have to do yourself.

In 1982, W. Edwards Deming authored *Quality, Productivity, and Competitive Position* [5]. Throughout his text he offered numerous references to the responsibility of management to embrace and get involved in quality control as a necessary function concerning the well-being of their plants. His frustration in attempting to get uninvolved management to do its job is apparent in the following quotes from his text:

> There is much talk about how to get employees involved with quality. The big problem is how to get the management involved. [5, p. 62]

> An important obstacle is the supposition that improvement of quality and productivity is accomplished suddenly by affirmation of faith. Letters and telephone calls received by this author disclose prevalence of the supposition that one or two consultations with a competent statistician will set the company on the road to quality and productivity—instant pudding. "Come, and spend a day with us, and do for us what you did for Japan; we too wish to be saved."... One man actually wrote to me for my formula, and the bill therefore. [5, p. 65]

> The supposition is prevalent the world over that there would be no problems in production or in service if only our production workers would do their jobs in the way they were taught. Pleasant dreams. The workers are handicapped by the system, and the system belongs to management. [5, p. 68]

> It is difficult to understand how any economic upturn of importance can take place in the United States till products made there become competitive at home and abroad. The only possible answer lies in better design and greater productivity. Better management can

> bring improvement in both. The big question is, how long will it be until top management becomes active in their responsibilities? and then how long? [5, p. 87]

Regarding attempts in Japan to bring the importance of quality control to management, Deming states [5, p. 102]:

> It was vital not to repeat in Japan the mistakes made in America. Management must understand their responsibilities.

Ichiro Ishikawa, president of Kei-dan-ren (Federated Economic Societies) and president of JUSE (Japanese Union [Society] of Scientists and Engineers) sent telegrams inviting top industry leaders to meet with JUSE, which they did. In 1950, Dr. Deming presented an eight-day seminar for JUSE. The topics of the seminar were (1) How to use the Deming Cycle (Plan, Do, Check, Action), (2) Dispersion in statistics, and (3) Process control through control charts.

The Japanese have been very grateful to Dr. Deming for the part he played in the development of quality control in Japan. Their highest award—the Deming Award—was named in his honor. The Deming Award is awarded each year to the organization that demonstrates the greatest overall accomplishment in the practice of quality control.

Deming offers his following 14 points for management, slightly revised in his latest text, *Out of the Crisis* [6, pp. 23–24]:

> **1** Create and publish to all employees a statement of the aims and purposes of the company or other organization. The management must demonstrate constantly their commitment to this statement.
> **2** Learn the new philosophy, top management and everybody.
> **3** Understand the purpose of inspection, for improvement of processes and reduction of cost.
> **4** End the practice of awarding business on the basis of price tag alone.
> **5** Improve constantly and forever the system of production and service.
> **6** Institute training.
> **7** Teach and institute leadership.
> **8** Drive out fear. Create trust. Create a climate for innovation.
> **9** Optimize toward the aims and purposes of the company the efforts of teams, groups, staff areas.
> **10** Eliminate exhortations for the work force.
> **11 a** Eliminate numerical quotas for production. Instead, learn and institute methods for improvement.
> **b** Eliminate M. B. O. Instead, learn the capabilities of processes, and how to improve them.
> **12** Remove barriers that rob people of pride of workmanship.
> **13** Encourage education and self-improvement for everyone.
> **14** Take action to accomplish the transformation.

The recommended 14 points are valid, with the exception of number 11. It is agreed that expecting production accomplishment on a daily or hourly basis may be, and often is, ill-advised, depending on process conditions. However, it simply seems unrealistic to imagine a management control system that does not forecast, schedule, and track plant accomplishment over time. Work standards provide a means for management to do that on an engineered basis over a sufficient review period. It need not be oppressive, but it needs to be a part of the management control system. It is difficult to imagine managerial leadership that does not have the desire, intention, and ability to measure and track progress against plan.

In defense of Deming's admonition concerning the use of standards, however, it is recognized that, combined with ineffective management, inappropriate standards can limit accomplishment. The following example is offered to make the point.

While reviewing department production standards for a large southeastern U.S. plant some years ago, I noted that the sum of individual operation-engineered standards in a department, including allowances, represented approximately 70 percent of the existing department standard, including allowances. The industrial engineering manager confirmed that was the case, explaining that factors (expected additional downtime, slow maintenance response, etc.) had caused adjustments to the sum of individual standards over time. The more generous department standards provided for longer downtimes and other accomplishment-degrading factors that had crept into the department standards over the years. The departments were all experiencing department accomplishments, relative to the department standards, in the range of 95 percent to 110 percent. When I asked the question "If maintenance, engineering, warehousing, production control, material handling, and all other supporting departments were doing their job, would the sum of individual work standards constitute a valid standard to judge departmental accomplishment?" the answer was "Yes."

My recommendation was to utilize the sum of individual engineered work standards as the department standard. When this possibility was discussed with the general manager of the facility, he indicated that such an action would result in reported department accomplishments dropping from 100 percent to 70 percent on average, and that would cause considerable consternation at the department manager level. He accepted the recommendation, however, and chose to implement the revised reporting of departmental performance the following month. When he informed his production department managers and supporting department managers of his decision, they tried their best to talk him out of it, but the new reporting basis continued. In the first month reported, typical accomplishments were at the 70 percent level, but in the months that followed, after pressure was put on maintenance by the line organization to do maintenance, and engineering to do engineering, etc., improved departmental accomplishments followed, first in the upper 70s, then in the 80s; I do not know the final levels attained, having lost touch since then, but I would not be surprised to find that they are now in the same range of 95 percent and 110 percent as they had been previously. People want to be challenged. When properly and equitably challenged, appropriate accomplishments happen. But note that for a challenge to be equitable and, therefore, proper, engineered standards are needed to determine the level of accomplishment to be employed as a standard. Engineered standards are essential for establishing demonstrable and, therefore, credible challenges. Under-challenging is a serious problem in American industry today. Over-challenging can be equally ineffective.

Tracking accomplishment over time is critical in managing any plant. When under-capacity scheduling is embraced as employed in Japan, thereby allowing for line stops, daily production accomplishment is so much more predictable. Note that under-capacity scheduling typically also includes "stay until daily production is met" as well. When an American firm embraces both engineered standards as an aid to production planning and under-capacity scheduling, the production control department often feels as though they have died and gone to heaven. Meeting daily forecasted production schedules, thereby eliminating the end-of-month rush to make monthly shipments, is a significant "quality of working life" improvement in the lives of production and inventory control personnel, and production personnel as well, once they get used to it.

In addition, the author frankly neither understands nor agrees with Dr. Deming's insistence on eliminating management by objectives (MBO) goals and annual merit ratings, sometimes limited to managers and staff. The author does not view these as "robbing people in management and in engineering of their right to pride of workmanship"; on the contrary, it permits each employee, and manager, to receive the feedback and guidance he or she typically wants and expects to permit him or her to improve professionally over time.

Of course, statistically speaking, it seems unlikely that a statistician and an industrial engineer are going to agree on all 14 points anyway; their life experiences are simply too different to expect that level of agreement. The statistician (Deming) and the industrial engineer (the author) do substantially agree, however, that the 14 points, with possibly one or two exceptions, will light the way for management emphasis in the future, if the United States is to successfully compete in the world economy of tomorrow. It seems appropriate to note, however, that we are agreeing on management basics more than on quality control theory.

Richard J. Schonberger, an industrial engineer in the Kawasaki Corporation, in his book *Japanese Manufacturing Techniques* [20] compares American and Japanese quality control concepts. Foremost in the total quality control approach is the placement of the responsibility for quality control in production. In the United States, quality control developed as a staff engineering function. As a staff function it could recommend, but if it got in the way of the line organization (i.e., production) it almost always lost politically. Plant managers have to support the line organization if they hope to meet their immediate production goals. American plants have traditionally been notorious for "waiving" quality problems in an often desperate attempt to meet production goals. It often meant shipping less-than-ideal product to the customer, with the resulting inevitable quality reputation of shipped American product that is now part of our history. The Japanese simply, and quite wisely, made production (i.e., the line organization) responsible for shipping only quality product. The staff quality engineering department then became an available technical resource to assist the line organization. The QC staff then devoted its resources to eliminating quality problems at the source, which is what they should have been doing all along, instead of wasting time in lengthy meetings that everyone involved knew would culminate in waiving marginally acceptable, or even marginally defective, product.

Consistent with comments made earlier concerning Crosby's conceptual approach to quality control, Schonberger states the following concerning TQC [20, pp. 50–51]:

> I believe that the statistical tools are not quite so important in the spectrum of concepts that constitute Japanese total quality control as are some of the more conceptual factors.

Figure 5-1 lists the primary TQC concepts, by categories, that Schonberger describes in some detail throughout his text. His list contains most of the primary concepts presently associated with TQC generally. The primary concept concerning organization is that quality should be a line responsibility, not a staff responsibility as it has traditionally been in the United States. This concept assumes that the quality control department, a collection of quality engineering and technical specialists, is an available resource to the line organization to assist them in meeting their quality objectives; it also assumes that quality plans are not enough—quality must be embraced by those who make the product.

The concepts contained in the Goals category, "habit of improvement" and "perfection," are in my judgment the largest cultural barrier for Americans to overcome. It is general knowledge that order is revered and culturally embraced by the Japanese and Germans. Japanese and German workers want to know how an operation should be performed so that they can then take pride in demonstrating to management how they faithfully and repetitively execute that operation exactly as instructed. Americans revere freedom, a desired societal benefit, which when embraced as part of an industrial process creates disastrous results. In the present

FIGURE 5-1 Total quality qontrol: concepts and categories. (From Schonberger [20, p. 51].)

TQC Category	TQC Concept
1. Organization	Production responsibility
2. Goals	Habit of improvement Perfection
3. Basic principles	Process control Easy-to-see quality Insistence on compliance Line stop Correcting one's own errors 100 percent check Project-by-project improvement
4. Facilitating concepts	QC as facilitator Small lot sizes Housekeeping Less-than-full-capacity scheduling Daily machine checking
5. Techniques and aids	Exposure of problems Foolproof devices $N = 2$ Analysis tools QC circles

vernacular, it is called "Do your own thing." It is one of the major sources of process variability and subsequent process waste, equipment downtime, and low yields in American industry today.

A similarly unfortunate cultural attribute of Americans is called "good enough." This phrase is an admission that the product is not what it ought to be, but if someone will agree that it is "good enough" and sign a waiver, we can ship it. For the A&E, it may be the final report with typos and sentence structure and grammar problems that does not present the A&E corporate image one would choose to leave with a client. TQM education can significantly reduce such deficiencies.

The concepts in the Basic Principles category are fairly self-explanatory; they are concerned with applying statistical process control, charts at each workstation to display quality control conditions, and yellow and red lights at each station so that a production line worker can indicate that he is either getting into trouble or in sufficient trouble to require shutting down the line. Operators correct their own line problems. The Japanese believe in 100 percent, preferably automatic, inspection of every item produced at a workstation, for instant removal of scrap and resetting of equipment. Each plant typically has two large lists of projects, on a wall visible to workers, showing quality problems previously solved and identified quality problems yet to be solved.

The Facilitating concepts include the quality control department as an available resource. Emphasis is on reduction of setup times, which permits small economic production lot sizes (i.e., Just in Time), which permits immediate feedback and correction of any developing quality problems. Housekeeping is an obvious desirable trait, as is daily machine checking. Less-than-full-capacity scheduling is a commitment to produce daily production requirements, even when it is necessary to shut down the line during the day to meet quality requirements or put in some overtime to meet schedule. If time is left over, quality circles meet to discuss opportunities to improve quality or efficiencies on the line. Compared to the American method of attempting to ship production requirements on the last day of the month and then resting the first week of each month to recover from last month's effort, thereby falling behind, the rewards in terms of predictable, uniform production output are obvious.

The Techniques and Aids concepts include exposure of problems, foolproof devices (production and inspection), last-piece inspection as well as first-piece inspection (i.e., $N = 2$), all available analysis tools (e.g., the Ishikawa fishbone cause and effect diagram, to be described shortly), and quality circles.

One concept missing from Schonberger's list is the participative management approach. A typical TQC effort today will likely involve functional and cross-functional team efforts as well as additional internal and external professional, ad hoc team efforts. Such an effort will also probably include, at minimum, a management steering committee.

A fundamental, yet monumental, difference in perspective under TQM as compared to TQC is that in TQM *all* corporate policies, procedures, and practices, including management and supervisory, are placed under scrutiny. Such a philosophy of management requires that all management policies, procedures, and practices be under continuous review and be made accountable in terms of their ultimate effects, potentially causing a needed revolution of the management culture in a corporation.

Probably the one TQC concept that best distinguishes Japanese quality control and culture from American quality control and culture is "habit of improvement." The Japanese have a name for it—*kaizen*. At the end of the day, if an American foreman has met his production goal for his department, he goes home with a feeling of total accomplishment. If a Japanese foreman in a similar department has had the same result at the end of the day, he may well go home feeling he has failed that day. At no time during the day did he spend any time improving his operation. All he accomplished for the day was production. He may well enter the plant tomorrow and discuss his concern with his employees so that they will endeavor to complete their production requirement in sufficient time to devote some time to discussing their efforts to improve their operation.

One TQM concept not mentioned in Fig. 5-1 is the elimination of receiving inspection, as Schonberger states [20, p. 65]:

> One type of inspection that Western [American] QC departments do a lot of and Japanese try to do none of (except for new suppliers) is receiving inspection.... By taking care of quality at the source [suppliers], Japanese industry has, over the years, been able to phase out the need for receiving inspections.

The Japanese simply do not want defective material contaminating their production operations. They first attempt not to make defective product, and if any defective product is made, they want to discover it as soon as possible and remove it from the process. They have a preference for automatic part-by-part inspection at the machine immediately following and integrated with its production.

The field of quality control started with the invention of the quality control chart by W. A. Shewhart of Bell Laboratories in the 1930s. Statistical process control (SPC) became widespread in the United States following the introduction of engineering texts such as Grant's statistical quality control text mentioned in Chapter 3.

Kaoru Ishikawa and some of his associates became involved in quality control in 1949. Ishikawa is to Japanese quality control what Shewhart, Grant, Deming, and Juran are to American quality control. He is one of the primary founders of modern quality control today.

In 1981, Ishikawa wrote *What Is Quality Control? The Japanese Way* in Japanese. In 1985, David J. Lu provided an English translation entitled *What Is Total Quality Control? The Japanese Way* [11]. The foreward material to the book contains comments from some of Ishikawa's consulting clients, such as Ford Motor, Bridgestone, Komatsu, and Cummins Engine. H. A. Poling, CEO of the Ford Motor Company, in his comments indicated that "In 1981, Ford Motor Company began a very intense effort to improve product quality to achieve 'Best-in-Class' levels in all world automotive markets." Poling states the following [11]:

> Our first involvement with the Japanese Union of Scientists and Engineers (JUSE) was in March 1982 when Ford sponsored a statistical study mission to Japan. This began a

> long and fruitful relationship with Dr. Ishikawa which continues today. In 1983, Dr. Ishikawa was kind enough to visit Ford on several occasions and conduct top-management training seminars for senior Ford executives, including myself, and in various positions including general management, product engineering, manufacturing, and marketing.
>
> Our experiences with Dr. Ishikawa have helped us develop a whole new approach to quality.

Note in the above quote that Mr. Poling indicates that he attended the seminar. His attendance at the seminar lies at the heart of total quality control as envisioned by Dr. Ishikawa.

On the first page of his text [11], he states, "Total quality control, Japanese style, is a thought revolution in management." When one reads his text it becomes apparent that this is not as much a book on quality control as it is a book on management.

He goes on to say, "My wish is to see the Japanese economy become well established through QC and TQC and through Japan's ability to export good and inexpensive products world-wide. It will follow that the Japanese economy will be placed on a firmer foundation, Japan's industrial technology will become well-established, and Japan will be in a position to engage in the export of technology on a continuous basis." Did his wish come true? Yes, it did. Do we in the United States wish the same for ourselves? Of course we do. Then it is time for American management to "get with it." Ishikawa further states, "To implement TQC, we need to carry out continuous education for everyone, from the president down to line workers" [11, p. 13]. In fairness to many managements, there is considerable movement in the United States today toward embracing modern quality control, but not as much as one might wish to see.

The author developed and annually teaches a one-week seminar for the Institute of Industrial Engineers entitled "Introduction to Industrial Engineering." Some years ago, at the completion of one of the seminars, a young lady walked up to me and said, "I have a problem, and I suspect other attendees do also. Everything you have taught me won't do me much good unless you teach a seminar to my boss." To this day, I have not yet determined how to get her boss, and all other bosses like hers, to attend the seminar. Teaching principles of better quality and productivity to engineers and technicians will produce only limited results if the managers they report to do not understand and appreciate the concepts involved.

Chapter 7 of Ishikawa's book is entitled "Dos and Don'ts for Top and Middle Management," and begins with the following list of topics to be covered in the chapter [11, p. 121]. Consider whether they are quality control issues or management issues.

1 If there is no leadership from the top, stop promoting TQC.

2 QC cannot progress if policy is not clear.

3 Organization means clarified responsibility and authority. Authority can be delegated but responsibility cannot.

4 QC cannot progress without attacking middle management.

5 Strive to become a person who does not have to be always physically present at the company, but become a person who is indispensable to the company.

6 A person who cannot manage his subordinates is not half as good as he is supposed to be. When he is able to manage his superiors, then he can be called an accomplished person.

As was the case with Deming's 14 points previously discussed, these are basic management issues. As stated at the beginning of this chapter, TQM has more to do with basic management than quality control. The cover page for Chapter 3 of Ishikawa's text is equally descriptive of the management nature of the material covered in his text [11, p. 43]:

1 The first step in QC is to know the requirements of consumers.
2 Another step in QC is to know what the consumers will buy.
3 One cannot define quality without knowing the cost.
4 Anticipate potential defects and complaints.
5 Always consider taking appropriate action. Quality control not accompanied by action is mere avocation.
6 An ideal state of quality control is where control no longer calls for checking (inspection).

In a section of his text entitled "How to Proceed With Control" he states, "Dr. Taylor used to describe control with these words, 'plan-do-see'" [11, p. 59] and then he discusses Deming's PDCA Circle, which he then expands to the control circle shown in Fig. 5-2.

Dr. Ishikawa developed a chart that he called a cause and effect diagram, which has come to be called a fishbone diagram because of its appearance; it is useful in recording and classifying possible causes in a search for the source of a quality problem. It is typical of the very simple and yet very practical approaches Dr. Ishikawa has developed from over 40 years of technical experience. Figure 5-3 is a typical fishbone diagram from Schonberger's text [20].

One story, believed attributable to Ishikawa, tells of a plant situation in which sheet material coming out of a set of rollers in a process displayed a defect. Ishikawa and his engineers analyzed the defect in the material and determined that an adjustment was necessary a number of process steps back in the process. An adjustment was then made in the earlier process step and when the material went through the subsequent process steps the defect was no longer apparent. His engineers were jubilant and were rejoicing in their success, but Dr. Ishikawa informed them that they were only half through in solving the problem, from a total quality control (TQC) perspective.

He went on to explain to his engineers that their process was "on line" at the time, and if the process specification and testing procedures had been correct, they should not be seeing this type of problem at this point in the process development. He then directed their efforts to identifying what part of their process testing procedures was deficient in missing the process step adjustment that led to the defect they had just observed. This story, whether true or not, is an attempt to explain a fundamental underlying concept of total quality management. "Total" means that any influence on production, whether it be material selection, tolerances, design, tooling, process testing procedures, process specifications, standard operating pro-

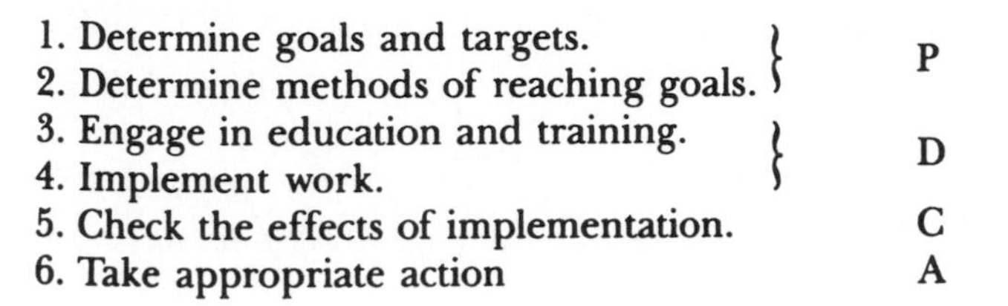
1. Determine goals and targets. } P
2. Determine methods of reaching goals.
3. Engage in education and training. } D
4. Implement work.
5. Check the effects of implementation. C
6. Take appropriate action A

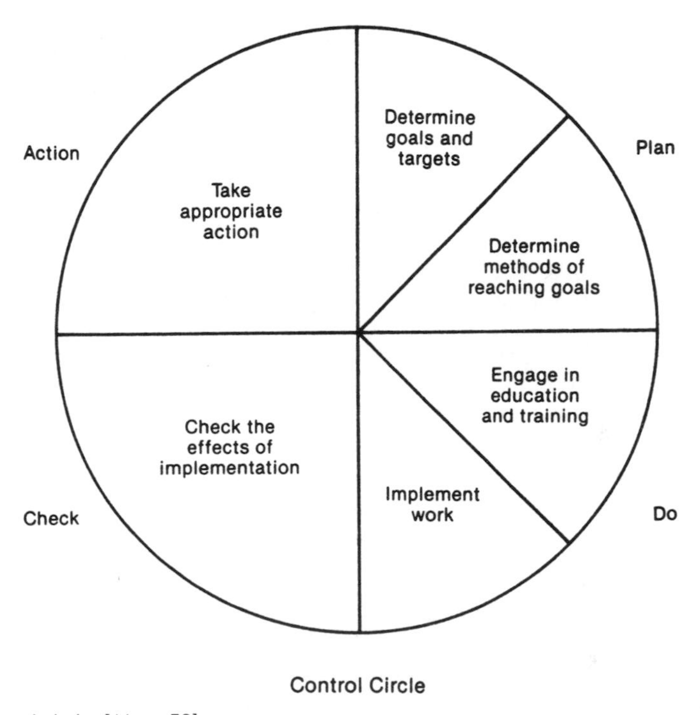

FIGURE 5-2 Control circle. [11, p. 59]

cedures, management operational procedures, management dictates, management policy, or restrictions and roadblocks, is open to reevaluation if it negatively impacts timely and cost-effective quality production.

For those process and design engineers who have had to live with numerous misguided management dictates in the past, which are common, this is a significant improvement. A total quality control operational philosophy gives them the right and duty to question all impediments to quality, cost, and schedule improvements. What a blessing! To a student reading this, it may not seem that important. To an engineer trying to do things right in an environment in which management dictates and associated management egos are major impediments to doing it right, this is very important. It says "You can do it right"; management gives you that right. You should also be cognizant that egos can be easily bruised, however.

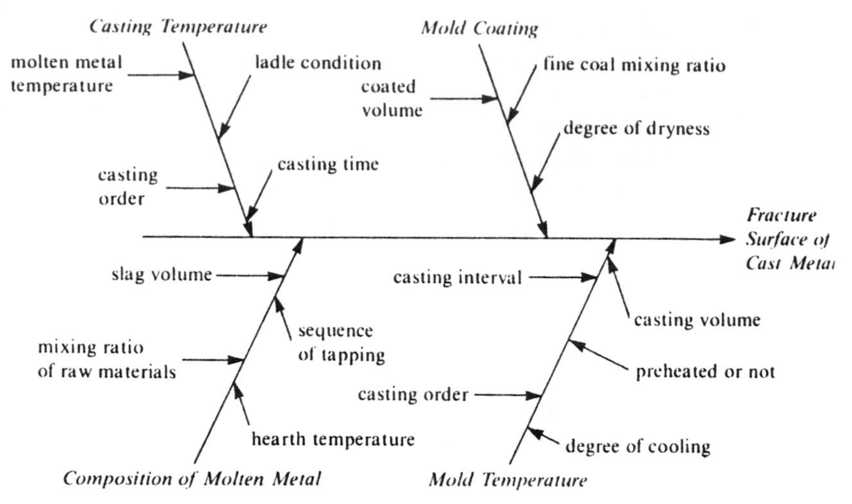

FIGURE 5-3 Fishbone diagram. (From Schonberger [20, p. 72].)

Much of what we know as total quality control (TQC), or the latest label, total quality management (TQM), comes from the efforts of Dr. Ishikawa and his colleagues. We owe them a considerable debt of gratitude.

Over the past 40 years, I have spent a considerable amount of time talking to plant people at all levels, and I have developed a sense of their concerns, their situation, their gripes, their frustrations concerning lack of support, etc. The Japanese authors mentioned thus far are "plant people." They know of what they speak.

American writers who write management texts predominantly are not plant people. Those that claim to be plant people are often not the plant people one meets in the back of the plant, but rather the people one meets close to the boardroom; these people are sometimes referred to as the "deep carpet folks." Those that are not front office plant people are more often than not present or past MBA school professors. When they visit the plant they typically come in the front door and go directly to the conference room in the deep carpet area. What they typically know about the plant is based on what they saw on a 20-minute tour, or what other front office people have told them. Could this be part of our problem?

Here is one other problem. Think about what an opportunity it represented, as painful as it must have been, for the Japanese to start over again at the end of World War II, including elimination of the *kaibatsus* (holding companies). Think of the opportunity it would represent in the United States, for example, if the U. S. Congress could be replaced—and it could be—and the primary, secondary, and high school educational bureaucracies of the United States, and many other corporate bureaucracies (e.g., General Motors), and simply start over again. It is truly an unfair advantage that Japanese society has had in recent years compared to American society.

There is another problem facing the United States in our present economic war with the other competing economies of the world, especially Japan. It was noted above that a foremen's society was started in Japan in the 1940s. Is the English translation of the name of that society "The Japanese Society of Foremen"? No. It is the Japanese Union [Society] of Scientists and Engineers. The reality today is that Japan graduates twice as many engineers as the United States and their population is half that of the United States (i.e., they produce four times as many engineers as we do per capita; we probably outproduce them in guitar players per capita). Many of those graduate engineers become the first-line foremen of Japanese industrial companies. What background does a typical first-line foreman in a U.S. plant have? History, music, English, education, agriculture, general studies, sociology, agronomy, recreational sports, hotel and restaurant management, literature, retired guitar players, etc. Do American foremen have a society, specifically for foremen, that permits them to improve their skills as foremen? Not to this author's knowledge. If there were a society, would any of these people join and actively participate in the society? Do they have the prerequisite skills to participate (e.g., math, physics, chemistry)? Would their respective managements pay for them to be members? Does American management invest in training for its first-line supervisors to keep them abreast of developments? The ugly truth, on average, is that they do not to the extent that they should; they are typically too fixated on how their quarterly profits will be viewed by the financial types on Wall Street. What does all this have to do with an introduction to industrial engineering and management? It portrays the environment in which industrial engineers and supervisors must practice; the more you know about it, the more prepared you will be to participate in it effectively. It was noted earlier that quality control has both technical and managerial components. Some QC professionals prefer to stay with the safe ground of technical aspects and have been known to denigrate the management component. The following quote from Ishikawa speaks to this issue [11, p. 18]:

> It is true that statistical methods are effective, but we overemphasized their importance. As a result, people either feared or disliked quality control as something very difficult. We overeducated people by giving them sophisticated methods where, at that stage, simple methods would have sufficed.

In 1986, Masaaki Imai wrote the book *Kaizen* [10]. Imai states the following [10, p. xxix]:

> Kaizen strategy is the single most important concept in Japanese management—the key to Japanese competitive success. Kaizen means improvement. In the context of this book, Kaizen means *ongoing* improvement involving everyone—top management, managers, and workers. In Japan, many systems have been developed to make management and workers Kaizen-conscious.
>
> Kaizen is everyone's business. The Kaizen concept is crucial to understanding the differences between the Japanese and Western approaches to management.

Imai shortly thereafter quotes a comment made by Toshiro Yamada, Professor Emeritus of the Engineering Faculty of Kyoto University, concerning a sentimental revisit to the River Rouge steel works in Dearborn, Michigan, which Yamada had first visited 25 years earlier. While shaking his head, he said, "You know, the plant was exactly the same as it had been 25 years ago." What does it say about the managerial improvement philosophy if a plant remains the same for 25 years? The River Rouge example is offered as an exact opposite of Kaizen management philosophy. How much ongoing investment does the typical American company make in education and upgrading the management or technical skills of its first-line management? Very little, on average, compared to the Japanese.

There are too many plants like the River Rouge plant in the American national economy today. More frightening than the physical plant remaining the same is the possibility that American industrial management philosophy has remained essentially the same as well.

In 1987, Dr. H. J. Harrington, then chairman of the board pro tem, American Society for Quality Control, and Project Manager, Quality Assurance, IBM Corporation, San Jose, California, wrote *The Improvement Process, How America's Leading Companies Improve Quality* [8]. In the preface, Dr. Harrington describes the problem of American industry as follows [8, p. xiv]:

> As we entered the 1980s, it was obvious to most managers that something was wrong with the present U. S. management system and the curriculum being taught in business schools around the country. Recession, inflation, foreign competition, government regulation, and taxes had put a tight squeeze on American business. We were well on our way to becoming a second-rate industrial power. The values that had made America great had seemingly shifted. Short-term profits were more important than long-term gain. Maximization of assets was taking priority over customer's needs. Management theory replaced experience, and company loyalty was a thing of the past.

Harrington is not alone in his indictment of business schools. The following quote from Deming speaks to the same issue [5, p. 63]:

> In respect to turnover in management, in and out, it may be worthwhile to point out that some schools of business lead students to suppose that for success they must be marketable, so that they may go into a company, perform miracles, and expect a call to go to another company for more miracles.
>
> Students learn little about their limitations and deficiencies. They are led to suppose that they are ready to manage.
>
> It is a fact that anyone might pass with high marks all the courses offered today in schools of business, and yet be helpless against the problems of low productivity and high costs that afflict American industry.

Harrington indicates that "our real purpose is to satisfy the customer's needs both today and tomorrow" [8, p. xiv]. "In today's buyer's market the customer is king. What your customer wants is value. Value—that's a key word. It stands for quality and reliability at a reasonable price" [8, p. 6]. In other words, the customer wants the right product (quality), at the right time (schedule), at the right price (cost). All are important, and all are simultaneously tied for first.

Of the numerous improvement approaches described in Harrington's book, the following 10 improvement activities best summarize the approaches reviewed in the book [8, p. 11]:

1 Obtain top-management commitment
2 Establish an improvement steering council

3 Obtain total management participation
4 Secure team participation
5 Obtain individual involvement
6 Establish system improvement teams (process control teams)
7 Develop supplier involvement activities
8 Establish a systems quality assurance activity
9 Develop and implement short-range improvement plans and a long-range improvement strategy
10 Establish a recognition system

Harrington does an excellent job of describing the opportunity that exists in embracing total quality management [8, p. xiii]:

> How would you like to increase profits by 50 percent? To turn the mounds of scrap that sit throughout your manufacturing facility into shippable product? To increase your corporate output by 20 percent without building or buying one new piece of equipment? To expand your R&D activities by 20 percent without adding one new engineer? To cut overtime in accounting from 30 percent to 2 percent? To see employees with smiles on their faces every day, rather than only at 4 P.M. on Friday?

Harrington's book is an excellent summary of total quality management approaches presently in use throughout America. What is needed is a far more universal implementation of such approaches throughout American industry.

In 1954, Dr. J. M. Juran accepted an invitation from JUSE to offer seminars in quality control to top- and middle-level managers. Ishikawa indicates in his text [11, p. 19] that Dr. Juran's visit represented a turning point in Japanese quality control efforts in elevating quality control beyond the factory to an overall concern throughout management. Dr. Juran's recent text *Juran on Planning for Quality* provides a prescription for installing quality control in organizations. The mission for the text is as follows [12, p. 2]:

> Create awareness of the quality crises; the role of quality planning in the crisis; and the need to revise the approach to quality planning.
>
> Establish a new approach to quality planning.
>
> Provide training in how to plan for quality, using the new approach.
>
> Assist company personnel to replan those existing processes which contain unacceptable quality deficiencies (March right through the Company).
>
> Assist company personnel to acquire mastery over the quality planning process, a mastery derived from replanning existing processes and from the associated training.
>
> Assist company personnel to use the resulting mastery to plan for quality in ways that avoid creation of new chronic problems.

The above list indicates that Juran endeavors in this text to provide a road map for initiating or overhauling an existing quality control program in an organization. His overall approach embodies what he calls the Juran trilogy. Figure 5-4 indicates the primary relationships of the trilogy. The three phases of the trilogy are (1) quality planning, (2) quality control, and (3) quality improvement. The first phase involves

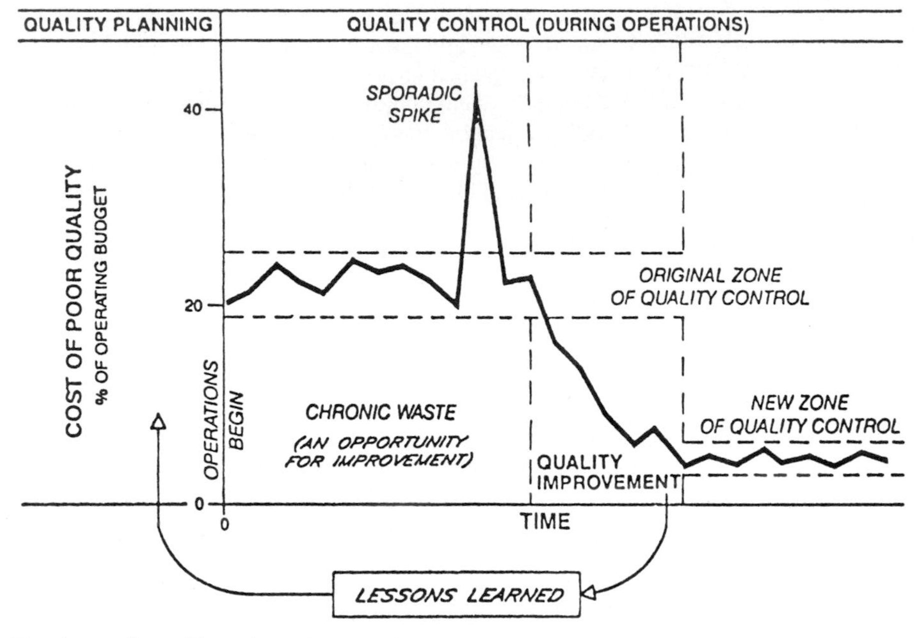

FIGURE 5-4 The Juran trilogy. (From Juran [12, p. 12].)

analysis of customer needs, products, and processes in preparation for implementing quality control. Note that in the second phase, the cost of poor quality is roughly 20 percent. This suggests that the process is a typical process that has not had the advantage of improved quality control analysis over time (i.e., it has the factory within the factory generating rework, as previously mentioned by Crosby [4]). In phase 3, quality improvement approaches identify sources of unacceptable variation, and adjustments are made in product designs, processes, procedures, etc., thereby eliminating sources of poor quality, such that the costs of poor quality can be greatly reduced (e.g., less than 1 percent defective as compared to 20 percent). The "new zone of quality control" translates to improved customer satisfaction, increased sales, lower unit material and labor costs, increased overall process capacity, a more motivated and secure work force with greater self-esteem, etc. As the improvement process progresses, lessons learned provide feedback to improve the planning process.

Juran's text follows the quality planning road map as indicated in Fig. 5-5. The remainder of the text describes in considerable detail the execution of each of the steps on this road map. It is an excellent guide for installing or revising a quality control program in a firm.

In 1988, Taiichi Ohno authored *Workplace Management* [15]. Ohno was primarily responsible for initiating the kanban "pull system" of production control in the Toyota Company. Ohno started with Toyota in 1943 as an assembly shop manager and worked his way up the management ladder. In 1970, he was promoted to

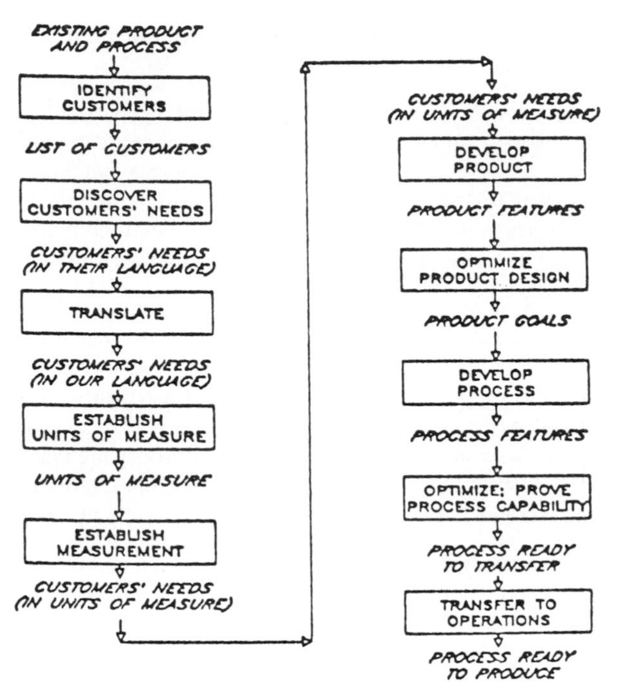

FIGURE 5-5 The quality planning road map. (From Juran [12, p. 15].)

executive director of Toyota. The title of the book, *Workplace Management*, was well chosen; it indicates the importance he places on the workplace. As mentioned earlier, it is clear from his writing that Mr. Ohno is "plant people"; he knows of what he speaks. Why are there so few books in the United States written on plant management by plant people who know from experience what they are talking about, and so many books on management authored by people who often are not plant people, and often have never served in a significant plant management staff or line capacity?

Last but not least in this sequence of sources concerning total quality management is a U.S. Department of Defense publication entitled *Bolstering Defense Industrial Competitiveness* [3]. This publication is a 65-page report to the Secretary of Defense of the United States, authored by the then Under Secretary of Defense (Acquisition) Dr. Robert B. Costello, and published in July 1988. The Defense Manufacturing Board, an advisory board to the Under Secretary of Defense (Acquisition), with members appointed from industry, labor, and academia, provided considerable input to the report. The report attempts to offer recommendations for improving the industrial competitiveness of American industry. Obviously, a strong industrial sector makes for strong national defense. The following are selected quotes from Chapter III of the report, entitled "National Policy Issues," which provides a summary of many of the problems facing our nation concerning industrial competitiveness today:

> Management issues consistently were identified by participants from industry and academia as the most important causes of declining American industrial competitiveness. There was a general consensus among these participants that American management culture and practices are less effective in the global marketplace than those of foreign firms. Industrial and academic participants in this effort identified numerous specific management practices they considered harmful to American competitiveness. [3, p. 14]
>
> There are historical reasons for current deficiencies. In the 1950s and 1960s, American industry dominated world manufacturing. American manufacturers could focus on quantity to the neglect of quality. American manufacturers were complacent, while other countries began building powerful new industrial infrastructures and developing superior process technology to manufacture easily-obtained American product designs and technology. Among the results of this period that persist today are many senior managers who continue to view the nature of markets as national, not international, and the nature of product requirements as *good enough,* not *world class.* [3, pp. 14,15]
>
> The Japanese do not invest heavily in defense research and development. Consequently, they are able to invest proportionally more in research and development for commercial products and processes. They also invest more in capital equipment. Relative to Japan, American firms are *underinvested.* The Japanese worker reportedly is supported by about $48,000 in capital investment in contrast to about $32,000 for the American worker. [3, p. 16]
>
> The American tax system at all levels of government places a heavy tax burden on American industry (for example: income, property, and labor taxes) but not on equivalent products manufactured elsewhere and sold in the United States. Many countries use a value-added tax to ensure that the products of both domestic and foreign producers are taxed equally and fairly, and, because of the value-added tax, are able to minimize other direct (and unequal) taxes on their domestic industries. [3, p. 17]
>
> During the decades of the 1960s and 1970s, American requirements for emissions, safety, and environmental controls imposed large *non-productive* costs on American manufacturers at the expense of additional investments in productivity improvements. [3, p. 17]
>
> Product liability laws and court awards are becoming a major issue in the United States. Test and evaluation requirements necessary to protect firms against lawsuits are becoming very costly. [3, p. 18]
>
> The American educational system does not produce the required numbers and skill levels of scientists, engineers, and technicians to support advanced manufacturing needs. Evidence suggests that the manufacturing workforce in some other countries may be better educated and trained than in the United States. For example, Japanese high school graduates appear to be much better educated in math, science, and technology than their American counterparts. [3, p. 19]
>
> Skill levels of many American high school graduates are not adequate, and firms often must invest in programs to upgrade basic reading and math skills. Such results suggest that a system of high-quality technical schools providing skills in applied mathematics,

machining, manufacturing methods and technologies, and fundamentals of technology management could be an effective means of providing highly skilled and motivated workers. Such a system might provide a constructive alternative for students who do not wish to or are unable to pursue a university education. [3, p. 19]

In large measure, the inability of American managers to achieve results in manufacturing equal to those of Japanese managers in the United States stems from management theory and practice, as taught in American universities (where for example, good management is management by financial control; good managers can manage anything; individual achievement is important, not teamwork; manufacturing is an unimportant function). Engineering schools in American universities also focus inadequately on manufacturing, training engineers for careers in product research and development. Few faculty members have industrial experience or expertise. Emphasis on specialization results in engineering professionals who are ill-equipped to understand total manufacturing systems. [3, p. 19]

Beyond the university level, American industry lacks adequate programs to provide continuing professional education and training to engineers and production workers. Continuing education and training programs in American industrial firms are often weak, ineffective, or non-existent. Stimulation of continuing education and training through tax incentives, Department of Defense contract incentives, and other Government efforts could be highly-productive and cost-effective. [3, p. 20]

There is a widely perceived failure of American institutions to instill basic skills in our citizens. The general lack of familiarity with foreign languages and cultures in the United States population detracts from American international competitiveness. [3, p. 20]

The absence of a national understanding (and Department of Defense understanding) that a healthy, productive manufacturing base is essential to our security greatly complicates efforts to develop and implement remedial measures. Manufacturing strength is needed to ensure that our armed forces can acquire the best weapons, and in quantities needed, to deter and defeat potential adversaries. Such strength encompasses commercial, as well as military production capacity. Commercial capacity adds the financial strength necessary to support research and development and capital investment and provides production resources that could be converted or diverted to military needs under emergency conditions. [3, p. 21]

The deeply ingrained adversarial relationships between Government and industry and between management and labor are major causes of declining American industrial competitiveness. The relationship between the Government and industry is characterized by Government constraints on industry behavior intended to protect the public good against profiteering and shoddy performance; and by industry performance *by the numbers* to stay within Government constraints and to document compliance. The relationship between management and labor is also adversarial. [3, p. 21]

These adversarial relationships undermine industrial efficiency, responsiveness, and technological innovation. This Government-industry relationship forces industry to operate within an extremely restrictive environment and discourages (or even penalizes) innovative behavior. Considerable industry effort is invested in satisfying Government paperwork requirements and responding to Government meddling in the manufacturing process. [3, pp. 21, 22]

> The management-labor relationship prevents cooperative efforts to identify and implement innovative processes and tends to hold labor productivity to some minimal standard, actively inhibiting any worker capability or desire to improve. Historical barriers to cooperation between management and labor are only beginning to fall, and at much too slow a rate. There is not yet a pervasive sense of shared interests and objectives for the common good. [3, p. 22]

> Other countries (Japan in particular) are much more effective than the United States in achieving industry/Government/labor cooperation on process and product development, and, through cooperation, are more effective in implementing new ideas to make manufacturing more efficient, responsive, and technologically advanced. [3, p. 22]

> American society, historically, has been action oriented and sharply focused on quick results. The short-term expectations that pervade American society have become a major impediment to the long-term planning necessary to compete effectively with other countries. [3, p. 22]

> The equity market is the major source of capital for American industry, in contrast to Japan, where commercial banks are the principal providers of capital to industry. The American stock market is driven by short-term expectations, whereas Japanese banks historically have supported long-term investment. [3, p. 22]

> The attitude in the United States toward manufacturing and manufacturing technology is somewhat negative. American universities have little to offer in these fields. Even within the manufacturing firm, research and design engineers are perceived to have more prestige than manufacturing engineers. One result is that the manufacturing function does not compete effectively for high-quality personnel. (Conversely, the Japanese have a high regard for manufacturing and are totally committed to innovation in both process and product.) These attitudes (and resultant rewards systems) toward manufacturing careers often prevent the best people from beginning or sustaining careers in manufacturing. [3, p. 23]

The above list is descriptive of many of the problems facing American industry. Surely the collapse of the Soviet empire, in time, will permit the United States to shift a much greater portion of its technical resources to domestic and commercial priorities. Figure 5-6 indicates that more than half of American research and development expenditures have been going into defense-related projects. It may be noted in Fig. 5-6 that defense R&D predominantly constitutes development expenditures as compared to basic and applied research expenditures. It seems reasonable to assume that when some portion of the reduced defense expenditures are shifted to nondefense expenditures, there will be an increase in basic and applied R&D in the United States relating to commercial products. This shift should greatly increase our ability to introduce new commercial technologies into the world economy, thereby greatly enhancing our international competitiveness.

Figure 5-7 indicates the enormous disparity in defense support provided by the United States as compared to Germany and Japan.

Both Japan and Germany have had a "free ride" as the result of post-war terms imposed by the western and Soviet powers following World War II. In not allowing Japan and Germany to rearm, we may have inadvertently provided them with an enormous opportunity and advantage in permitting them to focus their efforts in

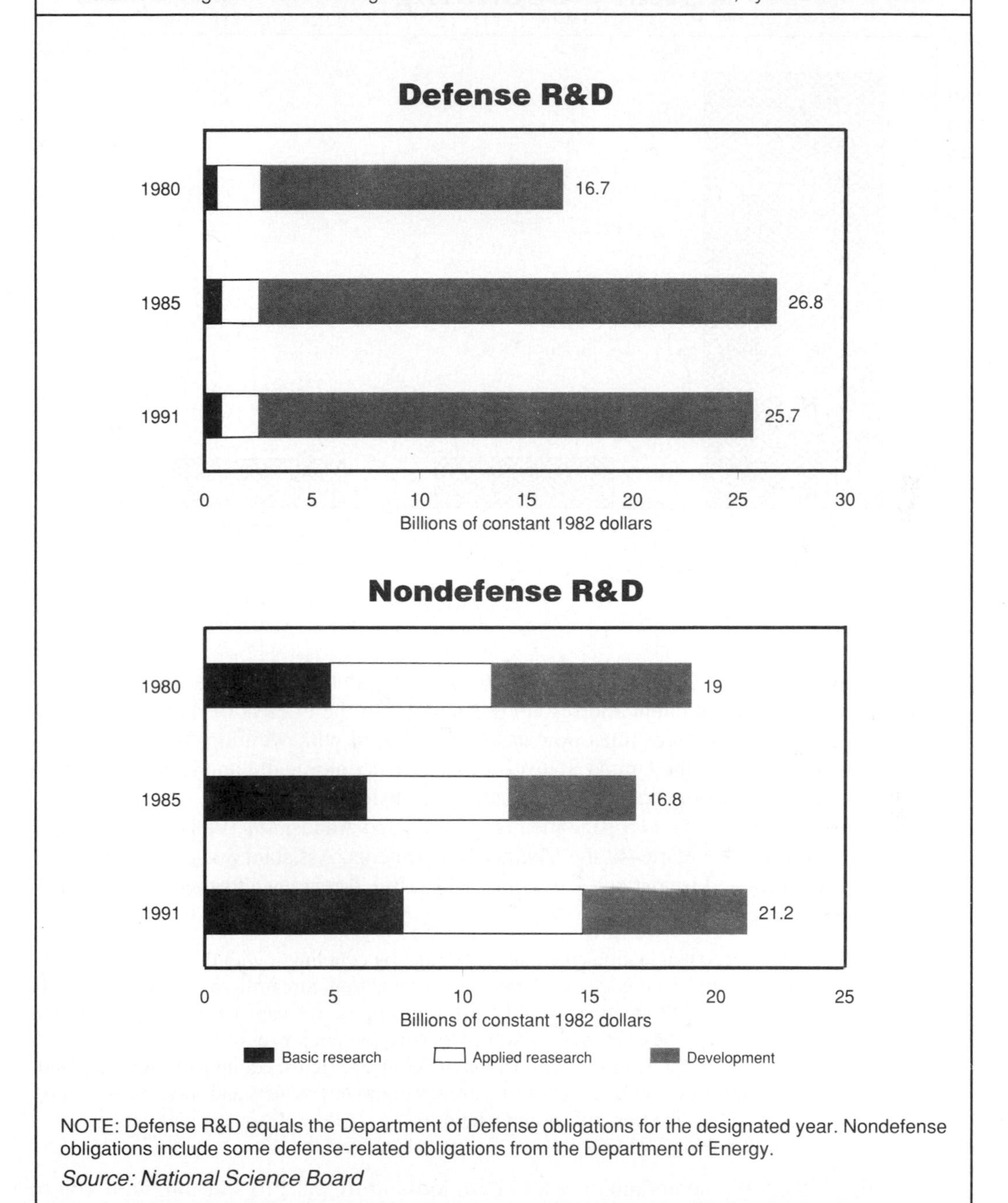
U.S. DEFENSE/NONDEFENSE R&D
Relative Changes in Federal Obligations for Defense and Nondefense R&D, by Character of Work
Defense R&D
1980
1985
1991
16.7
26.8
25.7
0
5
10
15
20
25
30
Billions of constant 1982 dollars
Nondefense R&D
1980
1985
1991
19
16.8
21.2
0
5
10
15
20
25
Billions of constant 1982 dollars
Basic research
Applied reasearch
Development
NOTE: Defense R&D equals the Department of Defense obligations for the designated year. Nondefense obligations include some defense-related obligations from the Department of Energy.
Source: National Science Board

FIGURE 5-6 U.S. R&D expenditures in 1991.

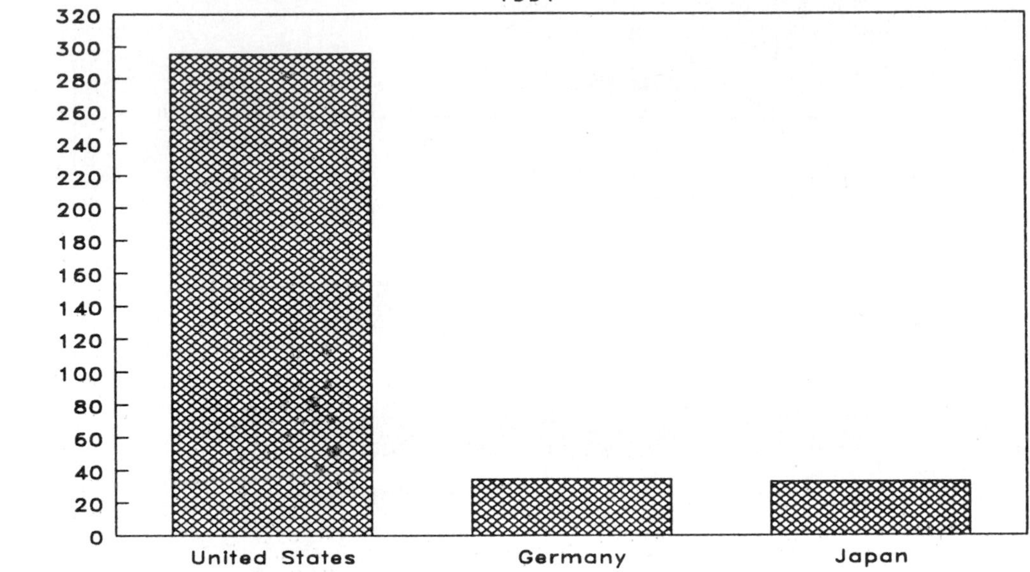

FIGURE 5-7 Military spending in 1991.

commercial R&D. The DoD has tied up much of the U. S. scientific and engineering technical talent and resources for years in the interest of national security. These finite resources must now at least be shared with winning the international economic war, if the United States is to survive economically and hopefully regain its prior international commercial leadership position.

Mr. Costello, in a two-page memorandum dated August 19, 1988 entitled "Memorandum for Secretaries of the Military Departments, Assistant Secretary of Defense (Production and Logistics), Directors of Defense Agencies," further states the following [2, pp. 1–2]:

> I am convinced that by implementing TQM, and by coupling it with the intensified application of such value-added strategies as Acquisition Streamlining, Transition from Development to Production, Could Cost, and others, we can achieve unprecedented improvements in the effectiveness of the DoD acquisition process.
>
> I want TQM applied to the acquisition of defense systems, equipment, supplies, facilities, and services to ensure continuous improvement of products and services being provided to, and by, the Department of Defense.
>
> ...I am making TQM success my primary objective.

He closed his memorandum with "I am looking forward to working with you to help achieve the extraordinary promise of TQM" [2, p. 2]. There is considerable effort under way at this time throughout the DoD to comply with his directive.

There are those who have their reservations about TQM. Somewhat typical is Tom Peters, a highly respected modern-day management guru, who exhibits his obvious skepticism about TQM in the following passage [17, p. 3]:

> A peck of past columns testify that I'm rabid about the need for better U. S. product quality. But some magic, infinitely flexible elixir called TQM is not the answer to all of America's vexing business problems. In fact, TQM often looks suspiciously like the latest act in a long-running farce called Revenge of the Number Nerds. The quantified "q-word"—quality—seems to be crowding out the far more important and messier "c-word"—customers.

This quote is confusing because it is not apparent to me that TQM is a "Number Nerd" approach, as compared, for example, to previous statistical quality control text content. Most TQM is composed of packaged commonsense management practices that improve quality, productivity, time to market, and management. TQM also places considerable value in knowing and accommodating customer needs.

Tom Peters, in his recent book *Thriving on Chaos* [18], concentrates on customer responsiveness, time to market, empowerment, leadership at all levels, and other primary issues in modern management philosophy—all of which are important issues. But they are only a part of what is needed.

The Malcolm Baldrige National Quality Award program is one of many efforts of late fostering much-needed improvement in quality in American industry. It is not clear whether it is part of TQM or simply complements it. The program has support from the National Institute of Standards and Technology (formerly the Bureau of Standards) and the American Society for Quality Control (ASQC), the national organization most uniquely associated with quality control development in the United States. Figure 5-8 lists the 1993 Baldrige Award criteria [14]. This list provides a long-term, comprehensive program to guide the development of a quality improvement program in a firm. Many firms who have limited interest in applying for the award are utilizing the criteria as an outline for guiding their quality improvement effort. If it serves that need, it is providing a considerable national service; it is the continuous quality improvement that is of primary importance, not the award.

Another incentive for improving quality is ISO 9000 certification, required today by many European Community (EC) country firms. ISO certification can be acquired at various levels, as is indicated in Fig. 5-9. Many U.S. firms are initially content to reach whatever level of certification demanded by their European customers (e.g., 9003 or 9002). The more progressive will pursue full ISO 9001 certification both to meet any and all future certification requirements and to have a displayable credential as an important aspect of their marketing program. He who "outcredentials the other guy," especially in the event of a tie, often gets the contract. In an analogous way, students should seriously reflect on this concept in considering engineering registration. The EIT examination, the hardest part of becoming a registered engineer, is so much easier to pass when you are still a student, and so much more difficult—impossible for some—five years after graduation (i.e., an opportunity lost).

1993 Examination categories/items			Point values
1.0	Leadership		95
	1.1	Senior Executive Leadership	45
	1.2	Management for Quality	25
	1.3	Public Responsibility and Corporate Citizenship	25
2.0	Information and Analysis		75
	2.1	Scope and Management of Quality and Performance Data and Information	15
	2.2	Competitive Comparisons and Benchmarking	20
	2.3	Analysis and Uses of Company-Level Data	40
3.0	Strategic Quality Planning		60
	3.1	Strategic Quality and Company Performance Planning Process	35
	3.2	Quality and Performance Plans	25
4.0	Human Resource Development and Management		150
	4.1	Human Resource Planning and Management	20
	4.2	Employee Involvement	40
	4.3	Employee Education and Training	40
	4.4	Employee Performance and Recognition	25
	4.5	Employee Well-Being and Satisfaction	25
5.0	Management of Process Quality		140
	5.1	Design and Introduction of Quality Products and Services	40
	5.2	Process Management: Product and Service Production and Delivery Processes	35
	5.3	Process Management: Business Processes and Support Services	30
	5.4	Supplier Quality	20
	5.5	Quality Assessment	15
6.0	Quality and Operations Results		180
	6.1	Product and Service Quality Results	70
	6.2	Company Operational Results	50
	6.3	Business Process and Support Service Results	25
	6.4	Supplier Quality Results	35
7.0	Customer Focus and Satisfaction		300
	7.1	Customer Expectations: Current and Future	35
	7.2	Customer Relationship Management	65
	7.3	Commitment to Customers	15
	7.4	Customer Satisfaction Determination	30
	7.5	Customer Satisfaction Results	85
	7.6	Customer Satisfaction Comparison	70
		TOTAL POINTS	1000

FIGURE 5-8 Malcolm Baldrige Award criteria for 1993. (1993 award criteria [14, p. 12].)

ISO 9000	Guidelines for Selection and Use
ISO 9001	Design/Development, Production Installation, and Servicing
ISO 9002	Production and Installation
ISO 9003	Final Inspection and Test
ISO 9004	Management and System Guidelines

FIGURE 5-9 Levels of ISO 9000 Certification.

Robert H. Schaffer and Harvey A. Thomson, in an article entitled "Successful Change Programs Begin with Results" in *Harvard Business Review* [19, pp. 80–89], describe their concern with activity-centered as compared to results-centered approaches. They describe a number of examples of large corporations who have initiated large improvement programs that have produced limited measurable results. The danger is that such programs will add to the bureaucracy without producing intended results. Numerous past quality circles programs produced less-than-satisfactory results and have since disappeared.

Schaffer and Thomson [19, p. 84] mention that of the 1000 points indicated in the criteria for the 1993 Baldrige Award (as indicated in Fig. 5-8), only 180 points relate specifically to results. In most sports, the team that scores the most points wins. With broad-based improvement programs such as the Baldrige Award, the team that prepares the best wins. Preparation is a worthy and necessary prerequisite, but it should not be confused with results.

The above comment is not intended to suggest that the Baldrige criteria are inappropriate as a guide to the development of a world-class quality program. It is offered merely to suggest that short-term goals and responsibilities for results must be identified and that timely accomplishment should be both measured and rewarded when it occurs, to prevent the program from "dying on the vine," which is so common a result with large, expensive, and potentially nonresponsive improvement efforts.

The most important attribute of any improvement program is its ability to produce results. When credibility is acquired through the accomplishment and recognition of some successful short-term projects, it is then typically time to plan for mid-term and eventually long-term improvements that often produce the more dramatic major improvements over the long haul. Support must be continually justified and rejustified by a series of accomplishments that demonstrate program payoff along the way.

Total quality management, a part of the overall management philosophy restructuring taking place in the United States today, has a great deal to offer in terms of improved management (what America needs most).

SUMMARY

Inspection came first, and then quality control (QC), quality assurance (QA), product assurance (PA), total quality control (TQC), and finally total quality management (TQM); each has a different connotation. In each progression, the role of quality control was broadened to take on a greater responsibility. TQM includes any aspect of the product, process, product design, and the management and control systems that has any potential effect on product quality.

Pioneers such as Crosby, Deming, Ishikawa, and Juran stressed the need for better management for accomplishing improved quality. TQM in fact has only modestly extended the technology of quality control. What it has done is stressed the need to examine how we do business and how we manage our operations. TQM is more a revolution in management than a revolution in quality control.

Most of the original QC statistical techniques still apply, but they are employed in a more rational management context. Team-building participative approaches in production, and making production responsible for setting and reaching quality goals, has empowered direct labor workers and others to participate in a continuous-improvement approach that was long overdue. Quality control is now a management practice of inclusion, whereas in years past it was a plant activity of restricted inclusion—only quality control personnel were permitted to do quality control.

A few of the TQM concepts are summarized below:

1 *Adoption of kaizen (improvement) within our industrial culture.* No other change is needed more in American industrial culture than change itself. Change implies critical self-examination, which is much needed in American industry today. The good news is that this is one of the United States of America's greatest strengths—the ability to change. The fact that I can say what I want to say, and you can read what you want to read and believe what you want to believe, is one of our greatest strengths. The United States is not where it needs to be, however, to best compete in the world today, but it can get there from here.

2 *"Do it right the first time."* The U.S. industrial culture must get away from the "do something and fix it later" approach and adopt the "do the proper analysis upfront to permit us to do it right (or at least near right) the first time" approach.

3 *Recognition that management* is *the problem and needs to be addressed.* Management improvement across the management spectrum is needed. Management improvement at all levels is needed: the top, middle, and first-line supervisory levels of management. If any one of those levels is left out of the improvement process, it will not work as it should.

4 *Quality control is a production responsibility.*

5 *Discipline.* Throughout American culture, including our industrial culture, discipline is a key element in determining our future success as a society. In the industrial society the connotation of doing what we do "in a disciplined way" is one of the keys to success.

6 *Recognition that for the last decade in particular, while American management has been chasing "quality" as an end in itself, the Japanese industrial culture has been addressing quality* and *productivity improvement, and more recently, responsiveness.* The behaviorist movement over the past 40 years classified words such as "efficiency," "productivity," "work measurement," and "control" as bad words, and in so doing has delayed productivity improvement in the United States.

It is time to admit that these words, and more importantly, the concepts that they represent, are needed if we are to successfully compete in today's internationally competitive world.

There is reason to believe that TQM is not just a passing "silver bullet" and will, in fact, become a continuous improvement element of our future industrial and government culture in the United States. As an example, the General Accounting Office, in a report issued October 6, 1992, indicated that 68 percent of the federal government's 2800 installations are employing TQM. Representative Donald Ritter, Republican from Pennsylvania who ordered the report, stated that "poor quality—including unnecessary reports, rework and other waste—eats up 25% of what government spends. That's more [350 million] than the Defense Department doles out each year" [9, p. 1B]. Only 18 percent of government offices have achieved significant results to date, however, and only 13 percent of government employees are on TQM teams or learning about TQM to date. It is one means to curbing waste in our enormous federal bureaucracy.

Hopefully, TQM is here to stay.

REFERENCES

1 Barker, J. A.: *Discovering the Future: The Business of Paradigms,* Film, Filmedia, Inc., Minneapolis, MN, 1986.

2 Costello, Robert B.: "Memorandum for Secretaries of the Military Departments, Assistant Secretary of Defense (Production and Logistics), Directors of Defense Agencies," August 19, 1988.

3 Costello, Robert B.: *Bolstering Defense Industrial Competitiveness,* Department of Defense, July 1988.

4 Crosby, Philip B.: *Quality Is Free*, McGraw-Hill Book Company, Inc., New York, 1979.

5 Deming, W. Edwards: *Quality, Productivity and Competitive Position,* Massachusetts Institute of Technology, Center for Advanced Engineering Study, Cambridge, MA, 1982.

6 Deming, W. Edwards: *Out of the Crisis,* Massachusetts Institute of Technology, Center for Advanced Engineering Study, Cambridge, MA, 1986.

7 Feigenbaum, A. V.: *Total Quality Control,* McGraw-Hill Book Company, Inc., New York, 1961.

8 Harrington, H. J.: *The Improvement Process, How America's Leading Companies Improve Quality,* McGraw-Hill Book Company, Inc., New York, 1987.

9 Hilkirk, John: "Uncle Sam Begins Push for Quality," *USA Today*, Oct. 7, 1992.

10 Imai, Masaaki: *Kaizen,* Random House Business Division, New York, 1986.

11 Ishikawa, Kaoru: *What Is Quality Control? The Japanese Way,* Prentice Hall of Japan, Inc., Tokyo, Japan, 1981.

12 Juran, J. M.: *Juran on Planning for Quality,* The Free Press, New York, 1988.

13 Lehrer, Robert N.: *The Management of Improvement,* Reinhold Publishing Corporation, New York, 1965.

14 "1993 Award Criteria," Malcolm Baldrige National Quality Award, United States Department of Commerce, National Institute of Standards and Technology, Admin Bldg, Room A537, Gaithersburg, MD 20899.

15 Ohno, Taiichi: *Workplace Management,* Productivity Press, Cambridge, MA, 1988.

16 Peale, Norman Vincent: *The Power of Positive Thinking,* Fawcett Books, New York, 10022, 1987.

17 Peters, Thomas J.: "Beware Politically Correct TQM," *The Costco,* vol. 5, no. 10, 1991, p. 7.

18 Peters, Thomas J.: *Thriving on Chaos,* Harper & Row, New York, 1987.
19 Schaffer, Robert H., and Harvey A. Thomson: "Successful Change Programs Begin with Results," *Harvard Business Review,* vol. 70, no. 1, 1992, pp. 80–89.
20 Schonberger, Richard J.: *Japanese Manufacturing Techniques—Nine Hidden Lessons in Simplicity,* The Free Press, New York, 1982.

REVIEW QUESTIONS AND PROBLEMS

1 What connotation does "product quality" add to previous titles of the quality field?
2 A powerful motivator for embracing continuous improvement was mentioned early in this chapter. What is it?
3 What is the "factory within the factory" that Crosby [4] talks about in his texts?
4 How would you characterize Deming's likely opinion of top management in the United States based on his quotes in this chapter?
5 What does *JUSE* mean, and what is its equivalent in the United States?
6 What is the name of the highest award presented annually in Japan in the field of quality control?
7 Do Deming's [6] 14 points for management primarily involve technical quality control issues or management issues? Develop a summary of his points in 25 words or less.
8 The author disagrees with one of the 14 points of Deming [6]. Which one is it and why? Do you agree or disagree with Deming on this point, and why?
9 According to Schonberger [20] concerning total quality control, which organization in a plant should have primary responsibility for quality control?
10 The author refers to "the largest cultural barrier for Americans to overcome" concerning quality control. The Japanese have a name for it. What is it?
11 How does TQM differ from TQC?
12 Who started the field of quality control? Is it a new development?
13 Does top management need to be involved in the development of quality control in its operations?
14 Ishikawa [11] had a wish concerning the implementation of QC and TQC in Japanese industry. What was his wish and was it fulfilled?
15 What is the purpose of Ishikawa's [11] fishbone diagram?
16 Describe the typical constituent in JUSE (Japanese Union of Scientists and Engineers) in Japan; what similar organization is there in the United States?
17 What are the three elements of the Juran [12] trilogy?
18 In 1988, Dr. Robert B. Costello [3], Under Secretary of Defense (Acquisition), authored a DoD publication entitled *Bolstering Defense Industrial Competitiveness* [3]. Based on his quotes contained in this chapter, summarize his assessment of the quality of U. S. industrial management practice and education.
19 What adversarial relationships does Mr. Costello [3] mention, and what effect do they have on U. S. international competitiveness? Is this impediment to productivity improvement getting better or worse, and at what rate?
20 Has the U. S. Department of Defense been a supporter of TQM in recent years?
21 Explain the difference between activity-centered and results-centered quality improvement efforts. Why is the distinction important?

6

PRODUCTIVITY

Decreased business base increases overhead. Increased business base increases overhead. (Law Number XV)

Norman R. Augustine

The field of industrial engineering began in manufacturing, and in that environment, as mentioned in Chapter 1, Frederick W. Taylor invented the concept of manufacturing staff. Initially there were few staff members—a stock clerk or two, a maintenance man, a paymaster, an accountant, a salesman. Today, in the United States, it is not uncommon for the cost of nondirect and overhead labor—those individuals not directly involved in making a specific product—to be eight times the direct labor cost of an operation. A century ago, direct labor was often the primary focus of industrial engineering in reducing manufactured cost because it typically represented 50 to 80 percent of the total cost of the product; it is not uncommon today for direct labor to constitute only 3 to 8 percent of manufactured cost. Not only have nondirect and overhead labor costs soared relative to direct labor costs, but through such factors as higher levels of mechanization, fixed costs and overhead costs have also soared relative to variable costs (i.e., direct material and labor). Typically, as fixed costs increase, so do indirect costs (e.g., maintenance).

This chapter is about both worker productivity and process productivity—the two are highly interdependent. A common misconception today, given that direct labor may represent only 5 percent of the cost of manufacture, is that direct labor is no longer important. It must be considered that despite the low cost percentage of direct labor, by their actions they exercise control of what gets produced today. If low morale and inefficient methods cause them to produce at only half their capacity, the remaining 95 percent of manufacturing costs—which are overhead (typically fixed) costs, excluding material costs—must be prorated over only half the amount of product. It is not the *cost* of direct labor that is at issue here as much as the *cost effect* of direct labor. Every day, in every plant in the United States, direct labor, which represents a steadily decreasing number of people in such plants, determine by their actions and inactions and the conditions and psychological environment of their workplace what will be produced. For that reason alone, they are very important. It is management's job to ensure that direct labor has the equipment, tools, methods, and motivation to be as productive as possible—something they will often choose to be if given a productive environment in which to work.

Note that most behaviorist philosophy espouses the importance of motivation but often ignores equipment, tools, and methods under the assumption that the properly motivated direct labor employee, the "expert" for that operation, will identify all of the necessary physical needs. My experience, gained through observation over a number of years and often contrary to that popular belief, is that direct labor employees would rather continue using a familiar method unless they are encouraged to perform the operation differently. Direct labor employees often "reach an accommodation" with their tasks. It may involve standing in ice water up to their ankles, but if they are used to it and it is nonthreatening (i.e., they have learned to perform it successfully), they will typically resist any change. It is called the RC factor—resistance to change. Any new method involves making an adjustment, thereby risking failure or humiliation in the eyes of their peers. They may not want to admit that the method they were using was not the best.

The focus of Chapter 2 was on the design of production systems that would permit increased productivity of direct labor. The focus of this chapter is on increasing the productivity of nondirect labor. With the cost of nondirect labor becoming a much larger fraction of total labor cost (relative to direct labor cost) and a more costly resource per person, there is an increasing need to analyze effectively the productivity of both white-collar and blue-collar nondirect labor personnel.

NONDIRECT LABOR PRODUCTIVITY

Productivity Improvement in Production and Service Employment

Marvin E. Mundel, Ph.D., has been a major contributor in the field of industrial engineering for many years. I once asked Mundel if he had received the first Ph.D. ever in industrial engineering, to which he replied, "No, Ralph Barnes was the first; I was the second. I studied under Ralph Barnes at Iowa State."

In 1983, Mundel published *Improving Productivity and Effectiveness* [6]. It is an extension and revision of his 1975 text entitled *Measuring and Enhancing the Productivity of Service and Government Organizations* [7], which was published by the Asian Productivity Organization, a treaty organization of 19 Pacific Rim member nations organized in 1962 under United Nations sponsorship. Their objective as an organization is to assist the productivity of member nations in agriculture, industry, and services. Based on their development of expertise over the past 30 years, it may now be appropriate for North American countries to form a U.S.-sponsored organization that encourages Pacific Rim engineers and managers to come visit us to lecture as Mundel, Deming, Juran, Nadler, Niebel, and others did in the Far East.

Mundel draws attention to the need for nondirect labor (i.e., white-collar and indirect labor) productivity improvement; he indicates that only 26 percent of employment in the United States is in the manufacturing and mining industries [7, p. 4]. The remaining 74 percent is in service organizations, both private and public. He suggests that only half of the 26 percent is touch labor (making product), bringing the total employment by all services in manufacturing, mining, and service organizations to 87 percent of total employment. Industrial engineering has traditionally concentrated in the development of improvement techniques for direct labor (i.e., touch labor) operations. If improvement of the U.S. gross national product (GNP) is a goal of U.S. industrial engineering, the potential impact from improvements in service employment represents a major opportunity. Mundel is "leading the charge" in this important area of future opportunity.

Mundel's *Improving Productivity and Effectiveness* attempts to accomplish a broad and ambitious goal for a text. The following statement from the preface describes the book's purpose:

> This book is intended to provide a flexible methodology for applying quantitative management techniques to all industrial and government activities including services, staff, and indirect work. The necessary concepts, terms, and procedures are given in detail. The purpose is to facilitate the measurement and subsequent improvement of the effectiveness and the productivity of all parts of all organizations.
>
> Thirty cases are given, showing the application of the methodology and covering a wide range of activities including printing, shipbuilding, judicial procedures, hospitals, and banking, to mention a few. All of the cases presented are from real situations; none are hypothetical. [6, p. ix]

In another introductory section of his book, subtitled "The Purpose of This Book," he states:

> This book is dedicated to aiding the peoples of the world in recapturing the steady increase of productivity and effectiveness that characterized the world during both the industrial and the green revolutions, and into the 1960s. [6, p. 9]

(In a footnote Mundel explains that the green revolution refers to "the introduction of high-yield agricultural crops.")

Mundel's approach to improving effectiveness and productivity is represented by a *cycle-of-management* process, involving the sequential and circular flow of nine activities that constitute effective management control of organizations. Figure 6-1 describes this process. For Mundel's approach to be understood and utilized effectively, it is critical that terminology and relationships be well defined. The first three chapters of Mundel's text devote considerable attention to defining and explaining terms and relationships (e.g., effectiveness, efficiency, productivity, performance) as a prerequisite to explaining specific methodology. For example, whereas *efficiency* implies doing things well, *effectiveness* implies doing the right things. In most endeavors, the goal is to do the right things well, which requires both effectiveness and efficiency.

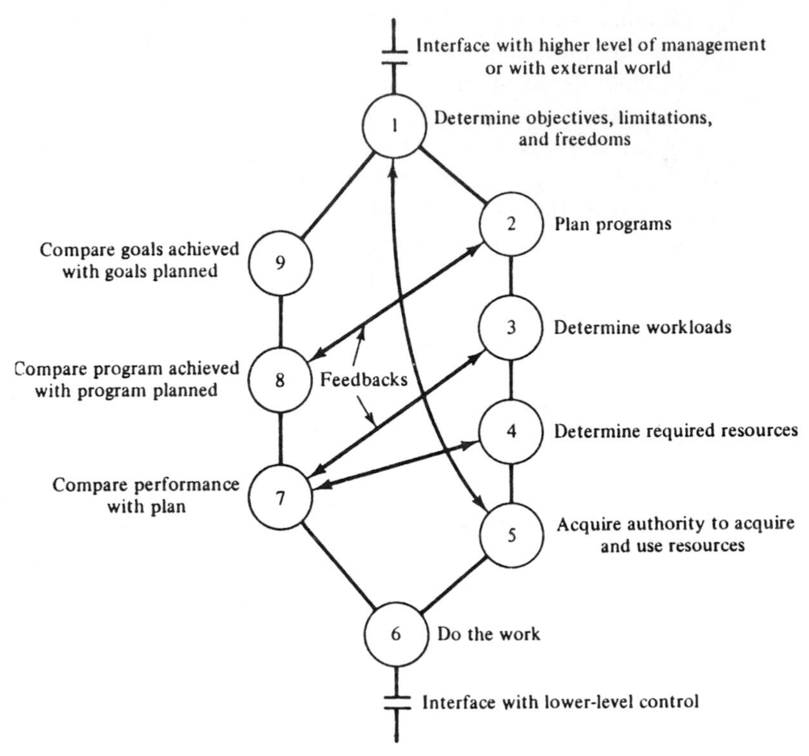

FIGURE 6-1 Cycle-of-management process. (From Mundel [6, p. 24].)

Mundel defines productivity as "the ratio of outputs produced per unit of resources consumed, compared to a similar ratio from some base period" [6, p. 46]. He offers the following description:

> Productivity is the ratio of the outputs produced for use outside an organization, with due allowances for the different kinds of products, divided by the resources used, all divided by a similar ratio from a base period. Hence it is an index; it has no dimension. Mathematically, the productivity index is:

$$(\text{A}) \quad \frac{(1)\ \dfrac{\text{AOMP}}{\text{RIMP}}}{(2)\ \dfrac{\text{AOBP}}{\text{RIBP}}} \times 100 \quad \text{or} \quad (\text{B}) \quad \frac{(3)\ \dfrac{\text{AOMP}}{\text{AOBP}}}{(4)\ \dfrac{\text{RIMP}}{\text{RIBP}}} \times 100$$

> where AOMP = aggregated outputs, measured period
> RIMP = resource inputs, measured period
> AOBP = aggregated outputs, base period
> RIBP = resource inputs, base period

> Both formulations produce an identical value, although the subordinate ratios have different meanings. Subordinate ratio (1) is called the *current performance index*. Ratio (2)

> is called the *base performance index*. Ratio (3) is referred to as the *outputs index;* (4) the *inputs index*.
>
> In formulation (A), the ratios of (1) and (2) may be computed from periods of different lengths without disturbing the meaning of the productivity computation or of the subordinate ratios. In formulation (B), if the subordinate ratios are to have meaning, all data must be from equal periods of time. Managers, operating in real time, will find formulation (A) more useful. [6, pp. 10–11]

Mundel also discusses such issues as distinguishing between "functions or activities" and "outputs." For example, "meetings attended" is not an output as far as the ultimate consumer is concerned; it is simply part of the overhead of the operation and is therefore an activity, not an output.

Mundel utilizes a necessarily elaborate structure called *work-unit analysis* (customer valued output products). He cautions that such units must be carefully developed when one is employing his approach:

> With service and government organizations, considerable effort is required to identify the outputs prior to any attempt to measure productivity. . . . Some method must be employed to avoid errors. Various erroneous applications have been:
>
> Weather bureau: Number of typhoons tracked per year.
> Food and drug inspection: Number of violations of law cited per year.
>
> Such applications do not instill much faith in productivity measurement. [6, p. 17]

Obviously, the weather bureau has no control over the number of typhoons that will develop in a year, and a food and drug inspector has no control over the level of unlawfulness that he or she must deal with. In a footnote, Mundel indicates that "in the case cited, the Weather Bureau tracked 26 typhoons in the base year; 17 in the subsequent year. The charge was leveled at the Weather Bureau that their productivity had fallen to 17/26 or 65 percent" [6, p. 17].

Mundel's work-unit structure is provided in Fig. 6-2. The work-unit structure is imbedded in the cycle-of-management control process as indicated in Fig. 6-3. Mundel's approach culminates in a master procedure containing the 13 following steps [6, p. 133]:

1 Make a general reconnaissance.
2 Develop a work-unit structure.
3 Select work measurement method.
4 Rough, tentative design of the staff resource budget system.
5 Tentative design of the ongoing staff resource control system.
6 Familiarizing affected organization with proposed systems and actions.
7 Apply selected work measurement techniques.
8 Work measurement data reduction.
9 Staff resource budget system and workload forecasting system development and pre-test.
10 Ongoing staff resource–workload management system design and pre-test.
11 Staff resource budget system installation.
12 Ongoing staff resource–workload management system installation.
13 Follow-up assistance provided.

Numerical Designation	*Name*	*Definition*
8th-order work-unit	Results	What is achieved because of the outputs of the activity.
7th-order work-unit	Gross output	A large group of end products or completed services of the working group, having some common affinity.
6th-order work-unit	Program	A group of outputs or completed services representing part of a 7th-order work-unit, but which are a more homogeneous subgroup in respect to some aspect of similarity.
5th-order work-unit	End product	A unit of final output; the units in which a program is quantified; a convenient-sized output which is produced for use outside of the organization and which contributes to the objectives of the organization without further work being done on that output.
4th-order work-unit	Intermediate product	A part of a unit of final output; the intermediate product may become part of the final output or merely be required to make it feasible to achieve the final output.
3rd-order work-unit	Task	Any part of the activity associated with, and all of the things associated with the performance of a unit of assignment by either an individual or a crew, depending on the method of assigning.
2nd-order work-unit	Element	The activity associated with the performance of part of a task, which it is convenient to separate to facilitate designing the method of performing the task or the determination of some dimension of the task.
1st-order work-unit	Motion	The performance of a human motion. This is the smallest work-unit usually encountered in the study of work. It is used to facilitate job design or dimensioning and never appears in control systems above tnis level of use.

FIGURE 6-2 Definitions of orders of work units. (From Mundel [6, p. 34].)

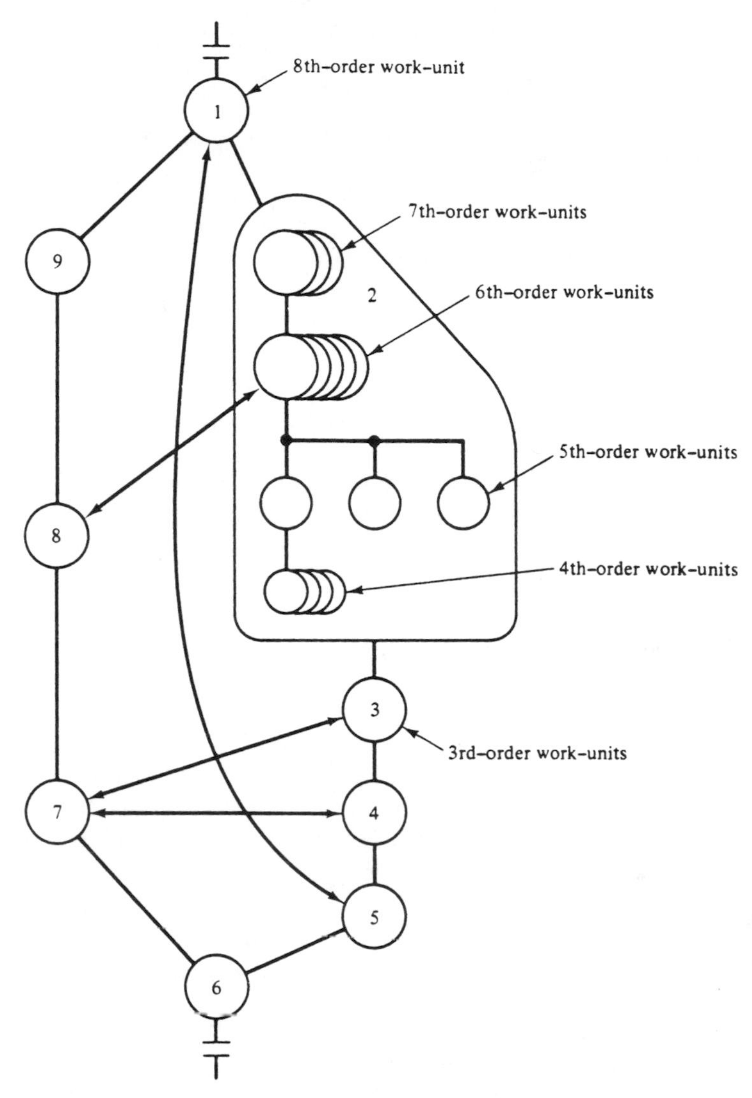

FIGURE 6-3 Cycle-of-management process, with work-order units. (From Mundel [6, p. 35].)

Mundel describes typical uses of his technique in the following quote:

The analyst must understand that the first improvement being sought is control: the matching of workload and staff. Control may subsequently be used as appropriate, to:

1 Support staff requests so as to match workload.
2 Measure effectiveness.
3 Measure productivity.
4 Reduce workload to feasible level.

5 Improve internal communications.
6 Improve schedule conformance.
7 Locate work where technological change would give sizeable benefits.
8 Reduce costs.
9 Improve quality of outputs, and so forth. [6, p. 135]

When you consider the above uses, it might be instructive to ask how such questions would be handled without such an approach. A few examples within my experience demonstrate how many managers presently deal with such issues.

For the first example, consider a manufacturer that makes the same product at five plants located throughout the United States. I visited a typical plant to assist in improving productivity. I noted that there appeared to be too many office people for the size of the plant, so I asked about the percentages of office personnel as a fraction of total personnel at each of the five plants. The percentages of office personnel were all 30 percent, except for one 31 percent value. That is *not* a likely coincidence. When I asked them, "Why 30 percent?" they said that "30 percent is all that corporate will allow." When I asked, "Why is one of them 31 percent?" they answered, "We let some people go in the plant, and we have not yet decided who in the office to let go."

Corporate managers "live and die" by ratios, but no one in any corporation I have ever visited can explain whether the ratio was objectively determined. Is this simply a problem with small companies? Not too long ago, General Motors Corporation, in a time of recession, announced that they were reducing their white-collar work force by 30 percent. Such a reduction suggests that they did not have control of their white-collar personnel costs. How long had they had 30 percent too many white-collar employees? Who authorized it? Why? It appears that in bad times large manufacturers often find themselves in a catch-22. If they admit they have excess personnel, they will be criticized for not managing more efficiently. If they do not cut their excess personnel, they have to suffer the resulting cost burden when they can least afford it. By using an approach such as that proposed by Mundel, they can have the right number of personnel all of the time, based on their true underlying personnel needs.

Consider a second example involving a large manufacturer that has numerous groups and divisions throughout the United States. In conversation with a group internal consultant, I asked, "If a plant manager calls you and says he needs an additional industrial engineer, two additional accountants, and three additional quality control engineers, how do you decide how many to give him, if any?" He eventually answered, "We don't give him any more than we have to." The inference was that such a circumstance represents "negotiation time." When plant-level management request white-collar resources, corporate managers in most corporations are "hard pressed" to know whether the resources are needed. Most U.S. corporations would benefit from developing and utilizing a technique such as that proposed by Mundel. A typical corporation cannot compete for long internationally if it has 30 percent too many white-collar personnel.

You might ask, "How well is management performing today in managing overhead and indirect labor control?" The answer is obvious. Recently, one major U.S.

corporation after another has indicated plans to implement a huge number of staff layoffs, which is undeniable evidence that they do not control overhead and indirect labor well.

I teach courses in industry as many as a dozen times each year, and I indicate that maintenance mechanics under a typical unmeasured maintenance system are reported to be productive only 50 percent of the time. Participants will often shake their heads in disagreement—they often think the percentage is more like 30 or 40 percent! That says something about indirect labor analysis and control in U.S. plants.

Improvement is an important issue today in American industry. Another way of considering an approach such as Mundel's, however, is simply from the perspective of good day-to-day management. Any management that claims to be managing must possess the means to know how many indirect and white-collar employees it needs now, six months from now, and a year from now in order to meet its corporate objectives effectively and efficiently; that is an ongoing management responsibility. It is an area of management that begs for dedicated attention.

To a sophomore engineering student, it may seem implausible that a plant manager would not, for the good of the company, actively pursue reduction of excess indirect and white-collar labor costs in his or her plant. The student may not fully understand the level of perceived risk that such a plant manager may feel in reducing his or her power base (i.e., the number of employees). A plant manager needs the support of those reporting to him or her to meet the monthly production goals. Becoming plant manager may have been that person's career goal for the last 30 years. If the subordinates feel alienated, they may choose to support the plant manager less, thereby threatening his or her continued employment as plant manager. This is typical of the very real personal choices that managers have to make outside of the realm of a university lecture on management practice or policy.

Mundel's approach requires considerable collection and analysis of data. His most recent text, *The White Collar Knowledge Worker* [8], with its accompanying floppy disk, provides computerized means for inputting and analyzing the data and providing appropriate reports. Mundel's approach to nondirect labor productivity improvement is leading the way in this important area today.

White-collar Productivity Improvement

In 1983, Robert N. Lehrer published *White Collar Productivity* [3], which provides a review and critique of the leading white-collar productivity improvement techniques in use. The techniques are summarized below in a sequence starting with the most clerical circumstances and progressing to the broadest white-collar application. The following descriptions summarize 11 of the techniques described by Lehrer.

Technique 1: Clerical Methods/Human Factors/Work Measurement This approach treats clerical productivity as analogous to industrial productivity, for an environment for which work content is considered to be well described by a limited number of finite and readily definable tasks, such as filing or typing. The human factors aspect deals with the determination of the arrangement assumed best for serving the needs of the employee and completing the work to be performed. The tasks to be performed are standardized and their work content measured. Productivity is then measured over time by means of a comparison of earned hours generated in relation to actual hours expended in performing the standardized tasks. This is the standard calculation for direct labor productivity measurement, as discussed in Chapter 2.

In many white-collar organizations, only a fraction of employees hold positions that fall within this readily measurable clerical category. For those employees (i.e., technical typist, preventive maintenance mechanic, janitor, etc.), the technique can be an effective means of measuring and reporting general labor performance.

Technique 2: Paperwork Simplification This approach applies logical paperwork analysis methods in order to identify a most effective paperwork system for communicating the necessary information in the most suitable modes. Such analyses, when performed effectively, typically identify unnecessary labor requirements that when eliminated or modified will significantly improve both labor performance and system products. In today's work environment, paperwork simplification and appropriate technology selection, as a combination of approaches, will produce even greater productivity improvement benefits. Participative approaches are being utilized today to improve the process at all levels.

Technique 3: Input/Output Ratios This approach utilizes macro work measurement and estimating techniques to establish ratios between input resources (e.g., labor by skill classes) and final outputs generated by organizations (e.g., weather reports for the United States Weather Service, meats inspected by federal meat inspectors). Manning standards presently employed by many organizations are not a preferred method because such standards typically lack a quantitatively derived basis for the ratios employed (i.e., strictly speaking, the "standards" are actually baseless baseline estimates).

Technique 4: Multiple Regression Multiple regression utilizes (1) an acquired database of associated variables and times for measured and normalized total operation accomplishments and (2) task input parameters; these elements form a basis for developing a linear multiple regression equation for any combination of those same task elements. Such an equation can be employed to predict the operation accomplishment time for any new operaticn that represents a different combination of the same task elements represented in the database.

An example would be a linear regression equation for estimating the time for a janitor to clean a classroom. The variables might be the square feet of rug to be vacuumed, the square feet of tile floor to be mopped, the number of furniture pieces to be dusted, the number of wastebaskets to be emptied, and the number of chalkboards to be cleaned. After the accomplishment times for a number of observed classroom cleanings have been normalized, the variables and normalized accomplishment times can be used to form a database that will permit determination of regression variable coefficients for the data submitted. The regression coefficients assigned to regression variables could then be employed to estimate classroom cleaning times for any additional classrooms for which regression variable values are known.

Multiple regression has been very successfully applied to operations in which a few well-defined tasks occur in various combinations and require varying times to be accomplished. Examples might include a utility crew installing a telephone pole, a trench-digging crew preparing a trench, or a painting crew painting a house. In years past, Warner Robbins Air Logistics Center developed and utilized a linear multiple regression equation for estimation of the required number of production-control-expediting person-hours based on the number of production control documents associated with a particular program to be undertaken at that facility. Each document generated required production control person-hours for accomplishing the required tasks.

Technique 5: Physical Resource/Technology Structure This approach concentrates technical effort on identifying and assessing the technological needs within the knowledge worker's and manager's workplace to allow selection of a best combination of feasible present technology for achieving optimum performance. It is generally recognized that knowledge workers and managers, at the upper two layers of the white-collar worker labor category, are often the most technologically underdeveloped. Two of the major factors in this underdevelopment are "ego protection" and reluctance to make the time investment necessary to acquire the skill improvements. It is likely that almost all managers and knowledge workers in most organizations would exhibit improved performance if they were to acquire at least minimal functional understanding and skill in, for example, time management, word processing, spreadsheet analysis, computer graphics, and database management capabilities. A psychologically engineered program that conveniently and sensitively encourages and provides the means of discreetly acquiring such skills would likely produce a significant productivity improvement within an organization. The program development requirement is at least as much psychologically and sociologically based (i.e., ego and image protection) as it is technically based.

Technique 6: Work-Unit Analysis This approach was reviewed earlier in this chapter in the discussion of Mundel's *Improving Productivity and Effectiveness.*

Technique 7: Management by Objectives (MBO) This approach is essentially a performance-appraisal aid for managing managers. It also typically assumes a zero-based budgeting basis. Such a system has been particularly helpful in assisting managers to develop a more robust and longer-term view of their subordinate's management responsibilities. Without such guidance, for example, managers have been known to focus on simplistic single-dimension objectives, such as profit. Such a short-term focus can deplete both physical and personnel assets of a corporation by maximizing short-term profits, for example, while limiting the long-term success of the firm.

Technique 8: Organizational Structure Analysis This approach, called INTROSPECT by General Electric, is utilized on an internal consulting review basis to determine the best organizational structure for an organization functional unit. Such factors as span of control, organizational layering, distribution of managerial workload, and logical grouping of technical personnel to eliminate or minimize duplicate capabilities are all part of the analysis. The goal of such a study is the reestablishment of an optimum organizational structure to most effectively and efficiently perform the overall mission of the function. It is a "back to the (management) basics" approach and therefore makes a lot of sense. GE has successfully developed and employed the technique in literally hundreds of functional organizations throughout its corporation. Management consulting firms have been performing such organizational analysis studies for many years. One might ask whether corporations perform such detailed analyses often enough. Such analysis can go a long way in providing insight and improving the effectiveness of both functions and organizations.

Technique 9: Operation Functional Analysis Lehrer describes operation functional analysis (OFA) as follows:

> OFA concentrates on analysis of the demands which require work within organizational units, and the interactive inter-organizational work flows associated with these demands. The primary aim is to identify how demands placed on each unit can be changed to reduce and eliminate unnecessary work, and to facilitate the work accomplishment associated with demands placed on other units. The approach is a participative one. [3, p. 203]

Technique 10: Overhead Value Analysis This approach utilizes a traditional value analysis approach for review, conception, evaluation, and redesign of overhead products (i.e., services) to best meet organizational needs in support of mission requirements. The analysis often results in the design of equivalent, but less costly, products for an organization's clients. An example would be a handwritten speed memo for requesting services, in which a list of alternatives serves as a checklist of available choices; this memo, when completed, constitutes the required sum of actions needed in a minimum time. For example, consider the convenient and forced choices on a standard Federal Express form. Such an approach is much more efficient than having a manager compose his or her thoughts concerning service needs in the form of a letter that is passed on to a secretary. The secretary must then prepare the letter, get approval and a signature, produce copies, and finally prepare an envelope.

Technique 11: Quality Circles This approach involves having employees meet to discuss the products or services they provide and any possible improvements that can be made. Such discussions may redefine the services the employees provide, and they may point out services that are needed to more effectively or efficiently accommodate the organizational mission. Quality circles focus attention on examining the day-to-day processes in place and reevaluating whether they need to be improved. Such examination of the process is healthy in revalidating present activities and procedures, as well as in serving as a motivational factor for those involved in performing tasks; the employees will then know that they have had an opportunity to evaluate and determine the process of which they are a part.

PRODUCTIVITY MANAGEMENT

In the practice of industrial engineering, a fundamental underlying goal is productivity improvement. Productivity improvement implies successful implementation, not merely a desire. In industrial engineering practice (as compared with academics, for example), talking about implementation is not enough; you must do it and do it successfully. Whereas meeting goals is typically the focus of operations, a primary responsibility of industrial engineering is to increase operational capabilities in order to permit an organization to meet even higher goals in the future. Therefore, knowing how to successfully implement productivity improvement in organizations is a critical requirement for the successful practice of industrial engineering. However, there are far fewer texts dealing with this specific requirement than there are texts concerned with analysis and design.

Implementation Strategies

In 1979, William T. Morris published *Implementation Strategies for Industrial Engineers* [5], one of few texts dealing specifically with implementation of industrial engineering. Morris makes the following statement concerning the professional development of industrial engineers:

> One way to view the professional development of the industrial engineer is to see it as a sequence of three stages.
>
> Stage I. Learning tools, techniques, methods, mathematical skills, experimental skills, and basic engineering concepts—the analytic stage
>
> Stage II. Learning synthetic skills, learning to design systems, learning creative skills, learning how analysis fits into design—the problem solving stage
>
> Stage III. Learning how to deal with clients, how to design change processes which will involve clients actively in the creation of new methods and systems, learning that involvement and collaboration are the essential features of successful change—the change process design stage, the client-centered stage
>
> This book presumes some progress through the first two stages and is concerned with the third, the client-centered stage. [5, p. v]

The traditional practice of industrial engineering has been based on a technical expert approach to solving problems. The following quote from Morris is descriptive of the approach:

> Define the problem
> Build the model and collect the data
> Obtain the optimal solution
> "Sell" the solution to the client

> It is a stance, a self-image, that comes very naturally to the IE whose training has often placed an overwhelming emphasis on the tools, techniques, methods, and models which are the technical core of the profession. [5, p. 63]

In years past, management was authoritarian in nature, and an authoritarian "technical expert" role for industrial engineering seemed consistent and compatible. However, the line organization did not always view the industrial engineer as an "expert" on "their" operations; operations people often would simply prefer to be left alone by management when trying to meet immediate goals, and industrial engineers were typically seen as an extension of management. Since operations' immediate interests typically lie in meeting *today's* production goals, the industrial engineer was often viewed as interfering with that goal. It is also not uncommon for industrial engineers to be viewed as "management spies" by operations people. The facetious statement "I'm from corporate, I'm here to help you" is well known in industry.

While speaking to a management group, I once said, "The practice of industrial engineering typically impedes present operations." My statement appeared to be received with some combination of amazement, if not doubt. I went on to explain that operations people are concerned with producing things today, whereas industrial engineers are concerned with producing things tomorrow. Any effort today to improve operations in the future will typically interfere to some degree with present operations (e.g., a highway construction sign that requests patience for any inconvenience illustrates the concept). For that reason, experienced industrial engineers are cognizant of operations personnel's focus on short-term goals (i.e., today's production) and the inconvenience that today's industrial engineering efforts impose on present operations. However, good operations personnel also know that today's operations are not good enough for the long run, and even if somewhat begrudgingly, they typically offer their assistance.

Morris states, "Failures to achieve implementation are a part of the professional history of many practicing IE's" [5, p. 8]. Management is often unsure about whether to accept an industrial engineering recommendation, as the following quote suggests:

> The relation of the management to staff is clearly not that of the wise doctor to whom the medically ignorant patient surrenders himself for care. Unless his own reasoning is involved, a manager is not likely to accept advice he considers within his own competence—in technological matters, perhaps, but not in running his business. [5, p. 13]

Therein lies a difficult problem for industrial engineers; what they do typically has a lot to do with running the business. The industrial engineer is often indirectly telling a manager that there is a better way of running the business, and such an inference, no matter how tactfully stated, is often received as inconsistent with the manager's self-image. Industrial engineers, as a result of their analyses and recommendations, often unavoidably bruise tender management egos. To further complicate matters, any change accepted by the management as a significant improvement is often treated as prima facie evidence that previous operations were not as well managed as they might have been. When accepted, industrial engineering recommendations have on too many occasions served as a catalyst for a management "witch hunt." There are two convenient, well-known, and commonly employed ways for management to prevent this from happening: (1) do not approve initiation of the study in the first place, and (2) if the study is made and the results are not favorably disposed toward present management, do not accept or simply ignore the study results.

My firm once performed a major study in which Phase I of the project was concerned with identifying and evaluating the opportunity for productivity improvement, and therefore cost improvement, of an operating system. At the conclusion of the Phase I effort, the results were enthusiastically supported at the middle-management-level briefings; however, they were later rejected at the corporate-level meeting, in which the study team was not permitted to participate, on the singular basis that "we don't manage that poorly." The study results had indicated that a $600 million annual operating budget could conservatively be reduced to $500 million (a 17 percent improvement) through the adoption of an industrial engineering consulting team's Phase II study effort over a period of a year and at an estimated cost of $1.2 million—a modest improvement investment. The in-house group that had contracted for the initial study was shortly thereafter disbanded, and the Phase II study was never initiated. Had the Phase I study team simply (and unethically) understated the available productivity improvement opportunity, indicating a 5 percent improvement opportunity in the annual operating budget, it is likely that the Phase II study would have been approved by corporate management. Seventeen percent was simply more than their egos could bear.

The above case was unfortunate. Equally unfortunate is the considerable national underutilization of industrial engineering as a productivity improvement resource simply because the managers who can approve improvement projects sometimes perceive an industrial engineering study as a potential threat to their self-image. It is therefore naive to assume that all managers are psychologically ready and willing to embrace analysis and evaluation of their operations. According to Morris,

> Self-evaluation is often seriously inhibited by:
>
> The need to defend past positions, past methods
> The threat of exposure as ineffective and even incompetent
> The resentment of measurement systems, standards, or criteria imposed from "the outside."
>
> A part of being open to change is being open to evaluation of present performance. This openness, unfortunately, is most frequently found among highly successful clients who are least in need of self-evaluation and self-change. [5, p. 51]

To be honest and fair to management, it needs to be stated that practicing industrial engineers ought to identify improvement opportunities and bring them to management's attention in such a way that the opportunity can be clearly understood by management. Not all practicing industrial engineers have been "stars" at meeting this need, nor has the profession stressed the need for industrial engineering education that prepares them to do so. This is the "bedside manner" part of successful practice that medical doctors are universally criticized for not having been adequately trained; the same criticism can and does apply to industrial engineers as well.

Morris, in describing the *client change process,* specifically the self-awareness phase, points out such psychological (i.e., ego-related) concerns:

> Bringing out data which can be interpreted as "showing people up" or implying criticism may, of course, harden attitudes, increase defensiveness, and simply build further the resistance that must eventually be dealt with if change is to occur. Protection of participants, maintenance of psychological safety, and appropriate assurances of confidentiality are of major concern in this phase. Evaluation, especially self-evaluation, comes later. Ideally, the IE hopes to cultivate the attitude among client personnel that the facts are almost always friendly, that change is exciting rather than threatening, and that ultimately there is no need to be apprehensive about an open appreciation of the way things are going. [5, pp. 49–50]

On numerous occasions in his text, Morris refers to ad hominem thinking on the part of the client (i.e., management); *ad hominem* may be defined as "appealing to a person's feelings or prejudices rather than his intellect" [10, p. 14]. An added complication, therefore, is the realization that managers are not perfect and exhibit normal human faults. An additional malady afflicting managers, often born of their possession of power, results from a vicious cycle of "success breeds confidence" and "confidence breeds success" that sometimes concludes with *ad hominem* decision making. There are few things that get managers into more trouble than the use of the expedient method of decision making commonly referred to as "winging it." Such an approach is nearly as bad as being so indecisive that the needed changes are never made.

If an organization is to achieve continuous improvement in its productivity over time, it must continuously change its behavior. Doing things the old way is simply not good enough. Without a permanent change in behavior, the improvement will simply not be permanent. In this context, the following quote from Morris concerning his proposed client change process should come as no surprise:

> Many of the professions engaged in helping or advising or changing behavior use a roughly common model of the phases through which the client passes as change takes place. Clergymen, psychiatrists, guidance counsellors, organization development specialists, and industrial engineers all typically assume that clients will move through similar processes. It is thus possible that what seems at first to be quite different efforts to bring about behavior change, to implement new methods, or to bring new systems into being may be illuminated by comparing them with the following simple model of the phases which seem to typify the client change process. [5, p. 40]

Morris's proposed client change process involves the following primary phases:

> *Scouting, Targeting, Diagnosis*—The client's attention comes to focus on a limited, although perhaps vaguely understood, problem area or source of difficulty and dissatisfaction. The client begins to refine feelings about "what is wrong" or where change is likely to have the best benefits-to-burdens ratio.
>
> *Stress Relief, Catharsis*—The client reduces the inhibiting effects of past frustrations, anxieties, and conflicts. Blame casting, fault finding, injustices, and so on are expressed,

set aside, and the client's attention is to some degree freed for concentration on the change process.

Self-Awareness—Data gathering, modeling the present system, and studies of "how we do it now" increase the client's self-perception and self-objectification.

Self-Evaluation—The client comes to make his or her own evaluation of present behavior. The client becomes clearer about goals, standards, and objectives and their relation to existing behavior. There is a growing appreciation of the need for change, the material and psychological costs of change, and the development of realistic expectations about the change process.

Self-Designed Change Strategies—The client or client organization plan or accept plans for new behaviors, methods, and systems. The greater the degree to which the client participates in the planning of change, the greater is the probability that change will actually occur. The phase also includes the planning of tests, experiments, and trials of the new behavior.

Trying Out the New Behavior—The client experiments with new methods and systems. Experiential learning leads to modifications and refinements.

Reinforcing the New Behavior—To be replicated, to be made a part of the individual's or organization's repertory of behaviors, it must be rewarded and reinforced. [5, pp. 40–41]

Morris proposes use of the fourth of the following four implementation strategies that he discusses:

Demonstration—actually showing the client that one's recommendations will work as promised.

Power, Politics—getting top management support for change and relying on the power of top management to order that it occur.

"Selling," Persuasion—telling the client what advantages are to be found in the proposed change or in the new system.

Involvement, Participation, Collaboration—acceptance is enhanced by making those whose behavior will be influenced by changes participants in the planning and design of those changes. [5, p. 10]

The second implementation strategy above is the traditional industrial engineering approach. Getting top management support, both in words and in action, is not always easy; even when support is given, it does not guarantee project success.

Morris proposes use of the *Nominal Group Technique* in developing improvements in organizations:

The Nominal Group Technique takes its name from the fact that it is a carefully designed, structured, group process which involves some participants in some activities as independent individuals rather than in the usual interactive mode of conventional groups. In its general form, it is a well-developed and tested method. [5, p. 126]

The following tasks and group session agendas, from a case that Morris describes concerning productivity measurement and improvement in an "ACI (ACI is the name of the firm) services data center" [5, p. 123] describe the group processes.

The Task I statement is "Please list below the important measures (indicators, criteria) of the productivity, effectiveness, or contribution of computing and information services from the viewpoint of your department or division" [5, p. 132].

The Session I agenda is as follows:

Session I Agenda

Consultant's Introduction	15 minutes
Silent Generation, Task I	10
Round-Robin Listing	20
Clarification	35
Ranking and Weighting	25
Discussion and Closure	15

Source: Reference [5], p. 130.

In Task II of Session II, "the group is first asked to spend ten minutes silently and independently generating the ways in which the eight most important measures could actually be made" [5, p. 138]. In Task III, "the group begins by spending ten minutes in the silent generation of the next steps they would be willing to undertake to continue the development and implementation of their productivity measurement system" [5, p. 138]. Both Tasks II and III are performed during Session II, as indicated in the following agenda:

Session II Agenda

Introduction	5 minutes
Silent Generation, Task II	10
Round-Robin Listing and Clarification	45
Introduction to Task III	5
Silent Generation, Task III	10
Round-Robin Listing and Clarification	15
Ranking and Weighting	15
Discussion and Closure	15

Source: Reference [5], p. 139.

The primary rationale for the approach that Morris offers for successful implementation of productivity improvement is suggested at the beginning of the first chapter of his text, entitled "Implementation and the New IE Role":

> Objectives of This Chapter: To define and illustrate the problems of translating industrial engineering knowledge into action. To introduce the concepts of alternative implementation strategies and multiple modes of professional functioning for the industrial engineer. To examine the basic concept of client collaboration as a necessary condition for implementation and the new role of the IE as a designer of change processes. [5, p. 1]

The key is client collaboration, involvement, ownership, and commitment, through group processes that unite the group toward a common goal. Does that mean that all industrial engineering efforts should be devoted to managing or administering group processes? More will be said about the utilization of industrial engineering resources in Chapter 10.

Role of the Industrial Engineer as an Improvement Consultant

In 1981, Peter Block published *Flawless Consulting: A Guide to Getting Your Expertise Used* [2]. Based on the above discussion of Morris's ideals, it should be obvious that "getting your expertise used" has something to do with implementing industrial engineering. A lot of people know that there are industrial engineering consultants, but are *all* industrial engineers consultants? The answer is yes! You cannot be an industrial engineer and not be a consultant. Some industrial engineers are paid directly by the company as employees (internal consultants), whereas other industrial engineers are paid as independent contractors (external consultants), but they are all consultants. All industrial engineers try through the results of their efforts to persuade their clients (i.e., typically management) to adopt their recommendations. The assumption is that the client will be better off for having done so, and therein lies the justification for remunerating the efforts of industrial engineers. An industrial engineer not caring about how to consult is somewhat analogous to a fish not caring about how to swim.

According to Block,

> The consultant who assumes a collaborative role enters the relationship with the notion that management issues can be dealt with effectively only by joining his or her specialized knowledge with the manager's knowledge of the organization. Problem solving becomes a joint undertaking, with equal attention to both the technical issues and the human interactions involved in dealing with the technical issues. [2, p. 21]

It is clear upon consideration of the above quote and the ideas of Morris that for one to be effective as an industrial engineer (in assisting a client to identify, evaluate, accept and then embrace productivity improvement), interpersonal skills are important. As described in Fig. 6-4, Block asserts that there are three primary skill groups that a successful consultant (e.g., industrial engineer) must master: (1) technical, (2) interpersonal, and (3) consulting.

If one compares the need for productivity improvement group processes as proposed by Morris [5] and the need for development of the three groups of consultant skills proposed by Block [2] with the education that industrial engineering students have traditionally received in preparing them for practice, the limited success that industrial engineers often have in getting their recommendations accepted and implemented is no surprise. More will be said about this educational deficiency in Chapter 10.

Organizational Improvement

In 1989, D. Scott Sink and Thomas C. Tuttle published *Planning and Measurement in Your Organization of the Future* [9]. Their text provides the basis for effecting organizational improvement over time. It is the kind of thoughtful, well-organized approach that manufacturing organizations have needed but too often in the past have not taken the time and effort to develop and embrace. The authors build on the prior group processes of Morris, particularly in their use of the Nominal Group

Technical skills	Interpersonal Skills	Consulting Skills
Specific to Your Discipline • Engineering • Project Management • Planning • Marketing • Manufacturing • Personnel • Finance • Systems Analysis	Apply to All Situations • Assertiveness • Supportiveness • Confrontation • Listening • Management style • Group process	Requirements of Each Consulting Phase *Contracting* • Negotiating wants • Coping with mixed motivation • Dealing with concerns about exposure and the loss of control • Doing triangular and rectangular contracting *Diagnosis* • Surfacing layers of analysis • Dealing with political climate • Resisting the urge for complete data • Seeing the interview as an intervention *Feedback* • Funneling data • Identifying and working with different forms of resistance • Presenting personal and organizational data *Decision* • Running group meetings • Focusing on here and now choices • Not taking it personally

FIGURE 6-4 The three required skill groups of a consultant. (From Block [2, p. 7].)

Technique. Figure 6-5 describes the Performance Management Process proposed by Sink and Tuttle.

Note that one of the earliest inputs to the process is "vision." Sink and Tuttle state:

> The creation of a vision is normally the role of the leader. In fact, the creation of a compelling vision is what separates the leader from the manager. Of course, what separates the successful leader from the dreamer is that the leader has the practicality to convert the vision to reality. [9, p. 56]

In a section of their text entitled "Implementing Visions Requires New Paradigms," Sink and Tuttle describe a 1986 videotape developed by Joel A. Barker [1] entitled *Discovering the Future—The Business of Paradigms* that does an excellent job of dramatizing the need for vision.

A dictionary definition of *paradigm* is "an outstandingly clear or typical example or archetype" [10, p. 830]. In this context, however, I prefer the broader Sink and

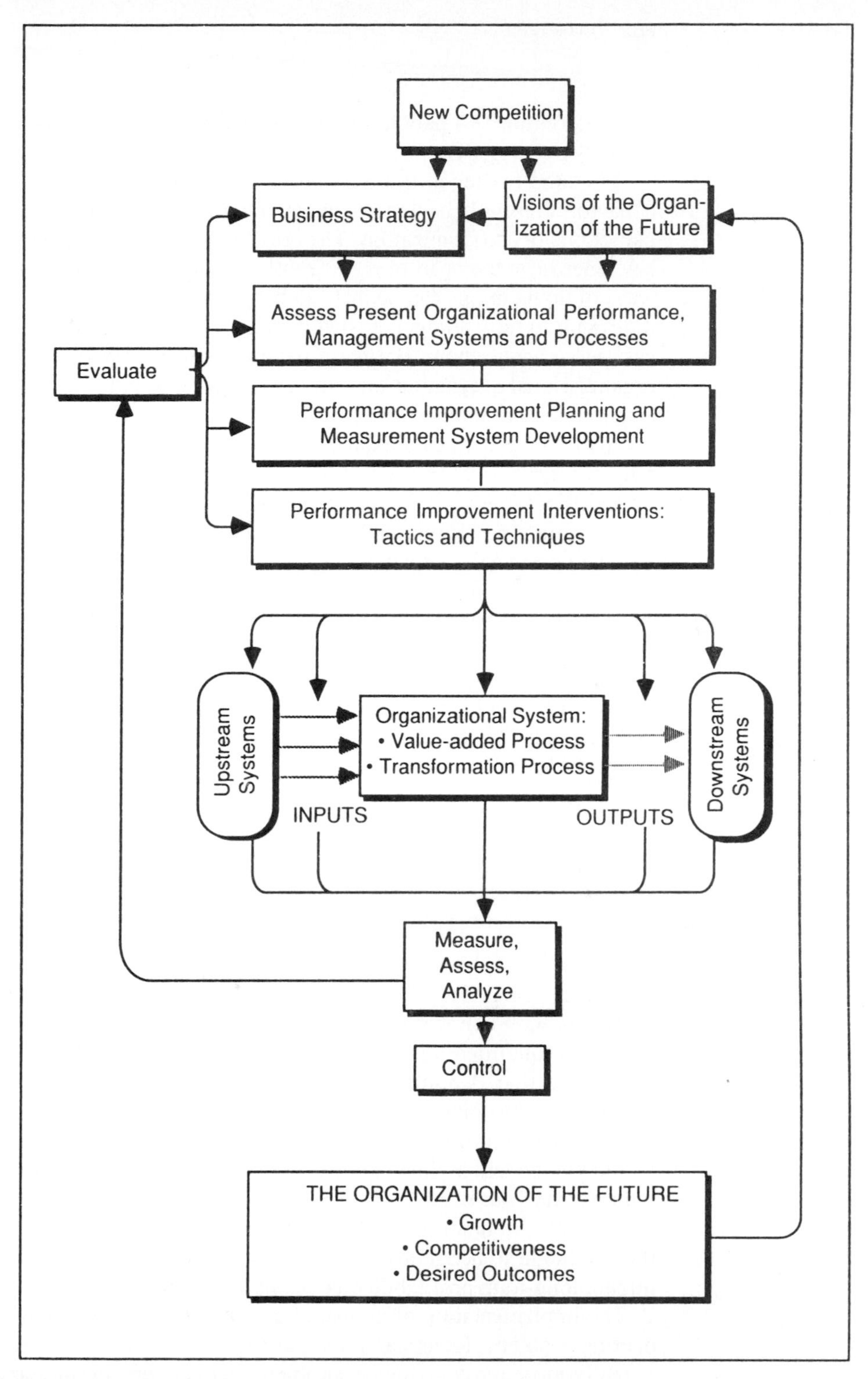

FIGURE 6-5 The Performance Management Process. (From Sink and Tuttle [9, p. 35].)

Tuttle definition: "A paradigm is formally defined as a pattern, example, or model" [9, p. 29].

As stated earlier in this section, continuous improvement requires change in behavior. Present behavior is contained within and can be described by the present paradigms of an organization. The breaking of old paradigms and the formation of new ones lie at the heart of significantly effective continuous improvement. Present ways of doing things can become so habitual that an existing organization may find it difficult to fully recognize what paradigms they presently utilize and then to successfully change them. Breaking paradigms in an organization is the opposite of organizational stagnation and arrogance in "maintaining the status quo," which ensure failure in today's fast-paced business world. The choice is to change or die.

Figure 6-6 describes Sink and Tuttle's proposed performance improvement planning process.

A difficult question to answer is "Where do visions come from?" A facetious answer might be "From leaders." But how do you know a good vision when you see one or a true leader when you meet one? Most great business decisions depend to some degree on faith—faith that the leader is truly a leader and that what he or she perceives to be true is true. Some things simply must be accepted on the basis of faith—there is no objective way of prejudging their value. If Wilbur and Orville's airplane or Edison's light bulb had been evaluated on the basis of ROI (return on investment), would there be airplanes or electric lights today?

What is the "bottom line" for an organization? Sink and Tuttle suggest that it is not profit:

> We also propose that the bottom line for any organization is not profitability or managing to budget. In our opinion, bottom line is whether or not the organization is achieving its visions of what it wants to or feels it must become. Bottom line, long term, is survival, growth, constantly improving performance, competitiveness, and behaving in accordance to your values and principles. If you do these things, profits follow. [9, p. 38]

In years past, I was occasionally asked to serve on an MBA oral examination committee. One of my favorite questions was "Do you believe in rewarding plant managers with bonuses based on profitability?" Candidates would often respond with an unqualified yes—only to regret not qualifying their response. It is not a simple question. A number of follow-up questions would lead them down the path; for example, "As plant manager, then, would you be less inclined to repair the boiler because it would impact this year's profitability and therefore your bonus? And would you be less inclined to train employees because that would also reduce profitability and your bonus?" The unacceptability of misusing assets to maximize this year's profit (in order to maximize management bonuses) would then typically lead the candidate to hedge and recommend a more qualified response, such as maximizing long-term profitability as one of many objectives.

The implementation of industrial engineering solutions is typically not an engineering problem. Technical industrial engineering is the easy part. Implementation is really more a psychological problem than an engineering problem; therefore, if industrial engineering is to be successfully practiced in the future, industrial engi-

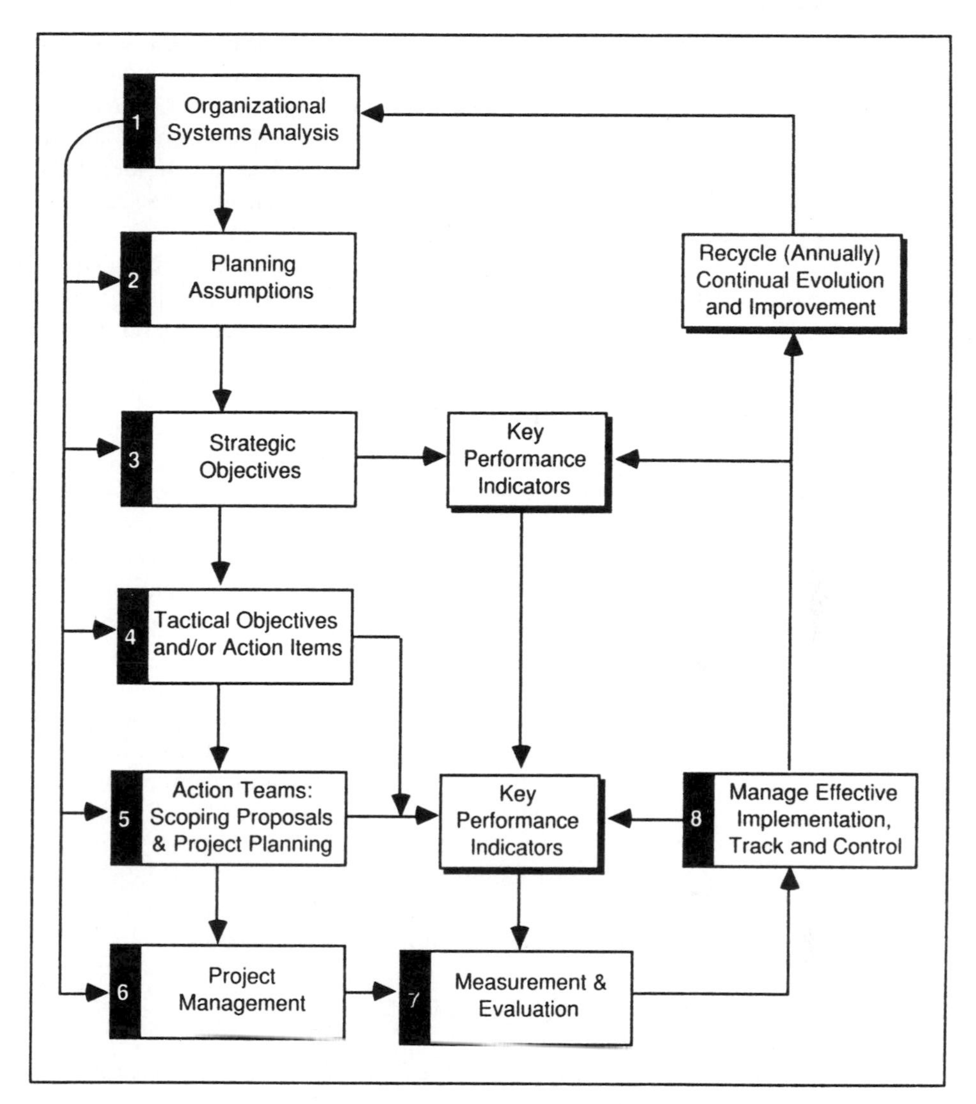

FIGURE 6-6 Performance Improvement Planning Process. (From Sink and Tuttle [9, p. 201].)

neers will need to be better educated in the "psychology of implementation." More will be said about this in Chapter 10.

SUMMARY

Industrial engineering has traditionally concentrated on reducing direct labor costs. When industrial engineering came into existence, direct labor was typically a major cost of doing business in manufacturing, and indirect and overhead costs were often minimal. Today it is just the opposite; it is not uncommon today for direct labor costs to represent 5 percent of manufactured cost and for indirect and overhead costs to represent 50 to 80 percent of manufactured cost. Industrial engineering emphasis and approaches have been slow in adjusting to this change in cost ratios.

One of the pioneers in bringing attention to this need and in developing improved techniques for addressing this need is Marvin Mundel, Ph.D. His approach will likely become increasingly employed in years to come as a means of gaining control of indirect and overhead costs.

Lehrer, ahead of his time, foresaw the need for a concentration of effort in productivity improvement, a concept called "continuous improvement" today. Much of the attention given today to productivity and quality improvement is consistent with his message of a quarter-century ago.

Morris provided one of few recent industrial engineering texts that address the important issue of successful implementation. The client change process he proposes identifies an important requirement of successful practice in industrial engineering—that a practitioner must approach a client in a way that will successfully change the client's behavior (i.e., industrial engineering is in the behavior modification business). Industrial engineering to date has been remiss in generally ignoring this requirement of successful practice. More attention (e.g., expanded curricula) needs to be given to such aspects if our goal is to enable future industrial engineers to maximize their accomplishments.

Group processes such as the Nominal Group Technique will be more commonly employed in the future as industrial engineering becomes increasingly practiced in work team and management consensus environments. Industrial engineers will need to be trained to both lead and function within team efforts of the future.

Block makes an excellent case for the need for all practicing industrial engineers to acquire three sets of required skills: technical, interpersonal, and consulting. All practicing industrial engineers are consultants. They are almost always trying to convince someone to do something; they basically make recommendations for a living. They need to learn how to do so effectively.

REFERENCES

1 Barker, J. A.: *Discovering the Future: The Business of Paradigms,* Filmedia, Inc., Minneapolis, MN, 1986.

2 Block, Peter: *Flawless Consulting: A Guide to Getting Your Expertise Used,* University Associates, San Diego, CA, 1981.

3 Lehrer, Robert N.: *White Collar Productivity,* McGraw-Hill Book Company, New York, 1983.

4 Maslow, Abraham H.: *Motivation and Personality,* Harper & Row Publishers, Inc., New York, 1970.

5 Morris, William T.: *Implementation Strategies for Industrial Engineers,* Grid Publishing, Inc., Columbus, OH, 1979.

6 Mundel, Marvin E.: *Improving Productivity and Effectiveness,* Prentice-Hall, Inc., Englewood Cliffs, NJ, 1983.

7 Mundel, Marvin E.: *Measuring and Enhancing the Productivity of Service and Government Organizations,* Asian Productivity Organization, Tokyo, Japan, 1975.

8 Mundel, Marvin E.: *The White Collar Knowledge Worker: Measuring and Improving Productivity and Effectiveness,* Asian Productivity Organization, Tokyo, Japan, 1989.

9 Sink, D. Scott, and Thomas C. Tuttle: *Planning and Measurement in Your Organization of the Future,* Industrial Engineering and Management Press, Norcross, GA, 1989.

REVIEW QUESTIONS AND PROBLEMS

1 If direct labor cost in a plant constitutes only 5 percent of manufactured cost, to what extent should production management focus its attention on direct labor?

2 Do direct labor employees typically express a desire to change the methods they are presently employing in performing work? Why or why not?

3 Who earned the first Ph.D. in industrial engineering, and from what university was it granted? Who earned the second Ph.D. in industrial engineering?

4 Were traditional industrial engineering techniques primarily developed to address direct labor or indirect labor cost reduction, and why is that important today?

5 Differentiate between efficiency and effectiveness.

6 At what level in an organization are knowledge workers typically least technologically developed for enhancing their effectiveness and efficiency in performing their daily tasks?

7 Describe the technique employed in General Electric called INTROSPECT.

8 Does the practice of industrial engineering get in the way of present operations; and if so, why do it?

9 What is ad hominem thinking?

10 Describe the step in Morris's client change process called "Stress Relief, Catharsis."

11 Of the four implementation strategies described by Morris, why is "Involvement, Participation, Collaboration" typically more effective than "Power, Politics."?

12 Why is Peter Block's book about consulting important to all industrial engineers?

13 What are the three skills that a successful industrial engineer must possess?

14 The United States has some cultural deficiencies; what are they?

15 What is a paradigm?

16 What do Sink and Tuttle propose as the bottom line (objective) for an organization?

LABORATORY EXERCISE 10: Quality and Production Problems and Nominal Group Technique (NGT)

The members of each toy train team should review their previous plant design and list and discuss possible quality and production problems and solutions. They should order the list with the greatest opportunity for improvement at the top of the list. This effort should take about half an hour.

All laboratory members should next form a single group to participate in a Nominal Group Technique discussion of possible improvements in the toy train product and process. If the group is larger than twenty students, it is recommended that the total group be subdivided so that no NGT group exceeds 20 students as a maximum.

The question to be addressed in this exercise is "What improvements can you suggest in product or process design that will improve either the quality, cost, or delivery (and "time to market" development) of this product to the customer?"

A nominal group technique analysis should then be directed by the instructor (or a team leader), employing the following steps, as described earlier in this chapter.

NGT SESSION AGENDA

Step	Task	Time (min)
1.	Instructor/Team Leader Introduction	10
2.	Silent Generation	10
3.	Round-Robin Listing	20
4.	Clarification	35
5.	Ranking and Weighting	25
6.	Discussion and Closure	15

In Step 1, the instructor/group leader describes the nominal group technique process.

Step 2 permits each participant to independently list possible improvement possibilities. This independent generation is considered a major strength of this technique.

In Step 3, the instructor/group leader asks each participant in sequence to list one idea that the team leader adds to a flip chart or writes on the board. In successive rounds, anyone who wishes to do so can "pass" if they do not wish to add any additional items at that time. Rounds are continued until all improvement ideas are submitted.

In Step 4, the list is reviewed to determine if some items are so close to other items that they can be combined into one idea. Each person who offers an idea should make the decision as to whether or not to combine his or her idea with another idea. Each resulting idea should be numbered on the final list.

In Step 5, each participant should be handed five to seven 3 × 5 cards—five if the list is 15 items or less, seven if 30 items or more, and six if between. Assuming five cards are issued to each participant, each participant should select the five items that he or she believes represent the best opportunities for improvement. In the upper left corner of the first card, the number of the item believed to offer the best opportunity for improvement should be recorded. The idea name should be listed in the center of the card, and the number 5 should be placed in the lower right corner of the card. On a second 3 × 5 card, based on the five best ideas selected, the number of the idea considered to be the "least best" of the five should be recorded in the upper left corner, with the idea name in the center and the number 1 in the lower right corner. The third card is for the second best idea, and the fourth card is for second "least best" idea of the five. The remaining card is scored 3 in the lower right corner. All five cards should then be handed to the group leader, and it is then time for a break.

The group leader summarizes the results from all participants, listing the scores recorded for each idea listed.

In Step 6, as soon as the summary is completed, the group is reconvened and the results discussed.

LABORATORY EXERCISE 11

Based on the results of Laboratory Exercise 10, each toy train team should reevaluate their toy train product and process and determine to what extent they would modify their product design or layout to accommodate improvements. They may utilize the Nominal Group Technique again in these deliberations. Each team should make the changes to their product design or plant layout accordingly.

The last hour of this laboratory should be utilized to prepare for Laboratory Exercise 12. (See Laboratory Exercise 12 for instructions.)

LABORATORY EXERCISE 12

It is suggested that the instructor invite three practicing industrial engineers from the community to participate in this laboratory exercise. At least one of the visitors should be a female, at least one should be a minority, and at least one should be from a service organization. The ages of the visitors should range from limited to considerable experience. Prior to the last hour of Laboratory Exercise 11, each student should have turned in a list of three to five questions concerning industrial engineering they wish to ask the visitors. Prior to the Laboratory Exercise 11 meeting, all questions should have been collected and listed. In the last hour of Laboratory Exercise 11, all class participants should vote, using 3 × 5 cards, for their preferred five questions, as described in Laboratory Exercise 10; the results should then be summarized.

The questions should be asked of the visitors in the preferred sequence indicated by the previous voting.

7

OPERATIONS RESEARCH

If you torture data long enough, it will confess.

Anonymous

Operations research is one of the many advancing areas of industrial engineering. Those most unfamiliar with operations research (e.g., management) have often viewed it as a collection of techniques (linear programming, dynamic programming, queuing theory, etc.). Managers who hold this belief have occasionally failed in the past because techniques were tried in situations for which they were inadequate. Successful application often depends on either modifying an existing technique to overcome its inherent weakness in a given situation or developing a new technique. Unfortunately, it is far easier in practice to apply an existing technique than to improve one. It takes considerably more ingenuity—sometimes more than exists at a particular place and time—to devise an appropriate operations research solution to a complex problem. For this reason, although it is introduced at the undergraduate level in a typical industrial engineering curriculum, serious preparation for practice in operations research today typically occurs at the graduate level.

What follows is an attempt to illustrate a few basic operations research techniques that are in general use today. Those selected are few in number and are meant to be representative, and are certainly not exhaustive. They are also limited to those with mathematics that the reader of this text can handle.

LINEAR PROGRAMMING

The one technique most commonly identified with operations research is linear programming. Dantzig [2] would be many operations researchers' choice for "father of operations research." His development of an algorithm (the simplex method) for solving linear programming problems is probably the most significant development to date in operations research. The generalized linear programming approach, for which the simplex method provides a solution technique, is of such generality that it is useful in solving resource allocation problems (that fit its assumptions) across a broad spectrum of application areas.

Before illustrating the generalized approach, two special cases of linear programming, the assignment algorithm and the transportation algorithm, will be illustrated. These two algorithms are more efficient than the generalized approach for problems that possess the specific characteristics required for their use.

The Assignment Algorithm

The assignment algorithm provides a means of assigning *n* resources to *n* tasks so as to maximize or minimize the sum of effectiveness of all task assignments.

Before an assignment algorithm example is discussed, however, it is important to understand the difference between two words employed extensively in operations research. One word is *heuristic,* the other *algorithm.* A heuristic is a rule or procedure that when followed will provide a good answer, but not necessarily an optimum answer. For example, if you want a rat entering a rat maze to ultimately emerge from the maze, the rat can be taught to hug the right (or left) wall continuously; "hugging the right wall" is a heuristic.

An algorithm, on the other hand, not only implies a good procedure but guarantees an optimum solution as well (e.g., first rat out of the maze, if getting out first is the success criterion).

Consider the following example. A department head would like to assign each of four subordinates to four assignments. An estimate of the number of days each worker would spend on each assignment with equal ultimate task effectiveness is as follows:

		Workers			
		I	II	III	IV
Tasks	A	6	7	10	9
	B	2	8	7	8
	C	8	9	5	12
	D	7	11	12	3

Table 7-1 is a three-phase assignment algorithm. The first step in attempting a solution to the problem is the application of phase 1 of the algorithm. Row reductions for the above problem would produce the following:

		Workers			
		I	II	III	IV
Tasks	A	0	1	4	3
	B	0	6	5	6
	C	3	4	0	7
	D	4	8	9	0

TABLE 7-1 AN ASSIGNMENT ALGORITHM*

Phase 1

1 Subtract the smallest value in each row (or column) from all other elements in the row (or column) until there is at least one zero element in each column and row.

Phase 2

1 Examine rows successively until a row with exactly one unmarked zero is found. Mark (□) this zero, as an assignment will be made there. Mark (X) all other zeros in the column to show that they cannot be used to make other assignments. Proceed in this fashion until all rows have been examined.
2 Next, examine columns for single unmarked zeros, marking them (□) and also marking with an (X) any other unmarked zeros in their rows.
3 Repeat steps 1 and 2 successively until one of three things occurs:
   a There are no zeros left unmarked and we have a complete assignment.
   b The remaining unmarked zeros lie at least two in each row and column and we must complete by trial and error.
   c We do not have a complete assignment and must proceed to phase 3.

Phase 3

1 Mark all rows for which assignments have not been made.
2 Mark columns not already marked which have zeros in marked rows.
3 Mark rows not already marked which have assignments in marked columns.
4 Repeat steps 2 and 3 until the chain of marking ends.
5 Draw lines through all unmarked rows and through all marked columns.
6 Select the smallest element not covered by a line and:
   a Subtract its value from all uncovered elements.
   b Add its value to all assignments covered by two lines.
   c Repeat the previous value for all elements covered by one line.
   d Return to phase 2 with the resulting matrix.

*A modification of algorithms in Maurice Sasieni, et al., *Operations Research: Methods and Problems,* John Wiley & Sons, Inc., New York, 1959, pp. 186–192.

Column reductions would then produce

		Workers			
		I	II	III	IV
Tasks	A	0	[0]	4	3
	B	[0]	5	5	6
	C	3	3	[0]	7
	D	4	7	9	[0]

A complete assignment, one worker to each of the four tasks, can be made as indicated above, using only zero elements. The optimal assignment of workers to tasks, referring to the initial matrix, would be

Task	Worker	Task time, days
A	II	7
B	I	2
C	III	5
D	IV	3
		17

No other assignment of the four workers, one each to a task, can be made with a sum of task times of less than 17 days.

In matrices of a larger magnitude, a complete assignment may well exist using only zero elements as above, but the larger number of zeros requires a procedure for identifying the set of zeros, one in each row and column, that represents the complete assignment. In these cases, phase 2 of the algorithm can be used to attempt to identify a complete assignment from among existing zero elements.

The following reduced matrix, considering zero elements only, can be used to illustrate phase 2 of the algorithm:

	I	II	III	IV	V
A		0			
B	0	0			
C	0		0		0
D	0		0		0
E			0	0	0

Application of steps 1 and 2 of phase 2 would result in the following:

	I	II	III	IV	V
A		[0]			
B	[0]	~~0~~			
C	~~0~~		0		0
D	~~0~~		0		0
E			~~0~~	[0]	~~0~~

As indicated in step 3b of phase 2, arbitrary assignment of C to III and D to V results in the following optimal solution:

	I	II	III	IV	V
A		[0]			
B	[0]	~~0~~			
C	~~0~~		[0]		0
D	~~0~~		0		[0]
E			~~0~~	[0]	0

To illustrate phase 3 of the algorithm, assume a solution is needed for the following reduced matrix:

	I	II	III	IV	V
A	5	0	8	10	11
B	0	6	15	0	3
C	8	5	0	0	0
D	0	6	4	2	7
E	3	5	6	0	8

Application of phase 2 would produce

	I	II	III	IV	V
A	5	[0]	8	10	11
B	~~0~~	6	15	~~0~~	3
C	8	5	[0]	~~0~~	~~0~~
D	[0]	6	4	2	7
E	3	5	6	[0]	8

Note that this does not represent a complete assignment. Application of steps 1 through 5 of phase 3 results in the following:

	✓ I	II	III	✓ IV	V
~~A~~	~~5~~	~~0~~	~~8~~	~~10~~	~~11~~
✓ B	0	6	15	0	3
~~C~~	~~8~~	~~5~~	~~0~~	~~0~~	~~0~~
✓ D	0	6	4	2	7
✓ E	3	5	6	0	8

Note that element BV is the smallest element not covered by a line. The application of step 6 of phase 3 produces the following:

	I	II	III	IV	V
A	8	0	8	13	11
B	0	3	12	0	0
C	11	5	0	3	0
D	0	3	1	2	4
E	3	2	3	0	5

Application of phase 2 to this revised matrix produces

	I	II	III	IV	V
A	8	[0]	8	13	11
B	~~0~~	3	12	~~0~~	[0]
C	11	5	[0]	3	~~0~~
D	[0]	3	1	2	4
E	3	2	3	[0]	5

The complete optimal assignment then is

A–II
B–V
C–III
D–I
E–IV

The assignment algorithm shown above is a minimization algorithm. To produce an optimum as a maximum it is necessary only to replace element values with complement values determined by subtracting each element from the largest element value in the original matrix. The following example will illustrate the procedure.

Assume that the expected profit from assigning ice cream sellers to four districts is as follows:

		Sellers			
		I	II	III	IV
Districts	A	16	14	12	16
	B	12	18	10	15
	C	13	16	14	20
	D	10	12	15	14

To produce a maximum, each element is subtracted from 20, which is the largest element value in the matrix, producing complement values as follows:

	I	II	III	IV
A	4	6	8	4
B	8	2	10	5
C	7	4	6	0
D	10	8	5	6

Phase 1 row and column reductions would then produce

	I	II	III	IV
A	[0]	2	4	0
B	6	[0]	8	3
C	7	4	6	[0]
D	5	3	[0]	1

Therefore, the maximal assignment would be

A–I
B–II
C–IV
D–III

This assignment produces a total profit of 69. No other assignment of the four ice cream sellers to the four districts will produce a greater profit. The generality of the technique should be apparent at this point. The assignments could represent the sum of effectiveness of any one of the following types of assignments:

1. Four nurses to four stations in a hospital
2. Four traffic controllers to four different air traffic tasks
3. Four police cars to four crime districts
4. Four heavy equipment operators to four pieces of equipment
5. Four pieces of equipment to produce four different products
6. Four bank auditors to four different types of banks in a district

It should be apparent that this list could be essentially infinite because of the level of generalization of the technique.

As an example of the use of the assignment method, consider an enterprising restaurateur who intends to attempt the following new concept in the restaurant business. The restaurateur has four locations in a city (W, X, Y, and Z). Three kitchens are to be located at three of the four locations, with one location remaining empty. Four kitchens will be labeled M, C, P, or N to represent Mexican food, Chinese food, pizzas, or no food, respectively. The restaurateur intends to produce the three different types of food in the three separate kitchens and deliver the prepared foods from all three kitchens to all five of the restaurant locations (1, 2, 3, 4, and 5) throughout the city. The four available kitchen locations and five restaurant locations are shown in Fig. 7-1. Locations are shown with respect to the nearest street crossing to simplify distance calculations by employing unit block increments (i.e., there are three block distances between locations X and 1, and five between Z and 1). The relative number of trips that would be needed between kitchens and restaurants over some defined period of time is given in Table 7-2. The restaurateur would like to locate the kitchens, one at a location, so that the sum of delivery distances from all kitchens to all restaurants is a minimum.

This problem is analogous to one proposed by Moore [7] in locating new machines at candidate locations in a factory so as to minimize the material handling distance for all products flowing between existing and new machines. Figure 7-2*a* illustrates the matrix multiplication of trips and distances to produce a matrix that shows the effect on distance of locating each kitchen at each possible location. An assignment method solution of this matrix, as shown in Fig. 7-2*b*, identifies the best location for each kitchen to minimize the total distance traveled for all required deliveries (i.e., 473 unit block distances). No other assignment of kitchens to possible locations will result in a smaller total distance. This example suggests the generality of the technique with respect to potential applications.

FIGURE 7-1 Kitchen and restaurant locations.

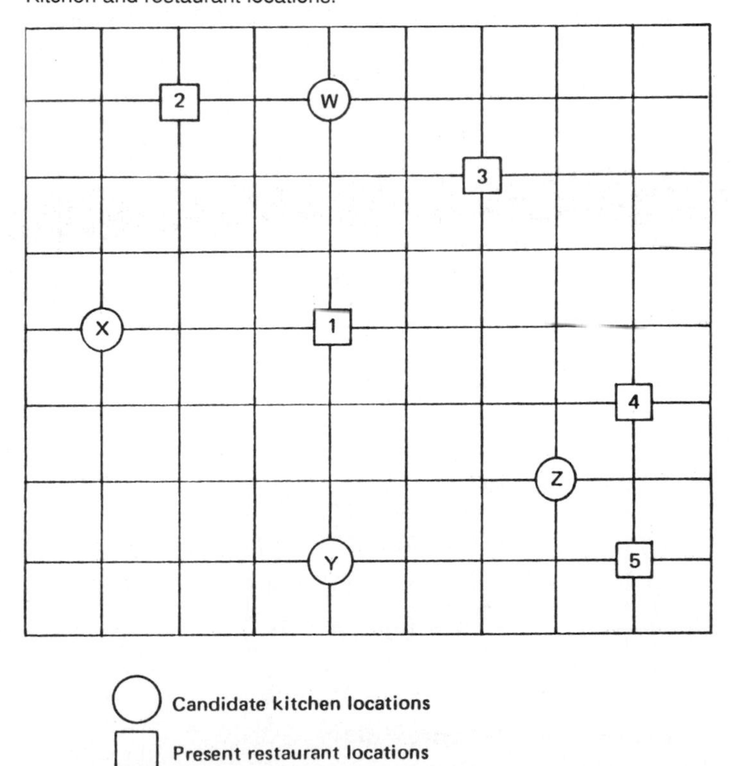

TABLE 7-2 RELATIVE NUMBER OF TRIPS REQUIRED BETWEEN KITCHENS AND RESTAURANTS OVER SOME DEFINED PERIOD

		Restaurants				
		1	2	3	4	5
Kitchens	M	10	12	4	8	6
	C	3	5	10	12	4
	P	7	3	5	4	8
	N	0	0	0	0	0

The Transportation Algorithm

Another very useful special case of the generalized linear programming problem is commonly referred to as the transportation problem. The transportation algorithm is employed when m sources are supplying n destinations, and individual cost coefficients per unit of flow between each source and destination are known and are linear as a function of the volume of flow. It is desired to allocate units from specific

FIGURE 7-2 Matrix multiplication of trips and distances and assignment method solution.

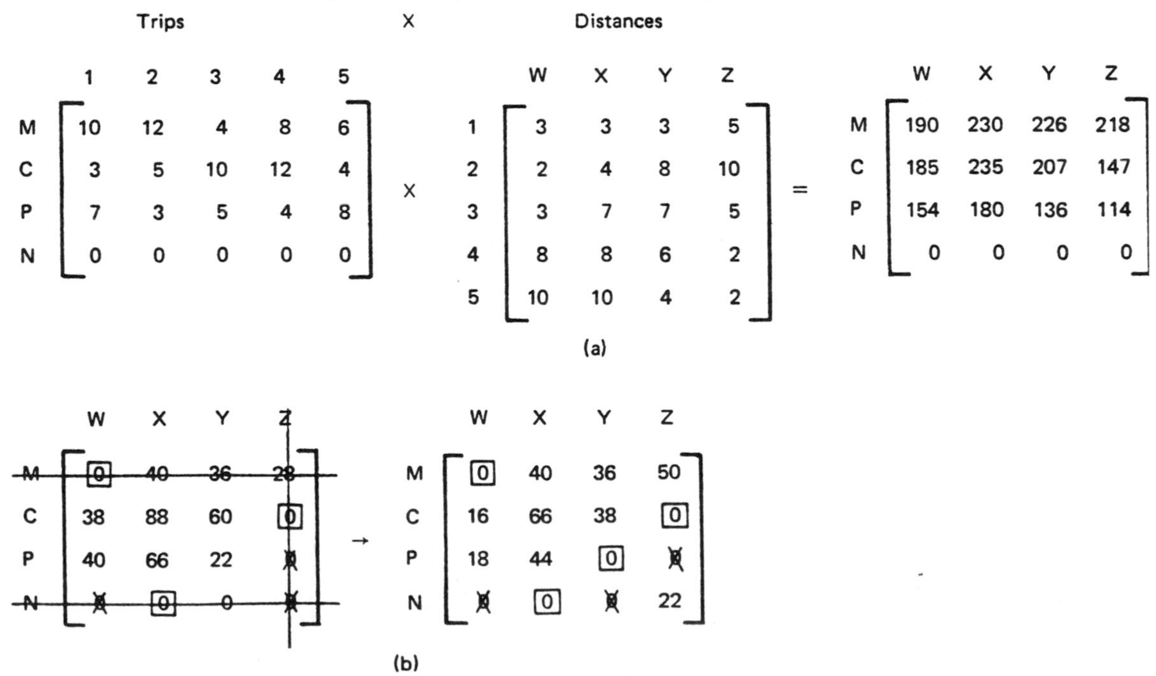

Total unit distance for all products = 190 + 147 + 136 = 473

sources to specific destinations to result in a least-cost solution for all products delivered. The following example illustrates a typical simple transportation problem.

How many tons should be shipped from each mine to each processing plant so that mine capacities are not exceeded, each plant receives its required tonnage input, and a least cost of delivered ore results? Shipping cost per ton between specific mines and plants, mine capacities, and plant requirements are assumed to be as follows:

Cost of shipping to plant, $/ton

		A	B	C	D	Mine capacities, tons
Mine	1	$3	$1	$4	$5	50
	2	7	3	8	6	50
	3	2	3	9	2	75
Plant requirements, tons		40	55	60	20	175

Table 7-3 contains the transportation algorithm that will be employed in solving this problem. Phase 1 of the algorithm is generally known as Vogel's approximation. It permits determination of a good starting point as a first step in searching for an optimal solution. Phase 2 permits ultimate determination of an optimal solution to any transportation problem.

The setup for the above problem employing phase 1 would be as follows:

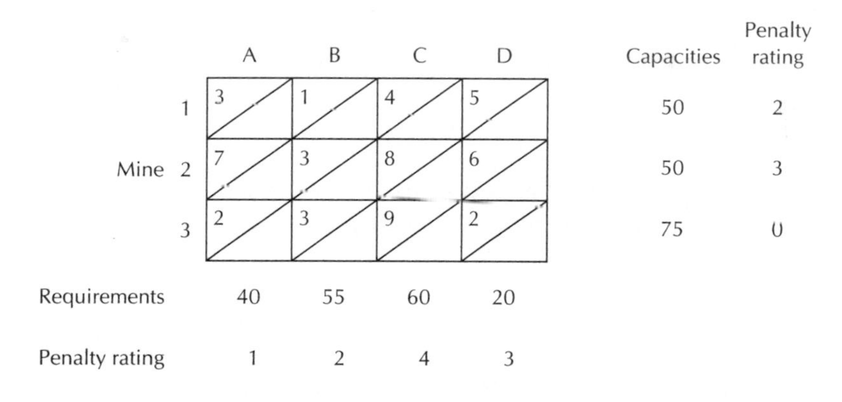

		A	B	C	D	Capacities	Penalty rating
Mine	1	3	1	4	5	50	2
	2	7	3	8	6	50	3
	3	2	3	9	2	75	0
Requirements		40	55	60	20		
Penalty rating		1	2	4	3		

Phase I maximum penalty rating is 4; therefore, as much material will be allocated to the lowest-cost cell in this column as possible. This heuristic simply allocates material in order to preclude suffering high penalty costs associated with second-best allocations. In this instance the difference between $4 per ton and $8 per ton makes it desirable to move as much material as possible at $4 per ton with respect to plant C. The limit of how much can be allocated to cell 1C is determined by the lesser of (1) the capacity of mine 1, 50 tons, and (2) the requirement of plant C, 60

TABLE 7-3 A TRANSPORTATION ALGORITHM

Phase 1 (Vogel's approximation)

1 Develop a penalty rating value for each row and column, which is the difference between the least-cost and the next-to-least-cost elements in each row (or column). Make the maximum possible assignments in decreasing order of the size of the penalty rating until all assignments are made.

Phase 2

1 Obtain an initial assignment by some prior method (e.g., Vogel's approximation).
2 Determine U_i and V_j values for rows and columns starting with $U_1 = 0$. Determine remaining U and V values by using the equation

$$U_i + V_j = C_{ij}$$

employing only cells in which an assignment has been made.
3 Determine t_{ij} values for all remaining cells by using U and V values developed in step 2 above as follows:

$$t_{ij} = U_i + V_j - C_{ij}$$

4 If all t_{ij} values are zero or negative, an optimal solution has been reached. If there are positive t_{ij} values, bring the largest positive t_{ij} into solution by employing step 5.
5 Identify a "transfer route" starting with the largest positive t_{ij} above, moving alternately up or down and left or right, on assigned cells only, returning to the largest positive t_{ij} cell above.
6 Alternately add and subtract a value θ from the assigned quantities in cells identified in step 5. The value θ is equal to the smallest quantity presently assigned to cells from which θ is to be subtracted. When θ is added or subtracted to cell quantities, one new assignment is created in the cell that possessed the largest positive t_{ij}, and one of the cells in which the quantity was reduced to zero should no longer possess an assignment.
7 Return to step 2.

tons. Therefore, the next matrix below shows no additional capacity remaining at mine 1, and a remaining requirement at plant C of 10 tons:

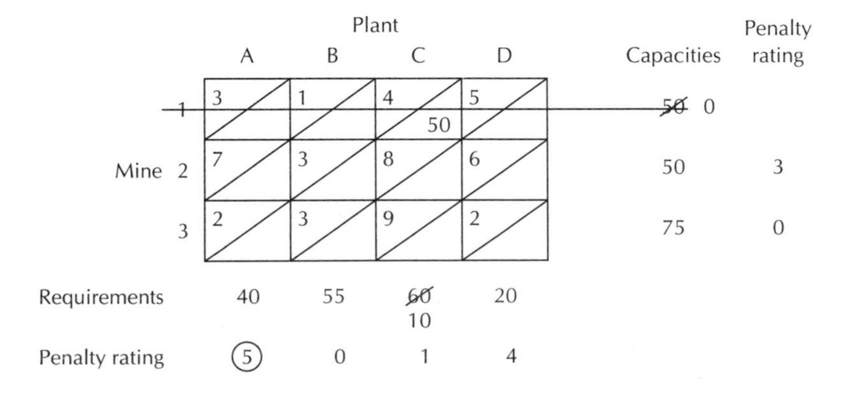

Recalculation of the penalty ratings, as shown on the above matrix, is necessary because row 1 has been removed from further consideration because of the allocation of all of the output of mine 1 to plant C. Because of the new penalty rating of 5 with respect to plant A, a maximum allocation is next made to this plant. Because 40 is less than 75, 40 tons are allocated to plant A from mine 3, reducing the remaining capacity from mine 3 to 35 tons and removing plant A from further consideration by reducing its remaining requirement to zero:

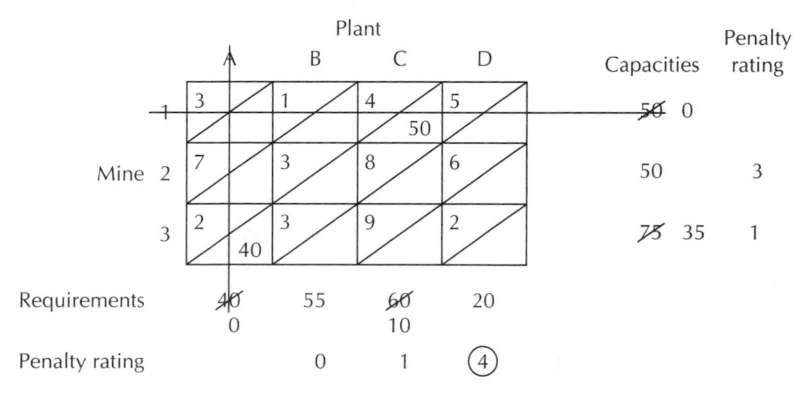

This process is continued until all mine capacity is allocated to meet the requirements of all plants.

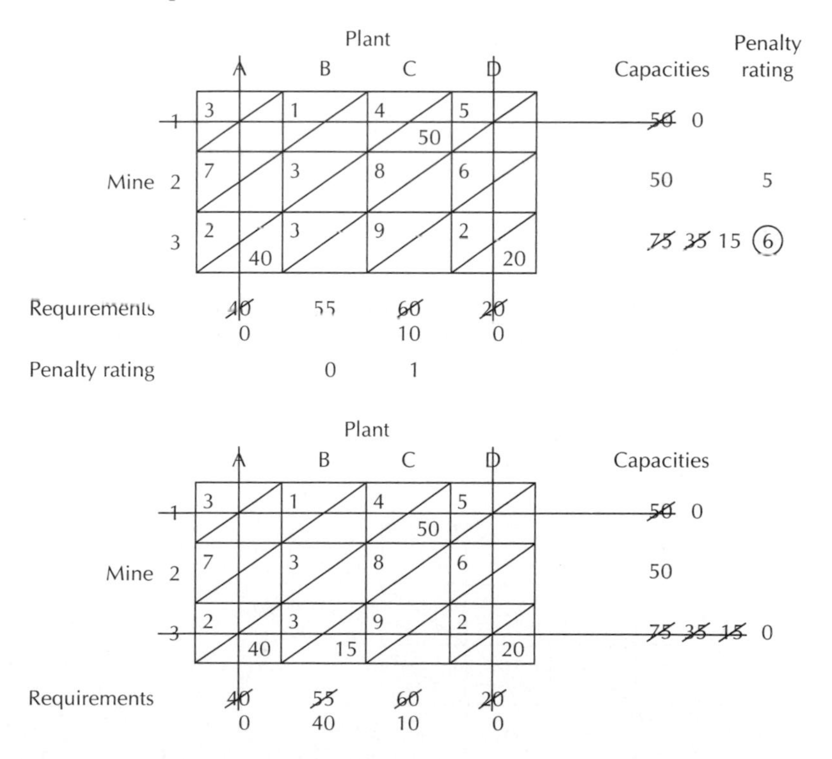

The only possible remaining assignments are 40 tons from mine 2 to plant B and 10 tons from mine 2 to plant C:

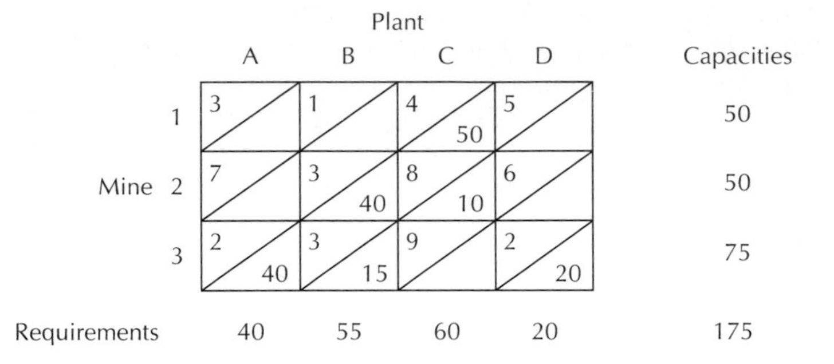

Plant
A
B
C
D
Capacities
Mine
1
2
3
3
1
4
5
50
50
7
3
8
6
40
10
50
2
3
9
2
40
15
20
75
Requirements
40
55
60
20
175

The total cost of shipping the 175 tons from the three (m) mines to the four (n) plants would be as follows:

Mine to plant	Cost per ton	Tons	Cost
1–C	$4	50	$200
2–B	3	40	120
2–C	8	10	80
3–A	2	40	80
3–B	3	15	45
3–D	2	20	40
			$565

The above assignment, obtained by employing Vogel's approximation, happens to be optimum. The test for optimality can be made by continuing to phase 2. The use of phase 2 will either prove that the application of phase 1 produced an optimal solution, as was the case above, or permit improvement in allocations until the optimum is reached.

Assume that phase 1 had produced the following allocation:

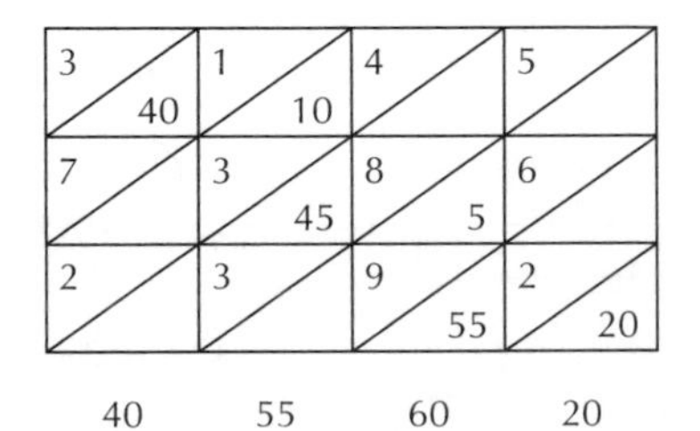

3
1
4
5
40
10
50
7
3
8
6
45
5
50
2
3
9
2
55
20
75
40
55
60
20

Then *Us* and *Vs* would be added as follows, employing steps 1 and 2 of phase 2:

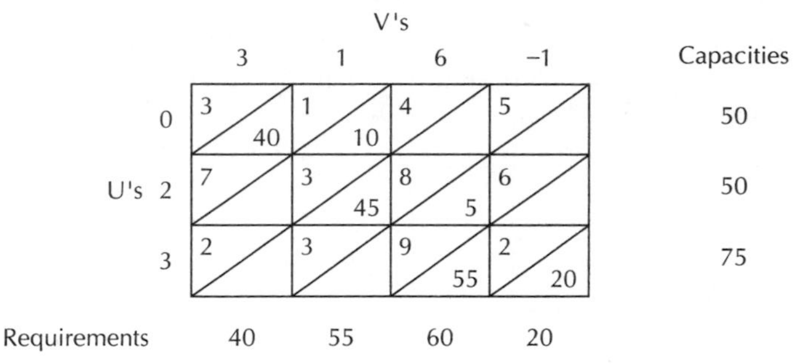

V's
3
1
6
−1
Capacities
0
U's 2
3
40
10
45
5
55
20
50
50
75
Requirements
40
55
60
20

The determination of t_{ij} for the remaining cells, as indicated in step 3 of phase 2, would add to the above matrix as follows:

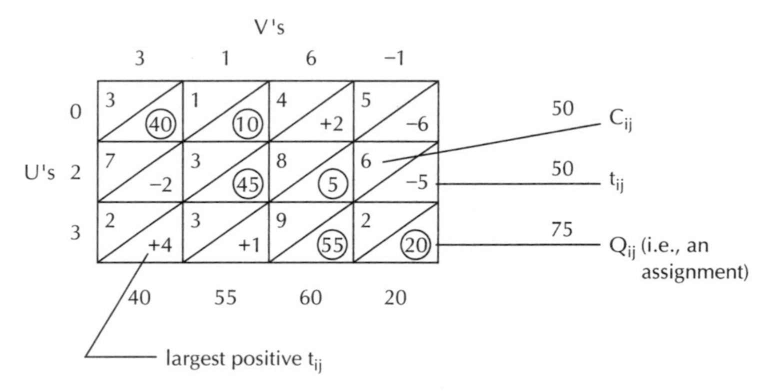

Step 4 of phase 2 is the test for optimality. There is a positive nonzero t_{ij} in the matrix above; therefore, optimality has not been reached and reallocation is necessary. Only assigned cells and the cell with the largest t_{ij} are considered in the reallocation, as indicated below:

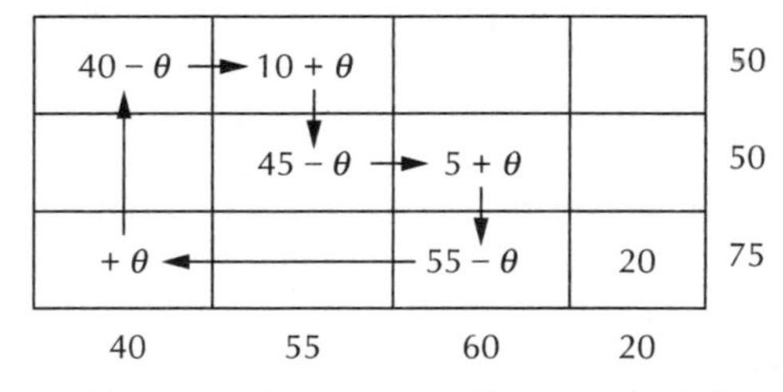

The quantity to be reallocated is θ. Starting by assigning $+\theta$ to the cell with the largest t_{ij}, one alternately adds $-\theta$ and $+\theta$ around the loop by successive vertical and horizontal transfers, returning ultimately to the t_{ij} cell. Thus, any quantity θ added to a row or column will be negated by a $-\theta$ quantity in the same row or column. Row 1 will still possess 50 allocated units, for example, regardless of the size of θ. Since reallocation of units will improve the solution as indicated by a positive t_{ij} in a candidate cell, a maximum reallocation is made to this cell. The maximum quantity is determined by the $-\theta$ cell in the loop with the smallest previous assignment. In this example, 40 is less than 45 or 55; therefore, 40 units will be reallocated. Obviously, if θ were larger than 40, the $40-\theta$ cell would become a negative quantity, which would represent a nonfeasible solution. The reallocation, therefore, would result in assignments as follows:

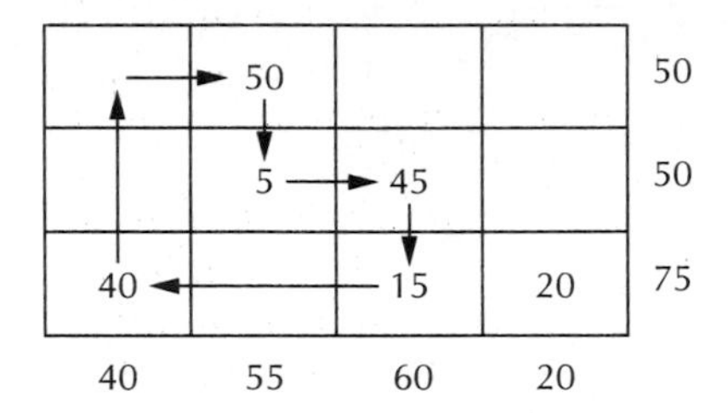

Applying steps 2 through 7 until an optimal solution is found produces the following sequence of matrix operations:

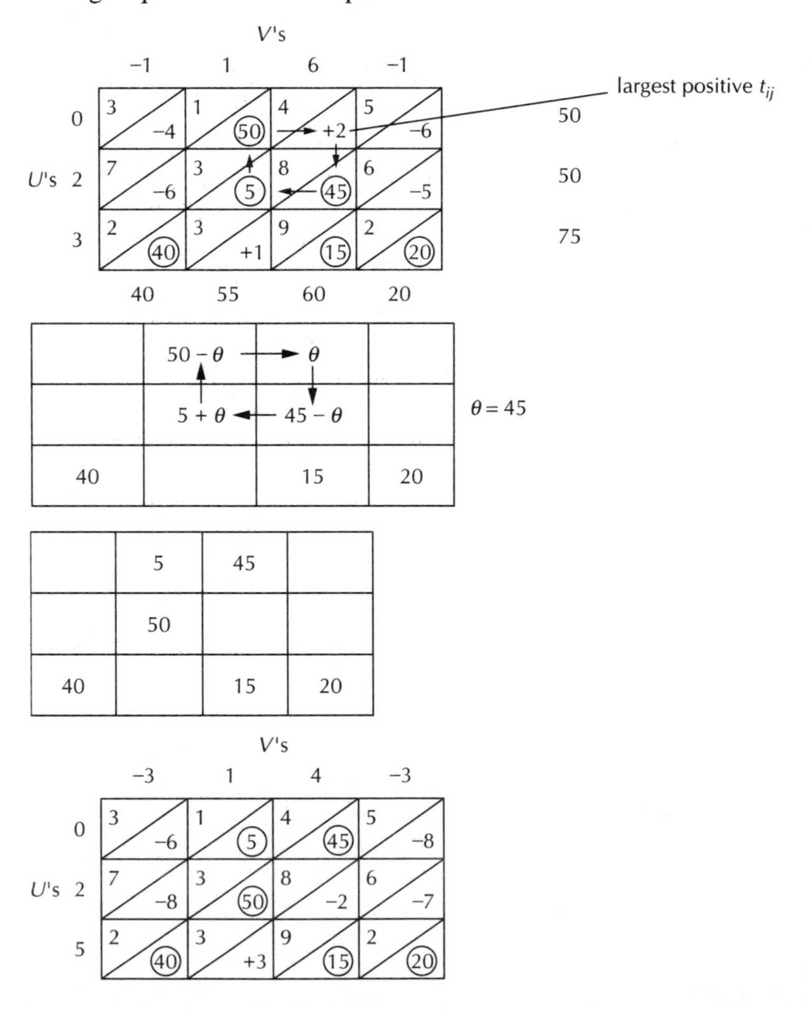

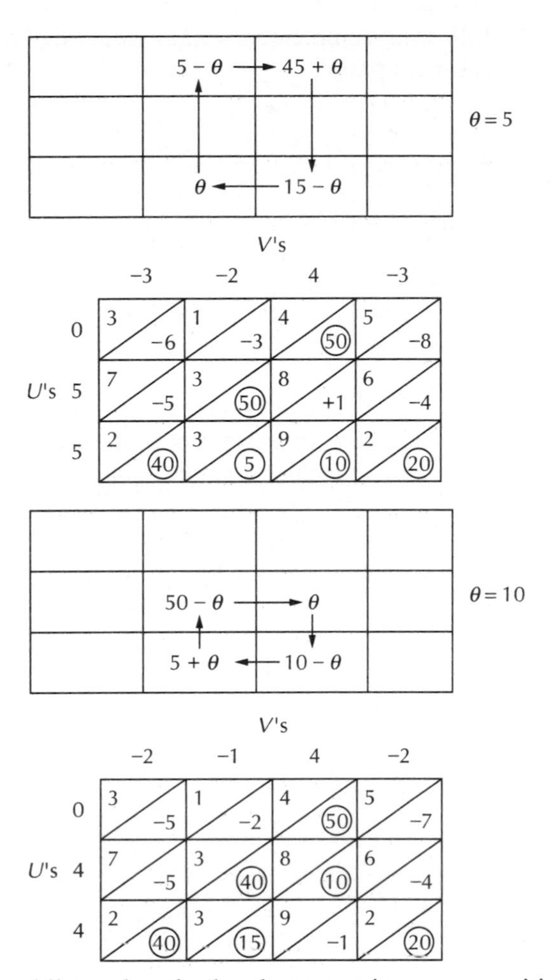

All t_{ij} values in the above matrix are nonpositive; therefore, an optimal assignment has been reached. A number of iterations were required in this example because the assumed initial assignment was a relatively poor one.

As in the case of the assignment algorithm, a minimization solution of complement cell values can be employed to produce a maximum solution. Assume, for example, that the profit received from sending a unit of product from source i to destination j was as follows:

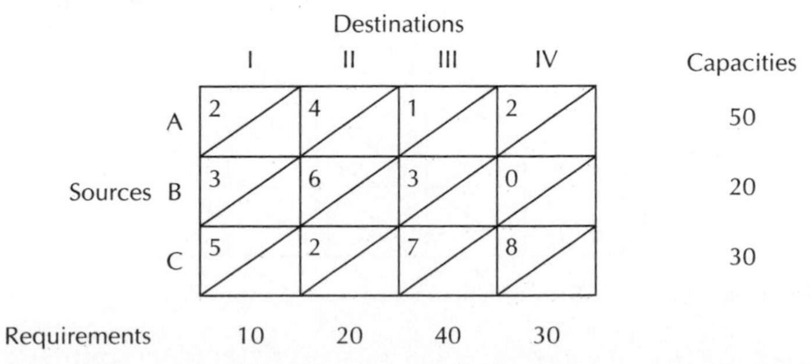

Minimization of the following complement values would determine the best allocation to produce a maximum profit:

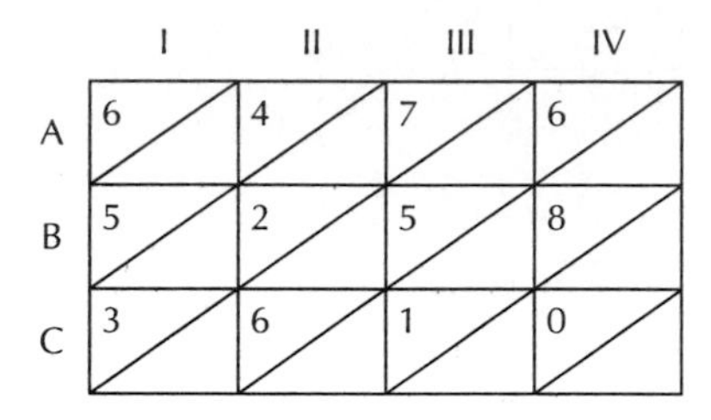

	I	II	III	IV
A	6	4	7	6
B	5	2	5	8
C	3	6	1	0

An alternative method is to bring cells into solution that have the largest negative t_{ij} until all t_{ij} values are nonnegative; this also produces a maximum assignment.

In some cases it is necessary to include a dummy (i.e., imaginary) source or destination in order to solve a transportation problem. Assume that, in the prior example concerning three mines serving four plants, the requirement for plant C was only 40 tons. The matrix would be as follows:

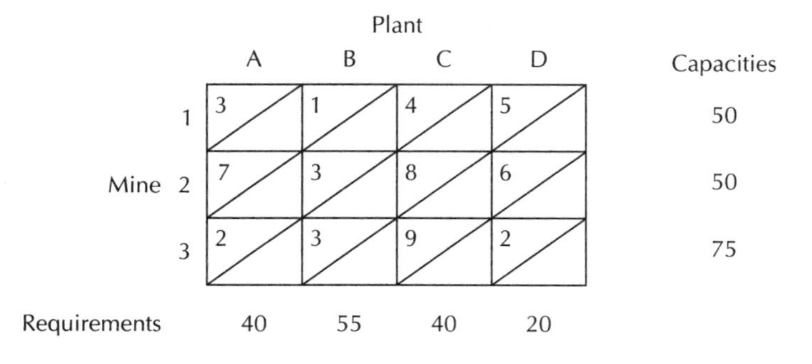

	Plant A	Plant B	Plant C	Plant D	Capacities
Mine 1	3	1	4	5	50
Mine 2	7	3	8	6	50
Mine 3	2	3	9	2	75
Requirements	40	55	40	20	

Note that the total capacity exceeds the total requirement by 20 tons. The transportation algorithm requires that the total requirement equal the total capacity; therefore, a nonexistent plant E is imagined to require 20 tons with shipping cost coefficients of zero for all mines. Actually, dummy plant E represents the 20 tons of capacity that will never leave the mines. The matrix would be as follows:

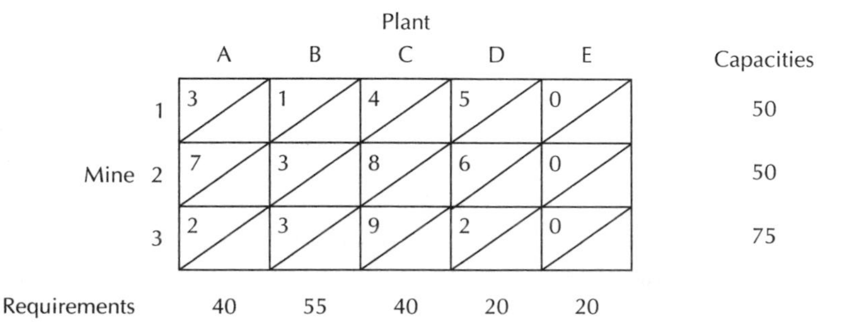

	Plant A	Plant B	Plant C	Plant D	Plant E	Capacities
Mine 1	3	1	4	5	0	50
Mine 2	7	3	8	6	0	50
Mine 3	2	3	9	2	0	75
Requirements	40	55	40	20	20	

Use of phase 1 produces the following assignment:

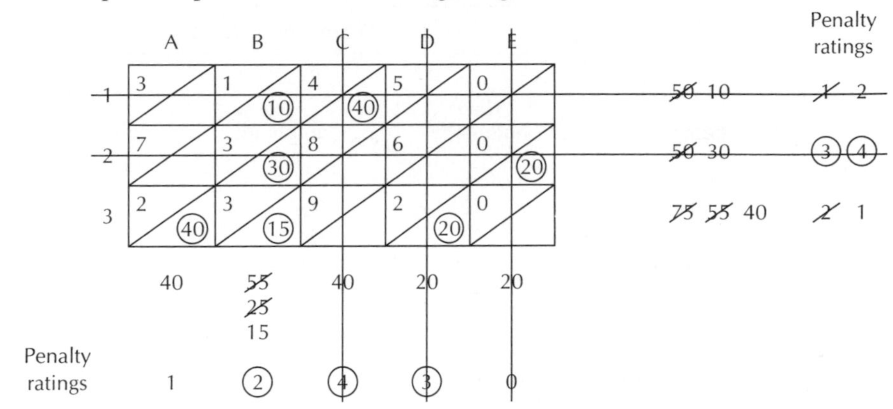

A
B
C
D
E
Penalty ratings
1
2
3
3 1 4 5 0
(10) (40)
7 3 8 6 0
(30) (20)
2 3 9 2 0
(40) (15) (20)
50 10
50 30
75 55 40
1 2
(3) (4)
2 1
40
55 25 15
40
20
20
Penalty ratings
1
(2)
(4)
(3)
0

The test for optimality, step 4 of phase 2, indicates that an optimum has been reached, as follows:

		V's 0	1	4	0	−2
	0	3 / −3	1 / (10)	4 / (40)	5 / −5	0 / −2
U's	2	7 / −5	3 / (30)	8 / −2	6 / −4	0 / (20)
	2	2 / (40)	3 / (15)	9 / −3	2 / (20)	0 / 0

The complete assignment would therefore be as follows:

Mine to plant	Cost per ton	Tons	Cost
1–B	$1	10	$ 10
1–C	4	40	160
2–B	3	30	90
2–E	0	20	0
3–A	2	40	80
3–B	3	15	45
3–D	2	20	40
			$425

The transportation technique is also a very general one, with a broad range of potential applications. For example, assume that dishwashing in a small hospital can be performed by employing any one of the three techniques shown in Table 7-4. The numbers of units normally requiring washing each day are indicated as

TABLE 7-4 DISHWASHER PROBLEM INFORMATION

	Worker-seconds per unit			
Items	New automatic	Old automatic	Hand dishwashing	Daily requirement, units
Pots	5	10	25	50
Plates	2	6	20	250
Silverware sets	5	12	25	200
Frying pans	10	13	30	20
Glasses	3	4	10	200
Cups	1	3	10	250
Capacity, units	550	350	70	970

requirements, and the capacities of the new and old automatic dishwashers are given as 550 and 350 units, respectively. Because the total requirement is for 970 units, 70 units are indicated as a capacity for hand dishwashing to equate total capacity to total requirement.

Assume that it is desired to minimize total worker-seconds associated with the dishwashing function. Figure 7-3 illustrates the solution to the assignment of the new

FIGURE 7-3 Assignment solution to the dishwasher problem.

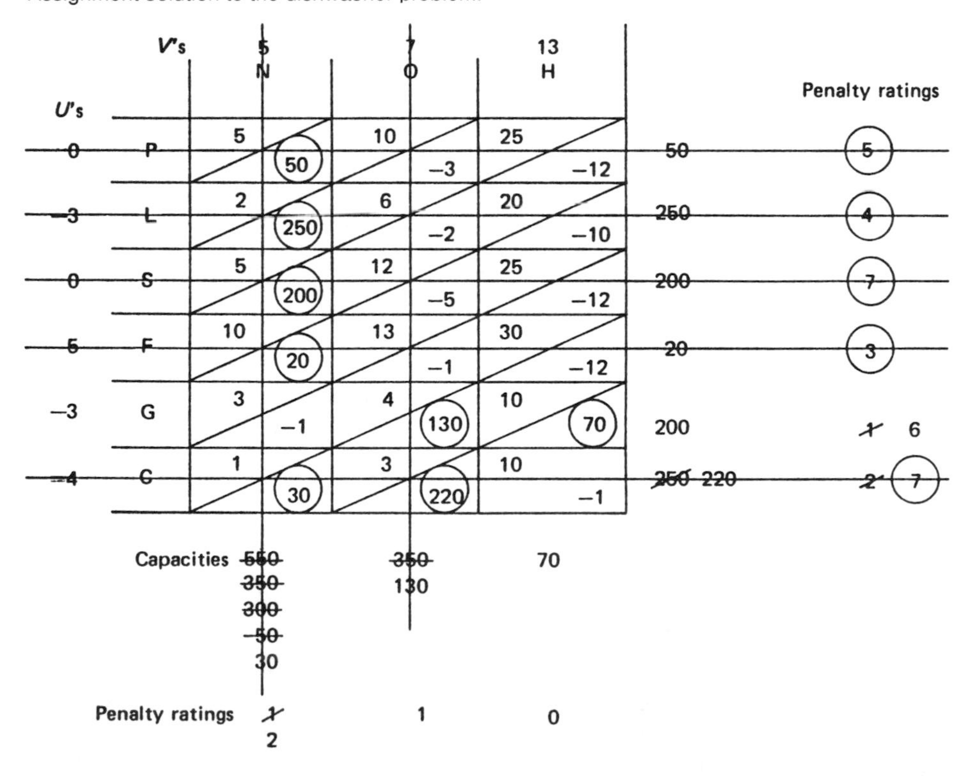

TABLE 7-5 DISHWASHING WORKER-HOURS REQUIRED

Total worker-seconds required	
$50 \times 5 = 250$	$30 \times 1 = 30$
$250 \times 2 = 500$	$130 \times 4 = 520$
$200 \times 5 = 1000$	$220 \times 3 = 660$
$20 \times 10 = 200$	$70 \times 10 = 700$
	3860 seconds
$\frac{3860}{3600} \approx 1.07$ hours	

dishwasher, old dishwasher, and the hand dishwasher for specific items requiring washing. This allocation of the available resources results in a minimum of approximately 1.07 worker-hours associated with the dishwashing function, as indicated in Table 7-5.

THE GENERALIZED LINEAR PROGRAMMING PROBLEM

There is a general class of problems that require maximization or minimization of an objective function represented by a sum of products of linear nonnegative variables and associated coefficients. The variables are also constrained by a set of linear equalities or inequalities relating them in a specific way. The general formulation of a linear programming problem can be explicitly stated as follows:

$$\text{Minimize (or maximize) } f(x) = C_1X_1 + C_2X_2 + \cdots + C_nX_n$$

subject to

$$a_{11}X_1 + a_{12}X_2 + \cdots + a_{1n}X_n \lesseqqgtr b_1$$

$$a_{21}X_1 + a_{22}X_2 + \cdots + a_{2n}X_n \lesseqqgtr b_2$$

$$a_{m1}X_1 + a_{m2}X_2 + \cdots + a_{mn}X_n \lesseqqgtr b_m$$

and

$$X_1, X_2, \ldots, X_n \geq 0$$

The following example will indicate the type of problem involved. Assume that products A, B, and C can be sold for \$5, \$10, and \$20 per unit, respectively. The unit costs of four possible resources, called inputs 1, 2, 3, and 4, for the manufacture of each product are \$2, \$1, \$0.5, and \$2, respectively, for the four inputs. Limitations on supply of inputs 1, 2, 3, and 4 are 100, 200, 400, and 100 units, respectively. Also, the following combinations of units of input are required to produce a unit of product A, B, and C:

		Products		
		A	B	C
Inputs	1	0	1	2
	2	1	2	1
	3	4	6	10
	4	0	0	2

The specific formulation of this problem would be as follows:

$$\text{Maximize } f(x) = C_1X_1 + C_2X_2 + C_3X_3 \qquad \text{(objective function)}$$

subject to the constraint

$$X_2 + 2X_3 \leq 100 \tag{1}$$

$$X_1 + 2X_2 + X_3 \leq 200 \tag{2}$$

$$4X_1 + 6X_2 + 10X_3 \leq 400 \tag{3}$$

$$2X_3 \leq 100 \tag{4}$$

and

$$X_1, X_2, X_3 \geq 0 \qquad (5, 6, 7)$$

The first coefficient in the objective function above (i.e., 2), denoted as C_1 in the general formulation, is the profit per unit of product A, where X_1 is the number of units of product A to be produced. It is determined as follows:

$$\begin{aligned} C_1 &= \text{product A unit profit} \\ &= \text{product A unit revenue} - \text{sum of product A unit input costs} \\ &= 5 - [1(1) + 4(0.5)] \\ &= \$2 \text{ profit per unit of product A sold} \end{aligned}$$

Coefficients for C_2 and C_3 ($3 and $6, respectively) were determined in a similar manner. The first constraint equation, Eq. (1) above, says that there is a limit to how many of products B and C can be made because each unit of product B requires one unit of input 1, and each unit of product C requires two units of input 1, and there are only 100 units of input 1 available. Equations (2) through (4) similarly constrain the use of inputs 2, 3, and 4; and Eqs. (5) through (7) are called nonnegativity constraints, requiring that the values of X_1, X_2, and X_3 be positive.

The simplex algorithm developed by Dantzig [2] is normally used to determine the values of $X_1, X_2, \ldots, X_n$, that meet the constraints imposed and maximize (or minimize) the objective function involved. In actual practice today, it is not necessary to manually employ the simplex method; a considerable amount of software is available for solving linear programming problems. It is only necessary to be able to construct the objective function and constraint equation set and input them into any one of a variety of available linear programming software programs, such as STORM [3], LINDO [8], and Quant Systems [11].

A two-dimensional problem can be graphed to illustrate the general way in which the simplex algorithm successively approaches a maximum (or minimum) solution. The following example illustrates this general procedure:

$$\text{Maximize } f(x) = X_1 + X_2$$

subject to

$$2X_1 + X_2 \leq 12 \quad (1)$$

$$X_1 + 2X_2 \leq 16 \quad (2)$$

$$2X_1 \leq 8 \quad (3)$$

and

$$X_1, X_2 \geq 0$$

Note that by assuming the constraint equations to be equalities, it is possible to plot them and then define the region included by the inequality, as follows:

Assume

$$2X_1 + X_2 = 12 \quad (1)$$

$$X_1 + 2X_2 = 16 \quad (2)$$

$$2X_1 = 8 \quad (3)$$

For constraint (1) above, if $X_1 = 0$, $X_2 = 12$; and if $X_2 = 0$, $X_1 = 6$, as indicated in Fig. 7-4. Linear constraint lines to represent constraints (2) and (3) can be determined in a similar manner and are also shown in Fig. 7-4. The shaded area represents the common feasible solution zone because the three equations are in fact "less than or equal" constraints. That is, any point in the shaded area of Fig. 7-4 represents a feasible solution with respect to all the constraint equations.

To determine a solution to the problem it is necessary to identify the point or points in the feasible zone (i.e., the shaded area) that will produce a maximum for the objective function.

To gain insight into how the simplex algorithm works, at least conceptually, lines can be added to Fig. 7-4 for values of the objective function $f(x) = 5$, 10, and 15 determined as follows:

if $f(x) = 5$ and $X_1 = 0$, $X_2 = 5$
and if $X_2 = 0$, $X_1 = 5$

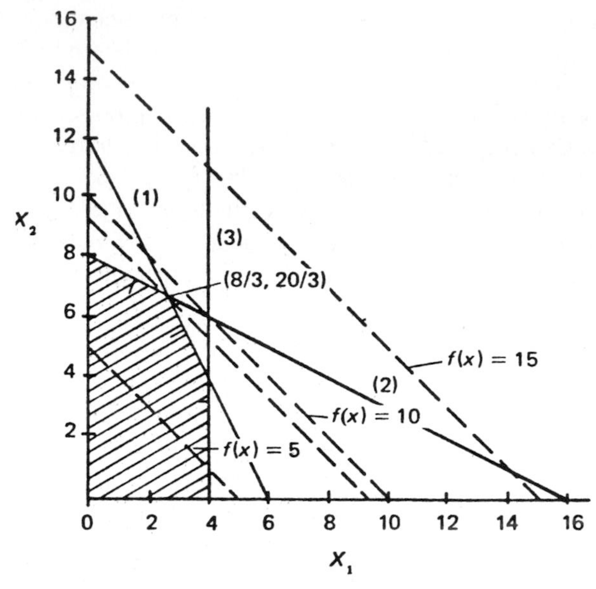

FIGURE 7-4 Graphical solution to a linear programming problem.

if $f(x) = 10$ and $X_1 = 0, X_2 = 10$
and if $X_2 = 0, X_1 = 10$

if $f(x) = 15$ and $X_1 = 0, X_2 = 15$
and if $X_2 = 0, X_1 = 15$

After plotting lines representing $f(x) = 5$, 10, and 15, it can be noted that they are parallel lines moving out from the origin for increasing values of $f(x)$. Note that a line parallel to the $f(x) = 10$ line would appear to possess a maximum objective function value of approximately 9 where constraints (1) and (2) intersect. Their intersection, then, determines the values of X_1 and X_2 that maximize the objective function. Solving these two equations simultaneously yields the following:

$$2X_1 + X_2 = 12 \quad (1)$$

$$\frac{2X_1 + 4X_2 = 32}{3X_2 = 20} \quad (2\times)(2)$$

$$X_2 = \frac{20}{3}$$

$$X_1 = \frac{12 - X_2}{2} = \frac{8}{3}$$

Therefore, the objective function is optimized if $X_1 = \frac{8}{3}$ and $X_2 = \frac{20}{3}$. The objective function reaches a peak feasible value of $9\frac{1}{3}$ [i.e., $f(x) = X_1 + X_2 = \frac{8}{3} + \frac{20}{3}$].

As can be noted in this example, the optimum occurs at constraint intersections. The simplex method is nothing more than an orderly search of n-dimensional intersections in an effort to identify the intersection farthest removed from the origin perpendicular to the plane of the objective function, yet still in the feasible zone, which produces a maximum value for the objective function.

In practice, recognizing the properties of a linear programming problem and knowing whether the linear assumptions are appropriate are the most difficult aspects of applying an existing operations research technique such as linear programming. As in all fields, good judgment is an essential ingredient in the successful application of a technique. Also, as in most fields, techniques rarely include all the effects that influence the outcome. Judgment plays a crucial part in sensing when an effect that cannot be included in a particular technique can be ignored because it will not have a significant effect on the outcome.

In the two-dimensional problem above it was possible to determine the solution by graphical means. In the n-dimensional space of more typical problems, graphical methods are inadequate. The simplex algorithm mentioned earlier is a procedure that utilizes matrix algebra methods to successively search improved constraint intersections on the surface of an n-dimensional convex polygon (i.e., the feasible solution space of an n-dimensional constraint equation set) to determine the point or points on this polygon that maximize the objective function.

The following example from Hadley [5, pp. 464–465] represents a more typical linear programming problem and suggests the need for some organized procedure for dealing with complex problems of allocation:

> A farmer has 100 acres which can be used for growing wheat or corn. The yield is 60 bushels per acre per year of wheat or 95 bushels of corn. Any fraction of the 100 acres can be devoted to growing wheat or corn. Labor requirements are 4 hours per acre per year, plus 0.15 hour per bushel of wheat and 0.70 hour per bushel of corn. Cost of seed, fertilizer, etc., is 20 cents per bushel of wheat and 12 cents per bushel of corn. Wheat can be sold for $1.75 per bushel and corn for $0.95 per bushel. Wheat can be bought for $2.50 per bushel, and corn for $1.50 per bushel.
>
> In addition, the farmer may raise pigs and/or poultry. The farmer sells the pigs or poultry when they reach the age of one year. A pig sells for $40. He measures the poultry in terms of one "pig equivalent" (the number of chickens needed to bring $40 at the time of sale). One pig requires 25 bushels of wheat or 20 bushels of corn, plus 25 hours of labor and 25 square feet of floor space. An equivalent amount of poultry requires 25 bushels of corn or 10 bushels of wheat, 40 hours of labor and 15 square feet of floor space.
>
> The farmer has 10,000 square feet of floor space. He has available per year 2000 hours of his own time and another 3000 hours from his family. He can hire labor at $1.50 per hour. However, for each hour of hired labor 0.15 hour of the farmer's time is required for supervision. How much land should be devoted to corn and how much to wheat, and in addition, how many pigs and/or poultry should be raised to maximize the farmer's profits?

This example was selected for two reasons: (1) it is indicative of the inherent complexity of many allocation problems, and (2) it should suggest the breadth of generality of the simplex algorithm. The farmer's problem above is basically one of inputs of production, constraints, and prices in relation to some objective function. There is some best number of bushels of corn and wheat and numbers of pigs and chickens to raise that will maximize the farmer's profits. For thousands of years farmers have grown what they have always grown, about the same way they did the previous year. Farmers who veered from tradition did it on their own. Sometimes what they tried worked and they were heros among their peers, but too often their curiosity cost them. The linear programming solution to the above problem, assuming linear constraints are appropriate and the data are correct, identifies the best combination of inputs to maximize profit.

Keep in mind that there is an infinite number of problems analogous to the one above. The following are a few that come to mind:

1 What products should be made in a steel mill to maximize profits?
2 What combinations of offensive weapons are needed to maximize destruction of an opposing force?
3 What combination of chemical pesticides should be used to control pests?
4 What combination of workforce inputs are needed to staff a corporation?
5 What assignment of products to machines minimizes costs of production?

Linear programming is only one of a set of related techniques generally referred to today as mathematical programming techniques. They include zero-one programming, nonlinear programming, integer programming, and dynamic programming, to name a few. They represent a very powerful arsenal of techniques when used by individuals who understand how and why the techniques work and what their limitations are. A further requirement, possibly the one most commonly violated, is the need for either extending or imaginatively employing a technique to meet the unique requirements of a particular application.

Having worked some of the earlier problems in this chapter, it should be apparent that working the above farmer problem by hand would be a time-consuming chore. Fortunately, working operations research problems by hand is no longer necessary. Of course, any technique when first learned should be done by hand in order to understand how the technique really works. Once the basics have been learned, however, more efficient means to solving problems can extend the learning experience.

As mentioned previously, a number of software choices are available for solving operations research problems. LINDO [8, pp. 8–9], Quant Systems [11], and STORM [3] are available for such use.

It has been said that a little education can be a dangerous thing. Operations research, and particularly mathematical programming, is a complex subdiscipline in which this adage can apply. Solving problems by hand is typically the best way to gain a thorough understanding of exactly how a quantitative technique really works. Once one understands how the technique works, and its limitations, a software aid can then obviously greatly expedite the generation of solutions. It is very important, however, that one understand the underlying limitations of the solutions that the software produces. Misinterpretation of software results is always a significant danger.

One of the difficulties in the practice of operations research today relates to the breadth of expertise it requires. Those who perform this work should be (1) well versed in operations research techniques, (2) skilled in collecting and analyzing data, and (3) sufficiently educated and capable of understanding the system environment they are attempting to improve. Operations research is being effectively applied in airline reservations systems, as an example, by industrial engineers who have become intimately familiar with their respective airline systems and have combined that knowledge with their thorough working knowledge of operations research.

One of the oldest techniques in operations research, and still one of the most useful, is queuing theory. *Queue* is not a common word in the United States, but it is in Britain. British people know that when they stand in line at the butcher shop they are in a queue.

QUEUING THEORY

Queues are so common that they are easily overlooked; in fact, they are almost impossible to avoid in the conduct of everyday life. For example, people form queues at the family bathroom (especially in the morning); at barber shops, service stations, and supermarkets; while buying theater tickets; before bank tellers; in hospital emergency rooms; and at highway toll booths. One learns very early in life to dislike waiting in line. A line represents a form of regimentation, a loss of freedom, and a waste of time. In industry, time is money; therefore, wasted time represents wasted money. To the $25-per-hour tool and die maker, standing in line for a blueprint or a tool seems like work—in fact, rather pleasant work, because it provides an opportunity to discuss the World Series while getting paid. To the industrial engineer, queuing represents a need for cost control and often a possibility for cost savings. Consider also, particularly in the rapidly expanding service industries, the displeasure created in those who have to wait to be served.

Consider for a moment the far-too-common supermarket that fails to provide a sufficient number of cashiers, which results in long queues. Does this limitation of service capacity save the store money (i.e., is it economizing)? No, the customers in queues must be served sooner or later. What happens in this type of situation is that the customer is led to consider the following question: "Based on the expected waiting time in this line in this store, and other factors, is it worth the loss of my time to shop here?" The modern neighborhood convenience stores came into existence because their higher prices were acceptable in light of (1) location, (2) longer hours, and, in particular, (3) shorter queues, in comparison to supermarkets. Some supermarkets seem to have realized this and are providing longer hours and shorter queues. The supermarket example was offered to stress the point that queuing is a fact of life, both on and off the job, in any industry. Of course, any service is wasting money when it provides more servers than are required. The optimum solution becomes one of staffing servers so that customers are served without waiting, while few enough servers are provided so that all are fully employed. Staffing must be supplied, therefore, to match demand.

Queuing theory was pioneered by the Danish engineer A. K. Erlang [1] in the telephone industry well over 50 years ago. Erlang drew attention to a family of probability density functions that he found useful in functionally expressing interarrival times of incoming telephone calls. This very useful family of mathematical functions bears his name today.

The study of queuing of calls in the telephone industry is at least partially responsible for today's efficient telephone systems. An analogous system is the flow of signals in a computer system. The study of these queues was an essential precursor to the development of the efficient computer systems of today. GPSS (General Purpose Simulation System [4]), which has been one of the most popular simulation languages for many years, was originally developed by IBM engineers to simulate the flow of information (i.e., electrical signal flows) in computer systems. In the development and use of the simulation languages it became apparent that the flow of cars in a bank drive-in system, for example, is quite analogous to the flow of signals in a computer. Consequently, only a minute fraction of simulations today are of computer systems; the bulk are concerned with a potentially infinite variety of actual flow or sequential event systems in everyday life.

In most productive systems, attention beyond process transformation steps is sooner or later directed toward inventories and queues. As was mentioned earlier, inventories serve as a buffer between unequal rates of flow, as indicated in Fig. 7-5. The inventory ensures that if the rate out is greater than the rate in for some defined period, an adequate supply of inventory at point B is available for continued utilization of the equipment at Point B.

It was noted in Chapter 3, however, that there is considerable interest today in employing a JIT (Just-in-Time) philosophy, which reduces inventories and produces other even more important effects. If the rate in is insufficient to provide product to keep the next machine utilized, the machine operator at point B may simply perform other duties required in the overall operation. Such actions limit the undesirable creation of in-process inventories that greatly lengthen product cycle times in the overall process. In years past, it was assumed that maintaining the availability of product at machines throughout the process, thereby creating high equipment utilizations, was an appropriate operational policy; JIT philosophy has proven such intuitively appealing past operating policies to be typically erroneous, in light of their overall effects, and therefore unacceptable.

One of the guiding principles of good plant design is that a dense design of the production system is typically superior. This results in minimum material movement distances, which translate to minimum material handling costs. Because mate-

FIGURE 7-5 A flow system.

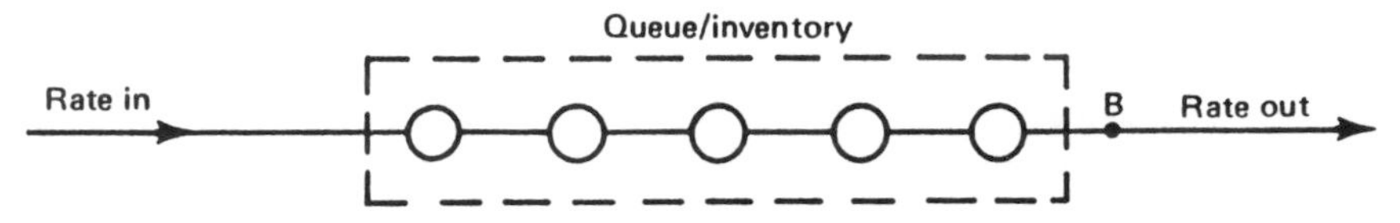

rial handling does not add "value in use" to a product, minimum material handling cost is probably the most important single criterion in plant design. Without belaboring the point further, it should be apparent that plant design, queuing theory, and inventory theory are interrelated. This is precisely why modern industrial engineering involves a blend of traditional industrial engineering, operations research, management science, computer science, and statistics.

Two characteristics of any queuing model that must be specified are the arrival and service distributions. Single-server queues with Poisson-distributed arrivals and exponentially distributed service times are the simplest to consider mathematically and normally represent the logical beginning point for queuing model development. The arrival (or service) distribution refers to the probabilistic description of the arrivals (or service times) in the form of a probability density function. The Poisson density function of Fig. 7-6 indicates that although 8 arrivals per hour is the most frequently occurring rate, the average arrival rate λ is 10 per hour, resulting in an arrival every 60/10 = 6 minutes on the average. The exponential density function of Fig. 7-7 has been employed in numerous queuing models to describe service times. The mean service time $1/\mu$ from Fig. 7-7 is 8 minutes.

If the arrival rate and service time distributions of Figs. 7-6 and 7-7, respectively, describe arrivals and service in a particular single-service system, it should be apparent that a queue should be expected to develop because arrivals on the average occur every 6 minutes, whereas the average service time is 8 minutes. As this system functions over time, the queue would be expected to continue to grow in size. Such a system has no steady-state characteristics; it continues to grow indefinitely because of its lack of service capacity.

Assume in another case, however, that the mean arrival rate is 6 per hour, resulting in a mean time between arrivals of 10 minutes. The time between arrivals is now greater than the mean service time of 8 minutes. In such a case (i.e., service

FIGURE 7-6 Poisson arrivals density function.

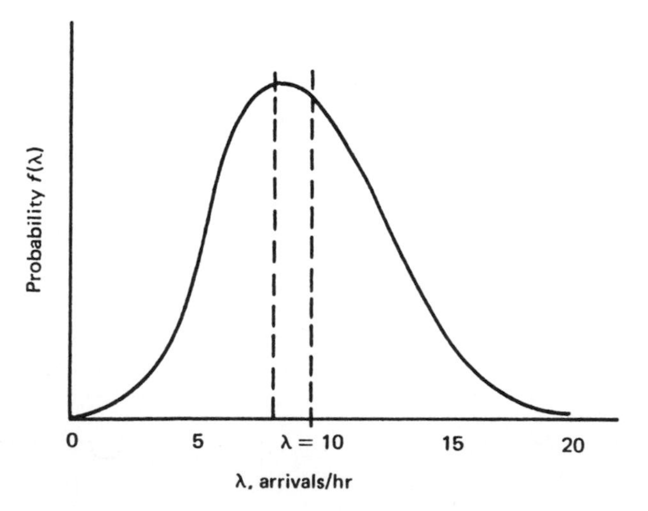

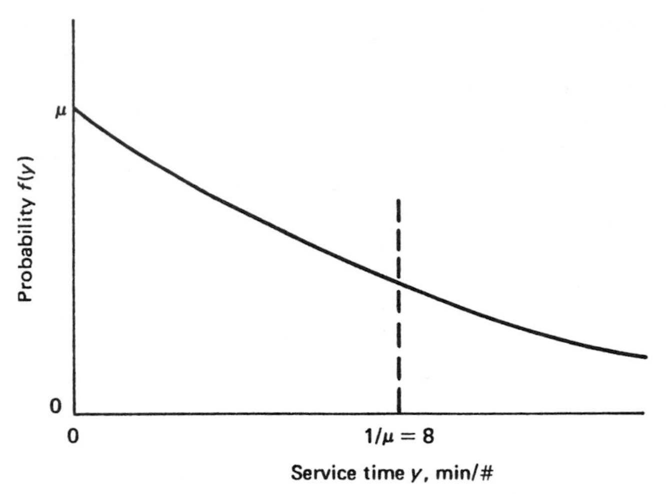

FIGURE 7-7 Exponential service times density function.

time < mean time between arrivals), steady-state queue statistics are expected. By "steady state" it is meant that the queue will not grow indefinitely but will vary in size, possessing some long-run mean size. Other characteristics of the system such as time in the queue and time in the system will also vary over time but have mean values in the long run. Obviously, if the mean time between arrivals is very large compared to the mean service time, the expected queue length will approach zero. However, if the mean time between arrivals is close to the mean service time, the expected queue length will be very large.

The following steady-state equations for Poisson-distributed arrivals and exponentially distributed service times will be offered without proof:

λ = mean arrival rate
μ = mean service rate
P_0 = probability that no units are in the system
P_n = probability that n units are in the system
m = queue length
n = number of units in the system
w = waiting time of an arrival
v = total time an arrival spends in the system
E( ) = expected (mean) value of ( )

$$P_0 = 1 - \frac{\lambda}{\mu}$$

$$P_n = \left(\frac{\lambda}{\mu}\right)^n P_0$$

$$\sum_{n=0}^{\infty} P_n = 1$$

$$E(m) = \frac{\lambda^2}{\mu(\mu - \lambda)}$$

$$E(m \mid m > 0) = \frac{\mu}{\mu - \lambda}$$

$$E(n) = \frac{\lambda}{\mu - \lambda}$$

$$E(w) = \frac{\lambda}{\mu(\mu - \lambda)}$$

$$E(w \mid w > 0) = \frac{1}{\mu - \lambda}$$

$$E(v) = \frac{1}{\mu - \lambda}$$

Assume, for example, that cars arriving at a bank drive-in window are Poisson-distributed with a mean time between successive arrivals of 5 minutes. Assume also that the service time of the window teller is exponentially distributed with a mean time of 2 minutes.

1 What is the expected queue length?

$$\lambda = \frac{1}{5} = 0.2$$

$$\mu = \frac{1}{2} = 0.5$$

$$E(m) = \frac{\lambda^2}{\mu(\mu - \lambda)} = \frac{(0.2)^2}{0.5(0.5 - 0.2)} = \frac{0.4}{0.5(0.3)} = 0.26 \text{ cars}$$

2 What is the expected queue length during the times when there is at least one car waiting to be served?

$$E(m \mid m > 0) = \frac{\mu}{\mu - \lambda} = \frac{0.5}{0.5 - 0.2} = 1.667 \text{ cars}$$

The expected queue length of 0.26 cars for question 1 as compared to 1.667 cars for question 2 indicates that for much of the time there is no queue. The averaging in of these zero queue lengths with the expected queues when they do exist accounts for the difference in the two different queue lengths.

3 What is the probability that an arriving car will have to wait before being served by the teller?

$$P(\text{an arrival will wait}) = 1 - P_0$$

$$P_0 = 1 - \frac{\lambda}{\mu}$$

$$\therefore P(\text{an arrival will wait}) = \frac{\lambda}{\mu} = \frac{0.2}{0.5} = 0.4$$

4 What is the expected waiting time of an arrival?

$$E(w) = \frac{\lambda}{\mu(\mu - \lambda)} = \frac{(0.2)}{0.5(0.5 - 0.2)} = 1.333 \text{ minutes}$$

5 What is the expected total time a car will spend in the system?

$$E(v) = \frac{1}{\mu - \lambda} = \frac{1}{0.5 - 0.2} = 3.333 \text{ minutes}$$

The difference in time between the total time in the system of 3.333 min. and the time waiting of 1.333 min. is 2 min. This is the time in service and agrees with that given in the statement of the problem.

The example above demonstrates how queuing theory can be employed to estimate the expected steady-state parameters of queuing systems. The next example illustrates how queuing theory can be employed to assist in making economic comparisons in special cases.

Assume that two different tools, A and B, can be purchased for use in repairing machines, and that the rate of breakdowns necessitating repairs is Poisson-distributed with a mean of 5 per hour. Assume also that downtime on a machine costs the company $4 per hour for the period of time the machine is out of service. Tool A can be leased at a rate of $3.50 per hour and repairs machines at an exponentially distributed rate having a mean of 6 per hour. Tool B can be leased at the rate of $17.00 per hour and repairs machines at an exponentially distributed rate with a mean of 10 per hour. Which tool should be leased?

The total hourly cost of employing tool A is the sum of the hourly leasing cost and the hourly downtime cost. The average number of machines experiencing downtime (i.e., the number of arrivals in the queue) is

$$E_A(n) = \frac{\lambda}{\mu_A - \lambda} = \frac{5}{6 - 5} = 5$$

The total cost of employing tool A is $3.50 + 5($4) = $23.50 per hour. The average number of machines experiencing downtime when tool B is employed would be

$$E_B(n) = \frac{\lambda}{\mu_B - \lambda} = \frac{5}{10 - 5} = 1$$

The total hourly cost for tool B would be \$17 + 1(\$4) = \$21.00 per hour. Therefore, tool B should be leased. Authors have referred to queuing systems as nonintuitive, and this example demonstrates that property. The \$17 per hour for tool B seems costly compared to \$3.50 per hour for tool A, yet tool B is cost-effective because of the queue lengths machine A produces, which are not obvious from the data given in the problem. It is only after calculation of comparative total costs that the best choice becomes obvious.

Hillier [6] reported a study performed at a Boeing aircraft plant that demonstrates the nonintuitiveness of queuing problems. In this case, supervisors were expressing concern that their mechanics were spending too much time waiting in line at the tool crib to receive tools. Management, concerned about overhead costs, was interested in not having more tool crib attendants than necessary.

A study of the tool crib indicated that the mean time between arrivals was 35 seconds, whereas mean service time was 50 seconds. Based on these data it would seem obvious that one clerk would not be able to keep up, but that two could do so without difficulty. Consider, however, that an idle mechanic is a great deal more costly than an idle tool crib attendant. If one assumes, as Hillier did at the time, that the mechanic is paid \$5 per hour and the clerk \$2 per hour, estimates of the total daily idle time are \$64.10, \$31.00, and \$40.00 for two, three, and four clerks, respectively. These results indicate that three clerks is the best choice and that even four clerks is a better choice than two clerks.

With four clerks, it is likely that management would question whether they had enough to do. Yet it is the ability of the fourth clerk to prevent the queue length from increasing significantly because of a combination of a temporarily high arrival rate and the service time involved that is of primary interest in this problem. The best solution in this type of situation is usually to provide excess service capacity when needed (e.g., three clerks) and secondary tasks, such as typing labels, assembling cardboard boxes, or other potentially intermittent duties, during slow periods in the tool crib.

Queuing systems are so common and diverse with respect to types and combinations of components that a system of classification of multiple-server queues was developed and is in common use today; the system was proposed in 1953 by Kendall and extended in 1966 by Lee. The Kendall-Lee system [10], detailed in Table 7-6, identifies the following six main characteristics essential to any multiserver queuing system, as illustrated in Fig. 7-8: (1) the arrival distribution, (2) the service distribution, (3) the number of parallel service channels, (4) the service discipline, (5) the number of units of flow permitted in the system, and (6) the source population.

The following problem situation illustrates the use of the system. Assume that the processing of frankfurters in a plant is to be modeled. Assume also that after the frankfurters are made they enter any one of five rotary cooking bins, which cook them for 5 minutes and then dump them one at a time onto a conveyor belt that carries them to a packaging machine. Assume that the queue that develops in cooking the frankfurters is the part of the process to be coded. A possible notation for this system might be (M, D, 5):(SIRO, 100, ∞). This notation indicates that this is a

TABLE 7-6 KENDALL-LEE MULTIPLE-SERVER QUEUE CLASSIFICATION SYSTEM

General Notation (a/b/c):(d/e/f)	
Specific notation	**Description**
a	Arrival or interarrival distribution
b	Leaving or service time distribution
c	Number of parallel service channels
d	Service discipline
e	Maximum number allowed in the system
f	Calling source
Code for a and b	
M	Poisson arrival (or equivalent exponential interarrival or service times)
D	Deterministic interarrival or service times
E_K	Erlangian* or gamma interarrival or service time distributions
GI	General independent distribution of arrivals or interarrival times
G	General distribution of leaving or service times
Code for d	
FCFS	First come, first served
LCFS	Last come, first served
SIRO	Service in random order
GD	General service discipline

c, e, and f in general notation above are finite or infinite numbers.

*The parameter K explicitly defines a unique member of the density family (e.g., if $K = 1$ the density function is exponential).

multiple queue system possessing Poisson-distributed arrivals, discrete service time, five multiple servers, random entry into service, a maximum of 100 units in the system, and an essentially infinite source of arrivals. This example suggests the diversity of potentially different multiple-server queuing systems.

FIGURE 7-8 Multiserver queuing system components.

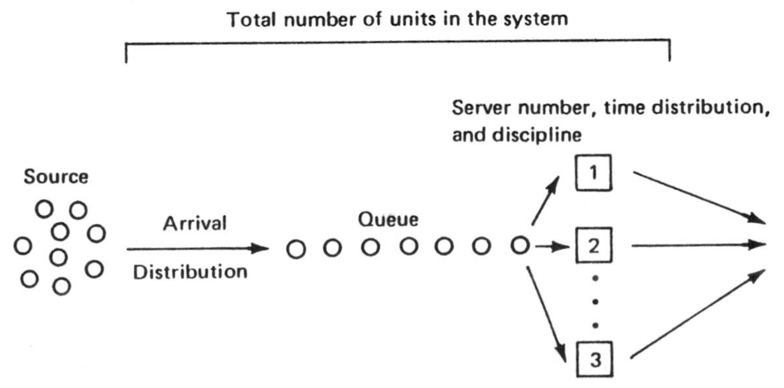

As would be expected, queuing formulations for multiple-server queues are more complex than those for single-server queues. In the following equations r represents the number of servers in the system. If the combined service capacity of the r servers exceeds the arrival rate into the system (i.e., $r\mu \geq \lambda$) the queue will not grow over time but instead will have predictable steady-state characteristics. The following multiple-server equations, assuming Poisson-distributed arrivals and exponentially distributed service times, are offered without proof:

$$P_0 = \frac{1}{\left[\sum_{n=0}^{r-1} 1/n!(\lambda/\mu)^n\right] + 1/r!(\lambda/\mu)^r[r\mu/(r\mu - \lambda)]}$$

For $n < r$,

$$P_n = \frac{1}{n!}\left(\frac{\lambda}{\mu}\right)^n P_0$$

And for $n \geq r$,

$$P_n = \frac{1}{r!r^{n-r}}\left(\frac{\lambda}{\mu}\right)^n P_0$$

$$E(m) = \frac{\lambda\mu(\lambda\mu)^r}{(r-1)!(r\mu - \lambda)^2} P_0$$

$$E(n) = \frac{\lambda\mu(\lambda/\mu)^r}{(r-1)!(r\mu - \lambda)^2} P_0 + \frac{\lambda}{\mu}$$

$$E(w) = \frac{\mu(\lambda/\mu)^r}{(r-1)!(r\mu - \lambda)^2} P_0$$

$$E(v) = \frac{\mu(\lambda/\mu)^r}{(r-1)!(r\mu - \lambda)^2} P_0 + \frac{1}{\mu}$$

Assume, as an example, that students instructed to see any one of three counselors in a high school arrive for counseling on a first-come, first-served basis at a Poisson-distributed arrival rate of 10 per hour, or on the average every 6 minutes. Assume also that all three counselors spend 15 minutes on the average distributed exponentially with each student. A check should first be made to determine whether service is adequate to expect steady-state characteristics.

$\lambda = 10$ per hour
$\mu = 4$ per hour per counselor
$r\mu = 3(4) = 12 \geq 10$

Therefore, there is sufficient service capacity and the steady-state equations are applicable. Assume that one wishes to know the percentage of time when there are no students either waiting to see or seeing any of the three counselors.

$$P_0 = \frac{1}{\left[\sum_{n=0}^{2} 1/n!(10/4)^n\right] + 1/3!(10/4)^3[3(4)/(3(4)-10)]}$$

$$\sum_{n=0}^{2} \frac{1}{n!}\left(\frac{10}{4}\right)^n = \frac{1}{0!}\left(\frac{10}{4}\right)^0 + \frac{1}{1!}\left(\frac{10}{4}\right)^1 + \frac{1}{2!}\left(\frac{10}{4}\right)^2 = 1 + \frac{10}{4} + \frac{100}{32} = \frac{212}{32}$$

Therefore,

$$P_0 = \frac{1}{212/32 + 1{,}000(12)/3(2)64(2)} = 0.0449 \cong 4.5 \text{ percent}$$

Queuing theory is very useful as an area of study for developing a theoretical understanding of how queuing systems function. Such a study will usually develop one's intuition at least to the extent of providing a qualitative understanding of the primary factors influencing the output characteristics of queuing systems.

Many practical systems that are of interest to industrial engineers represent flows of some resource. For this reason many systems of practical interest contain elements of queuing. Queuing systems are more often than not modeled by using simulation rather than the analytical queuing models discussed in this chapter.

SIMULATION

By far, the modeling technique most frequently used in industrial engineering is simulation. The applications of simulation are infinite. Some examples are people and baggage in airports, cars and people in outside teller window systems at banks, barges carrying coal down the Mississippi on the way to a power generation plant, a forest fire model that concludes with nutrients being added back to the soil, and a military engagement involving an enemy tank and a laser guided missile.

Simulations do not optimize, but they are a means for estimating outcomes. A good solution, if not an optimum one, is determined with simulation by modifying a representation of system structure or input parameters and observing the output until desirable output is discovered. It is hoped that the results of the simulation lead to policies that will produce the outcomes desired in the real system.

There are two basic types of digital computer simulations; one is called "discrete" and the other "continuous." Discrete means that the identities of individual units are maintained in the simulation. A discrete simulation of the work force in a plant, for example, would maintain the identity of individuals throughout the simulation. A continuous simulation of the work force, however, might show the number of direct employees in a plant to be 326.5 at the end of the month of March. The 0.5 employee may be rationalized a number of ways: (1) as a part-time employee, (2) as an employee considered to be only 50 percent productive, or (3) as a reflection of employee absences due to illness, etc.

Simulation is common today in applied industrial engineering. The following example is offered to provide some insight into how computer simulations work. The example involves the assumed unloading of ship containers at a container terminal (e.g., Sealand). A container is a metal box approximately 10′ wide × 10′ high × 20′ or 40′ long to hold cargo on a cargo ship. The container is placed on board a ship employing a crane and latched down to the deck prior to its journey at sea to its final port of call. A crane at the receiving end lifts the container off the ship and onto a truck (or tractor) beneath the crane. It is then typically stored at the container terminal for further processing (customs, agricultural inspection, consolidation with other containers, etc.).

The following simulation language demonstrated is GPSS, and the software is GPSS/PC [4]. The activity modeled is the arrival of ships at a container terminal and their subsequent unloading by a crane and put-away by tractors, as indicated in Fig. 7-9. Figure 7-10 is the simulation model (i.e., computer program) developed for this problem.

The left column is a line numbering means for referring to any specific line in the model. Line 100, the generate line, creates "transactions" (i.e., ships) in the model that arrive on average every 240 minutes (i.e., every 4 hours) as indicated in the A subfield. The B subfield value in line 100, "120," specifies that this nominal value of 240 minutes will be adjusted to as low as (240 – 120) = 120 minutes, or as high as (240 + 120) = 360 minutes. The standard GPSS distribution of values over this range is uniform (i.e., equally likely), employing a uniform distribution approach. Therefore, the B subfield introduces variability to ship arrivals, such that they arrive every 4 hours on average but, based on a specific random number value selection, can be any number between 120 minutes and 360 minutes, uniformly distributed.

The D subfield in line 100 is 10, which in GPSS modeling means that 10 transactions (i.e., ships) will be generated in this model. In summary, line 100 creates 10 ships arriving in sequence to a port on average every 4 hours; but the generation time for any one ship arrival following the previous arrival may be as soon as 2 hours later or as long as 6 hours later.

Every transaction (i.e., ship) arriving, created at line 100, immediately enters line 110, which is a "queue" (i.e., waiting for the dock). The queue statement is placed in the model to collect queuing statistics on ships entering this arrival queue. Arriving ships would typically prefer to tie up at the dock, but they cannot if another ship is using the dock. Placing the queue statement in the model at this point will permit determination of queue statistics on all ships waiting for the dock during the assumed time modeled in the simulation.

Line 120 is a "seize" statement. The seize statement is employed in GPSS when a transaction (e.g., a ship) is attempting to use a facility, in this instance the dock. The seize statement is used when only one transaction (a ship) can utilize a facility (the dock) at a time. All other generated transactions in line 100 must wait in the

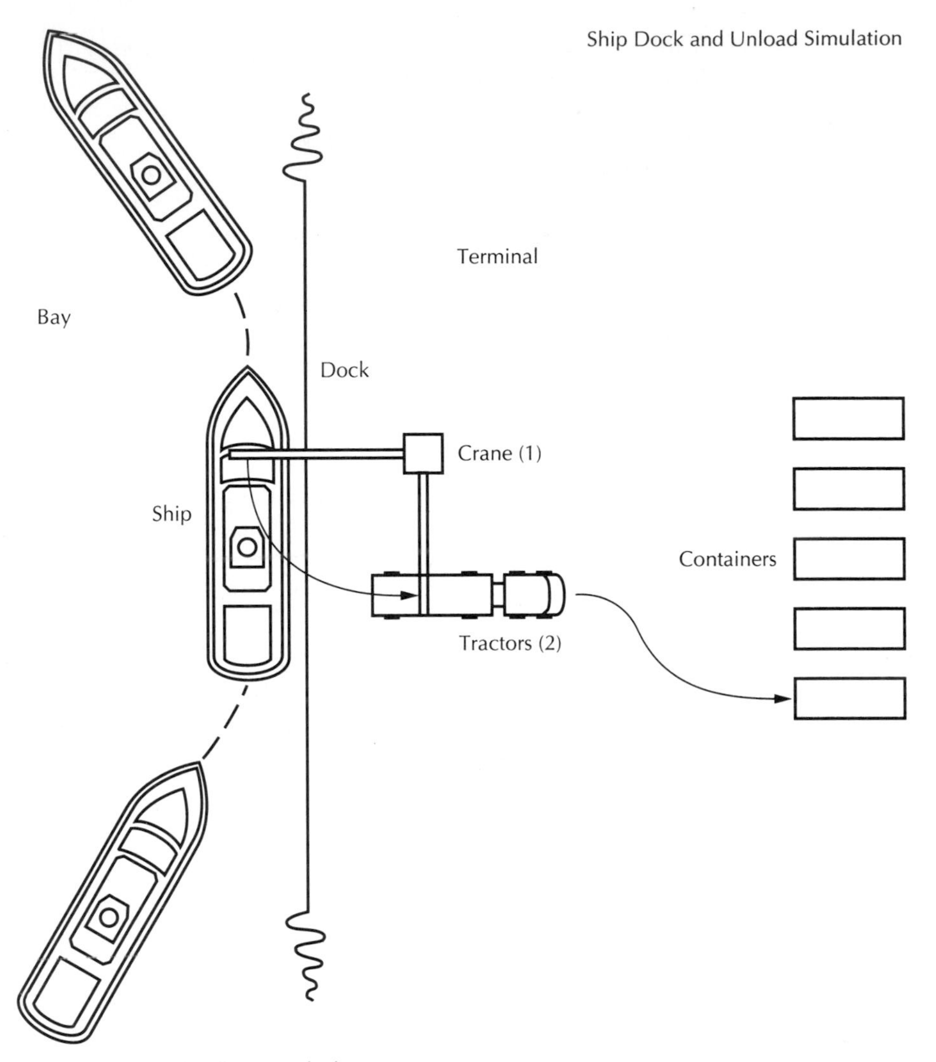

FIGURE 7-9 Container ship unloading at a dock.

queue in line 110 until the facility becomes available to handle them one at a time in line 120.

As soon as a transaction (a ship) seizes the dock, it has effectively left the queue (i.e., it is no longer waiting to get the dock). The "depart" block in line 130 is a complementary block with block 110, which records entry into a queue. The depart block records the time the transaction left the queue.

As soon as the transaction (i.e., ship) seizes the dock, it departs the queue, and it also expends a time indicated by the "advance" block in line 140. The "30" in subfield A of line 140 indicates that the "tie up and get ready to unload" time for the ship at the dock is assumed to be 30 minutes on average. Subfield B of line 140 indicates that the time may be 15 minutes less or more, however, for any one ship

```
; GPSS/PC program file SHIPSIM. (V 1.1, # 30890) 01-14-1993 12:45:52
10 TRACTOR STORAGE  2
*************************************************************************************************************
Ship Arrive, Dock, and Unload Simulation
*************************************************************************************************************
100          GENERATE     240,120,,10
110          QUEUE        DOCK
120          SEIZE        DOCK
130          DEPART       DOCK
140          ADVANCE      30, 15
150          SPLIT        25,ABC
155          TRANSFER     ,BCD
168 ABC      QUEUE        CRANE
170          SEIZE        CRANE
180          DEPART       CRANE
190          ADVANCE      10,5
200          QUEUE        TRACTOR
210          ENTER        TRACTOR
220          DEPART       TRACTOR
230          RELEASE      CRANE
240          ADVANCE      20,5
250          LEAVE        TRACTOR
260 BCD      ASSEMBLE     26
270          ADVANCE      30,15
280          RELEASE      DOCK
290          TERMINATE    1
```

FIGURE 7-10 A GPSS/PC simulation model of container unloading.

being tied up. The actual time between 15 and 45 minutes is determined by a generated random number in the simulation.

Line 150, a "split" block, indicates that the transaction has now produced 25 additional transactions, which proceed to model location "ABC" indicated in subfield B. These transactions are now intended to be containers to be unloaded from the ship.

The parent transaction in line 150 proceeds to the next block, a "transfer" block, which then sends it to model location BCD in line 260. This parent transaction is now considered to be the unloading paperwork that will be employed subsequently in the model after all 25 containers have been unloaded from the ship.

At line 168, the 25 container transactions enter a queue for the unloading crane.

In line 170, when the crane is available, containers seize the crane on a FIFO (first in, first out) basis, leaving the crane queue, as indicated by entering the crane "depart" statement in line 180.

Line 190 indicates that the average crane time is 10 minutes, plus or minus as much as 5 minutes, uniformly distributed over the range.

When the container is removed from the ship by the crane it must be placed on a tractor. Therefore, the next line in the simulation is a "queue" statement to collect queue statistics for all containers attempting to enter tractors.

There is more than one tractor; therefore, tractors are not considered a facility in this simulation. Note that statement 10 at the very beginning of the simulation model was set at 2. This indicates that there are two tractors being used in the simulation and this type of resource is generically called a "storage" rather than a "facility," precisely because it contains more than a single resource.

In line 210, rather than "seize," which is the appropriate statement for gaining use of a facility resource, the statement for gaining use of a storage resource (e.g., one of two tractors) is "enter." If the two tractors are presently tied up with containers at the time a transaction arrives at line 210, it will simply wait in the tractor queue created at line 200.

When a container finally enters the "enter" statement at line 210, it is assumed that the container is being placed on one of the two tractors in the process of being stored, and as indicated in line 220, it also instantaneously leaves the tractor queue (i.e., waiting for the tractor).

Line 230 indicates that when the container goes on the tractor it has effectively left the crane, which "releases" the crane to unload other containers, one at a time, from the ship.

The advance block at line 240 indicates that it takes 20 minutes, plus or minus up to 5 minutes, for a tractor to store a container.

Line 250 indicates that as soon as the tractor completes the container storage, it is free to pick up another container waiting in the queue to use the tractors. The "leave" block (250) is a complementary block to the "enter" block (210) for the tractors, and effects the recording of the queue leave time for tractor queue statistics.

Recall earlier that the paperwork for the 25 containers to be unloaded from the ship was transferred from line 155 to line 260. It has been waiting there for the 25 containers to be unloaded by the tractors. When the paperwork transaction and the 25 container transactions all finally collect at the "assemble" block at line 260, thereby equaling the subfield parameter value of 26, one, and only one, transaction will be permitted to leave line 260 and enter line 270; the other transactions will be destroyed. This transaction is now considered to be the ship attempting to leave the dock.

Line 270 indicates that it takes the ship 30 minutes, plus or minus up to 15 minutes, to untie and prepare to leave the dock. As soon as the departing ship transaction leaves the dock it enters line 280, which releases the dock to another ship waiting in the dock queue at line 110, previously blocked by the seize statement in line 120.

The departing ship, after releasing the dock in line 280, then departs the simulation by entering the "terminate" block in line 290.

A "start" statement, not shown here, specifies the number of terminations to be generated before concluding the simulation and summarizing collected simulation statistics. An additional line such as

```
PLOT Q$CRANE,25,0,2880
```

could be added just prior to the START statement to plot a figure that indicates the crane queues that developed during the simulation. The statement says, "Plot the

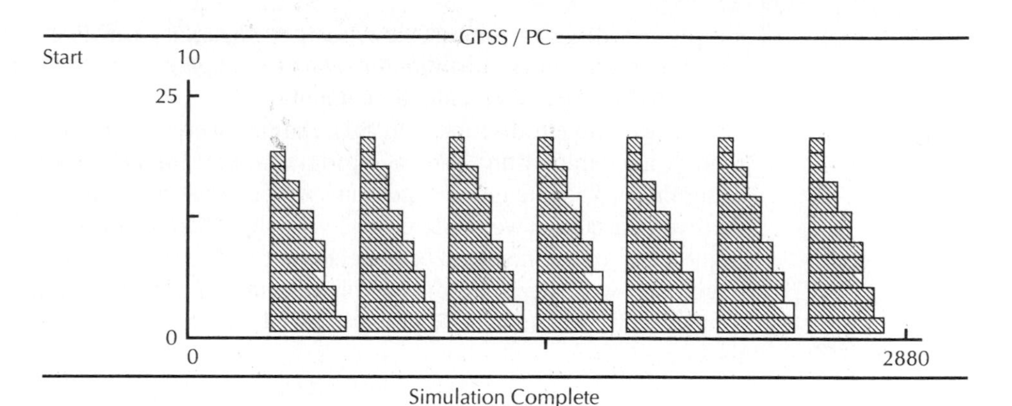

FIGURE 7-11 Plot of the crane queue statistics.

queue statistics for the queue whose name is 'crane' with a Y axis as high as 25, and an X axis that starts at 0, and is as high as 2880."

Figure 7-11 is a plot of the crane queue statistics for the above model. Many simulation packages today include both on-screen graphical representations of the simulation assets as well as animation of moving assets (e.g., material handling equipment). Figure 7-12 shows a typical representation of an AGV (automatic guided vehicle) following a path on the factory floor. The move increment in Fig. 7-12 is specified by the user and establishes how quickly the asset moves one increment forward on its path. Such animations permit visual evaluations of such aspects as facility clearances and densities of material handling devices within an area.

As interesting and entertaining as animation is—and it does have technical value—the primary technical value of simulations is in the statistics they generate. Animation is a fad today, whereas the primary underlying power of simulation is the considerable design data it generates as part of the engineering analysis of a system.

Simulation technical power has been considerably underutilized to date because so many decision makers in the real world are unfamiliar with its potential or are

FIGURE 7-12 Simulation animation. (From Sly [9, p.100].)

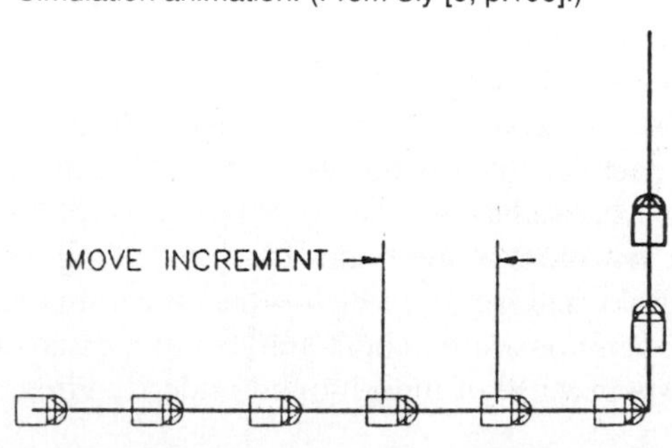

simply unwilling to commit the resources necessary for its use. It is an excellent example of the underutilization of industrial engineering that must be corrected if we are to become internationally competitive.

The techniques discussed in this chapter were linear programming, queuing theory, and simulation. Zero-one programming, integer programming, nonlinear programming, game theory, and many others were not considered. The techniques that were discussed were offered as a very small sample of operations research techniques. Of course, it should be understood that the examples offered were relatively trivial and were designed only to provide insight into the nature of typical operations research techniques. Real-world problems are a great deal more complex than those posed here.

This chapter has also been wholly technique-oriented. In years past there was a concern that operations research had become a collection of techniques from which the operations research specialist would draw one of the many predeveloped and categorized tools from his or her tool kit, on the assumption that the sum of such tools would suit all occasions. That thinking has matured today, viewing standard operations research tools as available where applicable, and recognizing the need for developing techniques to accommodate specific problem requirements. Operations research represents a powerful set of techniques for any practicing industrial engineer.

Drawing on what one has learned is not uncommon. Those doing the teaching were trying to pass on what they had learned, but they probably had high hopes that what they were teaching would be extended and adapted in innovative ways. Much operations research work is more analytical than is suggested by most of the examples discussed thus far. However, the restricted mathematical level of this text limits the use of analytic examples. This is unfortunate, because much of the best in operations research derives its value by forging new solution approaches to previously unsolved problems. The truly powerful operations researcher must be well versed in mathematics to be able to conceive and solve mathematical models appropriate to situations as they arise.

Unfortunately, as some operations researchers highly trained in mathematics have discovered, mathematical ability may be necessary but is certainly not sufficient. In the real world, it is necessary to find and understand problems buried in an environment replete with incomplete data, attitudes greatly influenced by motives of personal gain, semantic hurdles, managerial fears of risk, and support by average people (almost all of whom are imperfect) making average salaries and exhibiting average scientific enthusiasm.

Pure operations researchers may have an advantage over industrial engineers or management scientists because in their university educations they were able to devote more of their energies to the acquisition of mathematical skills. Having done so, however, they are at a disadvantage compared to more traditional industrial engineers because of the weakness of their education in dealing with physical systems and human and psychological factors, which is the case with management scientists as well. Of course, operations researchers are also at a disadvantage compared to management scientists because of their limited understanding and appreciation of the management environment, and this is also the case for industrial engineers to a lesser degree.

SUMMARY

A half-century ago, research first performed during World War I led to what was called operational research by the British, and later operations research in the United States. World wars involve the deployment of tremendous resources. Following the initial development and introduction of operations research, traditionalists resisted its use. There will always be those, with vested interests in present technology, trying to hold new technology in check; but if the new stuff works (i.e., if it has its niche) it will almost always prevail. Operations research typically involves the utilization of mathematics to resolve resource allocation problems. It is a quantitative approach that depends on structuring mathematical representations of variables to solve such problems.

The Desert Storm actions in Kuwait in the early 1990s provided an excellent example of the use of operations research in modern times. Barcode readers installed at unobtrusive locations throughout the area permitted the Allied forces to maintain effectively real-time inventory knowledge of their physical assets throughout the developing action. Every time a tank or personnel carrier passed a barcode reader in the field, its location was electronically relayed to a computer system maintaining real-time inventory, by location, of its massive technical resources.

In years past, there were OR zealots that falsely thought that OR would replace most if not all of traditional industrial engineering. Such has not proven to be the case. There is a place for traditional industrial engineering, as there is a place for operations research; they are not mutually exclusive resources. They are indeed very complementary, and any well-qualified industrial engineer today needs to be familiar with both of them.

One of the early developments in operations research was linear programming (LP). Special cases of LP began to take on their own lives and LP application subsets were developed, such as the assignment algorithm and the transportation algorithm.

Queuing theory was developed to deal with a day-to-day phenomenon that exists in all private and commercial life—waiting in line, what a British person calls standing in a "queue." Parts, pallets, and people in factories also wait in queues. JIT today is minimizing many of those queues. Mathematical formulations were developed to solve queuing problems, but over time the computer and a software approach, first called "Monte Carlo" simulation and later simply referred to as simulation, provided the means for handling extremely large and complex relationships in an effective way. Such simulations mimic the flow of thousands of items, often called transactions, through a time sequence analogous to the time effects the items will experience in the real world, providing the dynamic effects of item flows over time.

I recall simulating a ten-stacker crane automatic storage and retrieval system (ASRS) for bulk storage as part of a $40 million distribution facility design and installation that covered a space equivalent to 10 football fields. The simulation was developed from blueprints of the distribution facility to be constructed, and when built it operated much as the simulation had predicted. Simulation is one of the stars of operations research.

Operations research is not going to eliminate traditional industrial engineering, nor is traditional industrial engineering going to eliminate operations research. Both are viable and complementary technologies—all are part of the developing discipline of industrial engineering, which needs both tools in its tool kit to accommodate the vast variety of problems that exist in the real world today. Future industrial engineers need to understand both of them, and when to use each of them.

REFERENCES

1 Brockmeyer, E., H. L. Halstrom, and Arne Jensen: "The Life and Works of A. K. Erlang," *Trans. Dan. Acad. Tech. Sci. No. 2,* Copenhagen, 1948.

2 Dantzig, George B.: "Maximization of a Linear Function of Variables Subject to Linear Inequalities," in *Activity Analysis of Production and Allocation,* Chap. 21, Monograph no. 13, John Wiley & Sons, Inc., New York, 1951.

3 Emmons, Hamilton, A. Dale Flowers, Chandrashekar M. Khot, and Kamlesh Mathur: *STORM: Quantitative Modeling for Decision Support, Personal Version 3.0,* Simon & Schuster, Englewood Cliffs, NJ, 1992. Chapters include Linear Programming, Integer Programming, The Assignment Problem, The Transportation Problem, Distance Networks, Flow Networks, Project Management, Queuing Analysis, Inventory Management, Facility Layout, Assembly Line Balancing, Investment Analysis, Forecasting, Production Scheduling, Material Requirements Planning, Statistical Process Control, Statistics, Decision Analysis, and Decision Trees.

4 GPSS/PC Users Manual, Minuteman Software, Version 1.1, P. O. Box 171, Stow, MA 01775.

5 Hadley, G.: *Linear Programming,* Addison-Wesley Publishing Co., Reading, MA, 1962.

6 Hillier, Frederick S.: "The Application of Waiting Line Theory to Industrial Problems," *Journal of Industrial Engineering,* vol. 15, no. 1, 1964.

7 Moore, James M.: "Optimal Locations for Multiple Machines," *Journal of Industrial Engineering,* vol. 12, no.5, September-October 1961.

8 Schrage, Linus: *LINDO,* The Scientific Press, 651 Gateway Boulevard, Suite 1100, South San Francisco, CA, 94080-7014, 1992 Catalog, pp. 8–9.

9 Sly, David P.: *FactoryCad,* Release 2.01, CIMTECHNOLOGIES Corp., Ames, IA, 1991.

10 Taha, Hamdy: *Operations Research: An Introduction,* The Macmillan Company, New York, 1971.

11 Yih-Long, Chang, and Robert S. Sullivan: *Quant Systems, Version 2.0,* Prentice Hall, Englewood Cliffs, NJ, 1991. Chapters include Material Requirements Planning, Linear Programming, Integer Linear Programming, Linear Goal Planning, Quadratic Programming, Transportation and Transshipment Problems, Assignment and Traveling Salesman Problems, Network Modeling, Critical Path Method, Program Evaluation and Review Technique, Dynamic Programming, Inventory Theory, Queuing Theory, Queuing System Simulation, Decision and Probability Theory, Markov Process, Time Series Forecasting, Facility Location, Facility Layout, Aggregate Production Planning, Production Line Balancing, Job Shop Scheduling, Flow Shop Scheduling, Uncapacitated Lot Sizing, Quality Control, and Learning Curve and Work Measurement.

REVIEW QUESTIONS AND PROBLEMS

1 What is the difference in meaning between a "heuristic" and an "algorithm?"

2 Linear programming is one of the earliest techniques developed in operations research. Who developed the method most commonly employed in this general type of problem and what is the name of his method?

3 A nurse must assign five nurse's aides to one of each of five tasks. If the relative task times of each aide on each task with equal effectiveness are as given below, which task should be assigned each of the five aides to minimize the sum of task times?

		Tasks				
		I	II	III	IV	V
Aides	A	3	2	4	6	7
	B	1	2	4	3	5
	C	5	6	3	8	9
	D	2	3	5	7	1
	E	5	5	6	3	8

4 The costs for having five fire engines respond to five different locations simultaneously are given below. If the least-cost response indicates least cost to the property owner and less risk, which fire engine should be sent to which location to minimize losses?

		Locations				
		I	II	III	IV	V
Fire engines	A	24	25	23	22	26
	B	26	25	24	23	23
	C	23	22	22	27	26
	D	24	25	26	27	30
	E	24	25	25	26	23

5 If an auctioneer can sell only one item to each person, and the prices five customers are willing to pay for each item are as indicated below, to which individuals should the auctioneer sell which items?

		Customers				
		A	B	C	D	E
Items	I	6	4	2	3	4
	II	8	7	5	3	4
	III	3	7	5	4	4
	IV	6	2	3	7	7
	V	3	2	5	4	0

6 A sawmill company has three mills supplying four retail lumberyards. The costs to ship lumber in multiples of \$1000 per 10,000 board feet of lumber from each mill to each retailer are given below. Multiples of 10,000 board feet of lumber per month required by each retailer and available from each mill are also given. How many 10,000 board-foot loads should be shipped from each mill to each retailer to minimize shipping costs for the month?

		Retailers				
		A	B	C	D	Mill capacities, 10,000 board-feet loads
Mills	1	\$3	5	5	7	40
	2	5	8	3	6	30
	3	4	9	3	2	100
Retailer requirements, 10,000 board-feet loads		50	60	40	20	170

7 A commercial painting company has three paint crews. One crew uses an old sprayer (OS), another uses a new sprayer (NS), and the third paints by hand (H). The profits per 100 square feet painted by the three crews for four different buildings are as follows:

		Buildings			
		I	II	III	IV
Crews	OS	\$6	5	5	4
	NS	7	7	8	2
	H	3	3	3	3

Buildings I, II, III, and IV are 60,000, 30,000, 20,000, and 40,000 square feet, respectively. The old sprayer and new sprayer can paint 40,000 and 60,000 square feet, respectively. The owner can employ as much labor as he needs to hand-paint up to 70,000 square feet per month. Which buildings should be painted by which crews to maximize profit? Assume that any combination of crews can be used on the buildings and that crews paint in increments of 1,000 square feet.

8 Assume that the following specify a generalized linear programming problem:

$$\text{Maximize } f(x) = 2x_1 + x_2$$

subject to

$$x_1 + x_2 \leq 6 \qquad (1)$$

$$x_1 \leq 3 \qquad (2)$$

$$2x_1 + x_2 \geq 4 \qquad (3)$$

$$x_1, x_2 \geq 0$$

Graph this linear programming problem, identifying the three constraint equation lines and the feasible zone common to all of them. Plot dotted lines for values of 3, 6, 9, and

12 for the objective function *f(x)*. What appears to be the highest feasible value of *f(x)*, and for what values of x_1 and x_2 does it occur?

9 Assume that the mean arrival rate of moviegoers to a ticket window is three per minute and is Poisson-distributed. Assume also that the mean service rate of the ticket clerk is four per minute and that service times are exponentially distributed. After the system has reached steady state (e.g., at 11:15 A.M.):

a What is the probability that the ticket clerk has no one to serve?

b Is it more likely that one customer is buying a ticket with no one waiting in line, or that one person is buying a ticket and one person is waiting in line?

c What is the probability that there are at least two people waiting in line?

d What is the expected time an arriving customer will have to wait in line before being served?

e What is the expected number of customers waiting in line to be served?

10 Cars arrive at a state inspection center at a mean rate of 15 per hour. Assume that the arrival rate is Poisson-distributed. The inspection center has three parallel lines, and the mean service time in each line is 10 minutes. Assume service times are exponentially distributed. If cars form a single line while waiting, and cars enter any one of the three service cells when one becomes available on a first-come, first-served basis:

a Would this system be expected to exhibit steady-state characteristics?

b What is the probability that at 3:12 P.M. there are no cars in line or being inspected?

c What is the expected time a car will spend at the inspection station if it enters the system after steady state has been reached?

d Someone suggested that if the three service lines were distributed as individual lines about the city, each serving one-third of the cars, the time a car would spend at an inspection station would be less. Do you agree, and why?

11 Is simulation an optimization technique, and if not, how is it employed to provide "good" solutions?

12 What are the two basic types of digital computer simulations, and how do they differ?

13 What components are often missing in the education of pure operations researchers that may later negatively affect their performance in practice as compared to industrial engineers or management scientists?

8

DECISION SCIENCES

People can be divided into three groups: those who make things happen, those who watch things happen, and those who wonder what happened.

Anonymous

ENGINEERING ECONOMY

A considerable body of knowledge has been developed over the years concerning decision making. This topic area is generally referred to as the *decision sciences*. One subtopic of this body of knowledge, called *engineering economy*, is concerned with the financial analysis of engineering projects. As is sometimes the case, the title of a classic text—in this instance Eugene Grant's *Principles of Engineering Economy* [3] published in 1930—became so universally applied that it became the name of the body of knowledge initiated by it—much the same way that Xerox has become synonymous with a copier, or a Coke with a soft drink. Engineering economy has been and still is considered in engineering circles to be so essential to the practice of engineering that it is one of four required knowledge topics in which all engineers applying for engineering professional registration are tested.

In business schools, a similar body of knowledge is typically called *capital investment analysis* or *discounted cash flow analysis*.

Engineering designs are typically evaluated first for their relative engineering merits (e.g., Will the bridge stand up?). If it is deemed sound from a physical engineering perspective, a second evaluation often determines its relative economic cost (e.g., How much will it cost to build and maintain the bridge?). For an engineering design to go beyond design to construction and ultimate utilization, it typically must both represent sound physical engineering and be financially superior to other competing engineering designs as well. The client wants the function, but often at lowest attainable cost.

A typical engineering economy solution primarily answers one of two questions: (1) Which of the choices considered is best from a financial point of view? (i.e., Which equipment offers the required service at lowest cost?), or (2) What is the expected return on investment (ROI) in using this equipment? (i.e., If I were to purchase this equipment, would I make a satisfactory return for the money invested?).

Interest Calculations

Parents have been known to lend money to their children and later receive neither the simple interest nor the principal. In contrast, the business world generally employs compound interest calculations and expects repayment. Therefore, gaining an understanding of compound interest calculations is the typical starting point for teaching engineering economy. Consider the following four variables in an engineering economy problem:

$$P = \text{present worth}$$
$$F = \text{future worth}$$
$$i = \text{interest rate}$$
$$n = \text{number of interest periods}$$

Both present worth and future worth represent sums of money at points in time. The only requirement is that present worth precede future worth in time. If present worth is January 1, 1888, then future worth must follow January 1, 1888, some number of interest periods in the future (for example, February 1, 1888, one interest period later assuming monthly interest, or March 1, 1990, numerous interest periods later).

The interest rate is the rent charged on the money lent for a defined period of time. The problems in this section will be limited to annual interest, and interest will be stated as annual compounding interest. Interest is typically charged as a percentage of the principal lent (e.g., 6 percent). If I lend you a dollar at 6 percent annual compound interest, a year later you owe me \$1.06. Of this total, \$1 is principal and \$0.06 is interest (i.e., 6 percent of \$1).

The compound interest equation is

$$F = P(1 + i)^n$$

Assume that you are unable to pay me the \$1.06 you owe me at the end of the first year. Employing compound interest, the \$1.06 (i.e., all principal and previously accumulated interest) becomes the principal at the beginning of the next interest period (i.e., a year). Therefore, at the end of the second year you owe me

$$(\$1.06) \times (1 + 0.06)$$

The \$1.06 times 1 is the principal you owe me at the end of the second year, and the \$1.06 times 0.06 is the interest you owe me at the end of the second year. Of course, the total of what you owe me at the end of the two-year period is

$$\$1(1 + 0.06)^2$$

This amount, if not paid at the end of the second year, becomes the principal at the beginning of the third year. At the end of the third year, the amount due on the loan is

$$\$1(1.06)^3$$

Consider the following example:

Example 1:

I lent you \$500.00 three years ago at 6% interest. How much do you owe me now?

$$F = P(1 + i)^n = 500(1.06)^3 = \$595.51$$

Engineering economy problems have traditionally been solved employing interest tables, and their use is generally instructive. Note that values for four different interest rates are tabulated in Table 8-1. In the foregoing problem, the unknown F is on the left side of the equal sign and the known value P is on the right side of the equal sign. What follows P (i.e., $[1.06]^3$) can be found as a tabulated value in an interest table. To simplify the coding of this factor a convention is often employed of simply repeating the unknown and known letters of the applicable variables as a factor descriptor (i.e., FP). Because the interest rate and number of interest periods are two other required parameters in such a problem, the complete factor coding for the preceding problem is often designated as

$$(\text{FP},i,n)$$

Because the applicable interest rate is 6 percent and the number of interest periods is 3, the factor would be written as

$$(\text{FP},6,3)$$

The preceding problem then becomes one of finding an interest table value for FP,6,3 in Table 8-1 (1.191) that will make a present worth of \$500 become a larger amount three years hence, employing a 6 percent interest rate. The factor (1.191) is multiplied by the known value (\$500) that is to the right of the equal sign to produce the unknown value (F) that is to the left of the equal sign as follows:

$$F = P(\text{FP},6,3) = \$500(1.191) = \$595.50$$

TABLE 8-1 COMPOUND INTEREST TABLES

6% Compound Interest Table

n	FP	PF	AF	AP	FA	PA	PG
1	1.060	0.943	1.000	1.060	1.000	0.943	-
2	1.124	0.890	0.485	0.545	2.060	1.833	0.89
3	1.191	0.840	0.314	0.374	3.184	2.673	2.57
4	1.262	0.792	0.229	0.289	4.375	3.465	4.95
5	1.338	0.747	0.177	0.237	5.637	4.212	7.93
6	1.419	0.705	0.143	0.203	6.975	4.917	11.46
7	1.504	0.665	0.119	0.179	8.394	5.582	15.45
8	1.594	0.627	0.101	0.161	9.897	6.210	19.84
9	1.689	0.592	0.087	0.147	11.491	6.802	24.58
10	1.791	0.558	0.076	0.136	13.181	7.360	29.60
15	2.397	0.417	0.043	0.103	23.276	9.712	57.55
20	3.207	0.312	0.027	0.087	36.786	11.470	87.23

8% Compound Interest Tables

n	FP	PF	AF	AP	FA	PA	PG
1	1.080	0.926	1.000	1.080	1.000	0.926	-
2	1.166	0.857	0.480	0.561	2.080	1.783	0.86
3	1.260	0.794	0.308	0.388	3.246	2.577	2.45
4	1.360	0.735	0.222	0.302	4.506	3.312	4.65
5	1.469	0.681	0.170	0.250	5.867	3.993	7.37
6	1.587	0.630	0.136	0.216	7.336	4.623	10.52
7	1.714	0.584	0.112	0.192	8.923	5.206	14.02
8	1.851	0.540	0.094	0.174	10.637	5.747	17.81
9	1.999	0.500	0.080	0.160	12.488	6.247	21.81
10	2.159	0.463	0.069	0.149	14.487	6.710	25.98
15	3.172	0.315	0.037	0.117	27.152	8.559	47.89
20	4.661	0.215	0.022	0.102	45.762	9.818	69.09

12% Compound Interest Table

n	FP	PF	AF	AP	FA	PA	PG
1	1.120	0.893	1.000	1.120	1.000	0.893	-
2	1.254	0.797	0.472	0.592	2.120	1.690	0.80
3	1.405	0.712	0.296	0.416	3.374	2.402	2.22
4	1.574	0.636	0.209	0.329	4.779	3.037	4.13
5	1.762	0.567	0.157	0.277	6.353	3.605	6.40
6	1.974	0.507	0.123	0.243	8.115	4.111	8.93
7	2.211	0.452	0.099	0.219	10.089	4.564	11.64
8	2.476	0.404	0.081	0.201	12.300	4.968	14.47
9	2.773	0.361	0.068	0.188	14.776	5.328	17.36
10	3.106	0.322	0.057	0.177	17.549	5.650	20.25
15	5.474	0.183	0.027	0.147	37.280	6.811	33.92
20	9.646	0.104	0.014	0.134	72.052	7.469	44.97

20% Compound Interest Table

n	FP	PF	AF	AP	FA	PA	PG
1	1.200	0.833	1.000	1.200	1.000	0.833	-
2	1.440	0.694	0.455	0.655	2.200	1.528	0.69
3	1.728	0.579	0.275	0.475	3.640	2.106	1.85
4	2.074	0.482	0.186	0.386	5.368	2.589	3.30
5	2.488	0.402	0.134	0.334	7.442	2.991	4.91
6	2.986	0.335	0.101	0.301	9.930	3.326	6.58
7	3.583	0.279	0.077	0.277	12.916	3.605	8.26
8	4.300	0.233	0.061	0.261	16.499	3.837	9.88
9	5.160	0.194	0.048	0.248	20.799	4.031	11.43
10	6.192	0.162	0.039	0.239	25.959	4.192	12.89
15	15.407	0.065	0.014	0.214	72.835	4.675	18.51
20	38.338	0.026	0.005	0.205	186.688	4.870	21.74

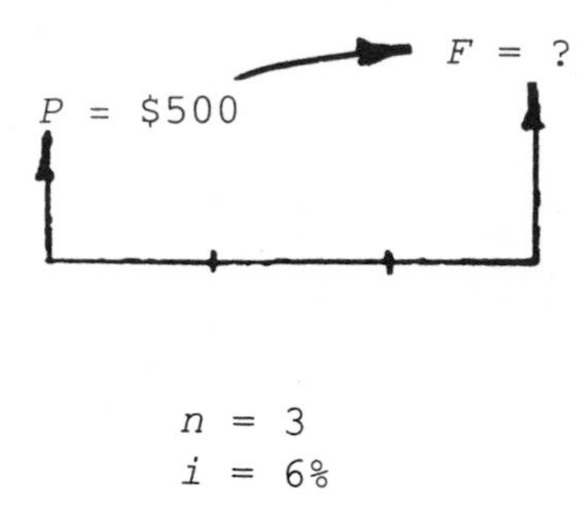

FIGURE 8-1 Cash flow diagram for P to F.

Of the \$595.50, \$500 is the original principal, and the remaining \$95.50 is accumulated interest over the three-year period.

Figure 8-1 is often called a cash flow diagram in engineering economy. It graphically displays the timing and amounts of cash flows in a problem.

The next problem essentially reverses the previous problem.

Example 2:

You paid me \$595.50 today for a loan made three years ago. If the interest charged was 6 percent, how much was the original loan?

In this problem F is known (i.e., \$595.50) and it is the prior present value P that is the unknown. The applicable interest rate is still 6 percent, and the number of interest periods is still three. The solution, therefore, is

$$P = F(\text{PF},i,n) = 595.50(\text{PF},6,3) = 595.50(0.840) = 500.22 \approx \$500$$

The cash flow diagram for this problem is shown in Fig. 8-2.

Another variable employed in engineering economy problems is commonly referred to as an *equivalent annual amount*. It represents a constant value (e.g., A) occurring at the end of each interest period over a continuous series of interest

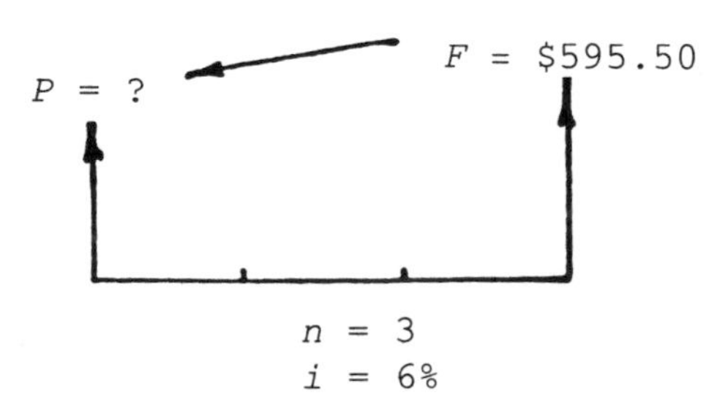

FIGURE 8-2 Cash flow diagram for F to P.

periods. This A amount occurring at the end of each interest period (e.g., a month) is well known to most Americans, examples being the monthly automobile or mortgage payment.

The following is representative of an equivalent annual amount problem.

Example 3:

If I put $1000 in a bank that pays 6 percent interest, what equivalent annual amount can I withdraw from this account at the end of each of the next five years?

The answer to this question is illustrated diagrammatically in Fig. 8-3. In solving any engineering problem one should always ask, "Is the answer reasonable?" in terms of its approximate size. If interest was 0 percent, $1000 paid out in five equal payments would be $200 (i.e., $1000/5). Since the applicable interest rate is 6 percent, some interest should be paid out in addition to the return of principal. The answer should therefore be greater than $200. The solution to this problem is as follows:

$$A = P(\mathrm{AP},i,n) = \$1000(\mathrm{AP},6,5) = \$1000(0.237) = \$237$$

It is assumed in reviewing the result that $200 is return of principal and the $37 difference represents accumulated interest payments.

The following example simply reverses the previous problem:

Example 4:

If I owe you $237 at the end of each of the next five years, how much would you be willing to accept as a lump sum payment now if the mutually agreed

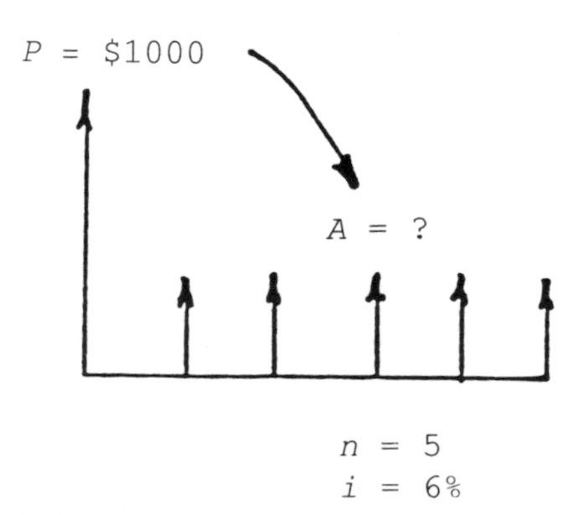

FIGURE 8-3 Cash flow diagram for P to A.

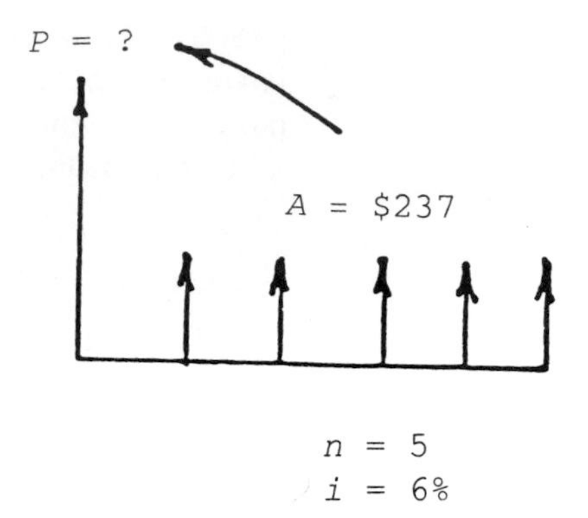

FIGURE 8-4 Cash flow diagram for A to P.

interest rate is 6 percent? This question, illustrated in Fig. 8-4, can be answered by employing the following equation:

$$P = A(\text{PA},6,5) = \$237(4.212) = 998.24 \approx \$1000$$

The resultant $998.24 is close to $1000 but is not exactly equal due to rounding error. Over the years I have noticed that when I say "close enough for government work," government seminar attendees often wince, which has the unfortunate effect of encouraging me to say it more often.

The next two examples convert A to F, and then the reverse problem of F to A.

Example 5:

If I put $500 in an account at the end of each year for the next four years, how much could I withdraw at the end of the four years as a lump sum payment if interest is 6 percent?

Let us consider this question, depicted in Fig. 8-5, using the equation:

$$F = A(\text{FA},6,4) = 500(4.375) = \$2187.50$$

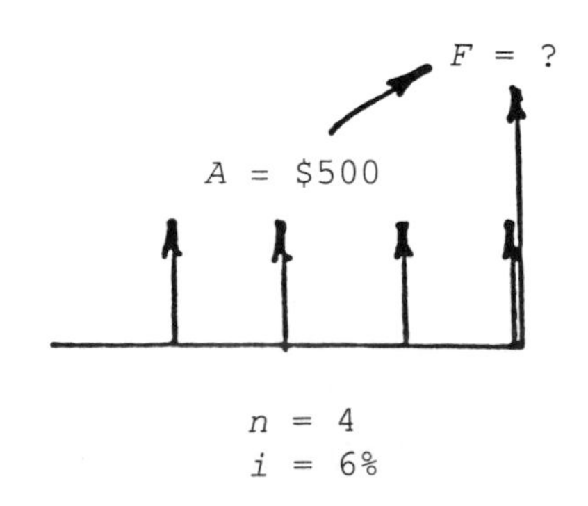

FIGURE 8-5 Cash flow diagram for A to F.

Is the answer reasonable? If interest was 0 percent, four equal payments of \$500 each would accumulate to \$2000. Therefore, it seems reasonable that the difference of \$187.50 may well be the accumulated interest.

Example 6:

If I need \$2187.50 four years from now to start a fortune cookie factory, what equal amount (i.e., equivalent annual amount) must I deposit in an account at the end of each of the next four years, if interest is 6 percent?

Using Fig. 8-6 and the following equation, we have

$$A = F(\text{AF},6,4) = \$2{,}187.50(0.229) = 500.93 \approx \$500$$

We find that A is approximately equal to \$500, which is correct to a first approximation. If the tabular values had contained more significant digits, the \$0.93 difference would have disappeared. In engineering work, approximate values are all we typically need anyway.

There are four commonly used methods for comparing the relative financial merits of competing alternatives in engineering economy problems: (1) annual cost, (2) present worth, (3) internal rate of return, and (4) payback. The first three are mathematically equivalent and, therefore, all three methods will produce the same final result (i.e., if one method chooses machine A as superior, the other methods will also). Payback, which is a historically popular method in general use, can and will provide the wrong answer, as will be demonstrated shortly. First let us consider the annual cost method, the most popular method employed in industry.

Annual Cost Method

In the next problem a few new terms will be introduced. *Equivalent annual cost* (EAC) represents an equal annual amount required to obtain the use of some service over a

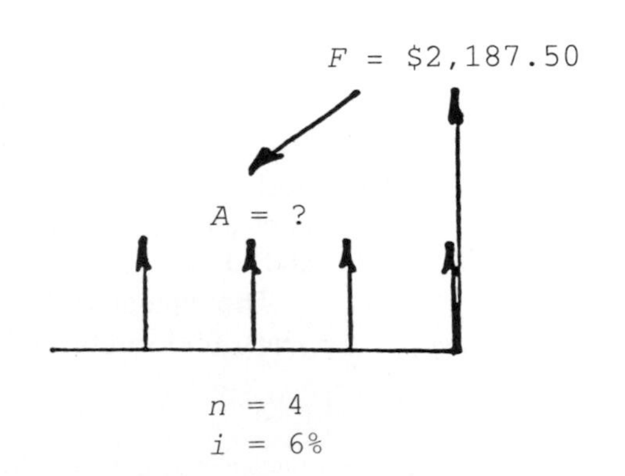

FIGURE 8-6 Cash flow diagram for F to A.

period of time, referred to as the *economic life* of the investment (e.g., equipment). Economic life as distinguished from *physical life* is often a shorter period of time, at the end of which time the intended use of the equipment is expected to end even though the equipment is physically able to continue functioning (i.e., its economic life has ended).

The EAC is somewhat equivalent to what the lease cost of the equipment would be. For example, instead of purchasing the equipment, providing net operating costs over its economic life, and obtaining salvage value at the end of its economic life, one may wish to make an equivalent annual payment over the same number of years to cover these costs and incomes (e.g., salvage value). *Annual operating cost* (AOC) refers to the assumed equal net outlays one is expected to make each year to support the investment over its economic life. Examples of such costs for a machine might be consumable expenses such as lubricating oil and gasoline, and maintenance costs associated with the replacement of V-belts, light bulbs, etc. *Net annual cost* implies that there may be some cash inflows during these periods that may effectively reduce these expenses.

Another terminology change will be to refer to the underlying interest rate i employed in an annual cost problem as the Minimum Acceptable Rate of Return (MARR). It represents the marginal rate of return (i.e., interest rate) a firm expects to make on its discretionary investments, typically based on its documented recent previous similar investments, adjusted for present relative investment opportunity compared with previous periods.

Consider the following annual cost method problem:

Example 7:

Machines A and B perform equally well. Machine A initially costs \$20,000, has estimated net annual operating expenses of \$3000, and has an estimated salvage value at the end of a 10-year economic life of \$4000. Machine B will cost \$25,000 initially, with annual operating expenses of \$2000, and a salvage value at the end of a 15-year economic life of \$5000. If your minimum acceptable rate of return (MARR) is 20 percent, which machine should you purchase?

The general cash flow diagram for this type of problem is presented in Fig. 8-7. Note the arrows going in opposing directions. As indicated before, cash flows may represent either costs or incomes (e.g., salvage value is income, therefore a negative cost). Because costs are more prevalent (welcome to the real world), costs will be shown as positive costs, and incomes will be shown as negative costs. To distinguish between costs and incomes, all costs in the cash flow diagram will be shown as "down arrows" and all incomes will be shown as "up arrows."

The typical annual cost method problem for machine service over a period of years, for example, has the following typical primary elements in its solution:

$$\text{EAC} = P(\text{AP},i,n) + \text{AOC} - F(\text{AF},i,n)$$

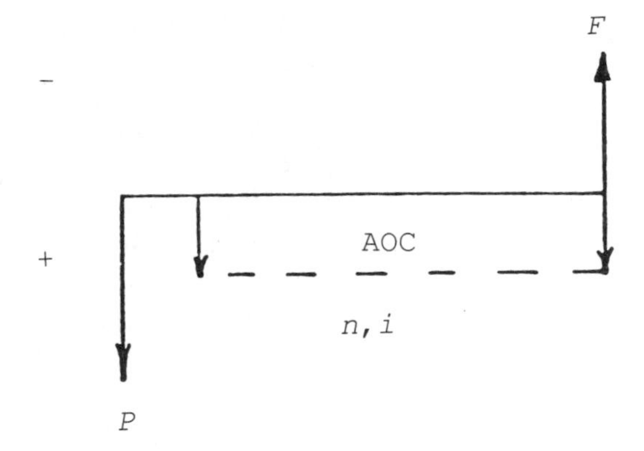

FIGURE 8-7 General annual cost method cash flow diagram for an investment.

The preceding elements can be generically considered as follows:

$$\text{EAC} = A_P + A_{\text{AOC}} - A_F$$

The first cost element on the right side of the equation, A_P, which stands for *annualized amount of the purchase*, employs the interest table symbolism of $P(\text{AP},i,n)$. What symbol would be expected on the left side of the equation to maintain consistent symbolism for this single element?

$$? = P(\text{AP},i,n)$$

The answer is obviously A. What A is it? The A can be interpreted as an annual lease cost equivalent to an initial lump sum purchase P. That is to say, this is an annual cost that is equivalent to an initial purchase investment. Therefore, A_P is the annual cost equivalent of the lump sum initial investment, that is, we have converted a beginning-of-period lump sum amount to an equivalent annual amount.

A_{AOC} is an annual amount, therefore, no annualization conversion is required.

In a similar fashion to the A_P factor above, what variable would one expect on the left side of the equation for the third element above?

$$? = F(\text{AF},i,n)$$

The answer is A again. What A is it? The answer is that this A is the annualized equivalent amount to receiving the salvage value at the end of the time series under consideration. A_F is the equivalent amount one might be paid at the end of each year rather than being paid the salvage value at the end of the series of years, for the applicable interest rate and number of years.

Therefore, the three elements of the equivalent annual cost are the annualized elements of the purchase cost (what the IRS calls *first cost*), the annual operating cost, and the annualized salvage value. The sum of these is the annual cost of providing the machine service over the economic life of the machine.

The IRS lets you claim a number of costs such as freight, engineering, installation, and others associated with placing the machine in service as legitimate elements of first cost. The total of all of these costs will then become the *capital investment* to be depreciated over a number of years.

Allowing firms to claim depreciation as a *business expense,* when in fact there may be no real cash flow expense, is a congressionally mandated tax inducement to business to purchase equipment that goes back to tax law passed in 1919. The fictitiously assumed depreciation expense, involving no real cash flow, reduces taxable income, thereby reducing corporate taxes due on the reduced reported income. The allowed fictitious expense rarely bears any close relationship to the reduction in asset value it is theoretically intended to compensate. In some instances, such as antique cars, one can claim depreciation expense when the asset value of the car is in fact increasing in value from year to year.

What is the equivalent annual cost for Machine A in Example 7? First let us write a cash flow diagram for Machine A (see Fig. 8-8).

$$\begin{aligned} \text{EAC} &= P(\text{AP},20,10) + \text{AOC} - F(\text{AF},20,10) \\ &= \$20{,}000(0.239) + \$3000 - \$4000(0.039) \\ &= \$7624 \end{aligned}$$

Likewise, we can write a cash flow diagram for Machine B (see Fig. 8-9).

$$\begin{aligned} \text{EAC} &= P(\text{AP},20,15) + \text{AOC} - F(\text{AF},20,15) \\ &= \$25{,}000(0.214) + \$2000 - \$5000(0.014) \\ &= \$7280 \end{aligned}$$

Since $7280 < $7624, Machine B is the better purchase.

FIGURE 8-8 Cash flow diagram for Machine A.

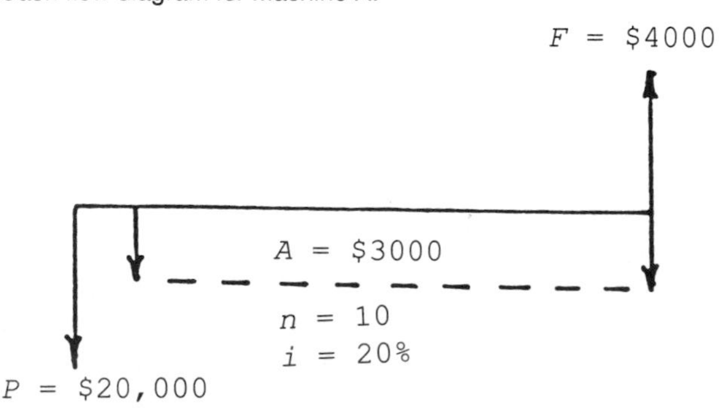

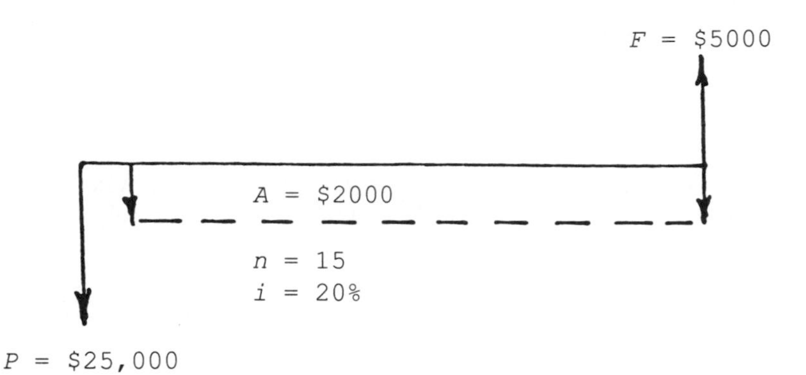

FIGURE 8-9 Cash flow diagram for Machine B.

Return on Investment (ROI)

Whereas an annual cost method solution attempts to select the financially best among choices (e.g., Machine A versus Machine B), an ROI solution determines the applicable underlying interest rate in a problem to answer the question "If I make this investment, will there be a sufficient return (i.e., interest) on the money invested?" The unknown variable, therefore, in such a problem is *i*, the interest rate.

Consider the following problem, shown as a cash flow diagram in Fig. 8-10:

Example 8:

If I invest $1 by depositing it in a bank paying compound annual interest and one year later I have $1.06 in the account, what is my ROI?

Employing the compounding interest formula, we have

$$F = P(1 + i)^n$$
$$1.06 = 1(1+ i)^1$$
$$i = 1.06 - 1 = 0.06, \text{ or } 6\%$$

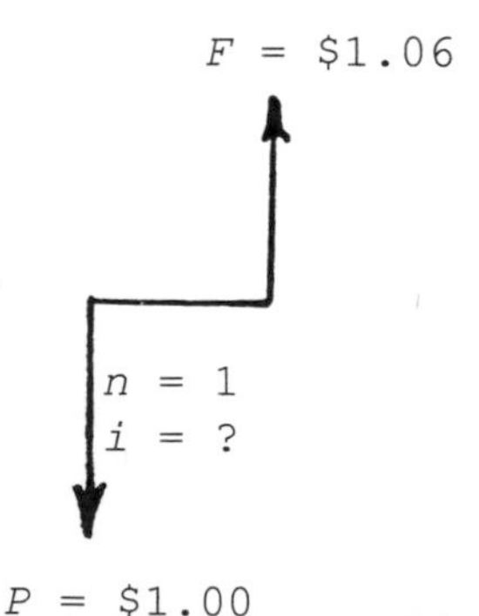

FIGURE 8-10 Cash flow diagram for a $1 investment for one year at 6 percent.

Therefore, if you put \$1 in a bank paying compound annual interest for a year, close out the account at the end of the year, and receive \$1.06, 6 percent must have been the interest rate.

In a similar fashion, to determine ROI for an investment, find an interest rate that will produce a present value for all future returns equal to the present investment.

For the preceding problem, consider Fig. 8-11. Employing the compounding interest equation again, this time employing interest table symbolism, we can write

$$P = F(\text{PF},i,n)$$
$$\$1 = \$1.06(\text{PF},i,1)$$
$$\text{PF},i,1 = \$1/\$1.06 = 0.9434$$

In reviewing the compound interest tables, note that the only interest rate table that has a PF,i,1 factor of 0.943 is the 6% table; therefore, the underlying interest rate in this problem is 6 percent. This "backing into the solution" by finding a factor in a table that produces the desired result is typical for ROI problems.

Consider the following problem:

Example 9:

You can purchase an antique car today for \$5070 that you estimate you can sell in six years for \$10,000. What return on investment (ROI) would you get from this project?

Employing the compound interest equation, we can write

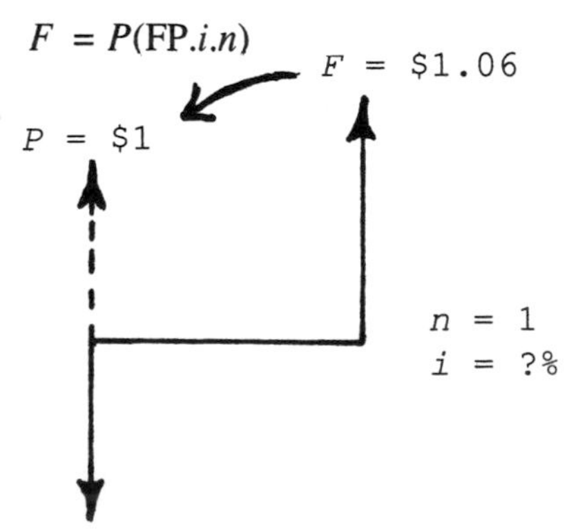

FIGURE 8-11 Cash flow diagram for determining ROI.

Note in the interest rate tables for an FP,i,6 factor that the 12 percent interest table has a factor value of 1.974, which (you guessed it) is "close enough for government work." The underlying interest rate, therefore, for the antique car problem is 12 percent. In reviewing Example 9, consider that if you put $5070 in a bank for six years, paying 12 percent interest, it will produce a lump sum future value of $10,000 at the end of that period.

Some years ago my uncle was concerned about his car, which was parked in the open next to his mobile home as a hurricane approached, so he moved it next to the carport of a relative. The hurricane picked up the carport and dropped it on my uncle's car. The best of plans do not always work out—but there is no excuse for not planning. The antique car project will produce a 12 percent ROI if the barn in which the car is stored does not fall on the car. There are no certain investments in the real world, only investments with differing degrees of risk.

Payback Method

The payback criterion for considering competing investments is simply this: *The alternative that returns the initial investment in the shortest time period is preferred.*

As stated, this criterion can produce unsound economic choices, and the following example is intended to demonstrate this underlying weakness.

Example 10:

	Machine A	Machine B
Initial investment	$10,000	$10,000
Net annual income	2,000	2,500

Machine A will pay back the original investment in five years, and Machine B in four years. If we employ the payback criterion, Machine B would be preferred.

The primary weakness in employing the payback criterion is in what it fails to consider, that is, returns after payback. Another weakness is that it does not consider the time value of money during the economic life of the equipment.

Assume that the economic lives of Machines A and B are eight and four years, respectively. What would be their underlying ROIs? Because it takes Machine B four years to pay back the original investment, and its economic life is four years, it cannot produce a return after payback, therefore, $i = 0\%$. Figure 8-12 is a cash flow diagram for Machine A.

$$\begin{aligned} P &= A(\text{PA},i,8) \\ \$10{,}000 &= \$2000(\text{PA},i,8) \\ \text{PA},i,8 &= 5.000 \end{aligned}$$

The tabulated value for PA,12,8 is 4.968, therefore, $i \approx 12$ percent.

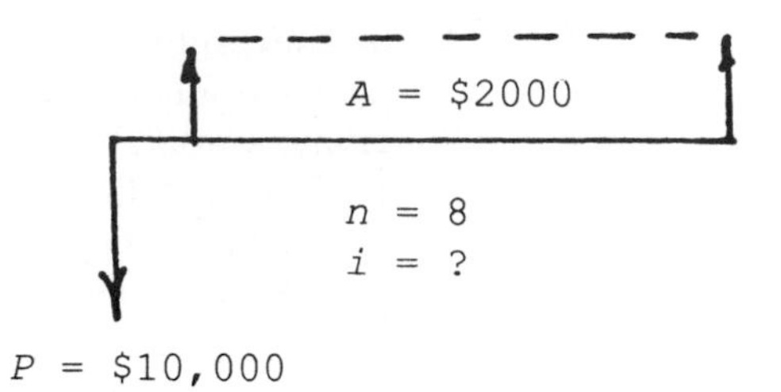

FIGURE 8-12 Cash flow diagram for machine A.

Assume for the moment that you are a stockholder in the corporation for which Machine A or Machine B will be selected. Which choice would you prefer? As a stockholder would you prefer that the management select a 0 percent or a 12 percent investment? The answer is obvious—a 12 percent investment is superior to a 0 percent investment. Therefore, Machine A is the better choice.

Because of the pressures exerted in the financial community in general and Wall Street in particular, through upper-level managers in U.S. corporations, far too often operations managers select the shorter payback investment choice over the maximum return investment choice. So much pressure is exerted to produce favorable annual and even quarterly returns on operations that short-sighted investing is far too prevalent, and it has contributed to significant erosion of the international competitiveness of American corporations. There is no valid excuse for such irresponsible management behavior—it needs to be fixed.

In comparison, Japanese and German management are generally known to make 10- and 15-year payback investments, whereas many managements in the United States restrict discretionary investments to one- or two-year paybacks. Is it any wonder that our plants cannot compete internationally?

Aftertax Analysis

The solutions to all the examples in this section thus far are for pretax problems. That is, that they do not consider taxes as a cash flow in solving the problem. The following example compares a pretax solution and an aftertax solution to the same problem.

Example 11:

The purchase of a minicomputer control system can reduce clerical labor required in controlling a process from $85,800 to $55,000 annually. The minicomputer control system can be purchased for $80,000 and will have an expected useful economic life of four years at which time it will have no salvage value. Assume straight-line depreciation, a 20 percent minimum attractive rate of return (MARR) before taxes, and an incremental tax rate of 40 percent. Should the minicomputer control system be purchased?

Pretax Analysis

Present System:	
Clerical labor	$85,800
Total	$85,800
Proposed System:	
Clerical labor	$55,000
Capital recovery ($80,000 × 0.386) [$P$ over $80,000; $AP,20,4$ over 0.386]	$30,880
Total	$85,880

Since $85,800 < $85,880, the present system should be retained.

Aftertax Analysis In performing the aftertax analysis, it is necessary to convert the pretax MARR to an equivalent aftertax MARR.

$$
\begin{aligned}
RR_{at} &= \text{the aftertax rate of return} \\
RR_{pt} &= \text{the pretax rate of return} \\
Tr &= \text{the marginal tax rate} \\
RR_{at} &= RR_{pt}(1 - Tr) \\
&= 0.20\,(1 - 0.40) \\
&= 0.12, \text{ or } 12\%
\end{aligned}
$$

Therefore, given a 20 percent pretax rate, or 20 cents on a dollar of marginal income, and a marginal income tax rate of 40 percent, or 8 cents on the dollar, the remaining 12 cents on the dollar, or 12 percent, is the aftertax equivalent rate of return.

Present System:	
Clerical labor	$85,800
Total	$85,800
Proposed System:	
Clerical labor	$55,000
Capital recovery ($80,000 × 0.329) [$P$ over $80,000; $AP,12,4$ over 0.329]	26,320
Labor tax effect 0.40($85,800 – $55,000)	12,320
Depreciation tax effect 0.40 ($20,000)	–8,000
Total	$85,640

Since $85,640 < $85,800 one should purchase the proposed system.

The pretax analysis indicates that the present system should be retained, and the aftertax analysis indicates that the proposed system should be purchased. Which is the more valid recommendation? The aftertax analysis is. From a purely business perspective, the alternative that produces the desired result after all other parties have been paid, including the IRS, is the final answer (i.e., the bottom line).

DECISION THEORY AND UTILITY

In the real world decisions are rarely made with certainty; most decisions involve risk. Managers are paid to make decisions encompassing risk. To remain managers they must, however, make enough good decisions to more than outweigh the bad ones, as compared with the relative capability of other available potential managers.

A common criterion for valuing risky decisions is the *maximum (or minimum) expected value criterion*. Under this criterion it is assumed that value can be estimated as the product of the value of an outcome and its probability of occurrence. For example, if the probability of receiving \$100 in a letter you have just received from a parent is 20 percent, the expected monetary value of the letter is $\$100 \times 0.20 = \$20$. Whereas we shall defer *utility* aspects of the problem until later, this criterion has been used extensively in what are called *decision tree* problems.

Assume, for example, that a contractor constructing a new building is installing wood flooring this week and also moving in certain pieces of equipment. One major piece of equipment will not arrive until next week. According to weather information, some change in the weather is possible this coming weekend. Although there is an 80 percent chance that the weather will be ideal, the forecast calls for a 15 percent chance of high humidity and a 5 percent chance of rain.

One choice available to the contractor is to install the permanent roof during this week and suffer an additional cost of \$3400 for removing a wall to install the piece of equipment coming next week. A second choice is to construct a temporary roof of polyethylene at a cost of \$1600. With the temporary roof, damage to the flooring and equipment is estimated to be \$7600 if it is very humid and \$22,600 if it rains, including the cost of the temporary roof in both instances. Another choice, and the most risky, is to install the permanent roof after the equipment is received next week. With no roof, if high humidity occurs, damage is estimated to be \$12,000, and if it rains, \$30,000. What should the contractor do? The choices are (1) install the permanent roof now, (2) install a temporary polyethylene roof now, and (3) do nothing and hope the weather prediction is wrong about the weather changing before the equipment arrives next week.

Figure 8-13 is a decision tree for this problem. To summarize, if the permanent roof is installed now, the contractor accepts a loss of \$3400 and need not be concerned further with the problem. If the contractor does not build the permanent roof and instead builds the temporary roof, there is an 80 percent, 15 percent, and 5 percent chance of losing \$1600, \$7600, and \$22,600, respectively. If no roof is built now, there is an 80 percent, 15 percent, and 5 percent chance of losing \$0, \$12,000, and \$30,000, respectively.

The expected losses for installing the temporary roof or not are as follows:

$$\text{Install temporary roof: } 0.8(\$1600) + 0.15(\$7600) + 0.05(\$22{,}600) = \$3550$$
$$\text{No roof: } 0.8(\$0) + 0.15(\$12{,}000) + 0.05(\$30{,}000) = \$3300$$

The *X* on the "Install temporary roof" path of Fig. 8-13 indicates that it should be dropped from further consideration. The expected value of \$3300 for the decision to

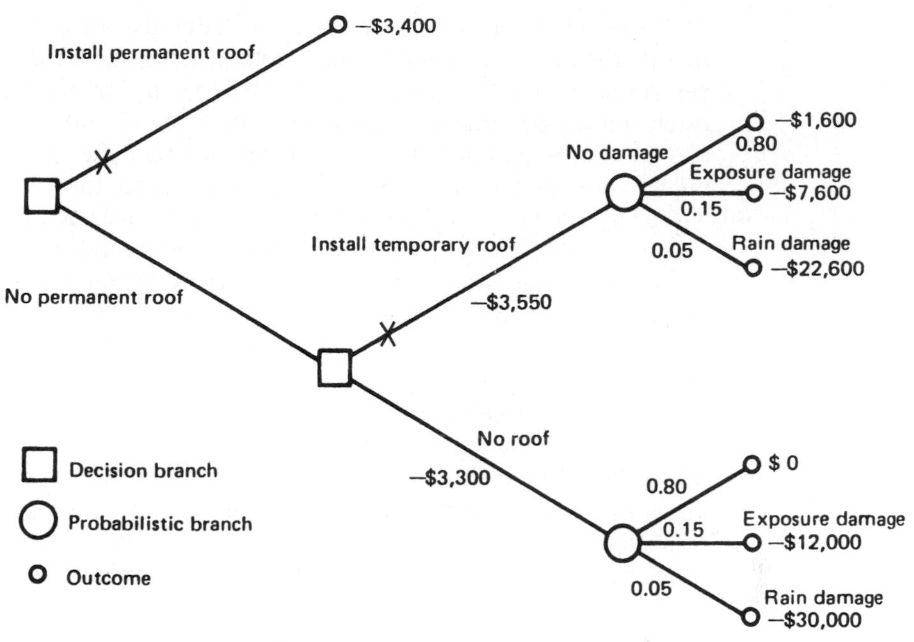

FIGURE 8-13 Decision tree for the contractor problem.

construct no roof at all is less than the $3400 for installing the permanent roof now, so the best choice using the expected value approach is to construct no roof and wait to see what happens.

No definitive meaning has been given to the word "value" thus far. Hall gives some insight into the complexities associated with determining value

> A fundamental problem in a general theory of value is to define value so as to include all its forms. One answer to this problem is to say that value resides in *any sort of interest or appreciation* of an object, event or state of affairs. Such appreciation involves feeling and ultimately desires, tendencies, or needs underlying the feeling. Therefore value *is* the feeling; *value and feeling of value are the same thing*. This is the psychological concept of value, and the propositions developed on this concept comprise the *psychological theory of value.*
>
> According to this theory, the measure of value is found in such conceptions as *intensity of feeling,* or the *strength* of the desire; such concepts are meant ultimately to reflect the importance of the psychological and biological needs and desires presupposed for establishing and maintaining the "good life." [4, p. 283]

Value is specific to individuals. It depends on a person's relative needs for security, pleasure, peer approval, aspirations, etc. In the previous contractor problem, value was assumed to be a linear function of dollars. This is typically not the case.

Assume that you and I have agreed to cut cards as a game of chance. Assume first that if I lose I pay you \$1, otherwise you pay me \$1. No difficulty exists as yet. Assume, however, that, having lost \$1 to you, I indicate that I have a train to catch and suggest that we cut one more time for \$20,000. Your expected gain is \$20,000(0.5) – \$20,000(0.5) = \$0. By employing the expected value criterion, it might seem that you would be as willing to cut cards for \$20,000 as \$1. What has slowed you down is consideration of the relative values of winning \$20,000 versus losing \$20,000. The pain of losing \$20,000 may be valued to be greater than the potential pleasure of winning \$20,000. Figure 8-14 represents a typical utility function for an individual. Not only do the psychological values of gains versus losses influence this nonlinearity, but the regressiveness of federal taxation adds another measure of nonlinearity.

Now let us assume that Fig. 8-15 is a utility function to be derived for the contractor with the roof problem. Then Fig. 8-16 is the decision tree for the contractor. It differs from Fig. 8-13 because it incorporates the added assumption of maximizing utility. Note that the dollar value of each outcome in Fig. 8-13 has been converted to a utility value by employing the utility function of Fig. 8-15. In terms of the contractor's utility, Fig. 8-16 indicates that it now appears more advantageous to install the permanent roof, accepting the loss of \$3400.

FIGURE 8-14 Utility function for an individual.

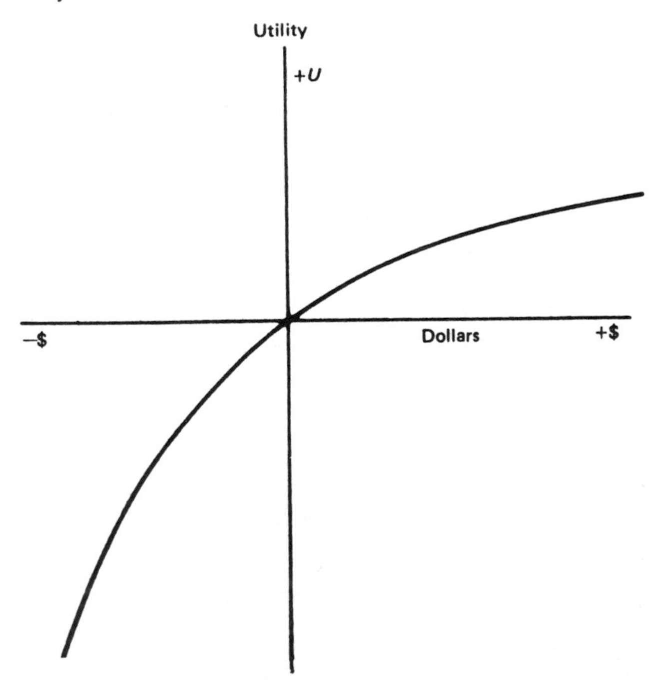

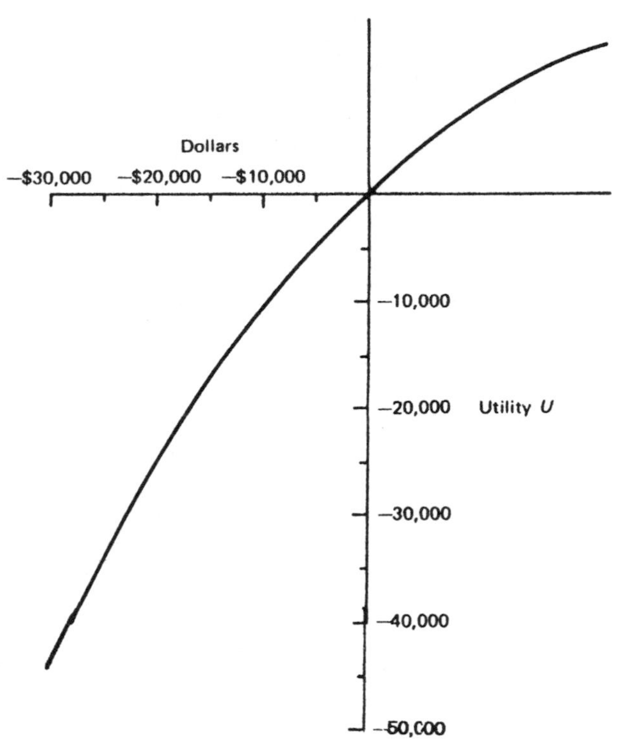

FIGURE 8-15 The contractor's utility function.

Of course, this method has utility only if we are able to derive a utility function for an individual. Note in Fig. 8-15 that the scaling of utility is such that –$30,000 is equal to –45,000 *U*s. The unit *U* is a measure of the advantages or disadvantages of a given outcome. If we use this as a starting point, it would be possible to ask what value the contractor would place on alternative B (i.e., the purchase of insurance) to equate it to alternative A in Fig. 8-17*a*. If the answer is $21,000, we can establish a point on the utility curve of Fig. 8-15 equating $21,000 to –27,000 *U*s. By a similar procedure, this point can be used as a basis for establishing a third point, as indicated in Fig. 8-17*b*.

Not only are utility functions specific to individuals, but individuals possess different utility functions for different resources. Added to these complexities is the fact that utility can vary significantly over a relatively short period of time, as is suggested by the utility curves in Fig. 8-18 for an individual named George. It should be apparent that the practical consideration of utility in industrial engineering problems is complex.

In the analysis of human systems it is not uncommon to need to specify a value or a functional relationship for a variable where data simply do not exist. In engineering circles, this is called making a *guesstimate*. A common way to provide an air of authenticity to a guesstimate is to have a group of people (i.e., a committee)

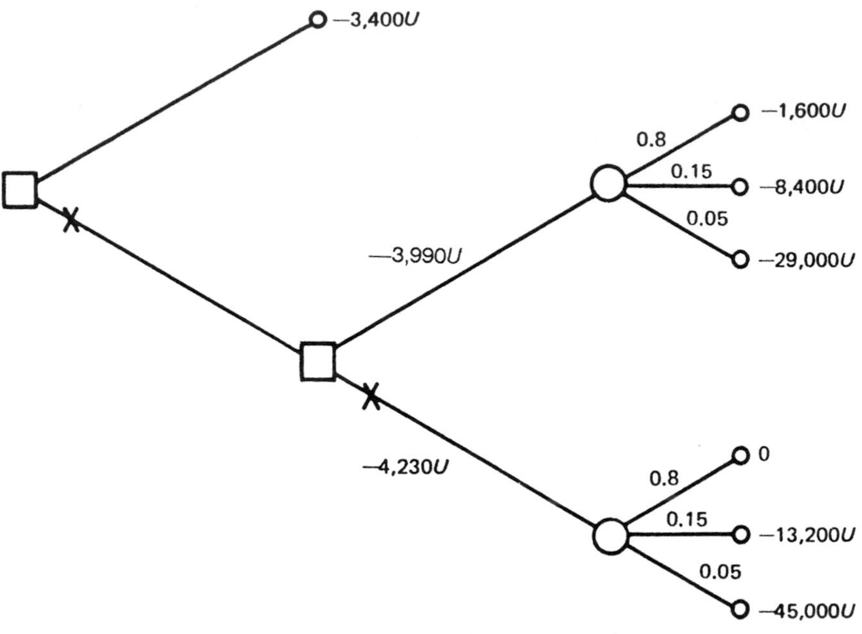

FIGURE 8-16 Decision tree employing utility for the contractor problem.

come up with it. Unfortunately, a group not so well informed as one individual in the group will typically produce a poorer estimate than the individual would have provided. In most committee environments, sociological and psychological factors are greater determinants of the outcome of the committee than logic or facts.

The need for expert opinion is often essential in systems studies, as indicated by Helmer:

> While model-building is an extremely systematic expedient to promote the understanding and control of our environment, reliance on the use of expert judgment, though often unsystematic, is more than an expedient; it is an absolute necessity. Expert opinion must be called on whenever it becomes necessary to choose among several alternative courses of action in the absence of an accepted body of theoretical knowledge that would clearly single out one course as the preferred alternative. This can happen if there is either a factual uncertainty as to the real consequences of the proposed courses of action, or, even if the consequences are relatively predictable, there is a moral uncertainty as to which of the consequent states of the world would be preferable. [5, page 11]

There are potential hazards in attempting to combine the preferences of individuals in the absence of adequate theory. One example, and a rather convincing one, is the voting paradox contained in Arrow's text:

> A natural way of arriving at the collective preference scale would be to say that one alternative is preferred to another if a majority of the community prefer the first

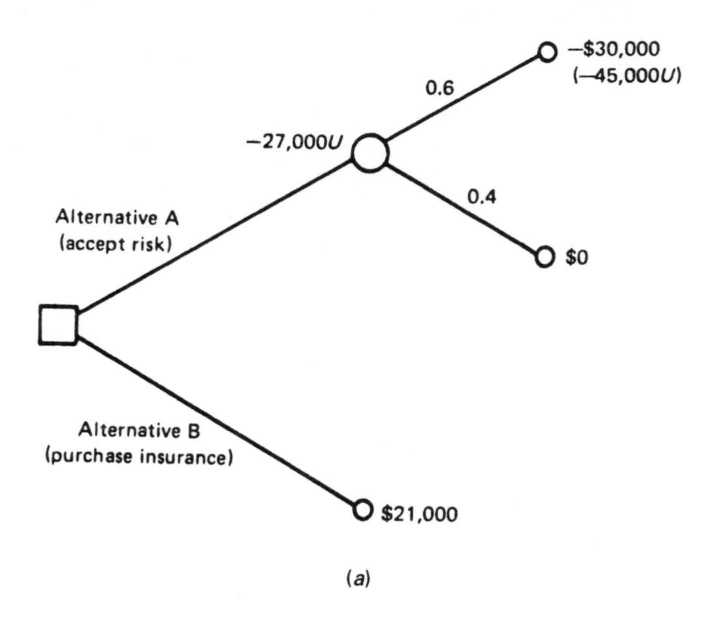

(a)

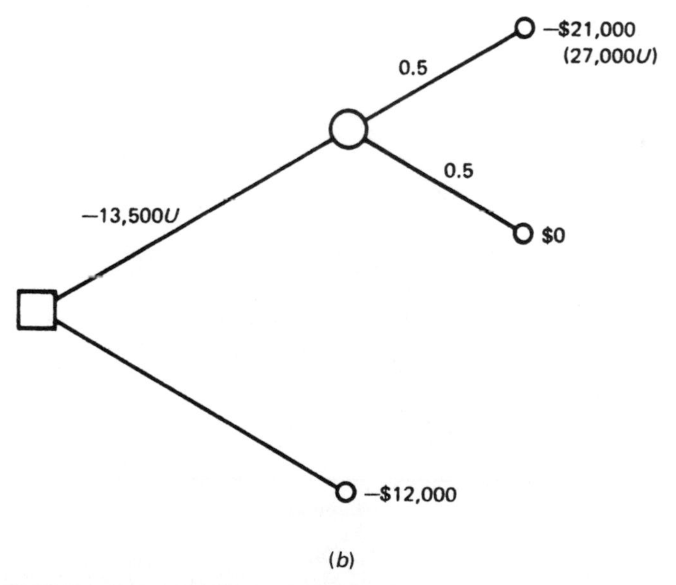

(b)

FIGURE 8-17 Decision tree derivations for a utility function.

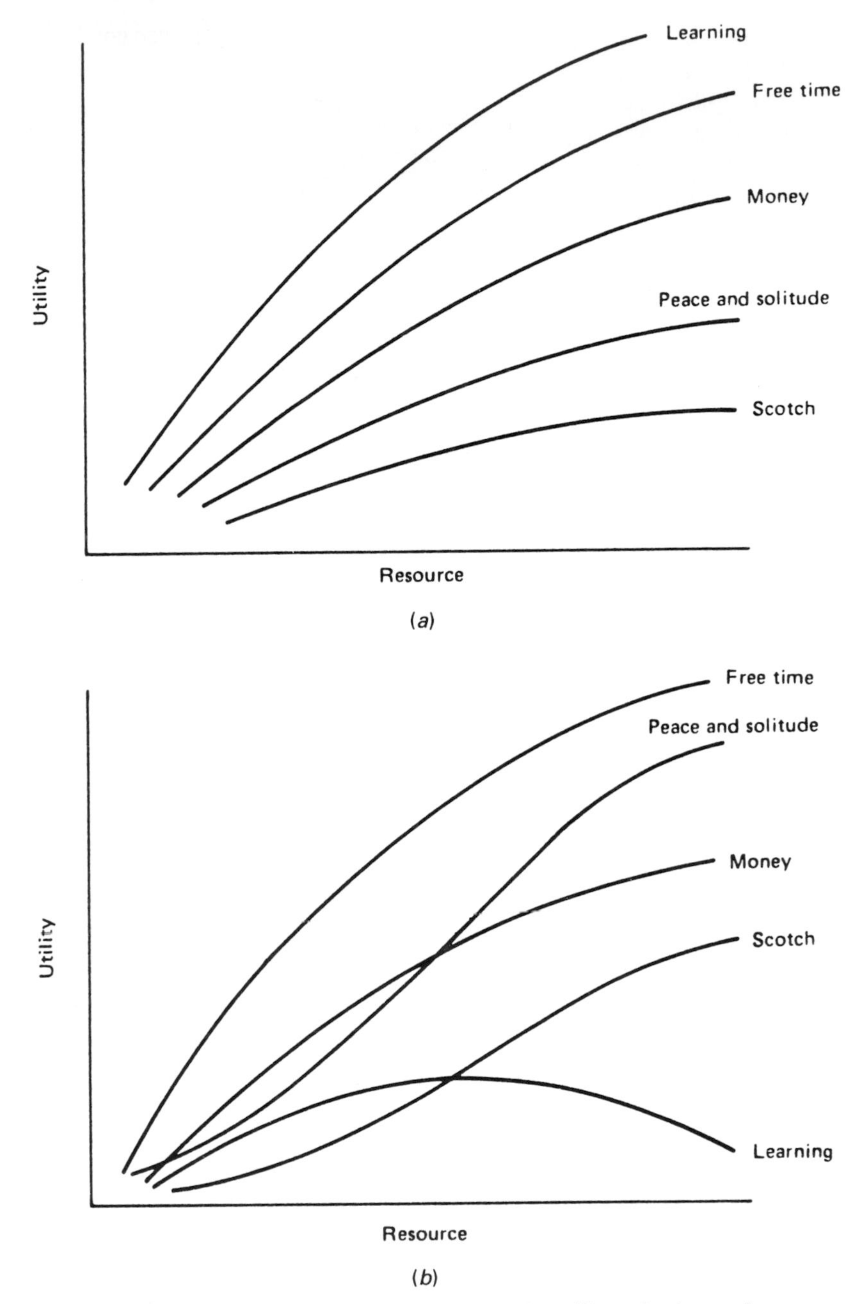

FIGURE 8-18 Utility curves for George: (*a*) incoming first-year student; (*b*) graduating senior.

> alternative to the second, i.e., would choose the first over the second if those were the only two alternatives. Let A, B and C be the three alternatives, and 1, 2 and 3 the three individuals. Suppose individual 1 prefers A to B and B to C (and therefore A to C); individual 2 prefers B to C and C to A (and therefore B to A); and individual 3 prefers C to A and A to B (and therefore C to B). Then a majority prefer A to B, and a majority prefer B to C. We may therefore say that the community prefers A to B and B to C. If the community is to be regarded as behaving rationally we are forced to say that A is preferred to C. But, in fact, a majority of the community prefer C to A. So the method just outlined fails to satisfy the condition of rationality as we ordinarily understand it. [1, page 3]

This would suggest that in systems study, as in life, things have a way of getting complicated.

COST CONTROL ENGINEERING

Inventory, production, and quality control are all concerned with developing procedures that ultimately affect overall cost control. Some control techniques can only be classified as cost control techniques, however.

Accounting is the language of cost control in manufacturing, and to be effective in appreciating cost effects as recorded in accounting one must know the language. Possibly as important as the language are the conventions embedded within the historical traditions of accounting. The recording and reporting of cost have long been a function of accountants, and only those familiar with their ways can properly decipher their results. Industrial engineers have traditionally been required to understand accountants and their methods, both in acquiring an education and in practice. Accounting is one of the dimensions in which industrial engineers have served in the role of translator from the accountant to technical management and vice versa.

As one example of accounting convention, it is common in cost accounting to prorate costs to products in some manner if costs were not originally recorded as required for each product in their consumption. Power cost in a plant, for example, may be paid as a single bill to the supplying electric utility by the accounts payable clerk within the accounting organization. In an effort to distribute this cost to reflect product costs for some period, a cost accountant would typically prorate this cost on some convenient basis. The square footage of plant space can readily be calculated for each product in the plant. Consequently, it is not uncommon for a product to bear its ratio of the power cost based on the fraction of plant space it occupies.

If products throughout the plant consume an essentially like amount of power per unit of floor space, proration on the basis of floor space may well be appropriate. If, however, as is often the case, some products require a much greater amount of power per unit of space than do other products, proration on the basis of space may distort the relative power cost allocation to products. In such a case, using clamp-on portable ammeters to measure the typical power demand in a product area could provide a much more equitable basis for power cost allocation.

In too many instances, accountants are too far removed from the production process to appreciate inequities they may have inadvertently introduced, and technical management may be too far removed from the allocation of product expenses to appreciate the basis for the product cost reports they receive. Industrial engineers often play an important role in bringing understanding between such disparate groups.

A recent development to improve the allocation of indirect and overhead costs is called activity-based costing (ABC). In past years there has been an overdependence on allocating costs as a function of a single variable, such as direct labor cost or plant square footage. In using ABC, costs are distributed based on appropriate "cost drivers," which correlate well with actual cost expenditures for each cost item (e.g., machine hours, floor space, number of setups, movement distances, size, amperage draw, etc.). With indirect and overhead costs representing the predominant costs of operation, a simplistic and, therefore, inaccurate allocation of such costs to products can greatly distort and degrade management decision-making ability. Product decisions need to be based on accurate cost information if a firm is to survive in an increasingly competitive world economy.

The following two examples are offered to demonstrate how standard accounting practice sometimes has to be overlooked or reevaluated in the best interests of the firm. The first example involves a company that had purchased a very expensive numerically controlled milling machine. The production demands on the machine and the associated operating costs were so great that it was decided that the cost of making any part on the machine first had to be compared with outside vendor quotations for producing the part. Only parts that could be produced more cheaply on the in-plant machine would be made on that machine, and all expenses for operating the machine would be borne by the parts being produced on the machine as an indirect charge.

A few years later, one part that constituted approximately one half of the load of the machine was suddenly discovered to be no longer needed. The remaining parts, which had constituted the other half of the loading on the machine, had to absorb the machine expense burden of the part no longer needed; consequently, the indirect portion of part cost doubled for the other parts. This so increased their calculated product cost that vendor-produced parts were now typically cheaper, resulting in lower and lower utilization of the machine and a higher and higher indirect cost burden for the remaining parts being made on the machine. Ultimately, apparent costs became so high for the remaining parts that management decided to dispose of the machine. Fortunately, this occurred at the same time that another technical group had submitted a proposal to management to purchase a new numerically controlled milling machine. When management reviewed the situation, the loss of one part of the load of the machine was seen to have excessively reflected infeasible in-plant costs for the remaining parts. All that was needed to remedy the situation was to shift some marginal in-plant versus vendor parts back to in-plant production to provide an adequate base for distributing the indirect expense. The example demonstrates how a make or buy decision may have to vary to suit the specific requirements at a particular time.

The second example shows how a method known as direct costing can aid the profitability of a plant in the short run. Assume for the purpose of this example that product pricing for a plant is calculated in the following manner. To the direct costs of material and labor to produce a product, one half of the cost of labor is normally added to cover all overhead costs, and an additional 10 percent of the total of the preceding costs is added for profit. A typical product would be priced as follows:

Product A Costs

Unit material cost	$10
Unit labor cost	20
Overhead (50% of unit labor cost)	10
Total manufacturing cost	40
Profit (10% of total manufacturing cost)	4
Unit price	$44

On the basis of typical accounting methods, if a potential purchaser offers to purchase 1000 units of Product A, which could be produced next month for a unit price of $41, the order would not be accepted. If the plant is scheduled at full capacity for next month, this action would seem to be justified.

If we assume, however, that the plant is presently scheduled for 70 percent of normal capacity for next month; the order for 1000 units of Product A would bring the plant to 90 percent capacity; and the purchase price offered is less than $44. Assume also that your employees are highly skilled, and letting them go for lack of demand for one month would not be wise (i.e., by letting them go you might lose them and have to hire other less qualified employees and suffer the associated training and quality effects of hiring new employees). Should the offer be accepted?

To answer this question, let us assume first that the offer is $40 per unit. Why would we accept an offer that does not add to profit? The answer is that in accounting a term called *contribution to profit and overhead* is used. Even though such an order will not produce profits in the traditional sense, it will help pay some of the overhead expenses of the plant; in fact, to the amount of 1000 × $10 = $10,000.

To stretch the point considerably, assume that the offer is $15 per unit; could there be any logical reason for accepting the offer? The answer is yes. If the plant has been operating at 90 percent capacity, dropping to 70 percent capacity would suggest that we either have to lay off employees next month for lack of work or pay them for doing nothing. Many employees today are so highly skilled that an employer wants to maintain full employment; otherwise, the employer has to run the risk of losing trained employees and has to suffer the considerable cost of training new ones—if they can be found. If the employer intends to keep all the employees next month, the $15 unit purchase cost can be viewed as offering a $5 per unit *contribution to labor, overhead, and profit* to the extent of 1000 × $5 = $5000.

Of course, an order for 1000 units at a $15 purchase price would only be accepted to keep the plant at nearly full capacity over the short run. In fact, having the customer pay for the material (i.e., $10/unit) that employees would use while maintaining production momentum and employee morale and retaining experienced employees is not out of the question. The sales manager would likely make it clear to the lucky purchaser that this would be a one-time deal that might never happen again, and why. For the purchaser, it would be a little like Christmas in July. In the final analysis, they both could conceivably come out ahead on the deal.

Another example of standard accounting practices is the overdependence of cost accountants on variance accounts to guide their decision making. A consultant was once asked to assist the cost accounting department to determine why at year's end there was always an approximate $50,000 negative variance in a particular scrap account. The industrial engineering consultant investigated the reason for the typical year-end negative variance and discovered that the reason was a misinterpretation of scrap information as part of the scrap reporting system. As a result of this clarification, year-end negative variances should no longer occur. In the final report the consultant explained what the problem was and how it was corrected to report scrap operations properly in the future. More importantly, however, the consultant raised the question of why the absolute annual scrap cost was so unreasonably high. A study of scrap was undertaken and a considerable opportunity for scrap reduction was uncovered. Reliance on out-of-date variance account measures as a guide to efficient operations is an area of considerable opportunity for cost improvement in many firms. Absolute values have to be reevaluated on a regular basis. Sitting in the office looking at variance accounts is not the answer.

The concepts mentioned here are certainly not unknown to accountants. The purpose of this discussion is to demonstrate that normal or standard accounting approaches need to bc adjusted to fit the occasion, and that the industrial engineers who bridge the gap that sometimes exists between accounting or sales and technical management can sometimes see the need for adjustments because of their unique vantage point. This may partially explain why the industrial engineering departments in IBM manufacturing plants have been referred to in the past as the "business manager of the plant."

A much more thorough discussion of ABC costing, entitled "Activity Based Costing and Controls," by McCormick appears in Section 9, Chapter 3, of *Maynard's Industrial Engineering Handbook* [6].

SUMMARY

Engineering economy is a traditional and important part of engineering practice. All engineering projects have costs associated with their activities, and these costs must be appropriately analyzed and considered. Advanced techniques are available today to deal with complex engineering problems (such as multi-dimensional criteria and inflation) in texts such as Canada and Sullivan [2].

One particularly important area of opportunity for cost control today is ABC (activity-based costing). Such a technique requires cost accounting to get better estimators for the proper allocation of indirect and overhead costs than the simplest allocation means employed in the past (e.g., square footage used by departments). Under ABC, cost accounting identifies and utilizes cost drivers that provide a more accurate allocation of specific indirect and overhead costs.

REFERENCES

1 Arrow, K. J.: *Social Choice and Individual Values,* John Wiley, New York, 1951.
2 Canada, John, and William Sullivan: *Economic and Multiattribute Evaluation of Advanced Manufacturing Systems,* Prentice Hall, Englewood Cliffs, NJ, 1989.
3 Grant, Eugene L.: *Principles of Engineering Economy,* Ronald Press, New York, 1930.
4 Hall, A. D.: *A Methodology for Systems Engineering,* Van Nostrand, Princeton, NJ, 1962.
5 Helmer, Olaf: *Social Technology*, Basic Books, New York, 1966.
6 McCormick, Edmund J., Jr.: "Activity Based Costing and Controls," in W. K. Hodson (ed. in chief), *Maynard's Industrial Engineering Handbook,* 4th ed., McGraw-Hill, New York, 1992.

REVIEW QUESTIONS AND PROBLEMS

1 If you put $3000 in a bank today that charges no fees and pays a fixed annual compounding interest rate of 6 percent, what is the total amount that you can withdraw from this account 10 years from now?

2 You were supposed to have received $10,000, six years ago. Instead, you were paid $2,169 at the end of each of the past six years. What was the annual compounding interest rate implicit in these payments?

3 How long will it take for an investment to double in value if the annual compound interest rate is 8 percent?

4 An airplane will require $30,000 of maintenance each year for the next 10 years and in the fifth year will require an additional maintenance payment of $120,000 for a major overhaul. Assume amounts during a year become due at the end of that year. If these funds were to be provided by an up-front investment to be made in an account that pays 6 percent annual compounding interest, what initial investment would be required to cover these future costs?

5 You can purchase a forklift for $30,000 that will have a 10-year economic life, at which time it will have an estimated salvage value of $3000. Annual operating expenses for the forklift truck are $2000. An alternative is to lease the same forklift truck, including all operating expenses, for $8000 per year. If your MARR is 20 percent, should you buy or lease?

6 You bought a baseball card eight years ago for $100, and you were just offered $430 for the card. If you sell it now for that amount, what will your return on the investment be?

7 You bought an antique car 10 years ago for $30,000. You had hoped when you purchased the car that you could earn a 12 percent return on that investment. It has been in your uncle's barn for the past 10 years at no cost to you. Someone has just offered you $100,000 through a dealer, and his commission would be 6 percent of the purchase price. If you sell it at that price, will you have made a 12 percent rate of return on the investment before taxes?

8 Mr. Johnson is considering selling ice cream and popcorn at five summer camping areas for the summer. One choice is to use his company trucks to do so. The expenses associated with the

use of these trucks for the summer are estimated to be $1000. Another choice available is to rent trucks for $3000 for the summer. A third choice, unappealing because Mr. Johnson does not particularly like his brother-in-law, is to use his brother-in-law's trucks. He estimates that his brother-in-law would expect remuneration for the use of his trucks of approximately $1000. Mr. Johnson has estimated the income he would expect to receive from employing each of the three types of trucks as follows:

Weather	Probability	Company trucks	Brother-in-law's trucks	Rental trucks
Excellent	.3	$24,000	$22,000	$28,000
Good	.5	20,500	17,000	23,000
Bad	.2	12,200	8,000	12,000

The incomes shown for using his brother-in-law's trucks assume that Mr. Johnson's wife will not help him, for which he estimates there is a 70 percent chance. If she does help, he will save $4000 net, regardless of the weather because of her unpaid labor for the summer. Which choice should he make?

9 Ms. Wall intends to sell stock in a company in which she holds shares within a week because she expects the company to go into bankruptcy within a year. If she sells today she will lose $23,000. If she waits until the end of the week there is a 25 percent, 35 percent, and 40 percent chance that her loss will be $10,000, $18,000, and $30,000, respectively. After this week, Ms. Wall expects the value of the stock to decline until bankruptcy occurs. If Ms. Wall's utility function is identical to that of the contractor in Figure 8-15, which choice should she make to maximize her utility?

10 What is meant by direct costing?

11 What is activity-based costing and what advantages does it offer over traditional cost accounting methods?

9

SYSTEMS

All the flowers of all the tomorrows are in the seeds of today.

Chinese Proverb

The word *system* means different things to different people.To an IBM systems engineer, a system is a computer system. To an electrical engineer, a system may be a collection of electromechanical components, such as the flight control or fire control systems in an aircraft. For a mathematics professor teaching a course in matrix theory, a set of simultaneous equations may be referred to as a *system of equations*.

Another sometimes confusing aspect of systems is that all systems, with rare exception, are part of a larger system. The fire control computer in an airplane is part of the offensive and defensive systems of an airplane. The airplane is also part of the defensive or offensive striking power of some segment of a specific military service. This service is also part of our national defense system, which is a major component of our national economy. Our economy is a significant component of the world economy, and world governmental action may well determine the future of our planet as a component of our universe. How many universes are components of larger universes or cosmoses (or whatever words one chooses to use to describe conceivably larger systems)?

The film *Orders of Magnitude* dramatizes this limitless range of systems within systems [10]. The problem, familiar to systems engineers, is the need to establish the boundary of a system carefully; otherwise, one must seemingly model the world to consider all potentially related factors. Of course, the establishment of a system boundary is more a necessary simplification than a reality.

Churchman began his classic text *The Systems Approach* with the following: [3, pp. 3–4]

> Suppose we begin by listing the problems of the world today that "in principle" can be solved by modern technology.
>
> In principle, we have the technological capability of adequately feeding, sheltering and clothing every inhabitant of the world.
>
> In principle, we have the technological capability of providing adequate medical care for every inhabitant of the world.
>
> In principle, we have the technological capability of providing sufficient education for every inhabitant of the world for him to enjoy a mature intellectual life.
>
> In principle, we have the technological capability of outlawing warfare and of instituting social sanctions that will prevent the outbreak of illegal war.
>
> In principle, we have the capability of creating in all societies a freedom of opinion and a freedom of action that will minimize the illegitimate constraints imposed by the society on the individual.
>
> In principle, we have the capability of developing new technologies that will release new sources of energy and power to take care of physical and economic emergencies throughout the world.
>
> In principle, we have the capability of organizing the societies of the world today to bring into existence well-developed plans for solving the problems of poverty, health, education, war, human freedom and the development of new resources.[1]

If we can, why haven't we? Is it because we lack the technology to do so? If technology means equipment and production methods, the answer is no. If technology means "Could we appropriately implement solutions to these problems if we possessed the capital and equipment?" the answer is "maybe." Attaining results is what systems is all about. Stated another way, we simply have more gadgets today than we know what to do with. It is not so much a question of coming up with a better mousetrap. Rather, the more relevant question is, given a full range of available mousetraps, when, where, and under what conditions should mouse traps be employed to best serve the common good with the limited resources available?

The automobile, the airplane, the computer, and the nuclear reactor are typical systems of things. They represent engineering's best in the 20th century. The second half of the 20th century has seen a shift toward an additional interest in human systems (i.e., systems in which human skill and performance determine system success). For example, who picks up your garbage? When is it picked up? Why does the garbage collector pick it up? Is it a cost-effective system? How many systems are there in your everyday life that, for lack of technical analysis, have been passed down from one generation to the next with inadequate analysis as to how truly effective they are with respect to those they are intended to serve? Churchman puts it this way: "The optimistic management scientist looks forward to a 'systems era,' in which man at last will be able to understand the systems he has created and lives in" [3, p. 43].

[1]Excerpt from *The Systems Approach* by C. West Churchman.

The systems approach is by no means new; politicians have been promising it for a very long time. Churchman wisely stated:

> The idea of a "systems approach" is both quite popular and quite unpopular. It's popular because it sounds good to say that the whole system is being considered, but it's quite unpopular because it sounds either like a lot of nonsense or else downright dangerous—so much evil can be created under the guise of serving the whole. [3, p. 11]

For years, the emphasis of traditional industrial engineering has been on "efficiency." How can we do the same thing faster? What fixtures and tools and what sequence of fundamental hand motions will increase output at a workstation? Only recently has there been a shift toward creative system design. Of course, that is not to say that efficiency is no longer needed. Efficiency is necessary, but not sufficient, in solving the problems posed by Churchman. The purely authoritarian labor control techniques of the past are inadequate, as have been the permissive and naive purely "humanistic" approaches of the recent past. Between these extremes lie an appropriate combination and compromise.

SYSTEM OBJECTIVES

Most systems authors seem to agree that the most important step in systems analysis is the determination of objectives. Churchman stated, "The objectives of the overall system are a logical place to begin, because, as we have seen, so many mistakes may be made in subsequent thinking about the system once one has ignored the true objectives of the whole" [3, p. 30]. It is disheartening to students and to practicing engineers and scientists alike to discover that they have an excellent solution to the wrong problem. Hall emphasized this point: "Yet it is much more important to choose the 'right' objectives than the 'right' system. To choose the wrong objective is to solve the wrong problem; to choose the wrong system is merely to choose an unoptimized system" [7, p. 105].

The fundamental problem with establishing objectives is that objectives implicitly involve value judgments. Hall stated, "The operation of setting objectives (alternatively, the design of a value system) is truly the key to good systems engineering; it summarizes all the results of environmental research guides and sets the standards for all future work, provides the means for optimizing systems, and provides the rules for choosing among alternative systems" [7, p. 229].

The determination of objectives is sufficiently difficult that it is far too often given only cursory attention in systems studies. Hall identifies the source of difficulty as follows: "Unfortunately, there is no general theory of value, though many individuals and groups in all walks of life, including science and engineering, pretend to have a philosophy, system, code or a pipeline to God that permits them to make end judgements as simply as they make instrumental judgements" [7, p. 253]. The lack of a value theory does represent a significant factor limiting success in systems studies; however, complete avoidance of this aspect of the problem is not the solution. The first look at a systems problem usually uncovers a short list of primary objectives. It is not uncommon for there to be more than one primary objective. Additional consideration typically results in the identification of other secondary objectives as well. Rarely, for example, is money the only dimension to the problem. Churchman refers to typical secondary factors such as aesthetics, recreation, and health. The author recalls some associates instrumental in the establishment of a fried chicken fast-food outlet in Heidelberg, Germany. They eventually went out of business, but who is to say that they managed their business badly considering the many friends they made, the recreational opportunities available to them there, the European culture they enjoyed, or the phenomenal amount of delicious German food and wine they consumed while they were there. It really depends on what their objectives and their values were.

Profit provides an inadequate measure of success. Individuals responsible for profit typically have aspirations quite different from the assumed corporate goal of maximum profits. Rensis Likert, in his theory of supportive relationships described in Chapter 4, suggests that management should attempt to develop management objectives that are compatible with the objectives and aspirations of its employees [9, p. 103]. To the extent that incompatibility exists, conflict will diminish the desired outcome for both individuals and the firm. In a negative sense, the essence of this theory might be paraphrased as, "It is practically impossible to get anybody to do anything unless they want to." The management approach then is not to make people do things, but rather to provide an environment in which what employees want to do is consistent with what management wants them to do. Numerous studies have shown that people are concerned with such factors as peer respect, power, and personal satisfaction, with money not at the top of the list. In human–machine systems, it is important to optimize some combination of all of these motivating factors.

Concern with needs leads one to needs research. Human systems typically involve satisfaction of some combination of human needs. How does one determine what needs should be satisfied?

The optimum system often does not result from operating individual components in what appears to be the optimum fashion for each component. As an example, a public utility group funded the Battelle Institute in Columbus, Ohio, to perform a system study of the Susquehanna River basin to determine optimum future development of the basin [8]. Experts were engaged to study various aspects of the problem. The specialties represented included demography (population), water pollution, recreation, economic development, hydroelectric power, and others. Some time later these experts met and found that they had vastly different and often opposing views as to what was best for the river basin. An industrial dynamics simulation model became the unifying mechanism for combining the data from the various specialties represented. As is true with many systems studies, the optimum solution for the system was at best a compromise of all interests represented.

One of the key problems that arose in the preceding study, which is common to many systems studies, was the need to determine just who were the "customers" in the study. The determination of customers can be sufficiently difficult that some systems are best studied without making value judgments as to desired outcomes. The author was involved in a systems study in which it was considered appropriate to attempt to link expected outcomes to actions and to let the local political process determine which combination of actions and outcomes, in light of benefits and costs, was most desirable for the common good. The study group, therefore, concentrated on identifying expected outcomes based on assumed inputs to the system, which was a far more manageable problem than valuing outcomes.

In a typical systems study, as in the foregoing example, one must deal with a set of outcomes (e.g., multiple objectives). Table 9-1 is a detailed list of multiple objectives offered in a military policy planning problem in Quade and Boucher (12, p. 399). If a particular set of actions with respect to some system were to produce various levels of accomplishment in relation to these criteria, there are two immediate problems: (1) How do we measure with respect to these attributes? If we could measure, how would we define the scales upon which we would measure? and (2) How do we rank and combine values obtained with respect to individual criteria to provide some single useful measure of success with respect to the overall mission of the system?

The specific area of decision theory concerned with defining, determining, scaling, measuring, and combining multiple criteria measures is an important frontier today. Texts such as *Economic and Multiattribute Evaluation of Advanced Manufacturing Systems* by Canada and Sullivan [2], or Section 14, Chapter 7, entitled "Multi-Criteria Decision Making" in *Maynard's Industrial Engineering Handbook*, by Reeves and Lawrence [13] are clarifying and extending this critical body of knowledge.

TABLE 9-1 MULTIPLE CRITERIA IN A POLICY PLANNING PROBLEM

	Criteria
1 Destructive potential	How well can the force mix destroy targets?
2 Responsiveness	How rapidly can the force mix be ready for military action?
3 Deployability	How rapidly can the force mix move to different theaters?
4 Mobility	How rapidly can the force mix move in the theaters?
5 Supportability	How effectively can the force mix be supported and maintained?
6 Survivability	How vulnerable is the force mix to enemy actions?
7 Flexibility	How many different postures or capabilities can the force mix employ?
8 Controllability	How responsive is the force mix to command requirements?
9 Complementarity	How well does the force mix complement the forces of our allies?
10 Versatility	How effective is the force mix in a variety of military and politicomilitary situations and crises?
11 Deterrent capability	How much does the force mix contribute to our ability to deter aggression?
12 Expandability	How fast can additional capability be mobilized for the force mix?
13 National acceptability	How readily will the force mix be accepted domestically?
14 International acceptability	How readily will the force mix be accepted by other nations?

Source: Quade and Boucher [12, p. 399].

TABLE 9-2 A PROCEDURE FOR DETERMINING RELATIVE VALUES OF FOUR OUTCOMES

1 Rank the four outcomes in order of importance. Let O_1 represent the outcome that is judged to be the most important O_2 the next, O_3 the next, and O_4 the last.

2 Tentatively assign the value 1.00 to the most valued outcome and assign values that initially seem to reflect their relative values to the others. For example, the evaluator might assign 1.00, 0.80, 0.50, and 0.30 to O_1, O_2, O_3, and O_4, respectively. Call these tentative valuess V_1, V_2, V_3, and V_4, respectively. These are to be considered as first estimates of the "true" values V_1, V_2, V_3, and V_4.

3 Now make the following comparison:

$$O_1, \text{ versus } (O_2 \text{ and } O_3 \text{ and } O_4)$$

i.e., if there was a choice of obtaining O_1 or the combination of O_2, O_3, and O_4, which would the evaluator select? Supppose the evaluator asserts that O_1 is preferable. Then the value of v_1 should be adjusted so that

$$v_1 > v_2 + v_3 + v_4$$

For example: $v_1 = 2.00$, $v_2 = 0.80$, $v_3 = 0.50$, and $v_4 = 0.30$. Note that the values of O_2, O_3, and O_4 have been retained.

4 Now compare O_2 versus (O_3 and O_4). Suppose (O_3 and O_4) is preferred. Then further adjustment of the values is necessary. For example: $v_1 = 2.00$, $v_2 = 0.70$, $v_3 = 0.50$, and $v_4 = 0.30$.

5 In this case, the evaluations are completed. It may be convenient, however, to "normalize" these values by dividing each by Σv_i. which in this case is 3.50. These standardized values are represented by v'_i

$$v'_1 = 2.00/3.50 = 0.57$$
$$v'_2 = 0.70/3.50 = 0.20$$
$$v'_3 = 0.50/3.50 = 0.14$$
$$v'_4 = 0.30/3.50 = \underline{0.09}$$
$$\text{Total} \quad 1.00$$

Source: Churchman et al. [4, pp. 139–140].

One of the simplest approaches to determining relative values of outcomes, referred to as "Procedure 1" in Churchman et al. [4], is given for four outcomes as Table 9-2.

The foregoing procedure is suggestive of the types of procedures or techniques that must be developed, if only empirically, for valuing system outcomes.

At least qualitatively, computer company systems engineers offer a modern example of the necessity to view problems in a systems sense. In fact, a computer system salesperson must deal with the reality of a system within a system. The first system is that combination of computer components (i.e., a system of things) that meets the customer's needs. The second system, the human–machine system, must be understood with respect to such factors as (1) How will the initial investment be recovered? (2) How many people (if any) will it replace? (3) What will become of the people it replaces? (4) Will it be reliable? (5) If the power goes out, where does it leave us?

In sociotechnical human systems there are not only multiple criteria but also multiple customers. As will be shown later, there is a further complication in the sense that each individual possesses a unique set of "utility functions" with respect to the satisfaction of his or her needs. For example, consider the relative negative value of a $25 traffic ticket in two different situations. Assume in the first situation that a soldier is stopped on Sunday 400 miles from home after driving for a day on a three-day pass and has only $18 available. In the second case consider a millionaire stopped for the same violation. The first case may well represent a major setback, the second a temporary inconvenience.

In any case, multiple outcomes, multiple criteria, multiple customers, and multiple utility functions complicate systems analysis.

The following example from Quade and Boucher is presented early in their book to attempt to show some of the complexities that exist in the relationships between the major system elements: objectives, alternatives, costs, effectiveness scales, effectiveness, and criteria [12]. This rather short example—old but worthy—does effectively highlight some of these underlying complexities.

An Example: Selection of a New Aircraft Engine*

E. S. Quade
W. I. Boucher

As an example of how these elements of analysis figure in a relatively narrow decision problem, let us consider the selection of a new aircraft engine, and assume that the *objective* is simply to increase engine performance. Then the *alternatives* are obviously the various possible engine types that achieve this objective by such means as exotic fuels or novel design. The *costs* would be of two general kinds: the total capital resources (such as manpower and research facilities) that must be allocated to the research, and the time required to achieve a successful prototype. In this simple case, the *effectiveness scale* relates directly to the objective, and might be taken as the difference between the specific fuel consumption typical today and that achieved by further research, for fixed engine weight. The *effectiveness* of a particular alternative engine type would then be its estimated reading on this scale. The greater the difference, the better the engine, since we desire to decrease the specific fuel consumption by research. In general, the amount of improvement will depend upon the amount of effort expended upon research, so that estimated costs and effectiveness might be related as shown in Figure 1.

Such different levels of performance might result from a situation [to be discussed more fully later] in which alternative 1 corresponds to a very conservative improvement over operational engines, and alternative 2, to a larger state-of-the-art advance.

Note, however, that even if we assume that both these alternative research programs can be completed on time and are subject to essentially the same amount of uncertainty, we still could not decide between them. What is missing is some knowledge of why the improved performance is needed. Thus, although alternative 1 achieves only a modest level of effectiveness (E_1), it does so at one-third the cost

FIGURE 1 Cost and effectiveness.

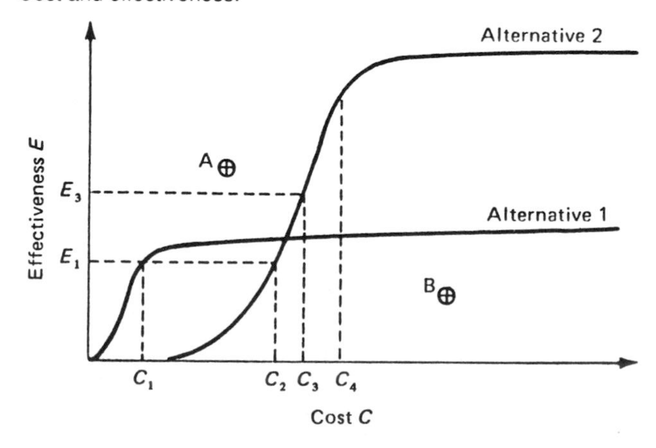

Source: Quade and Boucher [12, pp. 55–59].

of alternative 2. If the level E_1 is adequate, why not select alternative 1 and thereby minimize cost? Indeed, quite often cost will be limited by decree to some level such as C_2, in which case alternative 1 is the obvious choice. On the other hand, the goal of the research may be to achieve some minimal new level of effectiveness, such as E_3, no matter what the cost. Then alternative 2 is obviously the choice.

The point to be made is that, in general, it is not possible to choose between two alternatives just on the basis of the cost and effectiveness data shown in Figure 1. Usually, either a required effectiveness must be specified and the cost minimized for that effectiveness, or a required cost must be specified and the effectiveness maximized. Clearly, the results of the analysis of effectiveness should influence the selection of the final criterion. For example, if C_3 is truly a reasonable cost to pay, then the case for C_4 is much stronger, in view of the great gains to be made for a relatively small additional investment. As a matter of fact, this approach of setting *maximum* cost so that it corresponds to the knee of the cost-effectiveness curve is a very useful and prevalent one, since very little additional effectiveness is gained by further investment.

OVERSPECIFICATION OF CRITERIA

On the other hand, both required cost *and* effectiveness should not be specified; this overspecifies the criterion, and can result in asking for alternatives that are either unobtainable (point A in Figure 1) or underdesigned (point B in the same figure). An extreme case of criterion overspecification is to require maximum effectiveness for minimum cost. These two requirements cannot be met simultaneously, as is clear from Figure 1, where minimum cost corresponds to zero effectiveness, and maximum effectiveness corresponds to a very large cost.

MAXIMIZING EFFECTIVENESS/COST

Somewhere in the middle are criteria that apparently specify neither required cost nor effectiveness. One which is widely used calls for maximizing the ratio of effectiveness to cost. This seems to be a workable criterion, since, in general, we want to increase effectiveness and decrease cost. Nevertheless, as we can see by examining Figure 2, it has a serious defect. Since the effectiveness-cost ratio for either alternative is simply the slope of a line drawn from the origin to a given point on the curve for that alternative, and since, in this example, the ratio obviously takes on a maximum at the knee of the curve, our choice between the two alternatives seems to be settled at once. Thus, alternative 1 is clearly preferred with this criterion. However, if E_3 is the minimum level of effectiveness acceptable from a research program, then alternative 2 is the obvious choice. The point to be made here is that unless the decision maker is completely unconcerned about *absolute* levels of effectiveness and cost, then a criterion such as this, which suppresses them, must be avoided.

Theoretically, it is possible to escape this need for specifying either the required cost or effectiveness by expressing cost and effectiveness in the same units, such as dollars or equivalent lives saved. For if this can be done, then it is possible to

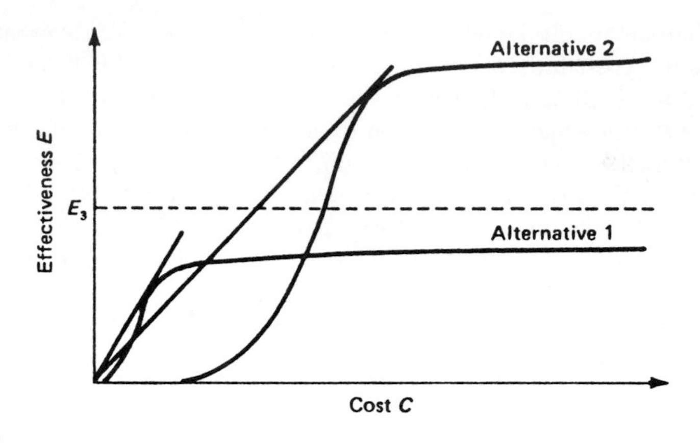

FIGURE 2

subtract cost from effectiveness, and take as the criterion the maximization of this difference. But seldom, if ever, can cost and effectiveness be expressed in similar units, and we may assume that the earlier description of a criterion applies.

DOMINANCE

Infrequently it happens that selection between alternatives is easy. An extreme case of this is shown in Figure 3, and occurs when an alternative such as 3 is more effective than any other at every cost. In such a case it is clearly advantageous to select alternative 3, which is said to *dominate* alternatives 1 and 2 at all levels of investment and effectiveness. Note that it is still not permissible to overspecify the criterion and require maximum effectiveness for minimum cost. For the situation of dominance only permits us to select alternative 3; minimum cost still corresponds to

FIGURE 3

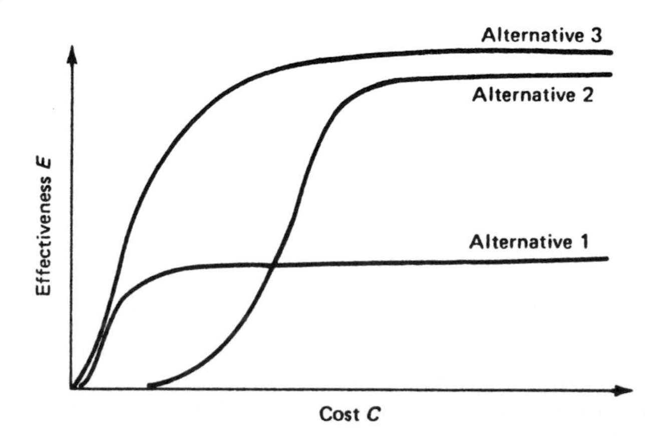

zero effectiveness for alternative 3, and so forth. Even though dominance designates alternative 3 as preferred, the required level of effectiveness must be specified before the preferred level of investment can be selected.

In this example of propulsion research, as in many others in advanced research or specific component design, the goal has been simple and obvious. Further, in such cases an appropriate scale of effectiveness is usually obvious and related directly to the goal. Finally, the measurement of effectiveness (that is, the location of an alternative upon that effectiveness scale) is straightforward in such cases.

The preceding article makes the point that selection between alternatives in a multiple criteria environment can be a complex task.

EXPERT SYSTEMS

Expert systems are one of many subsets of a broader field of technical inquiry called artificial intelligence (AI), as suggested in Figure 9-1.

The term *artificial intelligence* was coined by John McCarthy at a conference at Dartmouth College in 1956. Dr. Dell Allen of Brigham Young University defines artificial intelligence as follows: "AI is the combination of high performance processes and programs which, employing a knowledge base, can solve problems skillfully enough to equal or surpass the performance of human experts in a specific knowledge domain, while the knowledge base provides the information required to solve useful problems." [1]

Artificial intelligence has been a topic of intellectual discussion, and fruitful applications have been predicted for many years. One of the earlier examples of artificial intelligence that I recall was the creation of scripts for Class B Western movies. By viewing plots and deducing sequential events in such movies, and then generating discrete density functions concerning these sequential events, it becomes fairly easy probabilistically to generate conceivable movie plots. Employing a

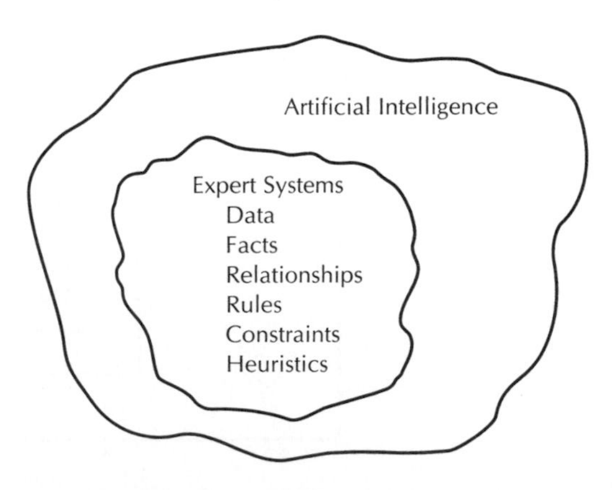

FIGURE 9-1 Expert systems and artificial intelligence.

random number generator, one can then let the computer generate Western movie scripts. In one study, actors were hired and the scripts were acted out and filmed. The lack of robustness in the system was excruciatingly obvious when during the movie some discontinuity would occur unexpectedly, for example, the hero might go from riding his horse to running alongside his horse in pursuit of the fair maiden's runaway stagecoach, with no explanation offered of how he got there. This type of script did provide an additional source of humor to the film, however.

FIGURE 9-2 Expert system development shell. (*From Parks* [11, p. 56]).

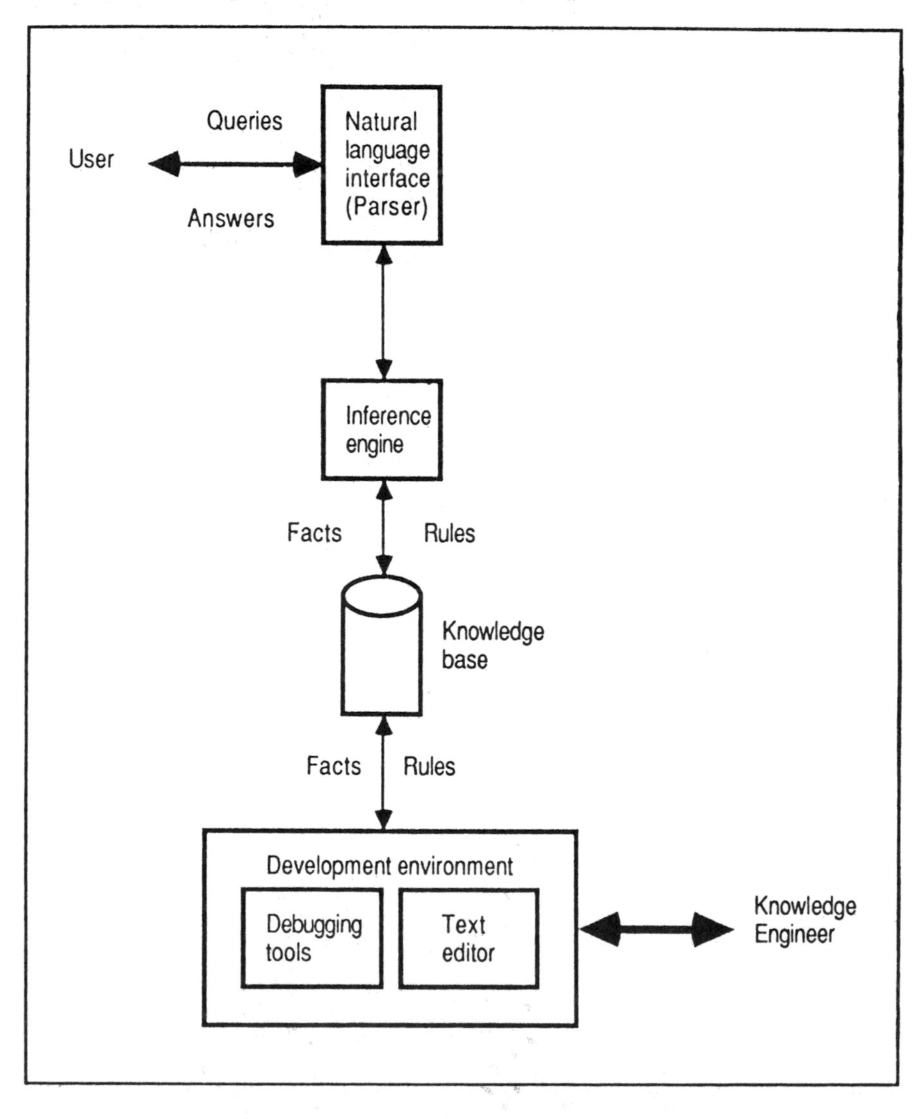

Expert systems are one of the subsets of artificial intelligence that is paying off today. It is the way of the future for dealing with manufacturing information and decision making. The following article by Michael W. Parks describes the use of an expert system to assist in the assembly of aircraft and is typical of expert systems in general [11]. Figures 9-2 and 9-3 graphically relate information flows in this system.

FIGURE 9-3 Aircraft assembly expert system. (From Parks [11, p. 57]).

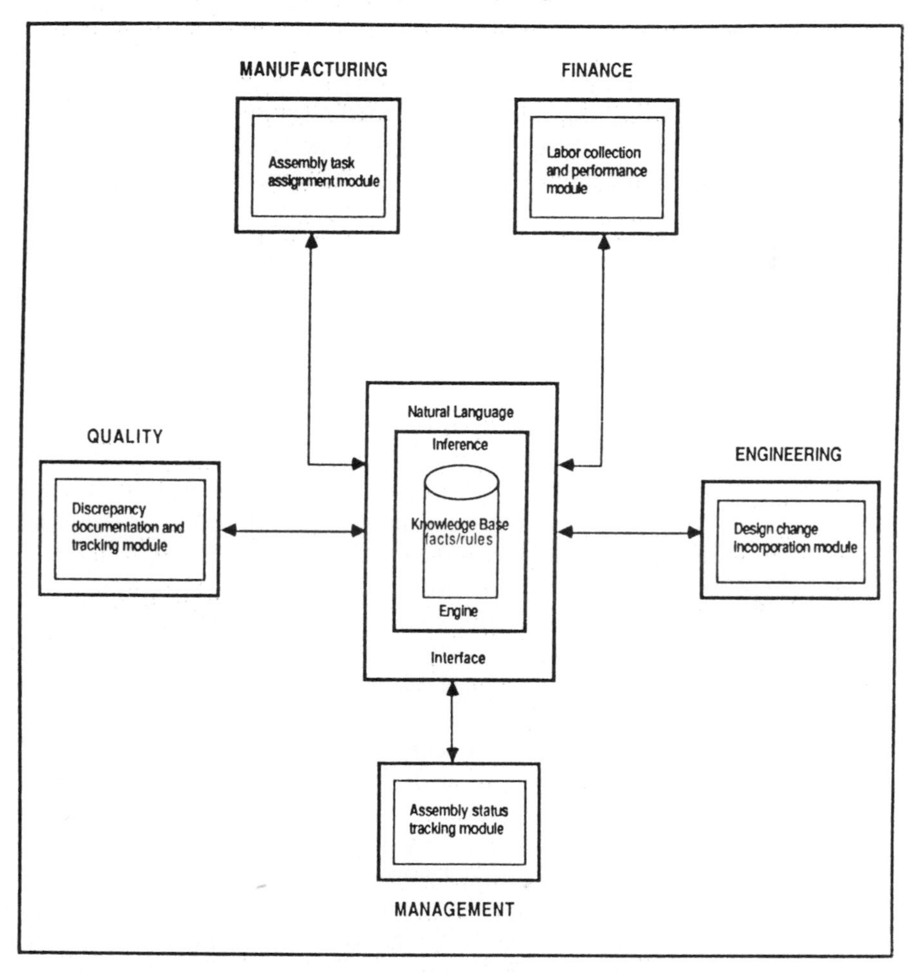

Expert Systems: Fill in the Missing Link in Paperless Aircraft Assembly

Michael W. Parks*
Lockheed Georgia Company

As with any manufacturing operation, aircraft assembly requires a multitude of documents containing the information necessary to structure, schedule, illustrate, control, and document the assembly process.

In most cases, the weight of the paperwork used to assemble the aircraft is greater than the total payload of the finished aircraft. The cost of generating and maintaining this mass of paper represents a significant portion of the total cost associated with aircraft manufacturing.

Even more significant, the inefficiencies associated with information lags/delays inherent in a manual paperwork system can easily represent the largest single deterrent to the improvement of manufacturing productivity.

TAKING THE LEAD

For this reason, major aircraft manufacturers are taking a leading role in the development of paperless manufacturing systems.

These systems provide manufacturing information to shop floor personnel using computer technology such as graphics terminals, laser printers, relational databases, and fourth-generation computer languages.

Shop floor personnel are provided convenient access to the most current information needed to build the aircraft, thus reducing rework and increasing quality.

The use of computer technology to reference assembly information also gives shop personnel the opportunity to update operations in the manufacturing assembly systems in a real-time fashion directly from the shop floor, thus eliminating intermediate key-input steps with corresponding data collection lags/delays.

By incorporating on-line feedback mechanisms for labor collection, job status and quality, management, and supervision can gain a much greater level of control over the assembly processes.

Today's paperless assembly concepts call for systems which dynamically assign job tasks to individual shop floor mechanics based on task precedence relationships and current shop conditions, collect and interpret quality data, track assembly status, measure effective assembly performance, and automatically incorporate changes into the manufacturing database.

The goal of paperless aircraft assembly is to drive management control of the assembly process to a level that is low enough to effectively highlight problems and plan their resolution at a level of detail that is not possible with manual paperwork systems.

Gaining this level of control in a complex assembly environment requires a decision-making capability beyond traditional data processing approaches. It is envisaged that a development of an expert system can provide such a capability.

*Source: Parks [11, pp. 55–62]

EXPERT SYSTEMS

An expert system (ES) is a computer program that is used to emulate the decision-making process of a human expert in a specific knowledge domain of limited scope. The expert system normally consists of two main parts—the knowledge base and the inference engine.

Knowledge Base (KB)

The knowledge base is a collection of the human expert's specific knowledge for the desired application domain. It is represented as both declarative and rule-based data. The following example shows how declarative type data might be represented in a knowledge base:

Wire number 10010 is red.
Wire number 10020 is blue.
Wire number 10030 is red.
Wire number 10040 is white.
Wire number 10050 is red.

Knowledge rules are typically in the form: IF condition clause THEN action clause. They are similar in principle to IF/THEN statements in conventional programs. The following rules provide an example:

IF wire number X is red THEN wire type is engine.
IF wire number X is white THEN wire type is instrument.
IF wire number X is blue THEN wire type is control.
IF wire number X is green THEN wire type is navigation.
IF wire number X is black THEN wire type is communication.
IF wire number X is tan THEN wire type is radar.

Notice that the declarative data is the KB which represent members of the knowledge domain and their attributes; i.e., wire number (member) and wire color (attribute). The rules in the knowledge base serve to structure the declarative data elements.

A knowledge base will also contain the goals to be reached by the system. A set of goals for our wire knowledge base might be written as follows:

1 The wire number is "ask user."
2 The wire is TYPE engine.
3 The wire is TYPE instrument.
4 The wire is TYPE control.
5 The wire is TYPE communication.
6 The wire is TYPE navigation.
7 The wire is TYPE radar.

This goal set can be interpreted to mean "Ask the user to input a wire type." This is a goal which our expert system could reach given the above KB data.

First, the system would prompt the user to input a wire number (goal number one). Notice that TYPE is not directly associated with wire number.

In order to determine wire type, the system will examine the color attribute for each wire number along with the rules which relate color to TYPE.

If the user inputs a wire number which is present in the knowledge base, the system can infer the wire type (from the rule set) because the color of the wire is known.

To further illustrate the example, suppose a user of our expert system (ES) were to input wire number 10040 in response to the prompt generated by the system as a solution to goal number 1.

In order to solve goals 2–7, the ES would examine the rule set in the KB and infer that color must be known in order to determine type. The system would then read the color attribute for wire number 10040, which is "white," from the database.

Given the wire color and the rule, "IF color of wire *X* is white then TYPE is instrument," the system has the necessary information to assign a type of "instrument" to wire 10040.

Now suppose that a user inputs an unknown wire number, say 10100. Our ES still infers that wire color must be determined first in order to determine wire type.

However, in this case the wire color is not already known by the system; i.e., data for wire number 10100 have not been entered into the KB. In this case the expert system will automatically prompt the user to input wire color.

Assuming that the color entered by the user matches a color in the given rule set (otherwise the system will display an "UNABLE TO REACH SOLUTION" message), the new wire number along with its color attribute will be added to the knowledge base. Our expert system has now "learned" the new wire number along with its associated color attributes.

The next time this same wire number is input by a user, the expert system will be able to automatically alter the wire type, because color is now known by the system.

SYSTEMS DIFFERENCES

A major difference between expert systems and traditional data processing programs is the sequence in which the rules are executed. A program written in a conventional programming language such as COBOL or FORTRAN has to be consciously structured by the programmer to ensure that IF/THEN statements are executed, or "fired," at the proper time during normal flow of execution.

An expert system, in contrast, can be programmed without thought as to the correct "firing" sequence of IF/THEN rules. In the above example, note that the sequence or location of rules in the KB does not affect the outcome of the goal search.

In addition, the expert system automatically determines what information is necessary to meet the goal(s) of the system and will prompt the user for input data as required to meet these goals.

Another important difference lies in the way that knowledge is stored in an information system. Traditionally, any inherent knowledge in the system about its application domain is embedded within the syntax of the programming language used to develop the control algorithms.

Thus, the traditional information system is very inflexible when changes must be made which affect its control structure (decisions, choices, conclusions, results). In an expert system, the control structure is represented as data (rules and goals) in the knowledge base, making it much easier to change and maintain.

In an expert system, the inference engine supervises the "firing" of knowledge-based rules by selecting the most appropriate rule to meet system goals.

Inference mechanisms use forward chaining (data driven) and/or backward chaining (goal driven) search processes to select the next rule to "fire."

In forward chaining, a change to data in the database will "fire" all rules whose condition clauses are met as a result of the change. This is why forward chaining is sometimes associated with the term "data driven." The rules are "fired" as the database is updated.

A backward chaining mechanism starts with the ACTION clause of a rule and searches the knowledge base for data which satisfy the CONDITION clause of the rule.

This is sometimes termed "goal driven" due to the fact that the backward chaining mechanism starts with a goal and searches the rules to find those whose consequent actions will achieve that goal.

Expert systems can be applied successfully to problems that have a relatively limited knowledge scope and take an expert a few minutes to a few hours to solve (the task is not too simple and not too hard).

There should be recognized experts who are available and willing to spend the time required to extract their expertise.

The task to be performed by the expert system should be primarily cognitive; that is, successful performance is not based on physical ability or common sense.

The following types of problems are good applications for expert system technology:

- Predicting—infer probable outcomes from given situations.
- Monitoring—examine real-time data and watch for developing malfunctions.
- Designing—configure objects on the basis of specifications and constraints.
- Instructing—provide problem simulation and decision checking and help individualize instruction by diagnosing learner weaknesses and prescribing remedial lessons.
- Diagnosing—infer system malfunctions from a set of observable symptoms.
- Interpreting—infer situation descriptions from sensor data.
- Controlling—interpreting, predicting, repairing, and monitoring system behaviors.
- Planning—evaluate possible future actions to determine the most logical series of steps leading to a desired goal.
- Repairing—combine the diagnosing and planning systems to create and execute plans for repairing faulty systems.
- Debugging—evaluate source code to identify syntax errors and predict errors in program logic based on defined goals.

SYSTEM SHELL

The term "expert system shell" is meant to describe a computer system written to facilitate the development of application-specific expert systems.

A shell normally consists of an inference engine or reasoning mechanism, an empty knowledge base, a knowledge engineering development facility, a front-end natural language interface for users and a workspace to keep track of how a problem is viewed by the system.

This type of development shell can be applied to many different expert system applications and is similar in concept to the popular microcomputer spreadsheet packages. The knowledge in a KB is analogous to the data in spreadsheet "cells" and to the formulas which tie them together.

There are numerous expert system shells available today which offer a wide range of price, capability, and performance. These systems are available to run on mainframe, mini-, and microcomputers.

Of the microcomputer-based systems, GURU from Micro Data Base Systems holds the current lead in dollar sales and is on the high end of both price and capability.

This is not surprising since GURU provides, in addition to an expert system shell, a completely integrated relational database, spreadsheet, report generator, text processor, business graphics, remote communications, procedural programming language, statistical analysis, and natural-language query system.

Another popular system, INSIGHT 2+ from Level Five Research, provides a great deal of capability in a lower-priced package. INSIGHT 2+ includes a complete, easy to use environment for the development of knowledge bases. PASCAL programs, which work in conjunction with INSIGHT's inference engine, can also be developed within this environment.

In addition, database files created with INSIGHT 2+ are fully compatible with files created using dBASE II and dBASE III.

SYSTEM LANGUAGES

Although there are numerous expert shells available today, many times an expert system will be written specifically for a particular application. The two most common languages for writing expert systems are LISP (LISt Processing) and PROLOG (PROgramming in LOGic).

LISP is the most common language used for expert system development in the United States, while PROLOG is more popular in Europe and Japan. In fact, PROLOG has been named the primary language in Japan's fifth-generation computer project.

A significant difference between these languages and traditional programming languages can be found in LISP's and PROLOG's ability to process symbols rather than numbers, and in their ability to handle parallel tasks. This is the primary reason for Japan's use of PROLOG in its fifth-generation computer project.

PROLOG is well suited to parallel computers, in which many processors tackle a problem simultaneously to provide performance several orders of magnitude beyond today's machines. Of equal importance is that LISP and PROLOG offer the programmer the ability to "extend" the kernel of the language.

It is this extensibility which allows both LISP and PROLOG to represent data/programs in more natural "symbolic" formats and, "grey" the distinction between "data" and "code" enough to allow for the existence of programs which are capable of getting "smart" (which rewrite themselves based on previous experience). The extensibility of a language is what essentially leads to its classification as an artificial intelligence language.

EXAMPLES OF SYSTEMS

Expert systems are being used today in a wide range of application areas such as medical diagnosis, equipment failure diagnosis, computer configuration, chemical data interpretation, financial decision making, mineral exploration, flexible manufacturing systems design, airline fare analysis, and production equipment setup.

AIRCRAFT ASSEMBLY

Today the shop floor is managed and supported by numerous individuals such as supervisors (both production and quality), manufacturing engineers, process planners, industrial engineers, and quality engineers. Each of these individuals is an expert in his or her area of responsibility, i.e., job task assignment, quality data interpretation, performance measurement, and change incorporation. A paperless computer system should be designed to duplicate and/or complement the decision-making process of the manufacturing "experts."

Expert systems can be used as "front-end" decision makers for assembly information systems, automatically deciding how to present screens and information to the current system user on a "need to know" basis.

This enables customization of system prompts depending on the identity of the user, resulting in fewer training requirements and better acceptance by shop floor personnel.

Expert systems can be used to analyze user inputs, allowing for less than complete responses to system prompts and providing a system which is more tolerant than typical data processing systems. Using an expert system in this manner enables the computer to "suggest" correct answers to prompts and automatically provide "help" instructions whenever necessary.

This tends to minimize system access time by encouraging the user to input/request the right information in the most efficient manner possible.

BACKGROUND DECISIONS

Expert systems can also make background decisions such as which task should be worked next and how to "work around" a part shortage or an absent mechanic. These decisions can either be made transparent to the user or the system can be asked "why" a decision was made, thus generating the rule path taken by the inference engine.

This is an important characteristic of a good expert system. That is, an expert system should be able to explain how a decision was reached, i.e., list the set of relevant knowledge rules in the sequence in which they were "fired" by the inference engine.

An important aspect of each shop supervisor's responsibility is to assign jobs to the most qualified individual to ensure greater overall quality and maximum learning curve effect. Many times the most qualified individual is the one with the most experience on a particular job.

The expert system can simulate the supervisor's reasoning process by "firing" the following two rules; IF mechanic A has the highest experience factor for job 1, THEN assign job 1 to mechanic A. WHEN mechanic A works job 1, THEN increment (mechanic A, job 1) experience factor by 1.

Of course, an experience factor would be a mechanic attribute in the knowledge base, along with the two stated rules. Remember that in an expert system the "firing" sequence of the rules will be handled by the inference engine, so the knowledge engineer does not have to be concerned about where the rules are placed in the knowledge base.

In a traditional computer program, the programmer would be responsible for making sure that the two rules are "fired" in the proper sequence and at the proper time.

ASSIGNING RESOURCES

A key component of paperless aircraft assembly is the ability to assign resources to specific work tasks based on precedence network relationships. This type of decision network provides the foundation on which the remaining functions are built.

In order to implement this concept in the real world, additional decision criteria must be built into the system. Space constraints, for example, must be considered so that an unrealistic situation will not be presented, i.e., three mechanics assigned to three different jobs located in an area able to fit only one individual at a time.

Supervisor preference should be captured by the system so that human factors can be incorporated into the decision rules; i.e., physical limitations, work preferences, individual abilities, etc. In addition, some tasks must always be worked in parallel.

Still other tasks can begin after their precedence task(s) have been partially completed. It is obvious that the decisions required to assign tasks to human workers require a multitude of decision rules and an extensive base of manufacturing knowledge, especially when these decisions are to be reached through computer technology.

Additionally, each assembly workstation requires a set of rules which will be similar but not necessarily identical to other workstations. In this situation, IF/THEN rules represent data to the system. An expert system is designed by the knowledge engineer to apply these rule sets which describe how knowledge in the system should interact.

The ES enables the capture of rules as data in a logical structure manner and allows for easy maintenance of rule sets in the knowledge base.

In a complex aircraft assembly, quality assurance becomes a critical function in order to achieve the high level of confidence necessary to maintain a successful aircraft manufacturing operation. Critical quality decisions are made continuously at a typical assembly workcell.

Each defect is classified based on the type and location of the problem, and the part or assembly involved. Corrective action is then based on this classification.

After classification the discrepancy must be routed through the proper departments for corrective action and approval. The proper routing and approval requirements are dependent on the discrepancy classification.

In addition, the discrepancy must be monitored in order to ensure that corrective action was taken and that the correct approvals were made and documented.

Expert systems are well suited for this type of problem. By prompting the quality inspector for information concerning the discrepancy's characteristics, the expert system can determine additional data that should be input by the inspector, the discrepancy category, and the proper routing sequence for corrective action.

The system can monitor the progress of the discrepancy record through the corrective process and request additional information or change the routing sequence as new data become available.

As with many CIM-related [computer-integrated manufacturing] systems today, the benefits resulting from expert systems are mostly intangible and therefore difficult to measure quantitatively.

One of the most common benefits specifically attributed to expert systems applies especially well in aircraft assembly processes—the capture of expertise gained by individuals through years of on-the-job experience.

The knowledge gained by experienced employees is a valuable company resource. As an expensive resource, it is a benefit to the company if an employee's knowledge can be captured prior to retirement or leaving the company.

An expert system allows this expertise to be captured in a knowledge base (KB), where it can be added to the knowledge of other experts in the same area, thus compounding the value of the stored knowledge.

Decisions that are made based on knowledge stored in the KB still benefit from being able to simultaneously draw on the combined knowledge of multiple experts.

Capturing the expertise of more experienced employees prior to their retirement is critical in aircraft assembly operations, due to the complexity of the task involved and the relatively small size of a production run (aircraft are typically built in lots of 10s or 100s, not 10,000s or 100,000s). This combination of a complex task environment and limited learning time provides an excellent opportunity for a high return on investment.

DOCUMENTING LOGIC

Using an expert system to capture the decision rules of shop experts provides an opportunity to formally study/document decision logic and evaluate/improve decision methodologies. This type of analysis will help to standardize decision rules, which will result in more consistent decision making throughout the organization.

Due to the complexity and magnitude of aircraft systems, the expertise required to build them is normally focused into areas of knowledge specialization such as cockpit wiring, exterior skin attachment, and inner wing hydraulic systems.

These focused areas of knowledge along with large scale operations provide excellent opportunities for the development and implementation of expert system technology.

HIGH PAYOFF

Large-scale operations result in high payoff potential and focused areas of knowledge provide finite domains for knowledge representation.

The documentation and standardization of manufacturing knowledge enables the development of a knowledge base which can be built as manufacturing plans evolve through different production programs.

It can be used for planning manufacturing requirements/costs for proposed contracts, training new employees and increasing awareness and understanding of the knowledge and expertise required in key manufacturing operations.

Many times, training a new employee requires the attention of an expert in a critical area of operation. Since these experts are usually in short supply, using an expert system as a training aid can be a valuable tool.

Remember that an expert system is able to "explain" how an answer was derived by listing the appropriate rules in the sequence they were "fired." An expert system can also be programmed to analyze a student's weak areas and adjust the instruction accordingly.

SERIOUS PROBLEM

As computer-based manufacturing systems evolve into a "paperless" environment, corresponding decision-support requirements will continue to become increasingly complex. At the same time, the quantity of available data on which to base decisions is growing at a tremendous rate. Systems such as shop floor data collection, CAD/CAM, bar coding, MAP [manufacturing resource planning], computer-aided process planning, time-and-attendance, and computer-aided testing are processing massive amounts of data in manufacturing organizations. As a result, the interpretation and utilization of available data are becoming a serious problem.

In most cases, human information experts (managers, supervisors, engineers, etc.) scan the reports generated by these manufacturing systems and make decisions based on various measurement "yardsticks," such as ratios, trend lines, and variances.

PAST SOLUTIONS

Many times criteria used to make decisions are probabilistic or "fuzzy" in nature and require the expert to spend valuable time analyzing the data before reaching a decision.

In the past, data processing has approached this problem by applying traditional programming solutions. However, as the volume of data increases along with the decision-making requirements, these traditional approaches will no longer be satisfactory.

As an example, software development presents the greatest challenge in implementing DOD's Strategic Defense Initiative, not the design of advanced hardware. This problem primarily results from the inability of today's programming practices to meet the complex decision-making requirements of the plan.

Expert systems provide an answer to this situation. They can provide an effective tool for quickly assimilating large volumes of data into meaningful patterns of information and applying inexact reasoning techniques to quickly arrive at the most likely solution to a problem.

The same kind of challenge faces manufacturing companies during the next few years. Continued growth in automation and just-in-time philosophies is increasing the need for complex/real-time decision-making capabilities on the shop floor. Expert systems will help manufacturing meet this challenge.

Examples of other expert systems that demonstrate the breadth of application, provided as an appendix in Parks' article (11), are as follows:

EXAMPLES OF EXPERT SYSTEMS

- XCON (eXpert CONfigurer)—This system was developed in a joint effort by Digital Equipment Corporation (DEC) and Carnegie-Mellon University. XCON is one of several expert systems used by DEC to configure VAX computer systems for customers. It contains over 4000 rules which enable XCON to correct mistakes in customer orders and correctly locate components in CPU cabinets.
- DELAT (Diesel–Electric Locomotive Troubleshooting Aid)—Developed by General Electric for the maintenance and repair of diesel–electric locomotives. To solve a problem, the system displays a menu of possible fault areas. When the user selects an area the system proceeds with a series of questions. At appropriate points in the session, the system explains "why" certain answers were given and/or displays CAD drawings and video disk sequences on a CRT screen to aid the user in locating specific components. Finally, when the troubleshooting system has determined the cause of the problem, repair instructions are generated. In addition, training film sequences are displayed if requested.
- MYCIN—One of the first successful medical diagnostic systems, MYCIN was developed at Stanford University to diagnose bacterial blood diseases. EMYCIN (Empty MYCIN), one of the first expert system shells, was developed by extracting the inference engine from MYCIN. The inference engine could then be applied to knowledge bases developed for other applications.
- DIPMETER—Schlumberger developed this system as an advisor for oil well drill sites. The system analyzes the readings of an oil well dipmeter and advises the

rig operator on the best way to approach the problem of freeing a drill which is stuck at some point in the well.

- STEAMER—Developed by Westinghouse, STEAMER is used to train machinist mates in shipboard repair and maintenance.
- FINANCIAL STATEMENT ANALYZER—Developed by Arthur Andersen for the SEC (Securities and Exchange Commission). Used to analyze and consolidate incoming financial statements. Due to the relatively unstructured nature of financial statements, FINANCIAL STATEMENT ANALYZER is required to interpret incoming source data by enabling accounting elements to "find" themselves in the data.

Our present information flows are rapidly becoming increasingly computerized. It is not an exaggeration to say in specific instances that *computer-integrated manufacturing* (CIM) has gone from a technical dream to technical reality. The following is a description provided by David L. Ellis-Brown of Martin Marietta's CIM system presently in use at the Electronic Systems Company facilities in Orlando, Florida [5]. Figure 9-4, APECS architecture, and Figure 9-5, APECS shop floor control system, provide graphical descriptions of the Martin Marietta CIM system. The shop floor control system is one of many subsystems of the APECS system. The APECS system is typical of computerized systems that have reached this higher level of development and integration.

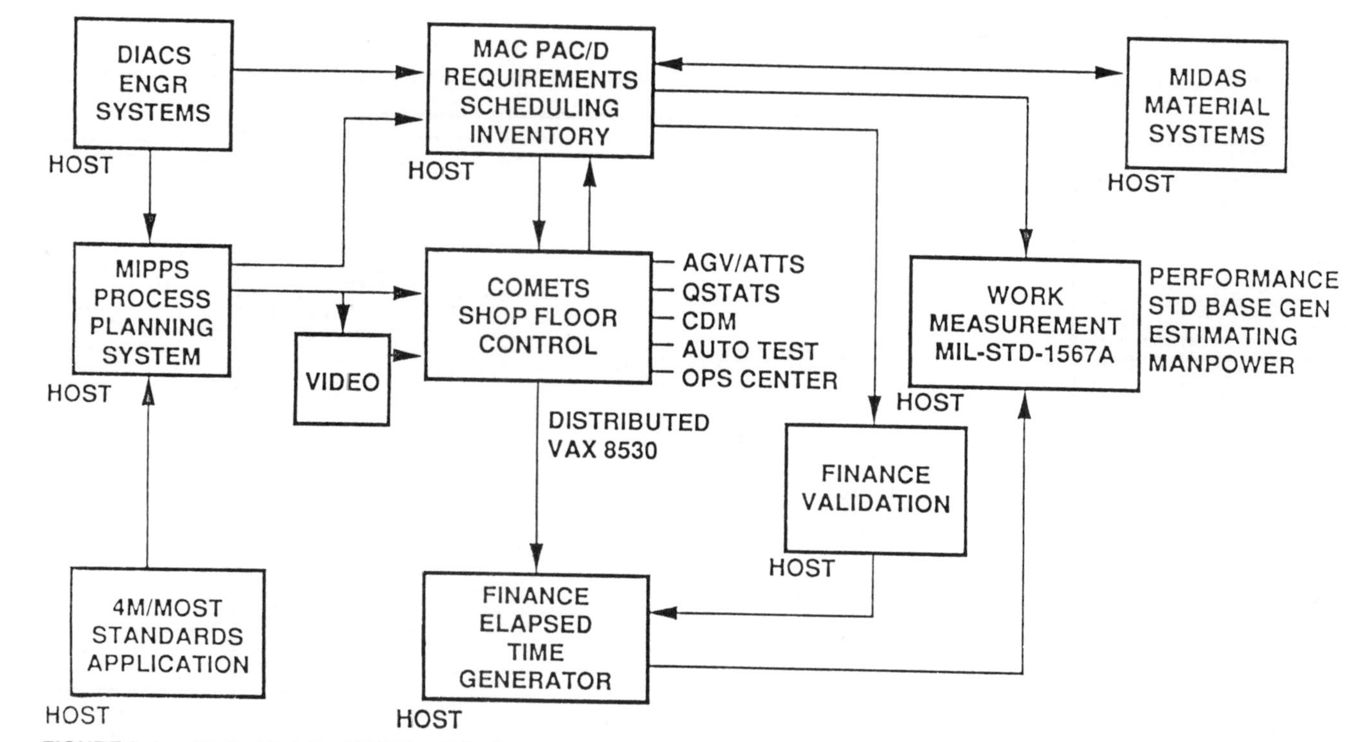

FIGURE 9-4 Martin Marietta APECS architecture.

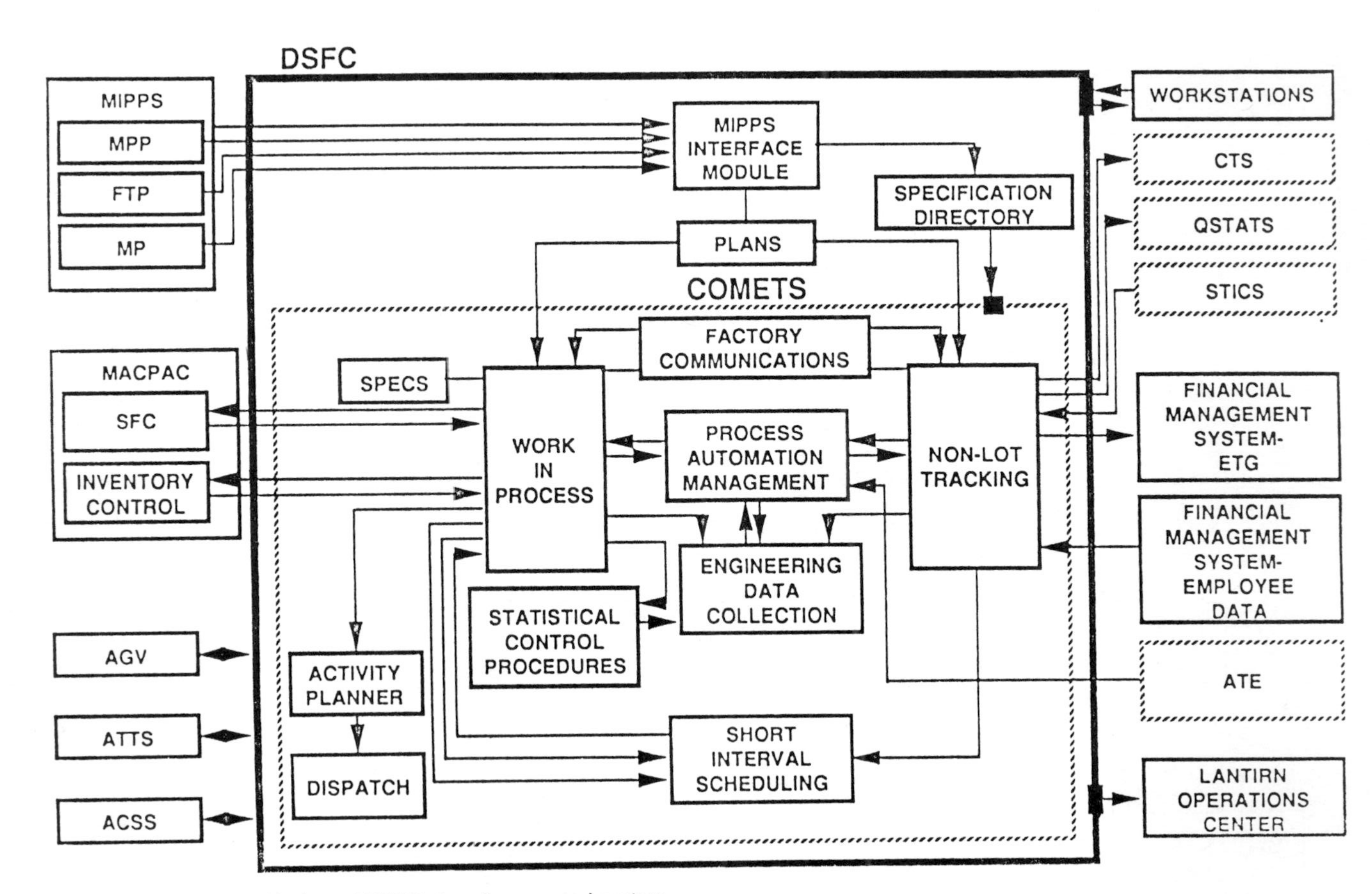

FIGURE 9-5 Martin Marietta APECS shop floor control system.

Martin Marietta's Computer Integrated Manufacturing System

David L. Ellis-Brown
Manager of Production, Planning and Control
Martin Marietta Electronic Systems Company
Orlando, FL

Martin Marietta Aerospace, Orlando, Florida, has developed, implemented, and is utilizing a totally integrated manufacturing system, called APECS (Aerospace Planning, Execution and Control System), which includes integrated modules of MAP II, DSFC (Distributed Shop Floor Control), computer-aided process planning, and a work performance module designed to support MIL-STD 1567A.

Also integrated into APECS are automated material handling elements which include AGV (Automated Guided Vehicle), ATTS (Automated Tote Transporter System), and ACSS (Automated Carousel Stocking System).

Due to the planned modularity of the system's design, a variety of configurations can be supported, including either "paperless" or "paper" shop floor control system, depending on program demands. This customization capability has provided the necessary flexibility to ensure that one basic system will be utilized by all programs at Martin Marietta Aerospace.

The Martin Marietta computer-integrated manufacturing system consists of the following modules with these basic functions:

1 An aerospace and defense MAP II (Manufacturing Resource Planning) module. This module is used to plan and schedule the manufacturing resources required to support contract requirements, both material and factory assets. This system module includes the following features:
- **A** Manufacturing Bills of Material and Part Master File
- **B** Inventory Control
- **C** Master Production Schedule
- **D** Shop Floor Control Interface
- **E** Procurement Interface
- **F** Capacity Requirements Planning (CRP) Interface
- **G** Financial Management System (FMS) Interface
- **I** Manufacturing Routing File Interface to MIPPS [manufacturing integrated production planning system] (a CAPPS [computer-aided process planning system] Systems Development by Martin Marietta)
- **J** Defense/Cross Contract MAP Netting Logic

2 A distributed shop floor control system module that is designed to:
- **1** Direct the efforts of the production floor personnel
- **2** Dispatch orders and material to the workstation level
- **3** Continually monitor the progress and status of the shop floor against the production schedules
- **4** Identify and isolate problems as they occur

- **5** Provide "on line, real time" data for problem resolution
- **6** Collect data via "touch screen and bar code" input
  - **A** Time and attendance
  - **B** Configuration data management
  - **C** Quality status data (q-stats)
  - **D** Variance clocking for performance measurement
  - **E** Work instruction (manufacturing process plan) display
- **7** Interface to MAP II, MIPPS, work measurement, and finance
- **8** Processes through interactive workstations
- **9** Operation control center interface
- **10** Online transaction validation
- **11** Online variance display
- **12** Directs AGV, ATTS and ACSS automated material handling systems
- **13** Real time validation
  - **A** Employee skill and certification validation
  - **B** Tool calibration and utilization validation

3 A computer-aided process planning system (CAPPS) called "MIPPS" (Manufacturing Integrated Production Planning System), that mechanizes the flow of product and process instructional information between design, manufacturing planning, and the shop floor. The following features are contained within "MIPPS":

- **1** Multiple shop floor instructional documents
  - **A** Manufacturing Process Plan (MPP)
    - **1** Work instructions
    - **2** Routing information
    - **3** History information
    - **4** Step level resources
    - **5** Step level references
- **2** Integrated text and graphics
- **3** Engineering drawing access
- **4** Sketch editor
- **5** Full screen editor
- **6** Flexible interfaces
  - **A** MAP II
  - **B** Bill of material validation
  - **C** Tool validation
  - **D** Work center validation
  - **E** Time standards
  - **F** Distributed shop floor
  - **G** Flexible document distribution
    - **1** CRT
    - **2** Distributed
    - **3** Paper
  - **H** Ad hoc query ability

4 A performance measurement system that is compliant with MIL-STD 1567A is being utilized that provides the following features:

1 Performance measured as labor efficiency at work center/department and part number levels.
2 Labor efficiency has the "off standard" work extracted by capturing selected variances code clockings.
3 Variance clocking is separated from operator time for measured work from non-measured work.
4 Formal variance report generation.
5 Allow accurate measurement of operator efficiency and learning improvements.
6 Allow identification/accumulation of realization factor data.

The APECS system is being utilized in a "paperless" environment at Electronic Systems Company. The normal paper documents of work orders folders, material pick lists, manufacturing process plans, configuration data collection, etc. have been eliminated.

All operator access to the APECS system is accomplished through the "wanding" of the barcode on the side of the "ESDS" [electronic systems data system] totes.

A production operations control center, similar to a missile control room has been interfaced to the APECS system with additional capabilities of teleconferencing, telecommunications and total system access. Large-screen displays have been incorporated to aid visual as well as systematic control of factory operations.

APECS at the Electronic Systems Company operates on a multi-platform network consisting of an IBM host based computer and a dedicated DEC distributed processor. The distributed system provides the shop floor operations with on-line, real-time data across a local area network (LAN) that is broadband based to provide moving video to the some 300 plus workstations. Automated material handling equipment has been integrated into the APECS system and is operational at the Electronic Systems Company. The following automated material handling equipment are being utilized in production operations:

1 Automated Guided Vehicles (AGVs) are being used to move electro-static protective totes of material between work centers. The move directions are being generated by the shop floor control system.

2 Automatic Tote Transporter System (ATTS), an automatic conveyor system, is used to direct the totes to specific workstations within the various work centers. Again the location selection is determined by the distributed shop floor control system.

3 Automated Carousel Stocking System (ACSS) is being used to stock and issue material. A "paperless" pick list is downloaded to the carousel controller and the system automatically cycles and indicates material location and material quantity to be picked for issue to work orders. A barcode label is printed for each component required. The pick list is generated from the manufacturing bill of material.

The foregoing description should remove any remaining doubt anyone might have as to whether we have reached a sufficient level of development today to claim that the age of computer-integrated manufacturing (CIM) systems has arrived. It has.

SUMMARY

Along with the development of operations research, came recognition of the need for systems engineering. Prior to the 1940s, industrial engineering was practiced at the micro level. A typical scenario in years past would involve an industrial engineering manager in a plant tasking industrial engineering problems deemed worthy of analysis to the engineers. The industrial engineer rarely questioned the system need; the engineer assumed the manager knew what were worthy problems. It was the engineer's job simply to solve them.

Somewhere along the way it was noted that far too often superb solutions to the wrong problems were developed. Systems engineering simply requires that the overall objective be fully understood before we charge forward.

One of the early and exceedingly able gurus in systems engineering development was C. West Churchman. He and others developed the *systems approach*. Key to any systems analysis is clearly defining objectives. It is not an easy task, as some articles in this chapter, I hope, have made clear (e.g., the Quade and Boucher article).

Two articles in this chapter—one by Parks [11] and the other by Ellis-Brown [5]—offer examples of expert systems and computer-integrated manufacturing systems (CIMS), respectively, that represent significant development areas in systems engineering today.

Systems engineering is a dynamic and important development area within industrial engineering today. It can improve one's chances of winning, but it may not lead to success, as is indicated in Fig. 9-6. Trying is important, however.

FIGURE 9-6 A princely failure. (From Quade and Boucher [12, p. 278]).

REFERENCES

1 Allen, Dell: "Knowledge Engineering," CAM-I Conference, Ft. Lauderdale, FL, November 1984.
2 Canada, John, and William Sullivan, *Economic and Multiattribute Evaluation of Advanced Manufacturing Systems,* Prentice Hall, Englewood Cliffs, NJ, 1989.
3 Churchman, C. W.: *The Systems Approach,* Bantam Doubleday Dell, New York, 1968.
4 Churchman, C. W., R. L. Ackoff, and E. L. Arnoff: *Introduction to Operations Research,* John Wiley, New York, 1957.
5 Ellis-Brown, David L.: "Martin Marietta's Computer Integrated Manufacturing System," Martin Marietta Electronic Systems, P. O. Box 628007 MP1601, Orlando, FL, 1988.
6 Forrester, J.: *Industrial Dynamics,* The MIT Press, Cambridge, MA, 1961.
7 Hall, A. D.: *A Methodology for Systems Engineering,* Van Nostrand, Princeton, NJ, 1962.
8 Hamilton, H. R., et al.: "A Dynamic Model of the Economy of the Susquehanna River Basin," Battelle Memorial Institute, Columbus, OH, 1966.
9 Likert, R.: *New Patterns of Management,* McGraw-Hill, New York, 1961.
10 "Orders of Magnitude," 16mm film, Charles Eames Studios, Venice, CA, 1973.
11 Parks, Michael W.: "Expert Systems: Fill in the Missing Link in Paperless Aircraft Assembly," *Industrial Engineering,* January 1987, Vol. 20, No. 1, pp. 55–62.
12 Quade, E. S., and W. I. Boucher: *Systems Analysis and Policy Planning,* Elsevier Publishing Company, Inc., New York, NY, 1968.
13 Reeves, G.R. and K.D. Lawrence: "Multi-Criteria Decision Making," in W.K. Hodson, ed. in chief, *Maynard's Industrial Engineering Handbook*, 4th ed., McGraw-Hill, New York, 1992.

REVIEW QUESTIONS AND PROBLEMS

1 Give an example of a system within a system within a system.
2 Give one example each of a system of things and a human–machine system.
3 Why is choosing the right objective more important than choosing the right system?
4 Give an example of a system problem for which there are at least five objectives, and state the objectives.
5 Assume that there are four objectives, A, B, C, and D, with respect to a systems problem. The lengths of the lines in Fig. 9-7 indicate the relative value of obtaining each of these four objectives. Employing Churchman's Procedure 1, by visual inspection only, determine normalized relative values for the four objectives.
6 Determine relative normalized values for the preceding line segments by measuring them and comparing these values with your results from Problem 5.
7 Who coined the phrase *artificial intelligence,* and how long ago was it coined?
8 What is a *guesstimate*?

FIGURE 9-7 Relative value by length.

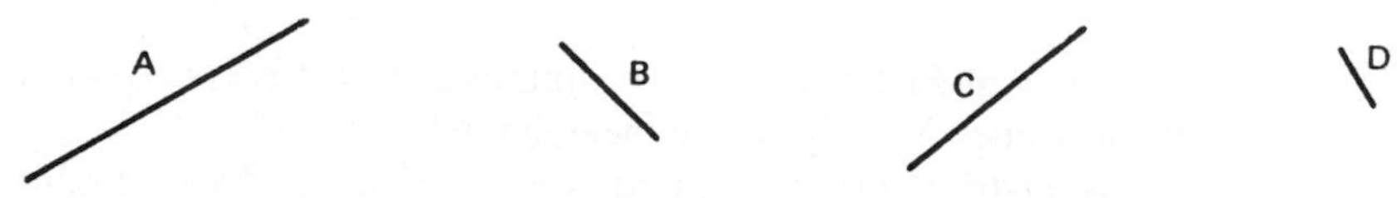

10

WHERE DO WE GO FROM HERE?

Progress, for the most part, only substitutes one partial and incomplete truth for another; the new fragment of truth being more wanted, more adapted to the needs of the time, than that which it displaces.

John Stuart Mill

Industrial engineering has been around for a century. How does industrial engineering differ from other engineering fields? What do industrial engineers do?

The following article by Smith [25], a former undergraduate student of mine, was an attempt to describe what industrial engineers do and how they differ from other engineers.

A Student IE's Answer to "What's an IE?"

Ronald Smith

"**Industrial Engineer.**" A very impressive title. Slides easily over the tongue. But what is one? Must be an engineer, but what type?

"Industrial" brings to mind bellowing smokestacks, mighty gears turning, the screaming of a steam whistle, forges flaring, millions of fellow working men building a stronger nation, the American way of life!... All very soul stirring, but what does an IE do?

First, let's see if we can find a niche for the IE in the vast engineering disciplines. There are four main fields: (In alphabetical order, for some sensitive types)

1 **Chemical:** Bubble-bubble, toil & trouble: If it is sticky, gooey, gummy, grubby, bubbly, they're in it.
2 **Civil:** Bridge builders, architects-to-be, road makers, stress and strain types.
3 **Electrical:** Zap! Crackle! Pop! AC/DC! Tiny black boxes with tiny black boxes within.
4 **Mechanical:** Clink-clank-clunk! What makes the world go round: gears, spindles, sprockets, & spanners.

(I'm sorry, Ag.E's, but I just could not bring myself to do it…)

Now, then, where were the IE's? We mixed some chemical together to make something, wired it up, put wheels & gears on it, and built a shed for it. What is left?

Some people consider IE similar to caulking compound—it fills the gaps between the other engineering fields—a catch-all for what the others don't care to use. Things like quality control, safety, & reliability. After the noble EE's, ME's, etc., have finished, the IE scuttles around tidying-up the process. "If you can't make it as a *real* engineer, you can always become an IE."

Others seem to consider the IE as a sadist whose biggest kick in life is making people work faster-for-lesser. "Efficiency Experts." The management's stool pigeons. "Young wet-nosed punks to tell me, after 10 years with the company, that I'm working too slow!"

These people are entitled to their opinions, but for those persons who have no idea at all about the difference between an IE and the other engineers, probably 90 percent of the people in the world (89 percent of whom have never heard of IE's), the following material may help. It is an attempt to describe some of the many different parts of being an IE, stressing the differences between IE's and those other engineers.

The IE brings a sense of **business reality** to the other engineers. The IE straddles the vast gap between the practical managers and the development engineers. He is the guy who tells the EE that his solid gold, platinum plated, lithium-lined relay contacts are being replaced by tin ones. "Is that third brace really needed?" This is where the IE's training in the various engineering fields comes in handy. He is a translator. The engineer can communicate with an IE when he may be unable to show the non-engineering boss what he is trying to do. And visa-versa, the IE can bring the holy word down from above to the engineers, reminding them that they are here to make money. Also, the IE has a sporting chance if the engineers try to snow him with technicalities; they do not know *what* he may know.

The IE is a **step backer.** He takes the proverbial one step back. He gets the big picture. Often people working intensively on the details of a project or design fail to see where they are going overall. While the ME is wondering if the right-torque framastan should have 5 or 6 spokes for the wombat model, the IE has decided that the fritler doesn't even need wombats.

An IE is a **people buffer.** Not only does he work with the bright, shining, purring machines, but also with their hairy, ham-fisted operators. Is Joe Blow working as fast as possible with the thingie cutter? How much should he be paid? Per piece or not? Call in the stop watch people, the IE's. They will help you to set up work standards for comparison, they will suggest ways of job evaluation, possible incentive plans to use, etc.

Joe Blow may complain at first, but the IE's help him, too (if, admittedly, he's not part of the dead wood trimmed away), (that is life, Joe), as more efficient work is usually easier work.

Most IE's really don't go around pulling wings off flies, kicking old ladies' crutches, etc. They are more or less human. They don't want Joe's job. They want to find faster, cheaper ways to do it. Also, safer ways, which brings up the next topic.

An IE is a **people protector**—safety. Keeping Joe Blow's paws out of the cutter's "field of authority." If not for Joe's sake, then it is at least more efficient. Many people consider safety as a joke—never take it seriously, like security. "Hah, the rope will hold; whatch'a worryin 'bout?" IE's make Mr. Blow safe if he wants to be or not. The IE's ears perk up at the sound of "famous last words..."

An IE can be a **plant planner.** Building a factory? Should the tinker-tuners go near the fudge-forgers? Should there be a moving belt or a passageway? How much how soon which way where when? Don't ask the CE—he's trying to keep the roof up; the ME's unsticking the door lift; the EE's in the John working on the lights; the ChemE is making smoke in the basement; the AgE is shoveling...well, anyway, it is the IE who's had the training, he is the dude who coordinates and plans. It is his neck if the feather fluffer winds up next to the molasses mixer. His training includes human factors as well as straight engineering knowledge.

The ME may see no reason why he shouldn't have the walls painted black to absorb heat, but the IE will also balance in the effect it will have on the employees. Also, where will you put the plant? Once the plant is located and built, it is very expensive to correct any major errors. An IE would be a valuable asset in assembling the assembly plant.

He can be a **sampler.** Who decides, out of 100 wombats, how many samples you must take to be 95 percent certain that 90 percent of the goods are good? A good IE can. He has been trained in statistics, probability, and reliability. Practical, applicable statistics, too, as opposed to what a math major would get. The other engineers? They may have had at most 3 hours of statistical theory. Not much of a confidence level there.

An IE is a **betterwayer**. He has "Is this the best way?" constantly rattling around in his head. Here, perhaps, the difference between an IE and the other engineers is less noticeable. All engineers try to come up with better methods of making something. The ME's, EE's, etc., have the advantage of technical knowledge in their own fields, but the IE may be able to pull something from, say, the ChemE's and apply it to an ME's problem. The IE's big advantage comes in after the advanced model is made. He knows the assembly line better than the other engineers, and can see a great use for the new No. 1 zoot sticker in the No. 3 wombat assembly line. The IE has ideas on what to do with the engineers' new ideas.

An IE is good at **maximizing.** If you want to ship your fresh wombats, and route A takes 3 hours at $4 per wombat, route B takes 6 hours at $2, and route C takes 4 hours at $3.50 each, and you lose $1 for each hour a wombat is outside, what is the best route? This example may be simple, but when you consider a plant's normal production of wombats per day, and all the various methods of transportation available, it gets very sticky indeed.

Many people would be surprised at the elaborate formulas IE's use to solve these types of problems (when they are not pulling wings off flies or winding their stop watches). You just cannot pull the answers out of your hat. The other engineers could possibly "plug and shove" to get the answer, but an IE is trained for it.

One main difference between IE's and the other engineers is **ladder climbability.** An EE can look forward to being head of the EE department someday, but that is usually about it. The IE finds it a comparatively easy transfer to management; he usually has a foot half-way into management to begin with. His background knowledge will be an asset. Also, he may be a bit more open minded than the EE, who would favor his old department. An IE's "old department" is the whole plant.

Another important difference is **survivability.** An IE is like a cat—he can always (well, more often than not) land on his metaphysical feet. His knowledge can be used in almost any field, from industry to hospitals, from the military to research projects. If the rubber market drops out, he can jump to, say, aeronautics. The ChemE, however, will be caught in a fairly specialized field, and he may go down with the rubber duckies.

Now then, many IE's are not going to like the fact that we may have brushed over lightly or even skipped their pet IE subject, but it is hard to cover IE'ism completely, which brings up the next characteristic of Industrial Engineering:

Indefinabilityness. One of the most frustrating characteristics of Industrial Engineering to some, but appealing to others, is that IE cannot be defined in a little capsule summary. There are no definite boundaries to IE. It is so wide, from management to time study, from quality control to design analysis, almost anyone can find his niche sooner or later.

An EE is an EE within fairly predictable limits, but when you say "I am an IE," people still haven't got you pinned down, pegged away in a mental cubby hole. You are an unknown factor, they don't known what to expect.

The world needs non-superspecialized engineers—people who can get an overall view, bring together the specialists, and handle new, presently unforseen and unpretrainable-for, events.

I hope this has given you at least some vague idea of what an IE is and how he differs from "them other slip-stick jockeys." I admit it is biased, if not a little aggressive, but then many an IE walks around with a large chip on his shoulder, because when he says "I'm an IE!" people still say "Whatzat?"

What engineering fields are there and how many graduates do they produce? Table 10-1 lists the number of degrees granted in the United States at the bachelor, master, and Ph.D. levels in 1990.

TABLE 10-1 DEGREES IN ENGINEERING BY FIELD, 1990

Field	Bachelor's	Master's	Doctorate
Aerospace	2,971	1,016	189
Agricultural	317	189	94
Architectual	375	33	0
Biomedical	695	310	103
Ceramic	348	80	38
Chemical	3,622	1,140	667
Civil	7,587	2,940	539
Computer	4,355	3,265	339
Electrical and electronic	21,385	7,691	1,262
Engineering, general	1,239	496	58
Engineering, science	1,045	701	239
Environmental	137	471	51
Industrial and manufacturing	4,306	2,489	200
Marine, naval architecture, and oceanography	475	146	21
Materials and metallurgical	857	671	392
Mechanical	14,969	3,994	900
Mining and mineral	168	192	79
Nuclear	264	236	115
Petroleum	286	162	54
Systems	362	692	69
Other	204	120	15
Total	65,967	27,034	5,424

Source: American Association of Engineering Societies, Engineering Manpower Commission, 1990.

Industrial engineering is fifth in the number of bachelor's degrees granted; industrial engineering is here to stay.

Accredited programs in industrial engineering are listed in Table 10-2. The American Association of Engineering Societies (AAES) is the primary accrediting organization for engineering programs in the United States.

Table 10-3 lists the largest producers of industrial engineering bachelor, master, and Ph.D. degrees in the United States.

INTERNATIONAL COMPETITIVENESS AND MANAGEMENT

In the early 1990s, the United States found itself in another economic recession. More important, the United States is falling behind in international competitiveness, and that fact is not less likely to go away by itself. Is this a fight we cannot win because we lack natural resources, or our standard of living is too high, or our work force is uneducated and lazy, or we lack competitive technology? No. Not at all. The United States simply needs to get its act together. Our only problem is our lack of focus and determination; we simply need leadership. Reread the quote at the beginning of the management chapter. There really is no one else to blame.

TABLE 10-2 ACCREDITED PROGRAMS IN INDUSTRIAL ENGINEERING IN THE UNITED STATES

Engineering Management
Missouri-Rolla, University of
Southern Methodist University
United States Military Academy

Industrial & Management Engineering
Montana State University
Rensselaer Polytechnic Institute

Industrial & Operations Engineering
Michigan, University of; Ann Arbor

Industrial & Systems Engineering
Alabama in Huntsville, University of
Florida, University of
Michigan-Dearborn, University of
Ohio State University
Ohio University
San Jose State University
Southern California, University of

Industrial Engineering
Alabama, University of
Alfred University, New York State
Arizona State University
Arizona, University of
Arkansas University
Auburn University
Bradley University
California Polytechnic State University, San Luis Obispo
California State Polytechnic University, Pomona
California State University, Fresno
California, Berkeley; University of
Central Florida, University of
Cincinnati, University of
Clemson University
Cleveland State University
Columbia University
Fairleigh Dickinson University, Teaneck Campus

Industrial Engineering (Cont.)
Florida International University
Georgia Institute of Technology
GMI Engineering and Management Institute
Hofstra University
Houston, University of
Illinois at Chicago, University of
Illinois at Urbana-Champaign, University of
Iowa State University
Iowa, University of
Kansas State University
Lamar University
Lehigh University
Louisiana State University
Louisiana Tech University
Miami, University of
Mississippi State University
Missouri-Columbia, University of
Nebraska-Lincoln, University of
New Haven, University of
New Jersey Institute of Technology
New Mexico State University
New York at Buffalo, State University of
North Carolina Agricultural and Technical State University
North Carolina State University at Raleigh
Northeastern University
Northwestern University
Oklahoma, University of
Oregon State University (Basic Option)
Pennsylvania State University
Pittsburgh, University of
Polytechnic University
Puerto Rico, Mayaguez Campus; University of
Purdue University, West Lafayette
Rhode Island, University of
Rochester Institute of Technology

Industrial Engineering (Cont.)
Rutgers-The State University of New Jersey
St. Mary's University
South Florida, University of
Stanford University
Tennessee Technological University
Tennessee at Knoxville, University of
Texas A&M University
Texas Tech University
Texas at Arlington, University of
Texas at El Paso, University of
Toledo, University of
Utah, University of
Washington, University of
Wayne State University
West Virginia University
Western Michigan University, Kalamazoo Campus
Western New England College
Wichita State University
Wisconsin-Madison, University of
Wisconsin-Milwaukee, University of
Youngstown State University

Industrial Engineering & Management
Oklahoma State University

Industrial Engineering & Operations Research
Massachusetts at Amherst, University of
Virginia Polytechnic Institute and State University

Manufacturing Engineering Option in Industrial Engineering
Kansas State University
Oregon State University

Operations Research & Industrial Engineering
Cornell University

In the last half-century living on credit has become a way of life. Credit cards have become so popular that wallets have been redesigned to hold them. The most colossal offender of the use of credit in this "shortcut to Valhalla" is the United States government. It seems evident now that the Great Society costs a great deal more than was originally anticipated.

TABLE 10-3 LARGEST PRODUCERS OF INDUSTRIAL ENGINEERING DEGREES, 1989–1990

Bachelor's			Masters			Doctoral		
	Institution	No.		Institution	No.		Institution	No.
1.	Georgia Tech	218	1.	Stanford	68	1.	Virginia Tech	16
2.	Purdue	178	2.	Northeastern	60	2.	Texas A&M	12
3.	Penn State	130	2.	Purdue	60	3.	Stanford	11
4.	Cornell	124	4.	Texas A&M	52	4.	Purdue	10
5.	Texas A&M	112	5.	AL - Huntsville	50	5.	UC - Berkeley	9
6.	N.C. State	103	6.	George Washington	48	5.	GeorgiaTech	9
7.	VirginiaTech	102	6.	SUNY- Buffalo	48	7.	Cornell	8
8.	Iowa State	101	8.	Arizona State	41	8.	Michigan	7
9.	Puerto Rico	100	8.	Georgia Tech	41	8.	Ohio State	7
10.	Michigan	97	10.	Virginia Tech	39	10.	Arizona State	6
11.	Wisconsin	94	11.	Penn State	39	10.	Columbia	6
12.	Ohio State	81	12.	Tennessee	38	10.	George Washington	6
13.	Tennessee	74	13.	Cornell	37	10.	Iowa State	6
14.	Illinois	71	14.	Michigan	33	10.	Penn State	6
15.	GMI	67	15.	Columbia	31	10.	Pittsburgh	6
16.	SUNY-Buffalo	63	16.	UC - Berkeley	30	16.	Massachusetts	5
17.	Florida	57	17.	Iowa	29	17.	Nebraska	4
18.	Cal Poly SLO	56	17.	New Mexico State	29	17.	Oklahoma State	4
18.	Lehigh	56	19.	Texas Tech	28	17.	Texas Tech	4
20.	Miss State	53	20.	San Jose State	26	17.	West Virginia	4
20.	Northwestern	53	20.	Southern Cal	26	17.	Wichita State	4
			20.	Wisconsin	26			

Source: American Association of Engineering Societies,Engineering Manpower Commision, 1990.

In addition, one out of five workers in the United States works for the government. The U.S. Navy has eight times as many captains as it has ships to captain. Parkinson's [21] "growth of the admiralty" is a present-day reality in the United States. These are merely samples, but they are indicative. Those who live on credit beyond their means must ultimately answer the "knock at the door": an individual finds that on the other side of the door is a bill collector; for an overexpanded government, on the other side of the door is inflation, recession, and a devalued currency.

Is the United States noncompetitive? Richard Wilson of North American Rockwell says [29, p. 9], "Today we are rapidly losing our competitive advantage. This is our national industrial economic disease—our inability to produce goods which can be sold at a competitive price in the world market."

One simple answer is that the United States has taught a lot of other people to be productive. Through foreign aid, we have given them the capital equipment to initiate modern industrialization. An achieving people are much less likely to want to go to war to meet their national objectives. We had our reasons; whether they were humanitarian or simply practical is a moot question at this time.

Through foreign aid and the exportation of U.S. technology we have helped our defeated enemies of World War II to the extent that we are now struggling to compete with them in international markets. One obvious advantage of countries with traditionally lower standards of living is lower labor costs. However, low labor cost is becoming less of an excuse for the noncompetitiveness of the United States, because many western nations and Japan pay salaries as high or higher than the United States. One disturbing turn of events is that we taught our competitors to be productive and then failed to implement for ourselves what we taught them. Much of what Taylor [26] espoused at the turn of the century we have taught others, and we have since moved away from many of our own teachings.

Of course, world conditions have changed drastically. During World War II, the United States and its allies, Britain, France, and the Soviet Union, vanquished their enemies (Germany, Japan, and Italy) and in the process decimated their industrial economies. The United States greatly developed its industrial base during that war, and it came out of the war unscathed and very well positioned for international trade. Our European allies had won the war but suffered considerable industrial and general economic damage. The war left the United States as the supplier of the world.

In time, with our and our allies' aid, our enemies rebuilt their economies and industrial bases. Ultimately, rebuilding provided them with a competitive advantage—completely retooled industrial bases. They were bringing up better equipment then we had at home. They had an opportunity to rethink how they would design, control, and manage their operations; the old guard was gone, and new ways could be tried. In the meantime, U.S. management could sell products worldwide, without much innovation, without concern for quality; times were good, and complacency set in. Companies ignored their customers; they often failed to hold the line with their unions (i.e., they let wages get too high and acquiesced to inefficient work rules); they let the pressure of Wall Street's preoccupation with short-term earnings affect their judgment, and they let their financial managers make product and plant decisions. The United States now finds itself trying to play catch-up with Germany, Japan, and many others.

The good news, though, is that we can and do change when we have to; our most valuable strength as a nation is our ability to change, to adjust. There are many industrial firms today that have not changed very much, however, and they are finding it harder and harder to compete. The United States does not have a lock on world customers anymore; those days are gone forever. Yesterday I stopped at a red light near our local university campus, and I counted the number of foreign and United States cars waiting for the light to change; if you are an American, you do not want to know the ratio. Try it some time. Those managements that are having difficulty competing today using the old traditional methods have two fundamental choices: (1) get with the program—read, listen, experiment, improve (continuously), or (2) get ready for retirement or a career change.

To a professional in the field, it can be exceedingly frustrating to see a firm that desperately needs industrial engineering help and not be able to get in the front door to help; management is happy with the way it is doing business, no one is requiring that it does better, and it may even be making a profit. As Jack Harrison, a JIT associate in Orlando, Florida, often says, "best is the enemy of better." He learned the phrase from another well-known consultant, George Plossl. If you (the management) think you are the best, I may not be able to help you (because you may choose not to listen to me). If you want to be better, I can help you (because you will at least listen to me). Learning to listen is so important; American management needs to learn to listen. As stated previously in this text, one of American managements' greatest barriers to their own improvement are their egos. Closely related as a counterproductive trait of top management in too many major American managements is arrogance. The recent movie *Roger and Me* makes the point quite convincingly. They have allowed themselves to become detached from outside influence, especially their customers.

To reiterate, our greatest strength as a nation is our ability to change, and yet there are times when we should have changed and did not. For example, the United States, Liberia, and Myanmar (formerly Burma) are the only remaining nations employing the British system of measurements (i.e., foot–pound–gallon); everyone else is using the far superior and more logical metric system. It makes no long-term sense to continue with our present measurement system, and yet our government, bending to pressure from our not-very-forward-thinking industrial "leaders," have delayed and delayed and delayed the change. Mechanics throughout the world, almost universally, have metric wrenches and rulers, and we sell products for which metric wrenches and rulers are essentially useless. Their CADD systems, manual drawings, formulations, and recipes are dimensioned and specified in metric dimensions. Our congress has not acted. Presidential candidates promise the American public that they will create jobs, but none have mentioned getting rid of the British system of measurements, which would significantly increase the number of American jobs.

And now, some good news—just recently, in meeting with a prime architectural and engineering firm to discuss our initiation of a project with NASA, I was handed a copy of the Metric Guide for Federal Construction, [19] First Edition, requiring that on this job we are to use metric only. The guide, in referring to the *Metric Conversion Act of 1975* as amended by the *Omnibus Trade and Competitiveness Act of 1988,* states, "It requires that, to the extent feasible, the metric system be used in all federal procurement, grants, and business-related activities by September 30, 1992." It is about time—the system does work!

Stated more directly, we simply lack leadership, in both government and industry. Our biggest problem by far today is leadership.

Deming [5] states the following:

> Most people in management are not aware that they are imprisoned by current practices of management…that these management practices are the cause of American corporate decline …
>
> Management was chasing phantoms, rewarding and punishing good workers, creating mistrust, fear, trying to manage people instead of transforming a flawed system …

Akio Morita, Chairman of the Sony Corporation, in a speech concerning developments in Japan delivered to the Malaysian Association of Bankers [17], stated the following:

> I cannot stress enough the importance of manufacturing. Nothing is more fundamental to any nation or economy than the ability to produce real products…

> The securities and banking industry, which used to play a supporting role to manufacturing and other production businesses, has become an end unto itself...
>
> A global brain drain of young talent to the financial sector was setting a dangerous trend that threatened the manufacturing industry.

In an *Orlando Sentinel* article entitled "Grad Students Blame Management for Problems" [13], 60 students in the Rollins College MBA program were asked a number of probing questions about management. It should be noted that the preponderance of Rollins MBA students are practicing managers attending Rollins College in the evenings; they have significant management experience to draw upon. The following are their responses to three of the questions posed:

Who's to Blame?	
Workers	2
Management	40
Society	11
All	2

What's the Answer?	
More recognition	23
More responsibility	8
More autonomy	5
Unspecified motivation	4
Better communication	3
More incentives	1
Tighter controls	2
Copy Japan	0
No Change	9

What's the Problem?	
Lazy Workers	2
Greedy workers	5
Disillusioned workers	29
Other priorities	17
No problem	2
Other	1

The survey results above speak for themselves. It is clear that those managers who participated in the survey believe that (1) management is the problem, (2) recognition is an important motivator, and (3) workers are disillusioned.

At a time when there is considerable agreement that United States management is the problem, it is ironic how arrogant and greedy CEOs too often appear to be, as compared to Japanese managers, who appear to be managing quite successfully, while presenting a far more humble demeanor. The newspaper cartoon in Fig. 10-1 makes the point. CEO salaries in the United States have become a national scandal. CEOs in major Japanese corporations typically receive salaries 20 times the lowest paid worker; in the United States, multipliers of 80 are not uncommon.

American management theorists would be well advised to revisit previous works in their own field for guidance. James Lincoln, of Lincoln Electric, in his 1961 book

FIGURE 10-1 American vs. Japanese management style. [28]

A New Approach to Industrial Economics, indicated that the common American management perspective needed to be addressed, as indicated in the following quote [15, p. 36]:

> The industrial manager is very conscious of his company's need of uninterrupted income. He is completely oblivious, evidently, of the fact that the worker has exactly the same need. The worker's fear of no income is far more intense than the industrialist's, since his daily bread and that of his family depend on his job. The industrialist would not miss a meal if his company should run at a loss for a length of time equal to that for which the worker was laid off because of lack of orders.
>
> In spite of these facts, the industrialist will fire the worker any time he feels that he can get along without him. The worker has no control over his future. His need of continuous income is far more urgent than that of management, yet he has no recourse. Only management is responsible for the loss of a worker's job. Only management can follow and develop a program that will bring in orders. The worker can't. Management, which is responsible, keeps its job. The man who had no responsibility is thrown out. Management failed in its job and had no punishment. The wage earner did not fail in his job but was fired. No man will go along with such injustice, nor should he. This is still true, in spite of custom which completely sanctions such procedure.

This does not mean that management should not have the right to terminate employees. However, some acceptance and acknowledgement of failure on their part may be appropriate. American managements should perform their duties with a demonstrated measure of compassion and humility, with a "kinder and gentler" approach than often appears to be the case.

There is good news, however. We have excellent natural resources. We are still by far the largest market in the world, and it will get even larger when implementation of the North American Free Trade Agreement (NAFTA) eliminates tariffs between the United States, Mexico, and Canada. In future years NAFTA will probably be broadened to include all of Latin and South America as well.

We have the infrastructure, even though in recent years it has begun to deteriorate from years of neglect. We have a well-educated work force; it is admittedly not as well-educated as the Japanese, but it is still correctable in the future. American workers are healthy and have a good work ethic. There are some unfair trade practices in Japan, but they do not represent the primary reason we are not competing well with them and others.

Most business students are taught that there are checks and balances in our capitalistic system, and that is one of the reasons it works so well. Two authors have recently shed some disturbing light on that subject, however. Lee Iacocca's book *Iacocca: An Autobiography* [11], raises some questions as to corporate directors as an effective check and balance. He quotes George Bennett, an outside director for Ford, as saying to him, "You know, if I'd had any guts I should have quit with you. But I handle a pension fund for Ford and I'd lose it in an instant if I followed you to Chrysler" [11, p. 132].

Iacocca states the following [11, p. 155]:

> When I went to my first board meeting, I began to understand the problem. Chrysler's board of directors had even less information than their counterparts at Ford—and that's saying a mouthful. There were no slides and no financial reviews. This was hardly the way to run the tenth largest corporation in the country.

Harold Geneen, retired chairman of ITT (International Telephone & Telegraph Corporation), in a chapter entitled "The Board of Directors" in his book *Managing* [7], states:

> Directors come in only once a month. What do they know? What have they got to back them up—intuition, feelings, gossip overheard, or a report read in the media? If one stubborn director continues, he is likely to be embarrassed by what he doesn't know. If he persists, he is casting himself in the role of a troublemaker, and no one likes a troublemaker. So what to do, except sit back in his chair, taste his cold coffee, and desist? [7, p. 259]

> A board of directors meeting is mostly a one-way line of communication. Management does 90 to 95 percent of the talking. Outside board members, who are not part of the management, sit there and listen; then they go to lunch, and then they go home and open the envelope which contains their fee. [7, p. 259]

> If the board of directors is really there to represent the interests of the stockholders, what is the chief executive doing on the board? Doesn't he have a conflict of interest? He's a professional manager. He cannot represent shareholders and impartially sit in judgement of himself. Nevertheless, in every corporation that I know of, the chief executive is a member of the board. In more than three quarters of the Fortune 500 corporations, the chief executive also sits as chairman of the board. As chairman, he is not only running the company, he is also running the board. [7, p. 261]

> Furthermore, the declining thrust and productivity of American industry in recent years can be attributed, at least partially, to what these boards of directors could have done and did not do. The stakes are enormous. The Fortune 500 corporations, after all, comprise the overwhelming share of the industrial output of this nation, and the archaic old-boy network no longer serves the true needs of the greatest industrial power in the world. [7, p. 267]

This is not to say that boards of directors of the preponderance of U.S. corporations do not serve as checks and balances to their management. But the observations of Iacocca and Geneen lead one to question, "Are American corporate managements in large corporations accountable to anyone but themselves?" This question must be raised because industrial engineers often assume that board policies, based on a review of relevant information, after due and thoughtful consideration and analysis, will bless proper actions of management—and therefore, the long-term best interests of the corporation are ensured. It appears that this may well be assuming too much.

Of course, the stockholders serve as another check and balance on directors. There is the familiar scenario of the widow, living off social security and the dividend from her one share of common stock, that stands up in the annual stockholders meeting, asks a superb and embarrassing question, and brings down the management. This usually happens after all the popcorn has been eaten—near the end of the movie. It can bring tears to your eyes, because she is so well justified at that point in the movie. Contrast that scenario with a short piece that was in the *Wall Street Journal* recently about the annual stockholders meeting of an American multinational corporation. During the meeting, worldwide operations for the year were reviewed, the meeting took 45 minutes, and it was concluded just in time for lunch. Annual stockholders meeting—give me a break!

In an effort to fill the void, some powerful shareholders have come forward recently to pressure boards of directors and professional management to perform more responsibly; one of the most powerful today is Dale Hanson, executive director of CalPERS, the California teachers retirement fund, with $73 billion in assets. In comparison to the one-share-of-stock stockholder described above, CEOs and directors roll out the red carpet when Mr. Hanson visits. As a *USA Today* article [12] indicates, "In November, James Robinson, then the embattled CEO of American Express, met with CalPERS officials in Hanson's office. As he was leaving, he bumped into Paul Lego, then the embattled CEO of Westinghouse. Both lost their jobs in January after bitter struggles against boards and stockholders."

Why am I bringing all of this up? Simply because management has a self-interest—a conflict of interest—in not wanting to expose present operations. An industrial engineer on a large white horse waiting in the wings to set the corporation straight because it's the right thing to do may wait a long time. I know a lot of industrial engineers who have been waiting a long time.

Pascale and Athos, in their book *The Art of Japanese Management* [22], describe Geneen's style of management as follows:

> Geneen's meetings were interrogatory, even adversarial [22, p. 70].
>
> The authors use the phrases "public spectacle," "personal confrontation," and "show trials" [22, p. 70].

> "Fear of humiliation is a powerful motivator. Part of what made Geneen's system work was fear" [22, p. 70].
>
> "For a guy in serious trouble," recalls one former IT&T group executive, "it was like watching a wounded rooster in a barnyard being picked to death by the others. If Geneen zoomed in on a guy, that was the cue for the staff to follow suit. Then he would sit back and watch" [22, p. 71].

Contrast the above with this quote (offered earlier in the management chapter of this book) from Pascale and Athos [22, p. 126]:

> The prime qualification of a Japanese leader is his acceptance by the group, and only part of that acceptance is founded on his professional merits. The groups harmony and spirit are the main concern.

When Japanese managements speak, I suggest that American managements listen. It is difficult to learn when talking; listening is much more effective. I recently listened when an industrial engineer in a Japanese-owned and -managed plant in the United States asked me to guess when the second shift in their plant began. Being accustomed to American plants I guessed 4:00 P.M. He smiled and said, "No. 8:00 P.M." I asked "Why 8:00 P.M.?" His reply was that his management believes that parents should be home with their children in the quality time period from 4:00 P.M. to 7:00 P.M. Japanese management feels strongly that parents should spend that time with their children. Therefore, the shift should start after most of the children have gone or are going to bed. I admit that I would not have thought of that one. I am from a different culture (the western culture) where management has not been truly concerned about what is best for its workers' children; but I am impressed, and I am learning. American management has a lot to learn; it needs to listen.

Robert H. Hayes and William J. Abernathy wrote "Managing Our Way to Economic Decline" [9], which appeared in the *Harvard Business Review,* in July-August 1980. It summarized much of what was wrong in American management then and now. In a section entitled "A Failure of Management" [9, p. 69], they state the following:

> In our judgement, the assumptions underlying these questions [involving risk-aversive management] are prime evidence of a broad managerial failure—a failure of both vision and leadership—that over time has eroded both the inclination and the capacity of U. S. companies to innovate.

In another section entitled "Corporate Portfolio Management" [9, pp. 70–71], they state:

> This preoccupation with control draws support from modern theories of financial portfolio management. Originally developed to help balance the overall risk and return of stock and bond portfolios, these principles have been applied increasingly to the creation and management of corporate portfolios—that is, a cluster of companies and product lines assembled through various modes of diversification under a single corporate umbrella. When applied by a remote group of dispassionate experts primarily concerned with finance and control and lacking hands-on experience, the analytic formulas of portfolio theory push managers even further toward an extreme of caution in allocating resources.

> "Especially in large organizations," reports one manager, "we are observing an increase in management behavior which I would regard as excessively cautious, even passive; certainly over-analytical; and, in general, characterized by a studied unwillingness to assume responsibility and even reasonable risk."

This sort of thinking results in lack of innovation; in addition, portfolio theory fails to deal with the fact that if you acquire a business you do not know anything about, you are probably in trouble. When a steel company buys an oil company because it likes the reported average return on investment in that industry (which one did) (i.e., the grass always looks greener in the other guy's yard) and then it tries to manage it (which it tried to do), the steel company might find that it is in trouble (which it did). Many American companies in the 1990s are divesting themselves of companies that they acquired during the "mergermania period" in the 1970s and 1980s. Management has been telling itself for some time that "If you know how to manage, you can manage anything." Many managements have since discovered that this concept has its limitations—painful limitations.

Hayes and Abernathy further state [9, p. 70]:

> The conclusion is painful but must be faced. Responsibility for this competitive listlessness belongs not just to a set of external conditions but also to the attitudes, preoccupations, and practices of American managers. By their preference for servicing existing markets rather than creating new ones and by their devotion to short-term returns and "management by the numbers," many of them have effectively forsworn long-term technological superiority as a competitive weapon. In consequence, they have abdicated their strategic responsibilities.

Some measure of the problem is indicated in a story related in James Lee's book *The Gold and the Garbage in Management Theories and Prescriptions* [14]. Lee relates the following concerning a corporate president who became aware of management by objectives (MBO) [14, p. 279]:

> Once viewed as a *program* or technique it became a fashion to be followed. Imagine four Fortune 500 corporation presidents who planned lunch together one day in 1965 at the Yale Club in New York. The three who arrived on time began discussing their MBO programs. Since the late president did not know what MBO stood for he ate in relative silence, but upon returning to his office he summoned his vice-president for Employee Relations and asked what MBO meant. The vice-president told him it referred to a program called Management By Objectives whereupon the president asked if they had such a program. When the vice-president told him that they did not, the president told him to hurry and get one going or he would become the laughing stock among his friends.

The above approach I refer to as "silver bullet" or "alphabet soup" management (e.g., zero defects, MIS, MRP, DFA, JIT, TQM, SPC, ABC, XYZ). A manager mentioned to me recently that his plant was employing a "brand-new technology" called SPC. I chose not to tell him that Shewhart started it in the 1920s; it is over 70 years old, which is hardly new. If his management knew that, the danger is that they might have stopped using SPC. The fact that there are as many silver bullets today as there are, and the voracity with which the latest one is pursued by management, whether it fits or not, says something about the quality of present-day industrial management. It is management grasping at straws. What is needed is rational performance improvement planning and implementation, as suggested by Sink and Tuttle [24], or by Tompkins in his recent text *Winning Manufacturing* [27], as examples. A hint of the problem is suggested in H. Ross Perot's *Fortune* magazine article entitled "How I Would Turn Around GM" [23].

My suspicion is that managers often either (1) do not want to expose their operations to critical analysis because they know they have significant unresolved problems in their areas of responsibility, or (2) falsely believe that they have excellent operations for which any attempt at significant improvement is unwarranted. Middle managers have learned through numerous work experiences that shining light on operations can get them in trouble, so they have learned to suppress the temptation to do so. Of course, such a management environment clearly indicates that the top managers are the problem. It is quite common for top management, in its actions (not to be confused with its words), to effectively discourage continuous improvement (i.e., experimentation). Progress never comes easily; there are always problems and setbacks along the way. When a temporary failure or temporary setback comes to light, those attempting change need to be rewarded with top management praise, with such statements as "We learned a lot from that one, keep trying. We are counting on you to change this place. If someone doesn't change the way we do business, we are history. Congratulations for failing on that one; keep up the good work and keep trying." Such management reassurances are uncommon in the real world. That needs to change.

The behaviorist approaches of Herzberg [10] and others have a lot to offer, but they are not *the answer*. What is needed is the combined potential of the knowing-where-we-are attributes of cycle time measurement, quality measurement, and productivity measurement, as examples, combined with the motivational attributes of behaviorist approaches. At the high levels of present-day investment per worker, it is essential that workers be interested in maximizing whatever throughput is possible from the equipment they control, directly or indirectly. Unfortunately, Japan now has a significantly greater investment in plant and equipment per employee as compared to the typical American worker.

The need for American management to gain an understanding of its own operations is well demonstrated in a story related in Sink and Tuttle's book *Planning and Measurement in Your Organization of the Future* [24, pp. 146–147] involving an experience that one of the authors gained on a summer job:

> When door panels were run that were painted white, the shaders—and I was one of the five shaders—basically did not have a job. So I would go over and stand by the spot where the overhead conveyor would bring the panels up from the first floor. I would watch for a ticket that would tell us which color to paint the next batch. As I waited for the tickets to come up, I noticed the inspectors and observed them using their red pens a lot. One day I went over and asked the inspectors what they were looking for when they decided whether to give a panel a red mark. They impatiently described the specifications they were instructed to apply. I asked them where most of the errors appeared to be coming from, and their reply was "the shaders." I then asked if they kept data on errors, and they shared the daily report sheets with me. I asked for these reports daily—while the white parts were run—and began to track good parts/total parts run (my measure of performance).

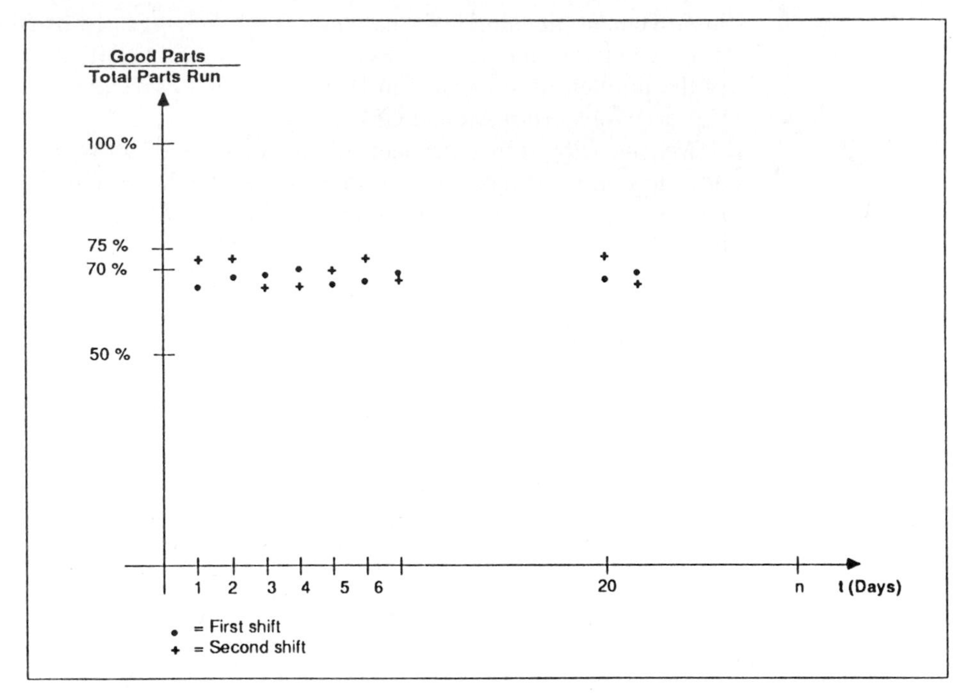

FIGURE 10-2 Run chart (before intervention). (From Sink [24, p. 147].)

Figure 10-2 presents the basic findings.

After about a week of tracking this data, I decided to post the results on a bulletin board (something in American organizations that could be used effectively as scoreboards if management only knew how) at the end of the paint booth work area. The data I presented was for the past four weeks, good parts to total parts painted, first shift and second shift, tracked day-by-day. Note that the paint line was apparently in steady state, performing at seventy percent, plus or minus a couple of percent. I explained to my team of shaders what the numbers meant and what the inspectors on the other side of the wall had told me they were looking for as they marked up charts. (Note: This was the first time, I learned, that the shaders had been told what the specifications were for the job. The training for the job had entailed placing a new shader in between two experienced shaders and saying, "Look to the left, look to the right—do what they tell you to do." I think they called this "on the job training.")

After a week, a little competition between shifts had set in, a little Hawthorne effect had undoubtedly set in, and performance (as measured, partially at least, by good parts to total parts painted) had improved substantially [see Fig. 10-3].

As you can see, performance increased from seventy percent to ninety-seven percent—plus or minus a couple of percent—and then settled down to about ninety-five percent after a couple of weeks. Performance stayed at ninety-five percent for almost three months. I kept the charts updated daily in my spare time as white parts ran on the line. One day, close to the end of my summer job period, our shift supervisor came up to me and asked me to explain the charts on the bulletin board. I proudly explained what I had done and explained the results. At the completion of my explanation, he shook his head

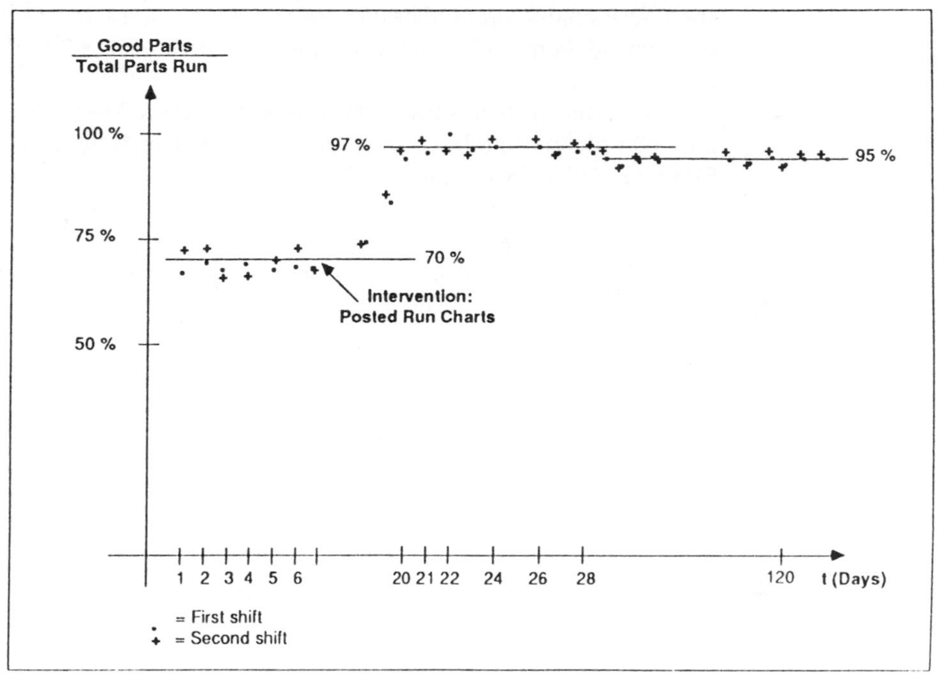

FIGURE 10-3 Run chart (after intervention). (From Sink [24, p. 147].)

and stated that he had severe reservations about my actions. First, he explained that I had created a major problem for him in that he now didn't have anything for his rework people to do! Second, he explained that it wasn't in my job description to analyze the past data. Third, and finally, he explained that the bulletin board was for official company business only and said that I should remove the charts from the board and cease spending time on the activity.

The type of analysis necessary to solve the underlying problem indicated in the above story can be simply described as "engineering the work." It is typical of the type of analysis necessary in plants today, and as performed by Frederick W. Taylor and his associates at the turn of the last century. The need is still there whether provided by a worker, by a foreman, or as basic engineering analysis of the process.

One of the obvious keys to success in operating a production system is the quality of first-line management, as evidenced by the example. If you do not have the first-line supervisor on board on a broad-based productivity improvement program, the program can only be partially effective, at best. With no words and a minuscule amount of body language, the first-line supervisor can instantaneously gain the support of his or her people, so that whatever you want is not what they want. They are not going to give you what you want just because it seems like a good idea to you. They are line, you are staff; they have work to do, and you are a temporary production interference. The most naive person in a plant can be the brand new industrial engineer who thinks that because he wears a shirt and tie, or she a fashionable dress, that the workers will recognize the inherent value of the recommendations and embrace them with all their hearts. It typically does not work that way.

A lot of the problems that hamper productivity in American operations represent a myriad of small problems that have been neglected by management in the past. Employees view the failure of management to deal with these issues as their justification for stating (on a regular basis) "If management doesn't care, why should I?" There is a name for the missing ingredient here—it is called "leadership."

The United States also suffers from some cultural deficiencies. We are a throw-away society. We have been taught to dispose of things if they do not suit us rather than repair or maintain them. "Buy something, use it, and if it doesn't suit you, throw it away and get another one." As mentioned in the total quality control section of this book, kaizen is an ingrained concept in the Japanese culture; Imai's text (see Chapter 5) bears *Kaizen* as its title. Kaizen means improvement. "Habit of improvement" and "perfection" are assumed goals of the Japanese total quality control approach. In years past, American firms have not done a good job in fostering continuous improvement, and only recently have begun to embrace the need to embed continuous improvement into the American industrial culture.

Management's focus in the past three decades has been toward espousing behavioral management while often continuing to practice Theory X, legal solutions, financial management, and sales. One wonders—if a typical American car manufacturer spent the money it spent on advertising and legal fees on making cars that work, maybe it would be much further ahead today. The focus has always been "get it off the lot," with far less concern about how far it will go down the street before it has to be hauled back to the dealer for repair. For example, I have a mother and two sisters in Ft. Lauderdale, Florida. A not uncommon Monday morning scenario in the past has been that my mother follows one of my sisters to the Cadillac dealer in her Volkswagen, and then returns my sister to the dealer later in the day when the repairs are completed. I do not ever recall one of my sisters following my mother to the Volkswagen dealer. Could this partially explain why the only Cadillac dealer in Orlando, Florida, with a great captive market, went bankrupt? If you want to know more, see H. Ross Perot's enlightening *Fortune* magazine article entitled "How I Would Turn Around GM" [23].

Industrial engineers have always respected the broad range of topics included in an MBA program, from marketing to accounting to human resources to management to international finance. Realistically, however, because of the broad range of topics to consider, only a limited amount of time can be devoted to each topic. Production system design and control typically represent part of one operations management course in an MBA program. There is barely enough time to learn the terminology, much less to acquire a detailed understanding of the technology. The typical industrial engineer spends much of five years gaining an in-depth understanding of production system design and control; that effort alone represents a series of a dozen or more junior- and senior-level courses, as well as many other supporting courses (e.g., probability theory, design of experiments, accounting). Business graduates and industrial engineers should both respect each others' potential contributions and work together as a team to conceive and develop the best production systems possible to accommodate the strategic plans of their corporation.

There has also been too much silver bullet and alphabet soup management in the past; there have been too many managers trying to catch up with too many management concepts they do not fully understand. Hopefully there will be a greater effort by managers in the future to draw upon the services of others (e.g., industrial engineers, manufacturing engineers, operations researchers, systems engineers, etc.) for which concentrated technical education has been acquired to assist in developing such systems of the future.

In too many instances there are so few, if any, industrial engineers in an operation that those that are in operations are overwhelmed with the work that needs to be done, often with little effective guidance, leadership, and support from above. If an operation has 1000 employees and one industrial engineer, he or she will never be able to accomplish what needs to be accomplished in that plant. A more commonly accepted ratio in the past in a typical industrial operation has been one industrial engineer for every 100 direct labor employees. As industrial engineers increasingly consider indirect and overhead operations, the ratio must be broadened beyond direct labor employees.

A railroad corporate office that I am familiar with has 70 auditors and no industrial engineers. The collapse of savings and loan organizations suggests that auditors have been only partially successful in years past. Most of their work is done in an office with little if any on-site understanding of the underlying operations they are auditing. If the railroad regional office above had 35 auditors in the office and 35 industrial engineers in the field, I suspect that a much greater ultimate accomplishment would be possible. Financial auditing is an established practice in American management practice. Operations auditing is not, and it needs to be.

To offer some sense of the possibilities, consider the following. The railroad mentioned above performs about 70 diesel locomotive overhauls each year at a cost of $800,000 per overhaul, at an annual locomotive overhaul cost of $56 million per year. Are the overhauls performed using the latest industrial engineering techniques, or by "good old boy" production supervisors who do it the same way they have always done it? Many of the top management people that I meet have no idea how it is done on the production floor; they assume that it is being done the right way. There is no connection between those doing the work and those responsible for the operation. The logical, cost-effective connection between the two is industrial engineering.

My firm recommended to an airline management some years ago that as an extension to a consulting assignment, we develop a new maintenance approach in the form of a pilot maintenance hangar to demonstrate to management how Level D (i.e., major teardown and inspection) maintenance could be done far more cost-effectively. At a cost of $50,000 a day for each additional day an airliner remains in maintenance, a CPM (critical path method)-developed technology that over time would provide probabilities of repairs by aircraft model and their effects on schedule could have produced dramatic scheduling effects. We were not given the assignment. The comptroller in a top management discussion of the project asked the question "Has any other airline done this?" (i.e., your classic "don't lead, always follow" type of risk-aversive, nonleading management, basically programmed for mediocrity, if not failure) and management choked. They are probably still waiting for some other progressive airline (e.g., American Airlines) to do it first. There are too many top management people in our economic system effectively acting as roadblocks to progress; where "do not rock the boat" is too often the name of the game, with no one requiring them to do otherwise.

Both the railroad and airline studies above would have been a candy store for industrial engineers, but these studies, and many more I could mention will never see the light of day because of risk-aversive, "maintain the status quo," non-continuous-improvement-oriented upper management. Allotting $100,000 to send all the sales engineers and their spouses to Hawaii is a repeatably acceptable investment in typical American corporations, but studying the way we do maintenance—something we are already supposed to know how to do—is far less acceptable and typically only marginally funded, if at all.

A survey entitled "The 1990 Productivity Survey" [20], performed for the American Electronics Association, and based on 262 responses to 3000 questionnaires in 1990, provides some understanding of the degree of relative acceptance and the rate of implementation of many of the approaches described in this text. Obviously, with such a low response rate, one would probably be correct to assume that those that did respond are more progressive than the average of the industry. Figure 10-4 indicates the level of response for what constitutes the top three factors concerning competitiveness for the respondents. Note that "quality" received the highest number of responses (i.e., 90). Ranges provided in figures from this survey are for responses from the various segments of the industry. Note that if "service," "technology," and "options" are considered as relating to quality, the three primary variables of "quality,""cost," and "delivery" are represented.

FIGURE 10-4 Competitive factors. (From *1990 Productivity Survey* [20, p. 6].)

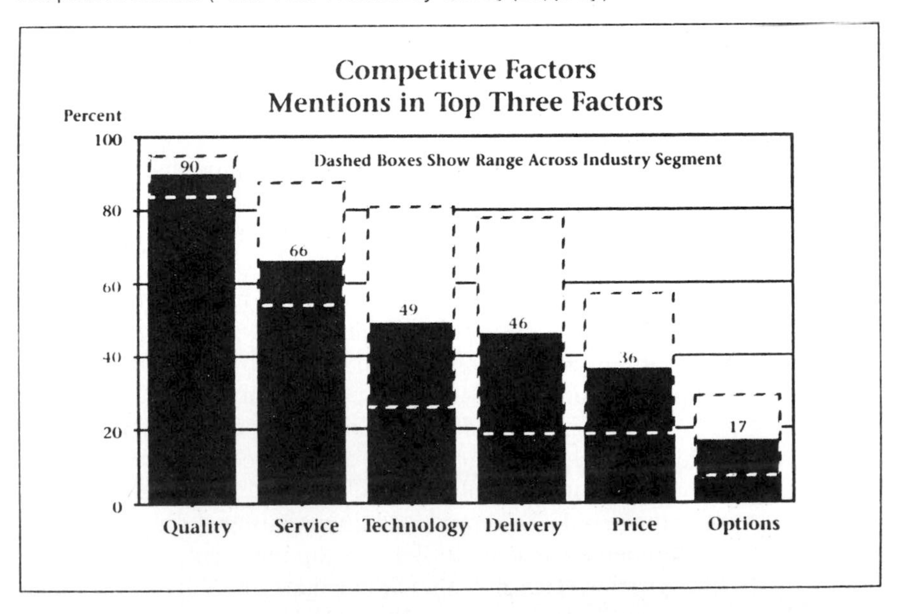

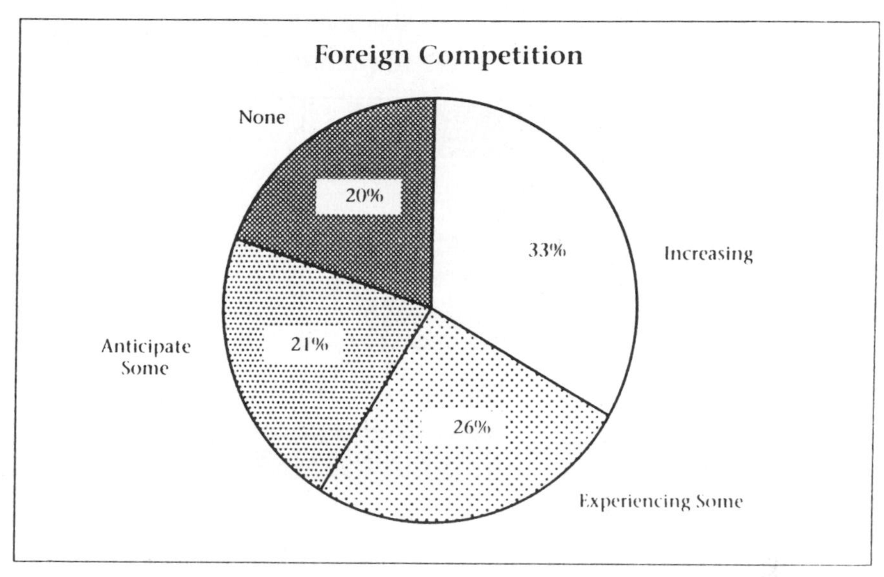

FIGURE 10-5 Foreign competition. (From *1990 Productivity Survey* [20, p.19].)

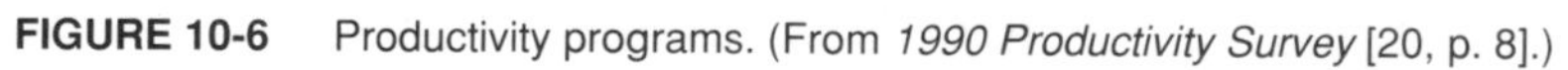

FIGURE 10-6 Productivity programs. (From *1990 Productivity Survey* [20, p. 8].)

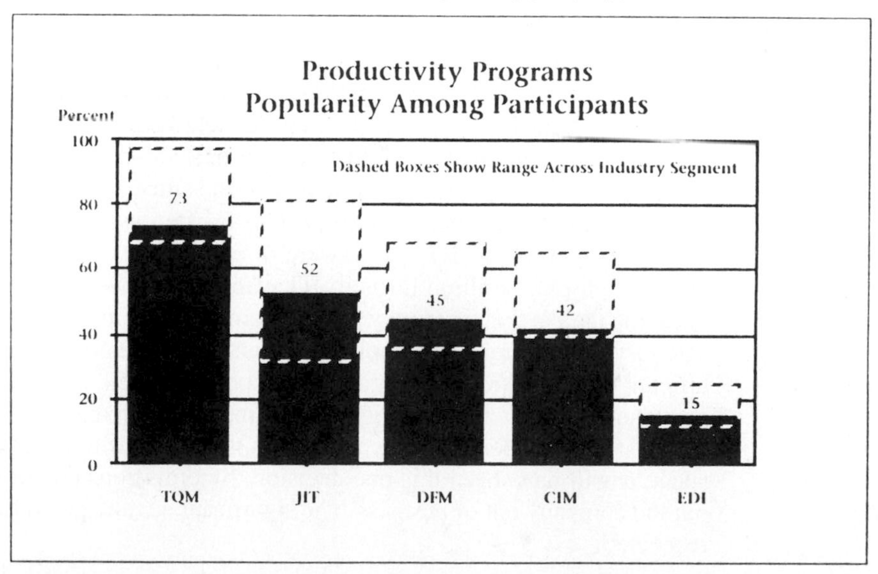

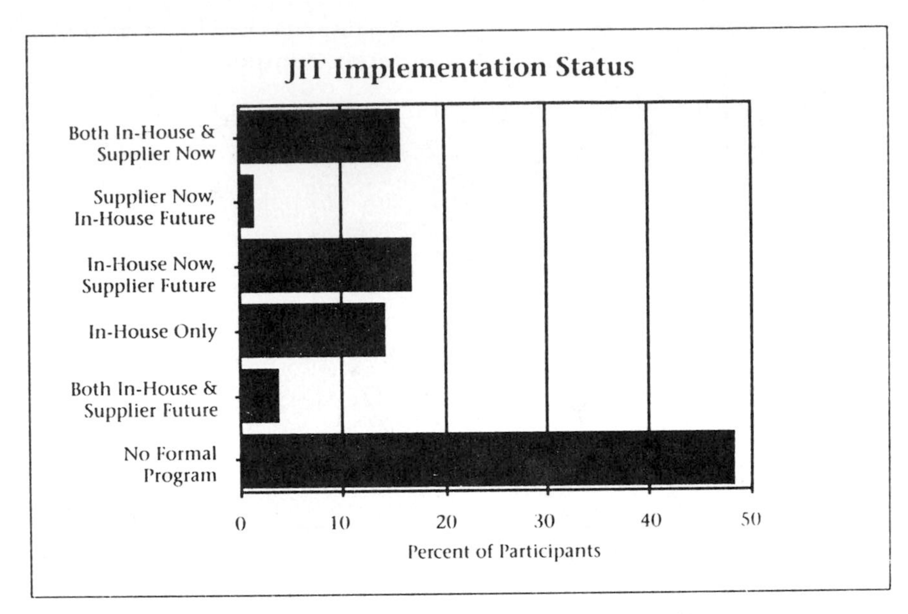

FIGURE 10-7 JIT implementation status. (From *1990 Productivity Survey* [20, p. 45].)

Figure 10-5 indicates the degree of foreign competition perceived by respondents. Note that 80 percent of the respondents either expect or are presently experiencing foreign competition; it is an integral part of a world economy. It highlights the need for "world-class manufacturing" capabilities if one hopes to survive and prosper.

Figure 10-6 illustrates the relative acceptance of five techniques discussed in this text: TQM (total quality management), JIT (just-in-time manufacturing), DFM (design for manufacturability), CIM (computer integrated manufacturing), and EDI (electronic data interchange). It may be noted that almost three-fourths of the respondents are in favor of TQM, and over half are in favor of JIT.

Figure 10-7 illustrates JIT implementation status. Note that almost half have "no formal program" for implementing JIT. It is often assumed that process industries are not well suited to JIT. I am aware of a steel firm that has recently reduced its inventory by $80 million through JIT efforts. The suspicion, however, in considering the data is that there may well be some light manufacturing operations in that category that are resisting implementation of a JIT approach. JIT offers such significant opportunities for improving an operation well suited to JIT that if such an opportunity is not embraced, it is likely that the company will have great difficulty successfully competing with competitors that have. It is conceivable that, in retrospect, it will have been this one decision, by omission of action or otherwise, that put the company out of business. The significant advantages JIT can offer need to be respected.

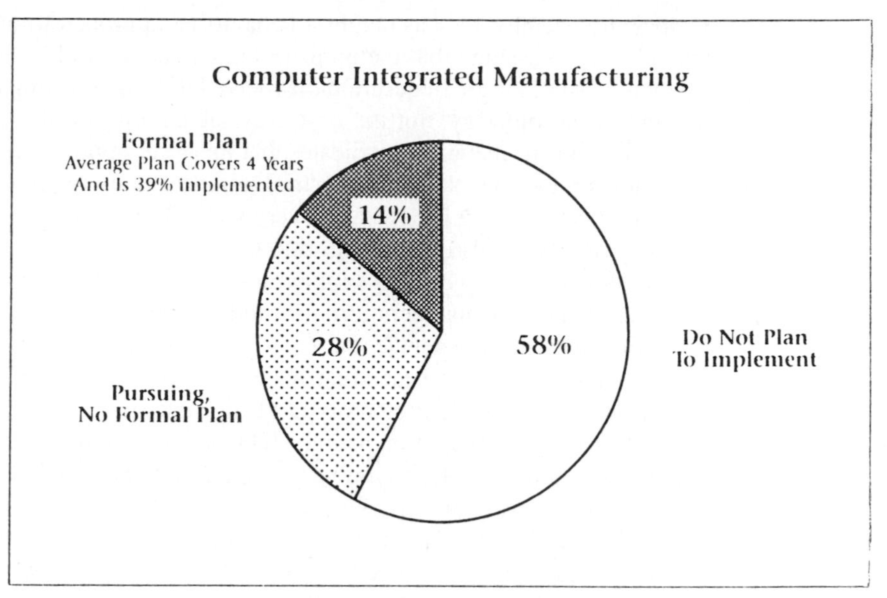

FIGURE 10-8 CIM implementation status. (From *1990 Productivity Survey* [20, p. 61].)

Figure 10-8 illustrates a general lack of interest in embracing CIM (computer-integrated manufacturing) in an industry in which computer use is familiar and prevalent. Note that only 14 percent of the respondents have a formal plan for pursuing computer-integrated manufacturing. Although the Allen-Bradley Corporation, for instance, has demonstrated success in developing a computer-integrated manufacturing facility, it appears that the preponderance of firms do not include that as an immediate objective. Computer-integrated manufacturing is more of an inevitable evolution than a program. For those of us who have survived the evolution from wearing slide rules on our belts to calculators to mainframes to PCs, computer-integrated manufacturing seems likely, warranted, desirable, and inevitable.

THE IE OF THE FUTURE

As mentioned in Chapter 4 of this book, Peter Block, in his book *Flawless Consulting* [2], stresses the need for three sets of skills: (1) technical, (2) interpersonal, and (3) consulting. It makes little sense to send graduating engineers out into the management "mine field" that exists today without some understanding of interpersonal and consulting skills. Industrial engineers by the nature of their work have to deal with the human element, and that human element typically represents both employees and management. Of the two groups, employees are typically the lesser problem. Dealing with management is typically far tougher.

Consider the following scenario. A recent graduate industrial engineer is attempting to convince an experienced and older present operations manager that things need to be changed in the operation. The manager may well assume that the industrial engineer has a different philosophy from his personal management philosophy because he was taught a behaviorist approach to management, and he may choose to assume that the industrial engineer is a holdover from Taylor management. The manager's education may have left him believing that industrial engineering is the old way, not the new way, of gaining productivity improvement. When the industrial engineer indicates that there are opportunities for improvement, the manager assumes that the industrial engineer does not think that he (the manager) has been running his operation very well. That may well be the case, and that may be why there is so much opportunity for improvement. The manager's ego is hurt. In defense of ego he may simply rationalize that the young industrial engineer understands neither his operation nor modern management philosophy. This may be sufficient for him to decide not to cooperate with the young industrial engineer.

The manager also assumes that his employees do not have a high opinion of industrial engineers; their stereotype of an industrial engineer is "someone who unfairly takes jobs away from people." The manager may also have some suspicion that his operation is not running very well, and he does not want managers above him to become aware of the deficiencies in his department, which he does not know how to fix. He therefore has a love/hate relationship with the industrial engineer. He would love to fix the problems in his operation, but he would hate to have to admit to his boss that he has problems. Because management people often vehemently dislike poor management that comes to light, but seem willing to accept poor management of which they are unaware, he may consider it simply best to "lay low in the grass." That is the management mine field mentioned above. Graduating industrial engineers need all the help they can get in interpersonal and consulting skills. The history of industrial engineering includes accumulated frustrations associated with failed attempts to implement industrial engineering. Any learning that will increase the probability of successful implementation is sorely needed.

As a member of the University of Central Florida Industrial Engineering Advisory Board, I had occasion in 1989 to conduct a local industry and government survey of the UCF undergraduate industrial engineering curriculum. A senior industrial engineer at each of 49 local industrial and government organizations was asked such questions as "If you could make one course change in the curriculum, what would it be? What course would you remove? What second course would you add? What second course would you remove?" I found the results to be both predictable and interesting. Ranked first for adding courses was communications, with 17 percent of the responses; second was computer applications, with 12 percent; and third, with 7 percent of the responses, was management and manufacturing philosophies. Number one above—communications—is consistent with Peter Block's [2] two major skill sets of interpersonal relations and consulting skills, which are predominantly communications skills. The respondents were saying that "If you can't communicate effectively, all the techniques in the world won't get the problem solved." By solved, I mean successfully implemented and then managed on a day-to-day basis.

References in management literature to Taylor [26] as being the way of the past, with phrases such as "no longer applicable in today's participative management environment," are both right and wrong. They are right in that the "expert" mode of instituting improvements is increasingly less applicable today, but they are wrong in that the preponderance of workstation improvement principles still apply, and industrial engineers are needed to teach these concepts to those who need the improvements in their workstations. Management is unaware of the lack of rational analysis and managerial discipline that is practiced on a day-to-day basis on too many production floors. I am certainly not against participative management; it is the way of the future. But it need not exclude industrial engineering input as part of these team efforts. A combined approach to quality and productivity improvement can be the best total solution. There is no reason why functional and cross-functional teams should not pursue operational quality and productivity improvements employing valuable industrial engineering input.

During a consulting assignment concerned with improving quality control operations in a plant, while reviewing production operations on the production floor, I became aware of what had been identified as the No. 1 quality-control problem in the plant in terms of defective product generated. The product is called an intraocular lens; it is inserted and attached in the human eye as a lens replacement for a lens removed in a cataract operation. The problem involved bulging of the artificial lens due to a hole for staking a lead to the lens being drilled too deeply during its manufacture, which caused optical distortion in the lens, as indicated in Fig. 10-9. The lead is used by the surgeon to attach the lens to the iris to hold the lens in place. Staking is nothing more than pressing down on the lens with a steel point to deform the plastic around the lead to hold it in place in the drilled hole. I noted that the lead extended five to seven lead diameters into the lens, beyond the stake point, and asked why. It seemed reasonable from a basic engineering point of view that any lead material more than about a diameter and a half or so beyond the stake point was not assisting the staking function, so I suggested that the drilled hole be reduced in depth to three diameters in length. A batch of product was then made that way and tested, and the No. 1 quality-control problem was solved. It was an obvious solution to the problem.

More important than the solution was the fact that a solution had not been implemented. Because it was their No. 1 quality problem for some time, I assumed that it must not have been as obvious to others, although it should have been. One might

FIGURE 10-9 Intraocular lens.

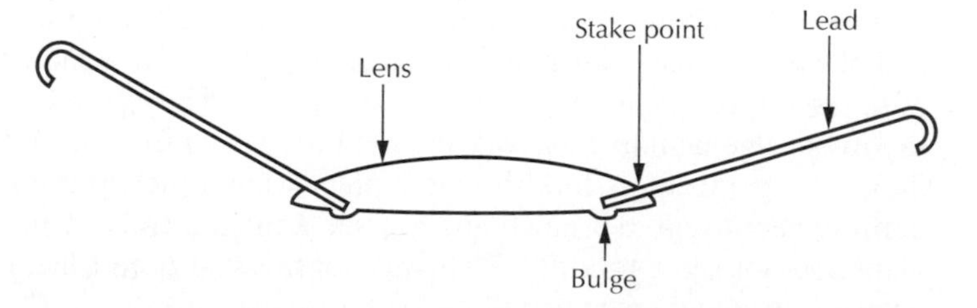

consider, however, that the focus of operations people is to get product out the door; they often do not have the time, or take the time, to improve their operations. It should also be noted that operations people, as good as they are at what they do, typically do not have the advantage of an engineering degree and follow-on related experience that provide a different view of the physical world (especially in the United States as compared to Japan). In my report to management, I asked why in six weeks I had not met another engineer on the production floor. The engineers were one flight up in the engineering office. I have no idea what they did, but I assumed it was important. The point is that there was a critical need for engineers on the production floor and they were not there.

In the final cleaning operation just prior to final inspection I asked a lead the following question: "The label on this box indicates a lot of 50. Does that mean that there are 50 lenses in the box?" His answer was that there could be more or less than 50 lenses in the box. It was my understanding that an operator cleaning the lenses would inspect a lens while cleaning it, and if the lens was obviously unacceptable, the operator would simply throw it away. I posed the following question to the lead: "If the box label indicates a lot of 50, but there are actually 60 lenses in the box, and the first 50 lenses that an operator cleans appear good, what does the operator do with the remaining 10 lenses?" His answer was, as incredible as it may seem, "We throw them away." When I asked why, he answered, "The computer [shop floor control terminal in his area] thinks we have 50 in the box and will not accept more than 50." You can imagine the reaction of the manager of quality control, when I related the story to him, and the director of quality control, when I related it to her. The manager of quality control spent some time with me on the production floor as I continued my investigation. Later he said, "Phil, how can you ask questions and get the answers that you do, and keep a straight face?" My answer was, "You get used to it."

Another case in point in another company involved a supervisor whom I asked why I had not seen an AGV (i.e., automatic guided vehicle) come by following the wire in the floor; she indicated that it had been shut off. When I asked why, she said, "Because it ran into boxes." That is really what she said. Obviously, if the AGV is going to run into boxes we should shut it off. A few minutes later, in a large rack storage area, I noted something was wrong but could not immediately put my finger on it. Then I noted that the two-way pallets in the racks were all turned 90 degrees, and the spacing on the support was such that they did fit that way. I asked, "How do you get pallets down from, or up into, the racks with a forklift." The answer was, "We don't have a forklift." They were placing material into the rack system, really using it as a shelving system, using portable stairs, carrying each carton up the stairs and placing it on the pallet, and taking cartons down the same way. My favorite terminology for this kind of thing is "bananaville." They asked me what they should say to management to get a forklift, and I suggested that they state the following to their management: "A forklift is to a plant what a pickup truck is to a ranch. You don't expect to go to a ranch and not see a pickup truck. One should not go to a plant and not see a forklift." That was not the first time I had offered those statements; I have them memorized.

Many workers have been trained to view the industrial engineer as the enemy (i.e., the stopwatch syndrome). I sincerely believe that if employees understood what an industrial engineer does and can do for them, they would realize that he or she is their best hope for a successful future in their company. Management and the industrial engineering profession have done a poor job of presenting this message. What do employees want most from a job? A job (i.e., a paycheck). Without the job, nothing else really matters. No one wants to go home and tell their spouse and children that they have become unemployed, with the possible rare exception of a lottery winner. Industrial engineers, in their quest to make plants and other operations more cost-effective, produce exactly what each worker really wants—job security. Job security does not come from a piece of paper, as some people have been led to believe, signed by management alluding to some measure of supposed job security. The management really cannot guarantee that for themselves, let alone for someone else. If management cannot guarantee job security, how does one attain it? The answer is quite simple—use industrial engineering and other essential business and technical disciplines. If the industrial engineers do their job properly, and others do their job as well, the plant will produce the right product at the right time at the right price. That *is* job security.

There is a considerable need in American plants for simple basic analysis of what workers are doing on the production floor. Even more important, at the systems level of analysis, there is considerable need for asking, "What are we trying to accomplish in this plant, now and in the future?" And then serious efforts need to be undertaken to conceive, develop, and implement a rational production system to meet those needs, develop appropriate control procedures, and then successfully manage the system on a day-to-day basis.

If you were to land on earth from outer space and read a typical trade magazine or technical society journal (e.g., *Modern Materials Handling* or *IIE Transactions*) or sit in on an industrial engineering lecture at a local university, you might come to believe that what we need in the United States is computer-integrated manufacturing (CIM) or nonlinear programming. The reality in much of industry is that what is needed is basic industrial engineering—engineering analysis of the production system. When the existing system is improved and is producing 50 percent more than at present, with a third or half as many workers, with 20 percent of the present amount of material on hand, in half the amount of space, maybe at that point it is time to think about computer-integrated manufacturing. CIM involves a considerable investment, and whether it will pay or not in a reasonable time period is a tough question to answer in many instances. Getting the existing production system to function much more cost-effectively than it presently functions often involves either no, low, or only modest investment. Over the long haul, major investments may well be required to attain other potential long-term improvements. In many cases, however, half of the improvements are simply training issues (e.g., don't throw the scrap on the floor because you will have to pick it up later; throw it directly into a proper, and properly located, scrap container at the machine). If all of Mildred's output goes to Margaret, move Margaret next to Mildred, or vice versa. I do not know how many times I have said, "Get the guy a decent chair."

As a beginning industrial engineering student, you might ask, "What is Hicks talking about? Haven't we done all that by now?" I suggest that you take a hard look at what is out there; you might be surprised. In the intraocular lens plant example above, management and I agreed that I would collect production forms along the way to see if there was opportunity for improvement there as well. When I got to the end of the process, I had collected more than 70 forms. Most of them were log books: "On March 20, at 9:05 A.M., I handed Harold a box of 55 lenses, type CSSR." So what? All of the time spent filling out forms takes time away from value-added production time. There is an analogous situation in teaching. The more time devoted to testing, the less time there is available for teaching. If you want to teach the most, test less. If you want to produce the most, fill out the least number of forms possible. In one meeting, over half of the forms were summarily eliminated. Over the long haul, management discipline is necessary to see that the unnecessary logs do not reappear in the future. At one point in the process, oven temperatures were being taken every half hour. When traced to its conclusion, the explanation was that the director of quality control had performed a "two-week" study two years ago, but the data collection had become institutionalized. This is not a rare occurrence; in too many instances this is the norm.

The magnitude of the problem in many U.S. plants is suggested in the following story. I was visiting a plant that was supposed to be further along in embracing participative management than other plants in the company, and I was there to help improve productivity. I was asking an operator about problems with her machine. She had mentioned a number of problems, such as a fixture that would stick, having to use a wrench because she had no nut runner, having to use shim material in the fixture because the tooling department had not fixed her fixture, etc. Quite suddenly she stopped, looked me straight in the eye, and said, "Have I told you too much?" Her statement was recorded in the executive summary of my report. Her question says volumes about the plant, and especially about the management. It shouts, "Employee input is not welcome here!" Why? Because management in that plant had a big ego, and they are in charge of deciding what will be improved and what will not, and they do not want help from employees. It went a long way to explain to me why productivity was so poor in that plant. A tremendous wealth of information—all the things the employees know need to be improved—is bottled up in their heads, because they have learned that employee input is not welcome. It is unfortunate that this kind of management is allowed to exist in this day and age.

Ineffective management exists far too often in the United States, and we have to get beyond it if we expect to compete internationally. Without effective management, nothing else matters (CIM or nonlinear programming, for example). The solution is in the details; it is time again to look at what workers are doing in their workstations, and ask how and why. Better yet, following training, have them look at their workstations. In the United States, particularly in the last 20 years, the biggest silver bullet to come along has been quality control. Can anyone be against quality control? Not really. Do we need improved quality? Of course we do! But should it be the singular, all-encompassing end in itself? When you reach ultimate quality, have you won? Not necessarily. It is still necessary to produce the right product (and quality control helps do that); but it is still also necessary to get it to the customer at the right time and at the right price. The triad of quality, cost, and schedule is inescapable. While we chase "quality," the Japanese are perfecting "productivity, quality, and time to market." We need to get focused on productivity and time to market as well, if we hope to catch up to them in the future.

It is interesting to note that some people react to "productivity" as if it were a bad word. It sounds like "efficiency," another word to which people have also been taught to react unfavorably. And yet the American farmer improved farming efficiency such that, whereas one farmer in 1960 fed 5 people in the United States, a farmer in 1990 now feeds 90 people. When I was a boy in the 1940s and 1950s, my father worked dawn to dusk, including Saturday. About the only time I saw my father was on Sunday, and he was often sleeping to get ready for the next week. I do not have a problem with a 36-hour work week. It represents a significant standard of living improvement as compared to leisure time available 50 years ago. If industrial engineers are successful in their endeavors and the work week is shortened to 18 hours a week, is that a problem? Not really. It really is not if we can produce as much product in 18 hours as we now produce in 40. A six-hour work day, three days per week, does not seem like a bad idea to me. Why not?

What, then, are the prospects for industrial engineering? They are considerable. Discussion throughout this text has been concentrated on problems, and every discipline has them. What was not discussed was the successes. Of course, there has not been as much success in the application of industrial engineering as some would like, but there have also been a lot of success stories. What is important, however, is that industrial engineering is making a greater and greater impact on the solution of problems of human systems. It seems likely that in the future, an almost inexhaustible supply of difficult human–machine systems problems will be available to challenge imaginative industrial engineers for a long time to come.

The greatest weakness of present industrial engineering graduates is likely to be an insufficient knowledge of psychological, sociological, and managerial environments, because so much of their curriculum is limited to engineering topics. This needs to be improved. Consider also that it is people who make organizations effective. Stephen R. Covey, in his *The Seven Habits of Highly Effective People* [4], provides a *must-read* text to assist any professional who wishes to focus his or her efforts toward a successful professional career. His seven habits, 1) Be proactive, 2) Begin with the end in mind, 3) Put first things first, 4) Think win/win, 5) Seek first to understand...then to be understood, 6) Synergize, and 7) Sharpen the saw, represent a powerful approach to personal effectiveness. From dependence to independence to interdependence (and team-based management) is a more complex and demanding management philosophy. When done right, however, team-based participative management is so much more effective than past management approaches. Covey offers persuasive arguments for developing and embracing a character ethic rather than a personality ethic. He invites you to review and reconsider the principles that constitute your personal managerial philosophy, which largely determine your professional behavior.

The participative style of management is becoming a fact of industrial life today, as it should be. Does that mean that industrial engineers will no longer design production systems and controls for production systems, and will simply be relegated to serving as facilitators for quality circle groups? I do not think so. With some exceptions, managements still believe that industrial engineers should develop production systems and the controls for such systems. What has changed is that the industrial engineer has taken on the additional responsibilities of assisting in both the planning of productivity and quality improvements and their implementation. It means that industrial engineers will therefore also assist in the development, management, and support of functional and cross-functional teams of employees in search of continuous improvement of their operations. There is every reason to have active team efforts so that everyone can participate and contribute to continuous improvement, and industrial engineers are well equipped to be deeply involved in such efforts.

Through participation, concepts that might have been rejected because of the NIH (not invented here) factor will likely enjoy higher acceptance in the future. Such approaches simply represent an approved understanding of what constitutes a better solution to the psychology of implementation. Industrial engineers must improve their ability to understand the psychology of implementation if they hope to improve their performance as a profession in the future. There is a considerable opportunity for closing the gap between industrial engineering and psychology, a gap that has been essentially ignored in the practice of industrial engineering to date, with the exception of Morris [18], Block [2], Sink and Tuttle [24], and a few others.

I suspect that many of my industrial engineering professor friends may be a little uncomfortable with my proposal that the industrial engineering curriculum needs to broaden to include more management and psychology. They know industrial engineering and that is what they want to teach. The fact, however, is that their students need more than traditional industrial engineering courses to be effective industrial engineers. The curriculum obviously should primarily reflect the students' needs, not the professors' teaching preferences.

This need for a broadening of the education of an engineer is well expressed in the following quote by the president of Georgia Tech, John Patrick Crecine [3], in the *Engineering Times*: "Kids get a terrific technical education, but often the organizational skills, the communication skills, skills that are required to operate in careers just haven't been there. So what we are trying to do is broaden the intellectual base of the institution."

I wish to leave you with one final thought. Repeatedly throughout this text I have discussed techniques whose sources were often outside the United States; in fact, quite often Japan. This is a turn-off to many audiences. Typical of the sentiment is the joke about the two men, one American, the other Japanese, on the way to the gallows. The sheriff explained that they would each be limited to one minute to speak before the hangman pulled the lever. The men flipped a coin to see who would go first. When the American won the coin flip, he said, "I elect to go first; I don't want to hear another short lecture on Japanese management." Such men have a lot of pride, as most of us have, and they would like to believe that we do not need help from outside the United States. My personal philosophy is that good advice should be embraced from whatever source. The source is irrelevant; the quality of the advice is what is important. If a source consistently gives good advice, does it not make sense to listen? I suggest that we swallow our pride and learn to listen.

SUMMARY

As one of the major surviving economies following World War II, the United States had a considerable competitive advantage in offering its products worldwide. Such a virtual monopoly made being successful easy and led to lack of attention to detail. The reality today is that there are numerous world-class economies competing with the United States to sell products worldwide—Japan, Germany, France, Korea, Malaysia, Hong Kong, Singapore, and China, to name a few. Today each country must compete on merit.

Throughout this chapter numerous references were made to management problems and a lack of industrial leadership in the United States. The Rollins MBA Program student survey, described earlier in this chapter, suggests that practicing managers are well aware of this problem as well. It is time to address the problem of management education in American MBA schools.

Symptomatic of the problem is the level of silver-bullet management being employed today. A never-ending wave of fads are attempted, often with insufficient analysis of whether they represent a good fit for a particular operation. Often those that do fit are later replaced by others that represent a step backward; too often, it is one step forward followed by one step backward.

One manager, recently asked if his organization intended to pursue TQM, indicated that they "did that." Responses to further questions indicated that they had had a meeting and discussed it.

There are also too many instances of managements attempting to technically lead production system design efforts for which they are technically unqualified, and unfortunately they do not know that. Many practicing managers were taught in their MBA programs that they were ready to go out and change the world, and they did. The question is, "Is it better now?" Many managements need to take a hard look at what their operations are and what they need to be, and then recruit the technical talent (e.g., industrial and manufacturing engineers) to develop and manage the installation of improved production systems, and the controls for such systems. The need is now.

Much of the work to be done by industrial engineers is modest- or no-investment stuff. In many cases they represent training issues. People on the production floor simply need to be told that there are principles concerning the performance of work that need to be followed. In many instances such principles simultaneously increase output, reduce fatigue, reduce risk of injury, and improve the quality of working life. A common theme in much of what needs to be addressed is simply an issue of discipline (i.e., doing things in a disciplined way). Go back and look at Fig. 2-63 (i) and ask yourself again, "Can anything of quality be produced productively in that workplace?" Workstations such as these should simply not be allowed, and an effective management would not allow such workstations to exist—those managers who do should not be managers.

What is the underlying message of this text? Simply, that in considering the logistical advantage that American manufacturers have in supplying the largest market in the world (i.e., the United States), there is absolutely no excuse for American manufacturers to be anything other than the best in the world for the products they represent. What is lacking is the commitment by top management to get it together. If there is a best silver bullet today, when it applies, it is probably JIT. In second place—and it applies everywhere—is TQM. In addition to these two approaches, all of the other improvements can and should come from across-the-board incorporation of all the techniques, approaches, and ideas mentioned in this text, to the extent they apply.

Management obviously needs to develop a relationship with its employees, suppliers, and customers to make it possible for them to solicit, consider, evaluate, develop, and implement the hundreds, if not thousands, of potential improvements in their operations; and they need to do it now. Any manufacturer today needs to give serious consideration to how it will approach the following requirements and then act: concurrent engineering, methods engineering, ergonomics, work measurement, production and inventory control, quality control, material handling, plant layout, maintenance, and many more. It might be noted that these are the titles of courses that industrial engineers take in acquiring a degree in industrial engineering. Manufacturing management should take advantage of this available resource.

To make the point that what was stated above is not just rhetoric, here are some quotes from the introduction to Edward J. Hay's text *The Just-In-Time Breakthrough: Implementing the New Manufacturing Basics* [8, pp. 6–8]. They demonstrate how a typical American manufacturer—Hutchinson Technology, Inc.—can compete successfully, both nationally and internationally.

> Though the JIT implementation effort was still underway at Hutchinson as of the end of 1986, the company had in just over a year realized dramatic improvements above and beyond those realized during the quality campaign. Specifically:
>
> Manufacturing lead times were reduced from 50 to 90 percent.
> Quality yields improved an additional 4 to 14 percent.
> Setup times were reduced by as much as 75 percent.
> Work-in-process was reduced 40 to 90 percent—that translated into 80 work-in-process inventory turns and 120 inventory turns in finished goods.

> These improvements in individual areas of manufacturing are only part of the Hutchinson picture, however. The big story is that at a time when the electronics industry was suffering, JIT helped Hutchinson remain competitive and maintain market share. The company that had issued the ultimatum to Hutchinson a year earlier—drop the price 67 percent or we will have our work done offshore—stayed with Hutchinson and is, in 1987, a major customer.

In addition, Hutchinson was able to respond to competitive pressures on its major product line (more than half of its total business). It maintained market share by lowering its price by 30 percent *while increasing its percent profit on each unit by one-third.* And the quality of Hutchinson's product is so good that the company is able to compete in the Far East, shipping some 30 percent of its product to Far East manufacturers.

There were two major factors in Hutchinson's success in implementing JIT and making it work. One was all the previous work the company had put into its quality program. The other, equally important factor was top management's dedication to the JIT concept.

"The top management team at Hutchinson is the JIT champion," Cingari said. From the beginning, top management thought through the idea conceptually and was totally committed to implementing it. And they passed that feeling on to workers at all levels of the company.

"Even before the effort was started, Hutchinson's management team had created a corporate culture for risk-taking and experimentation with new concepts. Each JIT team's charter was not to study JIT, but to experiment with it wherever the team felt it was appropriate. The teams were given the clear message from top management that experimentation was highly regarded. So there was no fear of failure."

This book, then, has two major functions and two major parts. One is an explanation of the technical aspects of JIT and how to make those technical aspects work. The second and more complex part is a discussion of the difficult management issues involved in implementing these technical changes. Some of these implementation difficulties are the changing of attitudes and the company climate with regard to manufacturing, getting the involvement of middle management and individuals on the shop floor, and rethinking the company's measurement and reward systems. These three elements are intimately involved with one another.

Slowly but surely the JIT ideal of manufacturing excellence is filtering through the world. This will lead to heightened competition; but in the long run, it will also lead to prosperity as resources are put into production rather than wasted.

Industrial engineers of the future need to be a new breed to effectively assist in such efforts. Working as an introvert, without interacting with those on the production floor, and developing and presenting sterile solutions without benefit of needed input from others is an ineffective and doomed approach. Industrial engineers in the future must, therefore, be team builders, team players, coaches, and facilitators. If they listen well and concern themselves with what is best for all parties, they should do quite well. There is a tremendous amount of good work out there yet to be done.

REFERENCES

1 American Association of Engineering Societies, 1111 19th Street, Suite 608, Washington, DC 20036-3690.

2 Block, Peter: *Flawless Consulting*, University Associates, San Diego, CA, 1981.

3 Chapple, Alan: "Georgia Tech to Leave Narrow Technical Path, Steer to Broad Offerings," *Engineering Times*, vol. 2, no. 2, February 1990, pp. 1, 10 (quote by John Patrick Crecine).

4 Covey, Stephen R., *The Seven Habits of Highly Effective People*, Simon and Schuster, New York, 1989.

5 Deming, W. Edward; quoted in Yates, Ronald E.: "U. S. Management Doomed to Failure," *Orlando Sentinel,* January 19, 1992.

6 Fein, Mitchell: "Motivation for Work," Monograph no. 4, Work Measurement and Methods Engineering Division, American Institute of Industrial Engineers, Norcross, GA, 1971.

7 Geneen, Harold: *Managing*, Avon Books, New York, 1984.

8 Hay, Edward J.: *The Just-In-Time Breakthrough: Implementing the New Manufacturing Basics*, John Wiley & Sons, New York, 1988.

9 Hayes, Robert H., and William J. Abernathy: "Managing Our Way to Economic Decline," *Harvard Business Review*, vol. 58, no. 4, July–August 1980, pp. 67–77.

10 Herzberg, Frederick: *Work and the Nature of Man*, World Publishing Company, Cleveland, OH, 1966.

11 Iacocca, Lee: *Iacocca, An Autobiography*, Bantam Books, New York, NY, 1984.

12 Kim, James: "CEO of CalPERS wields clout," *USA Today*, February 23, 1993, section B, p. 1.

13 Kuhn, Brad: "Grad Students Blame Management for Problems," *Orlando Sentinel*, May 21, 1992, section C, pp. 1, 6.

14 Lee, James A.: *The Gold and the Garbage in Management Theories and Prescriptions*, Ohio University Press, Athens, OH, 1980.

15 Lincoln, James F.: *A New Approach to Industrial Economics*, Devin Adair, Greenwich, CT, 1961.

16 "Listening to America: What Went Wrong," VHS video in 2 parts (LIBM102 and LIBM103), 58 minutes each, *Listening to America With Bill Moyers*, PBS Video, 1320 Braddock Place, Alexandria, VA 22314-1698, 1992.

17 Morita, Akio, in a speech to the Malaysian Bankers Association, from a Reuters press release, October 30, 1990.

18 Morris, William T.: *Implementation Strategies for Industrial Engineers*, Grid Publishing, Inc., Columbus, OH, 1979.

19 National Institute of Building Sciences, 1201 L Street N.W., Washington, DC 20005, 1991.

20 "1990 Productivity Survey," for the American Electronics Association, 5201 Great America Parkway, Santa Clara, CA, 95054-1120, 1990.

21 Parkinson, C. Northcote: *Parkinson's Law*, Houghton Mifflin Publishing Co., Boston, MA, 1957.

22 Pascale, Richard T., and Anthony G. Athos: *The Art of Japanese Management*, Simon and Schuster, New York, 1981.

23 Perot, H. Ross: "How I Would Turn Around GM," *Fortune*, February 15, 1988, pp. 45–50.

24 Sink, D. Scott, and Thomas C. Tuttle: *Planning and Measurement in Your Organization of the Future*, Industrial Engineering and Management Press, Norcross, GA, 1989.

25 Smith, Ronald: "A Student IE's Answer to 'What's an IE?'" *Industrial Engineering*, vol. 5, no. 7, July 1973, pp. 26–27.

26 Taylor, Frederick W.: *Scientific Management*, Harper and Row, New York, 1947.

27 Tompkins, James A.: *Winning Manufacturing*, Industrial Engineering and Management Press, Norcross, GA, 1989.

28 Wilkinson, Signe: *Philadelphia Daily News*, January 6, 1992.

29 Wilson, Richard K.: "Effectiveness in Industrial Engineering," *Industrial Engineering*, vol. 4, no. 4, April 1972, pp. 8–11.

REVIEW QUESTIONS AND PROBLEMS

1 What are the top five engineering bachelor's degree disciplines in the United States in terms of degrees granted annually, and in what order?

2 Is the high cost of direct labor the primary cause of international noncompetitiveness of the United States?

3 What is the likelihood that the nations of the world will adopt the British system of measurements? Should the United States continue to use the British system of measurements, and why?
4 What does "best is the enemy of better" mean?
5 Based on quotes in this chapter, what do you assume is Deming's opinion of present manufacturing management in the United States?
6 In the Rollins College survey of graduate students discussed in this chapter, which asks "who's to blame" in American industry, who is their top candidate?
7 A behaviorist management philosophy typically believes that "recognition" is an important motivator. Did the Rollins College survey indicate that recognition is important?
8 How do American CEO salaries compare with Japanese CEO salaries?
9 To what extent do you accept the following statement to be true, "If you know how to manage, you can manage anything."
10 To what extent is it necessary to teach first-line supervisors modern management techniques and approaches for running a manufacturing operation?
11 To what extent are financial and operations auditing established management practices in the United States?
12 Based on the results of the American Electronic Association's 1990 Productivity Survey, is foreign competition a concern of its member companies?
13 Based on the results of the American Electronic Association's 1990 productivity survey, is computer-integrated manufacturing rapidly becoming a reality in United States manufacturing?
14 In the University of Central Florida IE Advisory Committee survey discussed in this chapter, what type of additional educational content was proposed most frequently by the industry survey respondents?
15 To what extent is the answer to question 14 above consistent with Block's proposed skill requirements for consultants?
16 What do employees want most from a job?
17 Is there any connection between the practice of industrial engineering in a plant and providing job security for direct labor employees?
18 Does getting a manufacturing operation to become more cost-effective typically require a considerable capital investment?
19 What is one of the probable greatest weaknesses in the present education of industrial engineers?
20 Will participative management make it likely that future industrial engineers will no longer be involved in the design of production systems?
21 What would be the author's first choice for a silver bullet, assuming it were to apply in a given situation?
22 What do you believe to be the future prospects for industrial engineering, and why?

NAME INDEX

SUBJECT INDEX